AN INTRODUCTION TO NUMERICAL METHODS AND ANALYSIS

AN INTRODUCTION TO NUMERICAL METHODS AND ANALYSIS

JAMES F. EPPERSON
University of Alabama in Huntsville

JOHN WILEY & SONS, INC.
New York / Chichester / Weinheim / Brisbane / Singapore / Toronto

Senior Acquisitions Editor   *Kimberly Murphy*
Editorial Assistant   *Stacy French*
Marketing Manager   *Julie Lindstrom*
Senior Production Editor   *Norine M. Pigliucci*
Senior Designer   *Karen Kincheloe*
Production Management Services   *Publication Services*

This book was set in *Times Roman* by *Publication Services* and printed and bound by *Donnelley/Crawfordsville*. The cover was printed by *Lehigh Press*.

This book is printed on acid-free paper. (∞)

Library of Congress Cataloging in Publication Data:
Epperson, James F.
  An introduction to numerical methods and analysis / James F. Epperson.
    p. cm
  Includes bibliographical references.
  ISBN 0-471-31647-4 (cloth: alk. paper)
   1. Numerical analysis. I. Title.

QA297 .E568 2001
519.4–dc21

2001033012

Printed in the United States of America
10 9 8 7 6 5 4 3 2 1

To Frank Crosby and Ed Croteau, Peter Duren and Fred Gehring,
Richard MacCamy and (most especially) George J. Fix
—teachers all, mathematicians all.

PREFACE

This book is intended for introductory and advanced courses in numerical methods and numerical analysis, for students majoring in mathematics, sciences, and engineering. The book is appropriate for single-term survey courses or year-long sequences, where students have a basic understanding of at least single-variable calculus and a programming language. (The usual first courses in linear algebra and differential equations are required for the last four chapters.)

To provide maximum teaching flexibility, each chapter and each section begins with the basic, elementary material and gradually builds up to the more advanced material. This same approach is followed with the underlying theory of the methods. Accordingly, one can use the text for a "methods" course that eschews mathematical analysis, simply by not covering the sections that focus on the theoretical material. Or one can use the text for a survey course by covering only the basic sections; the extra topics can be covered if you have the luxury of a full-year course.

The objective of the text is for students to learn where approximation methods come from, why they work and sometimes don't work, and when to use which of the many techniques that are available, and to do all this in a style that emphasizes readability and usefulness to the beginning student. Although these goals are shared by other texts, the development and delivery of the ideas in this text, I think, make it different.

A course in numerical computation—whether it emphasizes the theory or the methods—requires that students think quite differently than in other mathematics courses, yet students are often not experienced in the kind of problem-solving skills and mathematical judgment that a numerical methods course requires. Many students react to mathematics problems by pigeonholing them by category, with little thought given to the meaning of the answer. Numerical mathematics demands much more judgment and evaluation in light of the underlying theory, and in the first several weeks of the course it is crucial for students to adapt their way of thinking about and working with these ideas, in order to succeed in the course.

To enable students to attain the appropriate level of mathematical sophistication, this text begins with a review of the important calculus results, and why and where these ideas play an important role in this course. Some of the concepts required for the study of computational mathematics are introduced, and simple approximations using Taylor's Theorem are treated in some depth, in order to acquaint students with one of the most common and basic tools in the science of approximation. Computer arithmetic is treated in perhaps more detail than some might think necessary, but it is instructive for many students to see the actual basis for rounding error demonstrated in detail, at least once.

One important element of this text that I have not seen in other texts is the emphasis that is placed on "cause and effect" in numerical mathematics. For example, if we apply the trapezoid rule to (approximately) integrate a function, then the error should go down by a factor of 4 as the mesh decreases by a factor of 2; if this is not what happens, then almost surely there is either an error in the code or the integrand is not sufficiently smooth. While this is obvious to experienced practitioners in the field, it is not obvious to beginning students who are not confident of their mathematical abilities. Many of the exercises and examples are designed to explore this kind of issue.

Two common starting points to the course are root finding and linear systems, but diving into the treatment of these ideas often leaves the students confused and wondering what the point of the course is. Instead, this text provides a second chapter designed as a "toolbox" of elementary ideas from several problem areas; it is one of the important innovations of the text. The goal of the toolbox is to acclimate the students to the culture of numerical methods and analysis, and to show them a variety of simple ideas before proceeding to cover any single topic in depth. It develops some elementary approximations and methods that the students can easily appreciate and understand, and introduces the students, in the context of very simple methods and problems, to the essence of the analytical and coding issues that dominate the course. At the same time, the early development of these tools allows them to be used later in the text to derive and explain some algorithms in more detail than is usually the case.

The style of exposition is intended to be more lively and "student friendly" than the average mathematics text. This does not mean that there are no theorems stated and proved correctly, but it does mean that the text is not slavish about it. There is a reason for this: The book is meant to be read by the students. The instructor can render more formal anything in the text that he or she wishes, but if the students do not read the text because they are turned off by an overly dry regimen of definition, theorem, proof, and corollary, then all of our effort is for naught. In places the exposition may seem a bit wordier than necessary, and there is a significant amount of repetition. Both are deliberate. While brevity is indeed better mathematical style, it is not necessarily better pedagogy. Mathematical textbook exposition often suffers from an excess of brevity, with the result that the students cannot follow the arguments as presented in the text. Similarly, repetition aids learning, by reinforcement.

Nonetheless I have tried to make the text mathematically complete. Those who wish to teach a lower-level survey course can skip proofs of many of the more technical results in order to concentrate on the approximations themselves. An effort has been made—not always successfully—to avoid making basic material in one section dependent on advanced material from an earlier section.

The topics selected for inclusion are fairly standard but not encyclopedic. Emerging areas of numerical analysis, such as wavelets, are not (in the author's opinion) appropriate for a first course in the subject. The same reasoning dictated the exclusion of other, more mature areas, such as the finite-element method, although that might change in future editions should there be sufficient demand for it. A more detailed treatment of approximation theory, one of the author's favorite topics, was also felt to be poorly suited to a beginning text. It was felt that a better text would result by doing a good job covering some of the basic ideas, rather than trying to cover everything in the subject.

The text is not specific to any one computing language. Most illustrations of code are made in an informal pseudocode, with more involved algorithms shown in a "macro-outline" form, and programming hints and suggestions are scattered throughout the text. The exercises assume that students have easy access to and working knowledge of software for producing basic Cartesian graphs.

A diskette of programs is *not* provided with the text—a feature that sets this book at odds with many others but that reflects the author's opinion that students must learn how to write and debug programs that implement the algorithms in order to learn the underlying mathematics. However, since some faculty and some departments structure their courses differently, a collection of program segments in a variety of languages is available on the text website so that instructors can easily download and distribute the code to their students. Instructors and students should be aware that these are program *segments*; none are intended to be ready-to-run, complete programs. Other features of the text website are discussed below.

Exercises run the gamut from simple hand computations that might be characterized as "starter exercises," to challenging derivations and minor proofs, to programming exercises designed to test whether or not the students have assimilated the important ideas of each chapter and section. Some of the exercises are taken from application situations; some are more traditionally focused on the mathematical issues for their own sake. Each chapter concludes with a brief section discussing existing software and other references for the topic at hand, and a discussion of material not covered in this text.

Historical notes are scattered throughout the text, with most named mathematicians accorded at least a paragraph or two of biography when they are first mentioned. This not only indulges my interest in the history of mathematics, but it also serves to engage the interest of the students.

The website for the text (`http://www.wiley.com/epperson`) contains, in addition to the set of code segments mentioned above, a collection of additional exercises for the text and, of course, information about any updates to be made in future editions. Colleagues who wish to submit exercises or make comments about the text are invited to do so by contacting the author at `epperson@math.uah.edu`.

A Note to the Student

This book was written to be read. Now, I am under no illusions that this text will compete with the latest popular novel for interest and thrilling narrative. But I have tried very hard to write a book on mathematics that could be read by the students. So do not simply buy the book, work the exercises, and sell the book back to the bookstore at the end of the term. Read the text, think about what you have read, and ask your instructor questions about the things that you do not understand.

Numerical methods and analysis is a very different area of mathematics, certainly different from what you have seen in your previous courses. It is not harder, but the differentness of the material makes it seem harder. We are concerned with different issues from those in other mathematics classes. In a calculus course you are typically asked to compute the derivative or antiderivative of a given function, or to solve some equation for a particular unknown. The task is clearly defined, with a very concrete notion of "the right answer." Here we are concerned with computing approximations, and this involves a slightly different kind of thinking. We have to understand what we are approximating well enough to construct a reasonable approximation, and we have to be able to think clearly and logically enough to analyze the accuracy and performance of that approximation. One former student has characterized this course material as "rigorously imprecise" or "approximately precise." Both are appropriate descriptions. Rote memorization of procedures is not of use here; it is vital in this course that the student learn the underlying concepts. Numerical mathematics is also *experimental* in nature. A lot can be learned by simply trying something out and seeing how the computation goes.

A Disclaimer

Each of the "topical" chapters—those chapters organized around a specific topic, such as root finding or numerical integration—begins with an illustration of how the computational problem considered in the chapter might arise in an actual application setting. The reader should be aware that the author has indulged his sense of humor and whimsy in concocting these vignettes. Some of the models and data are taken from real situations, and some are simply made up to illustrate how numerical calculations would arise in a typical setting.

Acknowledgments

Nobody writes a book like this without help, and I am no exception. This project would never have reached completion without the assistance of my editors at John Wiley & Sons, Barbara Holland, Mary Johenk, Kim Murphy, and a small army of faculty and student reviewers who provided helpful comments over the two years during which I was writing the book. I would particularly like to acknowledge my own students in several sections of MA 415 and MA 515 at the University of Alabama in Huntsville who used the text during the 1998–1999 and 1999–2000 academic years. My chairman, Dr. M.-H. Chang, has been very supportive, as have my other colleagues in the department. For help with LaTeX questions I would particularly like to mention Dr. Ken Howell.

Many of the biographies of mathematicians were constructed from material and sources on the website maintained at the University of St. Andrews, Scotland, `http://www-groups.dcs.st-and.ac.uk/~history`. Some additional information was gleaned from the collective resources and knowledge of the subscribers to the NA Digest, a mailing list for numerical analysts (`http://www.netlib.org/na-net/na_home.html`).

Finally, in addition to these professional acknowledgments, I have to personally acknowledge the support and perseverence of my family: wife Georgia, children Elinor and Jay, and even the family dog Bobbie Lee. It's over, kids; you can have your daddy back now, and I'll even get back to mowing the lawn.

Notation

Most notation is defined as it appears in the text, but we include here some commonplace items.

R	The real number line; $\mathbf{R} = (-\infty, \infty)$.
$\mathbf{R}^n$	The vector space of real vectors of n components.
$\mathbf{R}^{n \times n}$	The vector space of real $n \times n$ matrices.
$C([a, b])$	The set of functions f that are defined on the interval $[a,b]$, continuous on all of (a,b), and continuous from the interior of $[a,b]$ at the endpoints.
$C^k([a, b])$	The set of functions f such that f and its first k derivatives are all in $C([a,b])$.
$C^{p,q}(Q)$	The set of all functions u that are defined on the two-dimensional domain $Q = \{(x, t) \mid a \leq x \leq b, 0 < t \leq T\}$ and that are p times continuously differentiable in x for all t, and q times continuously differentiable in t for all x.
$\approx$	Approximately equal. When we say that $A \approx B$, we mean that A and B are approximations to each other. See Section 1.2.2.
$\equiv$	Equivalent. When we say that $f(x) = g(x)$, we mean that the two functions agree at the single point x. When we say that $f(x) \equiv g(x)$, we mean that they agree at all points x. The same thing is said by using just the function names, i.e., $f = g$.
O	On the order of ("big O of"). We say that $A = B + O(D(h))$ whenever $|A - B| \leq CD(h)$ for some constant C and for all h sufficiently small. See Section 1.3.
u	Machine epsilon. The largest number such that, in computer arithmetic, $1 + \mathbf{u} = 1$. This is architecture dependent, of course. See Section 1.3.
sgn	Sign function. The value of $\text{sgn}(x)$ is 1, -1, or 0 when x is positive, negative, or zero, respectively.

CONTENTS

CHAPTER 1

INTRODUCTORY CONCEPTS AND CALCULUS REVIEW

It is best to start this book with a question: What do we mean by "Numerical Methods and Analysis"? What kind of mathematics is this book about?

Generally and broadly speaking, this book covers the mathematics and methodologies that underlie the techniques of *scientific computation*. More prosaically, consider the button on your calculator that computes the sine of the number in the display. Exactly how does the calculator know that correct value? When we speak of using the computer to solve a complicated mathematics or engineering problem, exactly what is involved in making that happen? Are computers "born" with the knowledge of how to solve complicated mathematical and engineering problems? No, they are not. Mostly they are *programmed* to do it, and the programs implement algorithms that are based on the kinds of things we will talk about in this text.

Textbooks and courses in this area generally follow one of two main themes. Those titled "numerical methods" tend to emphasize the implementation of the algorithms, perhaps at the expense of the underlying mathematical theory that explains why the methods work; those titled "numerical analysis" tend to emphasize this underlying mathematical theory, perhaps at the expense of some of the implementation issues. The best approach, of course, is to properly mix the study of the algorithms and their implementation ("methods") with the study of the mathematical theory ("analysis") that supports them. That is the goal of this text.

When people speak of using a computer to design an airplane, predict the weather, or otherwise solve a complex science or engineering problem, they are talking about using numerical methods and analysis. The kinds of problems and areas of endeavor that use these kinds of techniques are ever expanding. For example, computational mathematics—another name for the material that we consider here—is now commonly used in the study of financial markets and investment structures, an area of study that does not ordinarily come to mind when we think of "scientific" computation. Similarly, the rapidly growing use of computer-generated animation in film production is based on a heavy dose of spline approximations, which we introduce in Section 4.8.

There are a number of different ways to break down the subject into component parts. We will discuss the derivation of the algorithms, we will discuss the implementation of the algorithms, and we will also analyze the algorithms mathematically, in order to learn how best to use them and how best to implement them. In our study of each technique, we will usually be concerned with two issues that often are competing with each other:

- *Accuracy:* Very few of our computations will yield the exact answer to the problem, so we will have to understand how much error is made, and how to control (or even diminish) that error.
- *Efficiency:* Does the algorithm take an inordinate amount of computer time? This might seem to be an odd question to concern ourselves with—after all, computers are fast, right?—but there are slow ways and fast ways to do things. All else being equal (it rarely is), we prefer the fast ways.

We say these two issues compete with each other because, generally speaking, the steps that can be taken to make an algorithm more accurate usually make it more costly, i.e., less efficient.

There is a third issue of importance, but it does not become as evident as the others (although it is still present) until Chapter 6:

- *Stability:* Does the method produce similar results for similar data? If we change the data by a small amount, do we get vastly different results? If so, we say the method is unstable, and unstable methods tend to produce unreliable results. It is entirely possible to have an accurate method that is efficiently implemented, and yet is horribly unstable; see Section 6.4.4 for an example of this.

1.1   BASIC TOOLS OF CALCULUS

1.1.1   Taylor's Theorem

Computational mathematics does not require a large amount of background, but it does require a good knowledge of that background. The most important single result in numerical computations, from all of the calculus, is Taylor's Theorem,[1] which we now state:

Theorem 1.1 *(Taylor's Theorem with Remainder)*   *Let $f(x)$ have $n+1$ continuous derivatives on $[a,b]$ for some $n \geq 0$, and let $x, x_0 \in [a,b]$. Then*

$$f(x) = p_n(x) + R_n(x)$$

for

$$p_n(x) = \sum_{k=0}^{n} \frac{(x-x_0)^k}{k!} f^{(k)}(x_0) \qquad (1.1)$$

and

$$R_n(x) = \frac{1}{n!} \int_{x_0}^{x} (x-t)^n f^{(n+1)}(t)\, dt. \qquad (1.2)$$

---

[1]Brook Taylor (1685–1731) was educated at St. John's College of Cambridge University, entering in 1701 and graduating in 1709. He published what we know as Taylor's Theorem in 1715, although it appears that he did not appreciate its larger importance and he certainly did not bother with a formal proof. He was elected a member of the prestigious Royal Society of London in 1712.

Taylor acknowledged that his work was based on that of Newton and Kepler and others, but he did not acknowledge that the same result had been discovered by Johann Bernoulli and published in 1694. (But then Taylor discovered integration by parts first, although Bernoulli claimed the credit.)

Moreover, there exists a point ξ_x between x and x_0 such that

$$R_n(x) = \frac{(x - x_0)^{n+1}}{(n+1)!} f^{(n+1)}(\xi_x). \qquad (1.3)$$

The point x_0 is usually chosen at the discretion of the user, and is often taken to be 0. Note that the two forms of the remainder are equivalent: the "pointwise" form (1.3) can be derived from the "integral" form (1.2); see Exercise 23.

Taylor's Theorem is important because it allows us to represent, exactly, fairly general functions in terms of polynomials with a *known, specified, boundable* error. This allows us to replace, in a computational setting, these same general functions with something that is much simpler—a polynomial—and yet at the same time we are able to bound the error that is made. No other tool will be as important in this course as Taylor's Theorem, so it is worth spending some time on it here at the outset.

The usual calculus treatment of Taylor's Theorem should leave the student familiar with three particular expansions (for all three of these we have used $x_0 = 0$):

$$e^x = 1 + x + \frac{1}{2!}x^2 + \frac{1}{3!}x^3 + \cdots + \frac{1}{n!}x^n + \frac{1}{(n+1)!}x^{n+1}e^{\xi_x}$$

$$= \sum_{k=0}^{n} \frac{1}{k!}x^k + R_n(x),$$

$$\sin x = x - \frac{1}{3!}x^3 + \frac{1}{5!}x^5 + \cdots + \frac{(-1)^n}{(2n+1)!}x^{2n+1} + \frac{(-1)^{n+1}}{(2n+3)!}x^{2n+3}\cos\xi_x$$

$$= \sum_{k=0}^{n} \frac{(-1)^k}{(2k+1)!}x^{2k+1} + R_n(x),$$

$$\cos x = 1 - \frac{1}{2!}x^2 + \frac{1}{4!}x^4 + \cdots + \frac{(-1)^n}{(2n)!}x^{2n} + \frac{(-1)^{n+1}}{(2n+2)!}x^{2n+2}\cos\xi_x$$

$$= \sum_{k=0}^{n} \frac{(-1)^k}{(2k)!}x^{2k} + R_n(x).$$

(Strictly speaking, the index on the last two remainders should be $2n + 1$ and $2n$, because those are the exponents in the last terms of the expansion, but it is commonplace to do as we have done here.) In fact, Taylor's Theorem provides us with our first and simplest example of an approximation and an error estimate. Consider the problem of approximating the exponential function on the interval $[-1, 1]$. Taylor's Theorem tells us that we can represent e^x using a polynomial with a (known) remainder:

$$e^x = \underbrace{1 + x + \frac{1}{2!}x^2 + \frac{1}{3!}x^3 + \cdots + \frac{1}{n!}x^n}_{p_n(x),\ \text{polynomial}} + \underbrace{\frac{1}{(n+1)!}x^{n+1}e^{c_x}}_{R_n(x),\ \text{remainder}},$$

where c_x is an unknown point between x and 0. Since we want to consider the most general case, where x can be any point in $[-1, 1]$, we have to consider that c_x can be any point in $[-1, 1]$ as well. For simplicity, let's denote the polynomial by $p_n(x)$, and the remainder by $R_n(x)$, so that the above becomes

$$e^x = p_n(x) + R_n(x).$$

Suppose we want this approximation to be accurate to within 10^{-6} in absolute error; i.e., we want

$$|e^x - p_n(x)| \le 10^{-6}$$

for *all* x in the interval $[-1,1]$. Note that if we can make $|R_n(x)| \le 10^{-6}$ for all $x \in [-1,1]$, then we will have

$$|e^x - p_n(x)| = |R_n(x)| \le 10^{-6}$$

so that the error in the approximation will be less than 10^{-6}. The best way to proceed is to create a *simple* upper bound for $|R_n(x)|$, and then use that to determine the number of terms necessary to make this upper bound less than 10^{-6}.

Thus we proceed as follows:

$$
\begin{aligned}
|R_n(x)| &= \frac{|x^{n+1} e^{c_x}|}{(n+1)!} = \frac{|x|^{n+1} e^{c_x}}{(n+1)!}, \quad \text{because } e^z > 0 \text{ for all } z \\
&\le \frac{e^{c_x}}{(n+1)!}, \quad \text{because } |x| \le 1 \text{ for all } x \in [-1,1] \\
&\le \frac{e}{(n+1)!}, \quad \text{because } e^{c_x} \le e \text{ for all } x \in [-1,1].
\end{aligned}
$$

Thus, if we find n such that

$$\frac{1}{(n+1)!} e \le 10^{-6}$$

then we will have

$$|e^x - p_n(x)| = |R_n(x)| \le \frac{1}{(n+1)!} e \le 10^{-6}$$

and we will *know* that the error is less than the desired tolerance, for *all* values of x of interest to us, i.e., all $x \in [-1,1]$. A little bit of exercise with a calculator shows us that we need to use $n = 9$ to get the desired accuracy. Figure 1.1 shows a plot of the exponential function e^x, the approximation $p_9(x)$, as well as the less accurate $p_2(x)$; since it is impossible to distinguish by eye the plots for e^x and $p_9(x)$, we have also provided Figure 1.2, which is a plot of the error $e^x - p_9(x)$; note that we begin to lose accuracy as we get away from the interval $[-1,1]$. This is not surprising. Since the Taylor polynomial is constructed to match f and its first n derivatives at $x = x_0$, it ought to be the case that p_n is a good approximation to f only when x is near x_0.

Here we are in the first section of the text, and we have already constructed our first approximation and proved our first theoretical result. Theoretical result? Where? Why, the error estimate of course. The material in the previous several lines is a proof of the following result:

Proposition 1.1   *Let $p_9(x)$ be the Taylor polynomial of degree 9 for the exponential function. Then, for all $x \in [-1,1]$, the error in the approximation of $p_9(x)$ to e^x is less than 10^{-6}, i.e.,*

$$|e^x - p_9(x)| \le 10^{-6}$$

for all $x \in [-1,1]$.

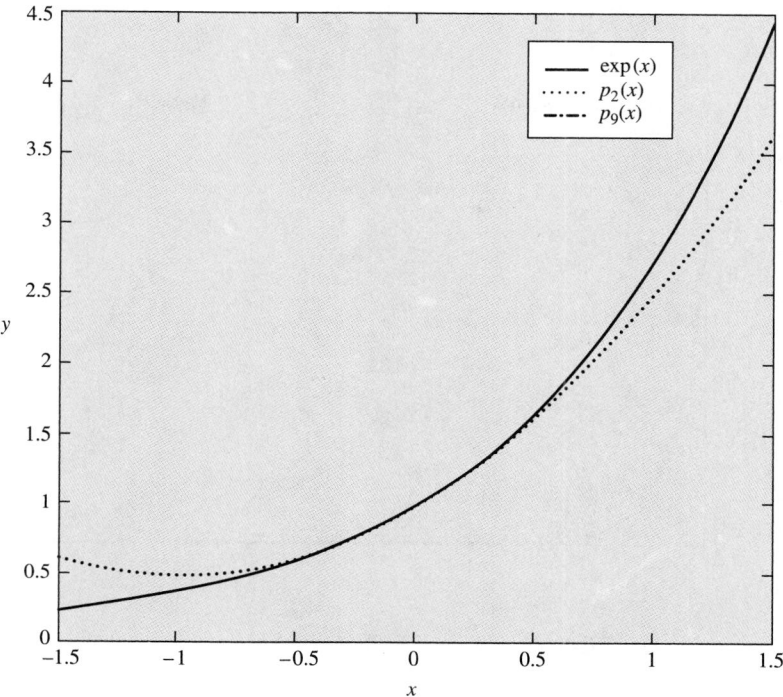

FIGURE 1.1 Taylor approximation: e^x, $p_9(x) \approx e^x$, and $p_2(x) \approx e^x$. Note that e^x and $p_9(x)$ are indistinguishable on this plot.

While this result is not of major significance to our work—the exponential function is approximated more efficiently by different means—it does illustrate one of the most important aspects of the subject. The result tells us, ahead of time, that we can approximate the exponential function to within 10^{-6} accuracy using a specific polynomial, and this accuracy holds for all x in a specified interval. That is the kind of thing we will be doing throughout the text—constructing approximations to difficult computations that are accurate in some sense, and we *know* how accurate they are.

EXAMPLE 1.1

Let $f(x) = \sqrt{x+1}$; then the second-order Taylor polynomial (computed about $x_0 = 0$) is computed as follows:

$$f(x_0) = f(0) = 1$$

$$f'(x) = \frac{1}{2}(x+1)^{-1/2} \quad \Rightarrow \quad f'(x_0) = \frac{1}{2}$$

$$f''(x) = -\frac{1}{2} \times \frac{1}{2}(x+1)^{-3/2} \quad \Rightarrow \quad f''(x_0) = -\frac{1}{4}$$

$$p_2(x) = f(x_0) + (x-x_0)f'(x_0) + \frac{1}{2}(x-x_0)^2 f''(x_0) = 1 + \frac{1}{2}x - \frac{1}{8}x^2.$$

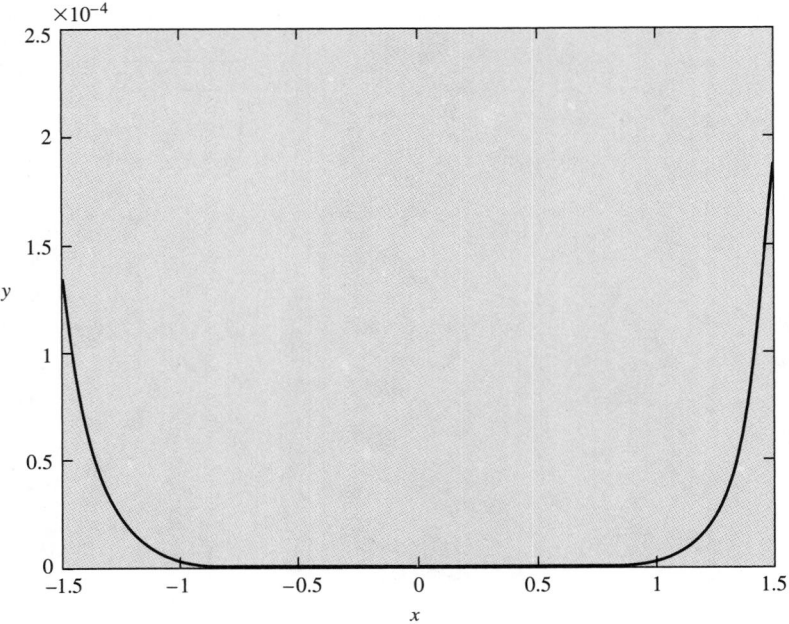

FIGURE 1.2   Error in Taylor approximation: $e^x - p_9(x)$.

The error in using p_2 to approximate $\sqrt{x+1}$ is given by $R_2(x) = (1/3!)(x-x_0)^3 f'''(\xi_x)$, where ξ_x is between x and x_0. We can simplify the error as follows:

$$|R_2(x)| = \left| \frac{1}{3!}(x-x_0)^3 f'''(\xi_x) \right|$$

$$= \frac{1}{6}|x|^3 \left| \frac{3}{2} \times \frac{1}{2} \times \frac{1}{2}(\xi_x+1)^{-5/2} \right|$$

$$= \frac{1}{16}|x|^3 |\xi_x+1|^{-5/2}.$$

If we want to consider this for all $x \in [0,1]$ then we observe that $x \in [0,1]$ and ξ_x between x and 0 imply that $\xi_x \in [0,1]$. Therefore,

$$|\xi_x+1|^{-5/2} \le |0+1|^{-5/2} = 1,$$

so that the upper bound on the error becomes $|R_3(x)| \le 1/16 = 0.0625$, for all $x \in [0,1]$. If we are interested only in $x \in [0,\frac{1}{2}]$, then the error is much smaller:

$$|R_2(x)| = \frac{1}{16}|x|^3 |\xi_x+1|^{-5/2}$$

$$\le \frac{1}{16}(1/2)^3$$

$$= \frac{1}{128}$$

$$= 0.0078125.$$

∎

EXAMPLE 1.2

Consider the problem of finding a polynomial approximation to the function $f(x) = \sin \pi x$ that is accurate to within 10^{-4} for all $x \in [-\frac{1}{2}, \frac{1}{2}]$, using $x_0 = 0$. A direct computation with Taylor's Theorem gives us (note that the indexing here is a little different, because we want to take advantage of the fact that the Taylor polynomial here has only odd terms)

$$p_n(x) = x - \frac{1}{6}\pi^3 x^3 + \frac{1}{120}\pi^5 x^5 + \cdots + (-1)^n \frac{1}{(2n+1)!}\pi^{2n+1} x^{2n+1}$$

$$R_n(x) = (-1)^{n+1} \frac{1}{(2n+3)!}\pi^{2n+3} x^{2n+3} \cos \xi_x.$$

Thus, the error in the approximation is

$$|f(x) - p_n(x)| = |R_n(x)| = \frac{(\pi x)^{2n+3}}{(2n+3)!}|\cos \xi_x|,$$

where ξ_x is between x and 0; this can be bounded for all $x \in [-\frac{1}{2}, \frac{1}{2}]$ in the following fashion:

$$|R_n(x)| = \frac{|\pi x|^{2n+3}}{(2n+3)!}|\cos \xi_x| \le \frac{(\pi/2)^{2n+3}}{(2n+3)!},$$

and some experimentation with a calculator shows that

$$|R_4(x)| \le 0.3599 \times 10^{-5}, \qquad |R_3(x)| \le 0.1604 \times 10^{-3},$$

from which we conclude that $n = 4$ will achieve the desired accuracy. Thus, we want

$$p_4(x) = x - \frac{1}{6}\pi^3 x^3 + \frac{1}{120}\pi^5 x^5 - \frac{1}{5040}\pi^7 x^7 + \frac{1}{362,880}\pi^9 x^9$$

as our approximation. Figure 1.3 shows the error between this polynomial and $f(x)$ over the interval $[-\frac{1}{2}, \frac{1}{2}]$; note that the error is much better than the predicted error, especially in the middle of the interval. ∎

While the usual approach is to construct Taylor expansions by directly computing the necessary derivatives of the function f, sometimes a more subtle approach can be used.

Consider the problem of computing a Taylor series for the arctangent function, $f(x) = \arctan x$, about the point $x_0 = 0$. It won't take many terms before we get tired of trying to find a general expression for $f^{(n)}$. What do we do?

Recall from calculus that

$$\arctan x = \int_0^x \frac{dt}{1+t^2},$$

so we can get the Taylor series for the arctangent from the Taylor series for $(1+t^2)^{-1}$ by a simple integration. Now recall the summation formula for the geometric series:

$$\sum_{k=0}^{n} r^k = \frac{1 - r^{n+1}}{1 - r}.$$

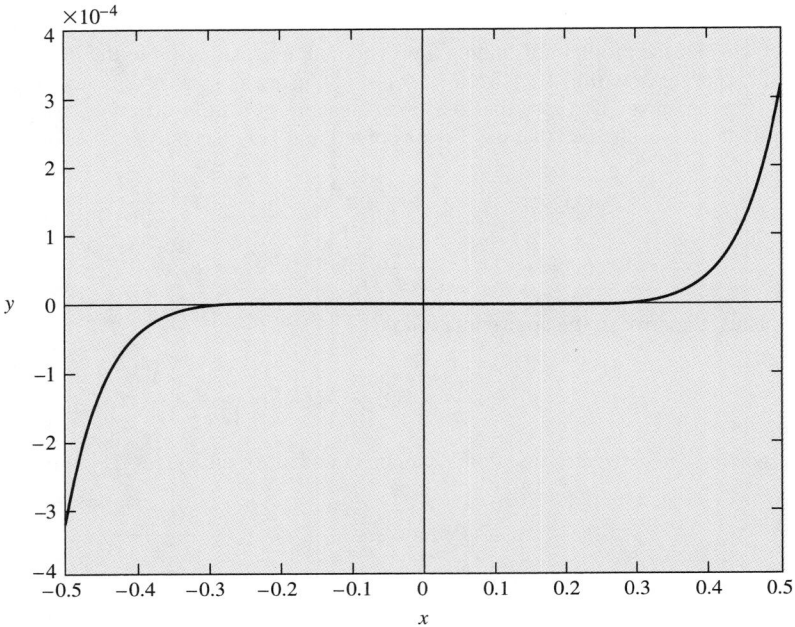

FIGURE 1.3 Error in Taylor approximation to $f(x) = \sin \pi x$ over $[-\frac{1}{2}, \frac{1}{2}]$.

If we substitute $r = -t^2$ into this and rearrange terms a bit, we get

$$\frac{1}{1+t^2} = \sum_{k=0}^{n}(-1)^k t^{2k} + \frac{(-t^2)^{n+1}}{1+t^2}.$$

Thus,

$$\arctan x = \int_0^x \frac{dt}{1+t^2}$$

$$= \int_0^x \left(\sum_{k=0}^{n}(-1)^k t^{2k} + \frac{(-t^2)^{n+1}}{1+t^2} \right) dt$$

$$= \sum_{k=0}^{n}(-1)^k (2k+1)^{-1} x^{2k+1} + (-1)^{n+1} \int_0^x \frac{t^{2n+2}}{1+t^2} dt.$$

This is known as Gregory's series[2] for the arctangent and was the basis for one of the early methods for computing π. (See Exercises 14–16.)

---

[2]James Gregory (1638–1675) was born in Aberdeen, Scotland, where he first went to university. He later (1664–1668) studied in Italy before returning to Britain. He published a work on optics and several works devoted to finding the areas under curves, including the circle. He knew about what we call Taylor series more than 40 years before Taylor published the theorem, and he might have developed the calculus before Newton did if he had been of a less geometric and more analytical frame of mind.

EXAMPLE 1.3

Let's use the Gregory series to determine the error in a ninth-degree Taylor approximation to the arctangent function. Since $2n+1 = 9$ implies that $n = 4$, we have

$$R_9(x) = -\int_0^x \frac{t^{10}}{1+t^2}\, dt,$$

so that (assuming $x \geq 0$)

$$|R_9(x)| = \int_0^x \frac{t^{10}}{1+t^2}\, dt,$$

and we can bound the error as follows:

$$|R_9(x)| = \int_0^x \frac{t^{10}}{1+t^2}\, dt \leq \int_0^x t^{10}\, dt = \frac{1}{11}x^{11}.$$

So for all $x \in [-\frac{1}{2}, \frac{1}{2}]$ the remainder is bounded above by

$$|R_9(x)| \leq \frac{1}{11}(1/2)^{11} = 4.44 \times 10^{-5}.$$

■

We close this section with an illustration of a special kind of Taylor expansion that will be used again and again.

Consider the problem of expanding $f(x+h)$ in a Taylor series, about the point $x_0 = x$. Here h is generally considered to be a small parameter. Direct application of Taylor's Theorem gives us the following:

$$\begin{aligned}
f(x+h) =& f(x) + ([x+h]-x)f'(x) + \frac{1}{2}([x+h]-x)^2 f''(x) \\
&+ \cdots + \frac{1}{n!}([x+h]-x)^n f^{(n)}(x) \\
&+ \frac{1}{(n+1)!}([x+h]-x)^{n+1} f^{(n+1)}(\xi) \\
=& f(x) + hf'(x) + \frac{1}{2}h^2 f''(x) + \cdots + \frac{1}{n!}h^n f^{(n)}(x) \\
&+ \frac{1}{(n+1)!}h^{n+1} f^{(n+1)}(\xi).
\end{aligned}$$

This kind of expansion will be useful time and again in our studies.

1.1.2 Mean Value and Extreme Value Theorems

We need to spend some time reviewing other results from calculus that have an impact on numerical methods and analysis. All of these theorems are included in most calculus texts but are usually not emphasized as much as will be necessary here.

Theorem 1.2 *(Mean Value Theorem)* *Let f be a given function, continuous on $[a,b]$ and differentiable on (a,b). Then there exists a point $\xi \in [a,b]$ such that*

$$f'(\xi) = \frac{f(b) - f(a)}{b - a}. \tag{1.4}$$

In the context of calculus, the importance of this result might seem obscure at best. However, from the point of view of numerical methods and analysis, the Mean Value Theorem (MVT) is probably second in importance only to Taylor's Theorem. Why? Well, consider a slightly reworked form of (1.4):

$$f(x_1) - f(x_2) = f'(\xi)(x_1 - x_2).$$

Thus, the MVT allows us to replace differences of function values with differences of argument values, if we scale by the derivative of the function. For example, we can use the MVT to determine that

$$|\cos x_1 - \cos x_2| \leq |x_1 - x_2|,$$

because the derivative of the cosine is the sine, which is bounded by 1 in absolute value. Note also that the MVT is simply a special case of Taylor's Theorem, for $n = 0$.

Similar to the MVT is the Intermediate Value Theorem:

Theorem 1.3 *(Intermediate Value Theorem)* *Let $f \in C([a,b])$ be given, and assume that W is a value between $f(a)$ and $f(b)$, i.e., either $f(a) \leq W \leq f(b)$, or $f(b) \leq W \leq f(a)$. Then there exists a point $c \in [a,b]$ such that $f(c) = W$.*

This seems to be a very abstract result; it says that a certain point exists, but doesn't give us much information about its numeric value. (Besides which, we might ask, why do we even care that c exists?) But it is this very theorem that is the basis for our first method for finding the roots of functions, an algorithm called the bisection method. (See Section 3.1.) Moreover, this is the theorem that tells us that a continuous function is one whose graph can be drawn without lifting the pen off the paper, because it says that between any two function values, we have a point on the curve for all possible argument values. Sometimes very abstract results have very concrete consequences.

A related result is the Extreme Value Theorem, which is the basis for the max/min problems that are a staple of most beginning calculus courses.

Theorem 1.4 *(Extreme Value Theorem)* *Let $f \in C([a,b])$ be given; then there exists a point $m \in [a,b]$ such that $f(m) \leq f(x)$ for all $x \in [a,b]$, and a point $M \in [a,b]$ such that $f(M) \geq f(x)$ for all $x \in [a,b]$. Morever, f achieves its maximum and minimum values on $[a,b]$ either at the endpoints a and b, or at a critical point.*

(The student should recall that a critical point is a point where the first derivative is either undefined or equal to zero. The theorem thus says that either $M = a$, $M = b$, $f'(M)$ doesn't exist, or $f'(M) = 0$, and similarly for m.)

There are other "mean value theorems," and we need to look at two in particular, as they will come up in our early error analysis.

Theorem 1.5 *(Integral Mean Value Theorem)   Let f and g both be in $C([a,b])$, and assume further that g does not change sign on $[a,b]$. Then there exists a point $\xi \in [a,b]$ such that*

$$\int_a^b g(t)f(t)dt = f(\xi) \int_a^b g(t)\, dt. \tag{1.5}$$

Proof   Since this result is not commonly covered in the calculus sequence, we will go ahead and prove it.

We first assume, without any loss of generality, that $g(t) \geq 0$; the argument changes in an obvious way if g is negative (see Exercise 26). Let f_M be the maximum value of the function on the interval, $f_M = \max_{x \in [a,b]} f(x)$, so that $g(t)f(t) \leq g(t)f_M$ for all $t \in (a,b)$; then

$$\int_a^b g(t)f(t)\, dt \leq \int_a^b g(t)f_M\, dt = f_M \int_a^b g(t)\, dt.$$

Similarly, we have

$$\int_a^b g(t)f(t)\, dt \geq \int_a^b g(t)f_m\, dt = f_m \int_a^b g(t)\, dt,$$

where $f_m = \min_{x \in [a,b]} f(x)$ is the minimum value of f on the interval. Since g does not change sign on the interval of integration, the only way that we can have

$$\int_a^b g(x)\, dx = 0$$

is if g is identically zero on the interval, in which case the theorem is trivially true, since both sides of (1.5) would be zero. So we can assume that

$$\int_a^b g(x)\, dx \neq 0.$$

Now define

$$W = \frac{\int_a^b g(t)f(t)\, dt}{\int_a^b g(t)\, dt}$$

so that we have

$$f_m \leq W \leq f_M.$$

By the Extreme Value Theorem, there is a point $M \in [a,b]$ such that $f(M) = f_M$, and similarly, there is a point $m \in [a,b]$ such that $f(m) = f_m$. Therefore,

$$f(m) \leq W \leq f(M),$$

and we can apply the Intermediate Value Theorem to establish that there is a point ξ in the interval defined by m and M such that $f(\xi) = W$; but this implies (1.5) (why?), and we are done.   □

This result will be useful for simplifying some error estimates for numerical integration rules in Chapter 5. A related result is the Discrete Average Value Theorem:

Theorem 1.6 *(Discrete Average Value Theorem)*   *Let $f \in C([a,b])$ and consider the sum*

$$S = \sum_{k=1}^{n} a_k f(x_k),$$

where each point $x_k \in [a,b]$, and the coefficients satisfy

$$a_k \geq 0, \qquad \sum_{k=1}^{n} a_k = 1.$$

Then there exists a point $\eta \in [a,b]$ such that $f(\eta) = S$, i.e.,

$$f(\eta) = \sum_{k=1}^{n} a_k f(x_k).$$

Proof   The proof is similar in spirit and technique to the previous one. We quickly have that

$$S = \sum_{k=1}^{n} a_k f(x_k) \leq f_M \sum_{k=1}^{n} a_k = f_M,$$

and similarly $S \geq f_m$, where f_M and f_m are defined as in the previous proof. Now define $W = S$ and proceed as before to get that there is a point $\eta \in [a,b]$ such that $f(\eta) = S$.   □

All three of the mean value theorems are useful in that they allow us to simplify certain expressions that will occur in the process of deriving error estimates for our approximate computations.

EXERCISES

1. Show that the third-order Taylor polynomial for $f(x) = (x+1)^{-1}$, about $x_0 = 0$, is

$$p_3(x) = 1 - x + x^2 - x^3.$$

2. What is the third-order Taylor polynomial for $f(x) = \sqrt{x+1}$, about $x_0 = 0$?

3. What is the sixth-order Taylor polynomial for $f(x) = \sqrt{1+x^2}$, using $x_0 = 0$? (Hint: Consider the previous problem.)

4. Given that

$$R(x) = \frac{|x|^6}{6!} e^{\xi}$$

for $x \in [-1,1]$, where ξ is between x and 0, find an upper bound for $|R|$, valid for all $x \in [-1,1]$, that is independent of x and ξ.

5. Repeat the above, but this time require that the upper bound be valid only for all $x \in [-\frac{1}{2},\frac{1}{2}]$.

6. Given that

$$R(x) = \frac{|x|^4}{4!} \left(\frac{-1}{1+\xi} \right)$$

for $x \in [-\frac{1}{2},\frac{1}{2}]$, where ξ is between x and 0, find an upper bound for $|R|$, valid for all $x \in [-\frac{1}{2},\frac{1}{2}]$, that is independent of x and ξ.

7. Use a Taylor polynomial to find an approximate value for $\sqrt{e}$ that is accurate to within 10^{-3}.

8. What is the fourth-order Taylor polynomial for $f(x) = 1/(x+1)$, about $x_0 = 0$?

9. What is the fourth-order Taylor polynomial for $f(x) = 1/x$, about $x_0 = 1$?

10. Find the Taylor polynomial of third order for $\sin x$, using:
(a) $x_0 = \pi/4$
(b) $x_0 = \pi/2$
(c) $x_0 = \pi/6$

11. For each of the following functions construct the third-order Taylor polynomial approximation, using $x_0 = 0$, and then estimate the error by computing an upper bound on the remainder, over the given interval.
(a) $f(x) = e^{-x}, x \in [0,1]$
(b) $f(x) = \sin x, x \in [0,\pi]$
(c) $f(x) = \ln(1+x), x \in [-\frac{3}{4}, \frac{3}{4}]$
(d) $f(x) = \ln(1+x), x \in [-1/2, 1/2]$
(e) $f(x) = 1/(x+1), x \in [-1/2, 1/2]$

12. Construct a Taylor polynomial approximation that is accurate to within 10^{-3}, over the indicated interval, for each of the following functions, using $x_0 = 0$.
(a) $f(x) = e^{-x}, x \in [0,1]$
(b) $f(x) = \sin x, x \in [0,\pi]$
(c) $f(x) = \ln(1+x), x \in [-\frac{3}{4}, \frac{3}{4}]$
(d) $f(x) = \ln(1+x), x \in [-1/2, 1/2]$
(e) $f(x) = 1/(x+1), x \in [-1/2, 1/2]$

13. Repeat the preceding, this time with a desired accuracy of 10^{-6}.

14. Since

$$\frac{\pi}{4} = \arctan 1,$$

we can estimate π by estimating $\arctan 1$. How many terms are needed in the Gregory series for the arctangent to approximate π to 100 decimal places? 1000? (Hint: Use the error term in the Gregory series to predict when the error gets sufficiently small.)

15. Elementary trigonometry can be used to show that

$$\arctan(1/239) = 4\arctan(1/5) - \arctan(1).$$

This formula was developed in 1706 by the English astronomer John Machin. Use this to develop a more efficient algorithm for computing π. How many terms are needed to get 100 digits of accuracy with this form?

How many terms are needed to get 1000 digits? (Historical note: Until 1961 this was the basis for the most commonly used method for computing π to high accuracy).

16. In 1896 a variation on Machin's formula was found:

$$\arctan(1/239) = \arctan(1) - 6\arctan(1/8) - 2\arctan(1/57),$$

and this began to be used in 1961 to compute π to high accuracy. How many terms are needed when using this expansion to get 100 digits of π? 1000 digits?

17. What is the Taylor polynomial of order 3 for $f(x) = x^4 + 1$, using $x_0 = 0$?

18. What is the Taylor polynomial of order 4 for $f(x) = x^4 + 1$, using $x_0 = 0$? Simplify as much as possible.

19. What is the Taylor polynomial of order 2 for $f(x) = x^3 + x$, using $x_0 = 1$?

20. What is the Taylor polynomial of order 3 for $f(x) = x^3 + x$, using $x_0 = 1$? Simplify as much as possible.

21. Let $p(x)$ be an arbitrary polynomial of degree less than or equal to n. What is its Taylor polynomial of degree n, about an arbitrary x_0?

22. The Fresnel integrals are defined as

$$C(x) = \int_0^x \cos(\pi t^2/2)\, dt$$

and

$$S(x) = \int_0^x \sin(\pi t^2/2)\, dt.$$

Use Taylor expansions to find approximations to $C(x)$ and $S(x)$ that are accurate to 10^{-4} for all x with $|x| \le \frac{1}{2}$. (Hint: Substitute $x = \pi t^2/2$ into the Taylor expansions for the cosine and sine.)

23. Use the Integral Mean Value Theorem to show that the "pointwise" form (1.3) of the Taylor remainder (usually called the *Lagrange* form) follows from the "integral" form (1.2) (usually called the *Cauchy* form).

24. For each function in Exercise 11, use the Mean Value Theorem to find a value M such that

$$|f(x_1) - f(x_2)| \le M|x_1 - x_2|$$

is valid for all x_1, x_2 in the interval used in Exercise 11.

25. A function is called *monotone* on an interval if its derivative is strictly positive or strictly negative on the interval. Suppose f is continuous and monotone on the interval $[a,b]$, and $f(a)f(b) < 0$; prove that there is exactly one value $\alpha \in [a,b]$ such that $f(\alpha) = 0$.

26. Finish the proof of the Integral Mean Value Theorem (Theorem 1.5) by writing up the argument in the case that g is negative.

27. Prove Theorem 1.6, providing all details.

28. Let $c_k > 0$ be given, $1 \leq k \leq n$, and let $x_k \in [a,b]$, $1 \leq k \leq n$. Then use the Discrete Average Value Theorem to prove that, for any function $f \in C([a,b])$,

$$\frac{\sum_{k=1}^{n} c_k f(x_k)}{\sum_{k=1}^{n} c_k} = f(\xi)$$

for some $\xi \in [a,b]$.

29. Discuss, in your own words, whether or not the following statement is true: "The Taylor polynomial of degree n is the best polynomial approximation of degree n to the given function near the point x_0."

1.2 ERROR, APPROXIMATE EQUALITY, AND ASYMPTOTIC ORDER NOTATION

We have already talked about the "error" made in a simple Taylor series approximation. Perhaps it is time we got a little more precise.

1.2.1 Error

If A is a quantity we want to compute and A_h is an approximation to that quantity, then the error is the difference between the two,

$$\text{error} = A - A_h;$$

the absolute error is simply the absolute value of the error,

$$\text{absolute error} = |A - A_h|; \tag{1.6}$$

and the relative error normalizes by the absolute value of the exact value,

$$\text{relative error} = \frac{|A - A_h|}{|A|}, \tag{1.7}$$

where we assume that $A \neq 0$.

Why do we need different measures of error? Consider the problem of approximating the number

$$x = e^{-16} = 0.1125351747 \times 10^{-6}.$$

Because x is so small, the absolute error in $y = 0$ as an approximation to x is also small. In fact, $|x - y| < 1.2 \times 10^{-7}$, which is decent accuracy in many settings. However, this "approximation" is clearly not a good one.

On the other hand, consider the problem of approximating

$$z = e^{16} = 0.8886110521 \times 10^7.$$

Because z is so large, the absolute error in almost any approximation will be large, even though almost all of the digits are matched. For example, if we take $w =$

0.8886110517×10^7, then we have $|z - w| = 4 \times 10^{-3}$—hardly very small, even though y matches z to nine decimal digits.

The point is that relative error gives a measure of the number of correct digits in the approximation. Thus,

$$\left| \frac{x - y}{x} \right| = 1,$$

telling us that not many digits are matched in that example, whereas

$$\left| \frac{z - w}{z} \right| = \frac{4 \times 10^{-3}}{0.8886110521 \times 10^7} = 0.4501 \times 10^{-9},$$

showing that about nine digits are correct. Generally speaking, using a relative error protects us from misjudging the accuracy of an approximation because of scale extremes (very large or very small numbers). As a practical matter, however, we sometimes are not able to obtain an error estimate in the relative sense.

In the definitions (1.6) and (1.7), we have used the subscript h to suggest that, in general, the approximation depends (in part, at least) on a parameter. For the most part, our computations will indeed be constructed this way, usually with either a real parameter h that tends toward zero, or with an integer parameter n that tends toward infinity. So we might want to think in terms of one of the two cases

$$\lim_{h \to 0} A_h = A$$

or

$$\lim_{n \to \infty} A_n = A.$$

In actual applied problems there are, of course, many sources of error: simple mistakes, measurement errors, modeling errors, etc. We are concerned here only with the computational errors caused by the need to construct computable approximations. The common terminology is *truncation error*, *approximation error*, or *mathematical error*.

1.2.2 Notation: Approximate Equality

If two quantities are approximately equal to each other, we will use the notation $\approx$ to denote this relationship, as in

$$A \approx B.$$

This is an admittedly vague notion. Is $0.99 \approx 1$? Probably so. Is $0.8 \approx 1$? Maybe not. We will almost always use the $\approx$ symbol in the sense of one of the two contexts outlined previously, of a parameterized set of approximations converging to a limit. Note that the definition of limit means

$$\lim_{h \to 0} A_h = A \quad \Rightarrow \quad A_h \approx A$$

for all h "sufficiently small" (and similarly for the case of $A_n \to A$ as $n \to \infty$, for n "sufficiently large"). For example, one way to write the definition of the derivative of a function $y = f(x)$ is

$$\lim_{h \to 0} \frac{f(x+h) - f(x)}{h} = f'(x).$$

We therefore conclude that, for h small enough,

$$\frac{f(x+h) - f(x)}{h} \approx f'(x).$$

Moreover, approximate equality does satisfy the transitive, symmetric, and reflexive properties of what in abstract algebra is called an "equivalence relation":

$$A \approx B, \quad B \approx C \quad \Rightarrow \quad A \approx C,$$
$$A \approx B \quad \Rightarrow \quad B \approx A,$$
$$A \approx A.$$

Consequently, we can manipulate approximate equalities much like ordinary equalities (i.e., equations). We can solve them, integrate both sides, etc.

Despite its vagueness, approximate equality is a useful notion to have around in a course devoted to approximations.

1.2.3 Notation: Asymptotic Order

Another notation of use is the so-called "big O" notation, more formally known as "asymptotic order" notation. Suppose we have a value y and a family of values $\{y_h\}$, each of which approximates this value,

$$y \approx y_h$$

for small values of h. If we can find a constant $C > 0$, independent of h, such that

$$|y - y_h| \le C\beta(h) \tag{1.8}$$

for all h sufficiently small, then we say that

$$y = y_h + O(\beta(h)), \qquad \text{as } h \to 0.$$

meaning that $y - y_h$ is "on the order of" $\beta(h)$. Here $\beta(h)$ is a function of the parameter h, and we assume that

$$\lim_{h \to 0} \beta(h) = 0.$$

The utility of this notation is that it allows us to concentrate on the important issue in the approximation—the way that the error $y - y_h$ depends on the parameter h, which is determined by $\beta(h)$—while ignoring unimportant details such as the precise size of the constant, C. The usage is similar if we have a sequence x_n that approximates a given value x for large values of n. If

$$|x - x_n| \le C\beta(n) \tag{1.9}$$

for all n sufficiently large, then we say that

$$x = x_n + O(\beta(n)), \qquad \text{as } n \to \infty.$$

The formal definitions are as follows.

Definition 1.1 *(Asymptotic Order Notation)   For a given value y, let $\{y_h\}$ be a set of values parameterized by h, which we assume is small, such that $y_h \approx y$ for small h. If there exists a positive function $\beta(h)$, $\beta(h) \to 0$ as $h \to 0$, and a constant $C > 0$, such that for all h sufficiently small,*

$$|y - y_h| \le C\beta(h),$$

then we say that

$$y = y_h + O(\beta(h)).$$

Similarly, if $\{y_n\}$ is a set of values parameterized by n, which we assume is large, such that $y_n \approx y$ for large n, and if there exists a positive function $\beta(n)$, $\beta(n) \to 0$ as $n \to \infty$, and a constant $C > 0$, such that for all n sufficiently large,

$$|y - y_n| \le C\beta(n),$$

then we say that

$$y = y_n + O(\beta(n)).$$

EXAMPLE 1.4

Let

$$A = \int_0^\infty e^{-2x}\, dx$$

$$A_n = \int_0^n e^{-2x}\, dx.$$

Simple calculus shows that $A = \frac{1}{2}$ and $A_n = \frac{1}{2} - \frac{1}{2}e^{-2n}$, so that we have $A = A_n + O(e^{-2n})$. Here $\beta(n) = e^{-2n}$. ∎

EXAMPLE 1.5

Another example—and many such, in fact—can be generated from Taylor's Theorem. We have

$$\cos x = 1 - \frac{1}{2}x^2 \cos \xi_x,$$

where ξ_x is between x and 0. Since $\cos \xi_x$ is bounded by 1 in absolute value, we easily have

$$|\cos h - 1| \le \frac{1}{2}h^2,$$

so we can write $\cos h = 1 + O(h^2)$. Similarly, we can write $e^h = 1 + h + O(h^2)$. In both these cases we have $\beta(h) = h^2$. ∎

The following theorem shows how the asymptotic order relationship can be manipulated.

Theorem 1.7   Let $y = y_h + O(\beta(h))$ and $z = z_h + O(\gamma(h))$, with $b\beta(h) > \gamma(h)$ for all h near zero. Then

$$y + z = y_h + z_h + O(\beta(h) + \gamma(h))$$
$$y + z = y_h + z_h + O(\beta(h))$$
$$Ay = Ay_h + O(\beta(h)).$$

In the last equation, A is an arbitrary constant, independent of h.

Proof   We simply show that each relationship satisfies (1.8) for some constant C. For example,

$$|(y + z) - (y_h + z_h)| \leq |y - y_h| + |z - z_h|$$
$$\leq C_1\beta(h) + C_2\gamma(h)$$
$$\leq C(\beta(h) + \gamma(h)),$$

where $C = \max(C_1, C_2)$. Thus $y + z = y_h + z_h + O(\beta(h) + \gamma(h))$. Moreover, since $b\beta(h) > \gamma(h)$, we also have

$$|(y + z) - (y_h + z_h)| \leq C(\beta(h) + \gamma(h))$$
$$\leq C(\beta(h) + b\beta(h))$$
$$\leq C(1 + b)\beta(h).$$

Also,

$$|Ay - A_{y_h}| = A|y - y_h|$$
$$\leq C_1 A\beta(h)$$
$$= C\beta(h),$$

so that $Ay = Ay_h + O(\beta(h))$.   □

A similar result holds for $u_n \approx u$ with $u = u_n + O(\beta(n))$, for $v_n \approx v$ with $v = v_n + O(\gamma(n))$, etc.

EXAMPLE 1.6

We close this section with a simple example that illustrates the utility of the O notation. Consider the combination of function values

$$D = -f(x + 2h) + 4f(x + h) - 3f(x),$$

where f is assumed to be continuous and smooth, and h is a (small) parameter. We can use Taylor's Theorem, together with the definition of the O notation, to write

$$f(x + h) = f(x) + hf'(x) + \frac{1}{2}h^2 f''(x) + O(h^3)$$

and

$$f(x+2h) = f(x) + 2hf'(x) + 2h^2 f''(x) + O(h^3).$$

Therefore,

$$
\begin{aligned}
D &= -f(x+2h) + 4f(x+h) - 3f(x) \\
&= -(f(x) + 2hf'(x) + 2h^2 f''(x) + O(h^3)) \\
&\quad + 4(f(x) + hf'(x) + \frac{1}{2}h^2 f''(x) + O(h^3)) - 3f(x) \\
&= (-1 + 4 - 3)f(x) + (-2h + 4h)f'(x) + (-2h^2 + 2h^2)f''(x) + O(h^3) \\
&= 2hf'(x) + O(h^3).
\end{aligned}
$$

If we then solve this for $f'(x)$, we get

$$f'(x) = \frac{-f(x+2h) + 4f(x+h) - 3f(x)}{2h} + O(h^2),$$

where we have used the fact that $O(h^3)/h = O(h^2)$ (see Exercise 10); thus we can use the expression on the right as an approximation to the derivative, and the remainder will be bounded by a constant times h^2. See Sections 2.2 and 4.5 for more on approximations to the derivative. This particular approximation is derived again, by alternate means, in Section 4.5. ∎

Note that we have rather quickly obtained an approximation to the first derivative, along with some notion of the error—it behaves proportionally to h^2—by using the O notation.

EXERCISES

1. Use Taylor's Theorem to show that $e^x = 1 + x + O(x^2)$ for x sufficiently small.

2. Use Taylor's Theorem to show that $(1 - \cos x)/x = \frac{1}{2}x + O(x^3)$ for x sufficiently small.

3. Use Taylor's Theorem to show that

$$\sqrt{1+x} = 1 + \frac{1}{2}x + O(x^2)$$

for x sufficiently small.

4. Use Taylor's Theorem to show that

$$(1+x)^{-1} = 1 - x + x^2 + O(x^3)$$

for x sufficiently small.

5. Show that

$$\sin x = x + O(x^3).$$

6. Recall the summation formula

$$1 + r + r^2 + r^3 + \cdots + r^n = \sum_{k=0}^{n} r^k = \frac{1 - r^{n+1}}{1 - r}.$$

Use this to prove that

$$\sum_{k=0}^{n} r^k = \frac{1}{1-r} + O(r^{n+1}).$$

(Hint: What is the *definition* of the O notation?)

7. Use the preceding to show that nine terms are all that are needed to compute

$$um_{k=0}^{\infty} e^{-k}$$

to within 10^{-4} absolute accuracy.

8. Recall the summation formula

$$\sum_{k=1}^{n} k = \frac{n(n+1)}{2}.$$

Use this to show that

$$\sum_{k=1}^{n} k = \frac{1}{2}n^2 + O(n).$$

9. State and prove the version of Theorem 1.7 that deals with relationships of the form $x = x_n + O(\beta(n))$.

10. Use the definition of O to show that if $y = y_h + O(h^p)$, then $hy = hy_h + O(h^{p+1})$.

11. Show that if $a_n = O(n^p)$ and $b_n = O(n^q)$, then $a_n b_n = O(n^{p+q})$.

12. Suppose that $y = y_h + O(\beta(h))$ and $z = z_h + O(\beta(h))$, for h sufficiently small. Does it follow that $y - z = y_h - z_h$ (for h sufficiently small)?

13. Show that

$$f''(x) = \frac{f(x+h) - 2f(x) + f(x-h)}{h^2} + O(h^2)$$

for all h sufficiently small. (Hint: Expand $f(x \pm h)$ out to the fourth-order terms.)

14. Explain, in your own words, why it is necessary that the constant C in (1.8) be independent of h.

1.3 A PRIMER ON COMPUTER ARITHMETIC

We need to spend some time reviewing how the computer actually does arithmetic. The reason for this is simple: Computer arithmetic is generally inexact, and while the errors that are made are very small, they can accumulate under some circumstances and actually dominate the calculation. Thus we need to understand computer arithmetic well enough to anticipate and deal with this phenomenon.

Most computer languages use what is called *floating-point arithmetic*. Although the details will differ from machine to machine, the basic idea is the same. Every number is represented using a (fixed, finite) number of binary digits, usually called *bits*. A typical implementation would represent the number in the form

$$x = \sigma \times f \times \beta^{t-p}.$$

Here σ is the sign of the number (± 1), denoted by a single bit; f is the mantissa or fraction; β is the base of the internal number system, usually binary ($\beta = 2$) or hexadecimal ($\beta = 16$), although other systems have sometimes been used; t is the (shifted) exponent, i.e., the value that is actually stored; and p is the shift required to recover the actual exponent. (Shifting in the exponent is done to avoid the need for a sign bit in the exponent itself.) The number would be stored by storing only the values of σ, f, and t. The standard way to represent a computer word containing a floating-point number is as follows:

σ	t	f

To keep things simple, we will use a base 2 representation for our floating-point examples here.

The total number of bits devoted to the number is fixed by the computer architecture, and the fraction and exponent are each allowed a certain number of bits. For example, a common situation allows 32 bits for the entire word,[3] assigned as follows: 24 bits for the fraction, 7 bits for the exponent, and a single bit for the sign.

---

[3] A "word" is the largest unit of computer storage. Usually a word consists of two or more *bytes*, which themselves consist of a certain number of bits, typically 8.

Note that this imposes limits on the numbers that can be represented. For example, a 7-bit exponent means that

$$0 \leq t \leq 127.$$

In order to allow for an equal range of positive and negative exponents, the shift should be taken to be $p = 63$, so that

$$-63 \leq t - p \leq 64.$$

Attempts to create larger exponents result in what is called an *overflow*. Attempts to create smaller exponents result in *underflow*.[4] The fraction is also limited in size by the number of bits available:

$$0 \leq f \leq \sum_{k=1}^{24} 2^{-k} = 1 - 2^{-24}.$$

In practice, most architectures assume that the fraction is *normalized* to be between fi^{-1} and 1; any leading zeroes would be dropped and the exponent adjusted accordingly.[5] Thus we actually have

$$\frac{1}{2} \leq f \leq \sum_{k=1}^{24} 2^{-k} = 1 - 2^{-24}.$$

The errors in computer arithmetic come about because the floating-point numbers are allowed only a fixed number of bits, and not every number can be exactly represented with a fixed number of bits. The common name for the error that is caused by this finite representation is *rounding error*.

Let's consider the simple addition of two numbers,

$$x = 0.1, \qquad y = 0.00003,$$

assuming that they are represented in the scheme outlined above. The exact answer, of course, is

$$z = x + y = 0.10003.$$

We first note that neither of these numbers can be exactly represented in our scheme. The best we can do is[6]

$$\tilde{x} = 0.00011001 \quad 10011001 \quad 10011001 \quad 100_2$$
$$= 0.0999999940395...$$

---

[4]Most modern computers adhere to the so-called IEEE standard for arithmetic, which uses a kind of "extended floating-point" number system. In addition to ordinary numbers, the IEEE standard allows for the results Inf (infinity) and NaN (not a number), and includes rules for manipulating with ordinary floating-point numbers and these special values. For example, x/Inf yields 0 as a result, while x/0 yields plus or minus Inf, depending on the sign of x. Most manipulations with NaN return NaN as their result. In older arithmetic schemes, if an overflow or division by zero occurred, program execution usually terminated.

[5]Some architectures take advantage of this assumption to avoid actually storing that leading bit—all the basic arithmetic algorithms are written to assume an extra leading 1—and thus they are able to get 25 bits of information into 24 bits of storage space.

[6]Note that we have used the subscript 2 to indicate that the number should be interpreted as a base 2 fraction.

and

$$\tilde{y} = 0.00000000 \quad 00000001 \quad 11110111 \quad 01010001 \quad 0000010_2$$
$$= 0.0000299999992421...$$

Thus we have (admittedly small) errors being made even before the addition occurs.

In our floating-point scheme these two numbers would be stored as

$$\tilde{x} = 0.11001100 \quad 11001100 \quad 11001100_2 \times 2^{60-63}$$
$$\tilde{y} = 0.11111011 \quad 10101000 \quad 1000010_2 \times 2^{49-63}.$$

Because the two numbers are of somewhat different sizes, a normalization is required in order to get equal exponents. One way to do this (the precise details depend on the particular computer architecture) would give us

$$\tilde{x} = 1100 \quad 11001100 \quad 11001100 \quad 11000000 \quad 00000000_2 \times 2^{-39}$$
$$\tilde{y} = 0000 \quad 00000000 \quad 11111011 \quad 10101000 \quad 1000010_2 \times 2^{-39},$$

and we can now add the mantissas to get the sum, which is first written as

$$\tilde{w} = 1100 \quad 11001101 \quad 11001000 \quad 01101000 \quad 1000010_2 \times 2^{-39}.$$

Note that the fraction here is too long to be stored with only 24 bits. What is done with the extra bits? Depending on the machine, they are either thrown away, regardless of their size (chopping), or the result is rounded up or down, depending on size (rounding). Rounding is more accurate, of course, but chopping is faster. If we round, we have

$$\tilde{z} = 11001100 \quad 11011100 \quad 1000011_2 \times 2^{-27}$$
$$= 0.11001100 \quad 11011100 \quad 1000011_2 \times 2^{60-63}$$
$$= 0.100029997527599...$$

and the error is

$$|z - \tilde{z}| = 0.24724... \times 10^{-8}.$$

Let's next assume that our machine uses chopping, in which case we end up with

$$\tilde{z} = 11001100 \quad 11011100 \quad 1000011_2 \times 2^{-27}$$
$$= 0.11001100 \quad 11011100 \quad 1000011_2 \times 2^{60-63}$$
$$= 0.1000299900770...$$

and a final error of

$$|z - \tilde{z}| = 0.992298... \times 10^{-8}.$$

Note that the chopping error is indeed larger than the error with rounding. Similar errors would, of course, occur with the other arithmetic operations.

The difference here—whether we chop or round—is indeed very small, and we might be tempted to ignore it as being too small to worry about. In fact, this is usually the case. But it is possible for the effects of different rounding errors to combine in such a way as to dominate and ruin a calculation. We can illustrate this point with simple decimal arithmetic so long as we insist on using only a small number of digits.

Consider an 8-digit approximation to $a = e^{-(1/100)^2} = 0.99990001$ and a similar approximation to $b = e^{-(1/1000)^2} = 0.99999900$. By construction, both of these numbers are accurate to 8 decimal digits. What about their difference, $c = a - b = -0.00009899$? How many accurate digits do we have here? The answer is, only 4. Because we were subtracting two nearly equal numbers, we lost a great deal of accuracy. This phenomenon is called *subtractive cancellation*. If we had started with more accurate approximations, then the difference would contain more accurate digits; try this by looking at the 16-digit values

$$a = 0.9999000049998333$$
$$b = 0.9999990000005000$$
$$c = -0.0000989950006667.$$

The result c is now accurate to 12 digits.

To see how subtractive cancellation can destroy almost all the accuracy in a calculation, consider the quantity

$$D = (f(x_1) - f(x_2)) - (f(x_2) - f(x_3)),$$

where $f(x) = e^{-x^2}$ and $x_1 = 999/10,000$, $x_2 = 1/10$, $x_3 = 1001/10,000$. Then the calculation, as organized above and done in 8-digit arithmetic, yields $D = -0.1 \times 10^{-7}$. But when we do it in 16-digit arithmetic, we get $D = -0.194049768 \times 10^{-7}$. The 8-digit calculation had *no* accurate digits. Subtractive cancellation is therefore something to avoid as much as possible, and to be aware of when it is unavoidable.

Sometimes the problem with rounding error can be eliminated by increasing the precision of the computation. Traditionally, floating-point arithmetic systems used a single word for each number ("single precision") by default, and a second word could be used ("double precision") by properly specifying the type of data format to be used. Most languages now use double-word arithmetic by default. Sometimes the entire extra word is used to extend the length of the fraction; sometimes the length of the exponent is extended as well. *If changing the precision of a calculation dramatically changes the results, then it is almost certain that the computation is being seriously affected by rounding errors.*

Another example of subtractive cancellation occurs with the evaluation of the function

$$f(x) = \frac{e^x - 1}{x}$$

for values of x near zero. L'Hôpital's Rule[7] can be easily used to show that

$$\lim_{x \to 0} f(x) = 1,$$

---

[7]Guillaume François Antoine Marquis de L'Hôpital (1661–1704) was not educated as a mathematician, but took it up after resigning from the military due to poor eyesight. He studied with Johann Bernoulli in the 1690s, and in 1692 published what is generally regarded to be the first calculus textbook, *Analyse des infiniment petits pour l'intelligence des lignes courbes*. What we know as "L'Hôpital's Rule" first appears in this book, and is probably actually due to Bernoulli.

and Taylor's Theorem can be used to show that

$$f(x) = 1 + \frac{1}{2}x + O(x^2)$$

for small x, but the evaluation of f for x near 0 will exhibit subtractive cancellation that is amplified by the division by x (since x is small precisely when the subtractive cancellation is worst). Table 1.1 shows the results of computing f using single-precision and double-precision floating-point arithmetic. Note that the error in the single-precision results increases dramatically for $x \leq 2^{-12} = 1/4096$, which is not that small a number. A second threshold of inaccuracy is reached at around $x = 2^{-24}$. Note that the use of double-precision arithmetic defers the onset and lessens the severity of the error, but does not eliminate it entirely, as the last few rows show. (Even though the limiting value of f is 1, the error in the computed value for $x = 2^{-30}$ is still nonzero; $f(2^{-30}) - 1 = 0.4657 \times 10^{-9}$. While this error would be acceptable in single-precision arithmetic, it is not acceptable for double-precision arithmetic.) How do we fix this?

One approach would be to use Taylor's Theorem:

$$
\begin{aligned}
f(x) &= \frac{\left(1 + x + \frac{1}{2}x^2 + \frac{1}{6}x^3 + \cdots \frac{1}{n!}x^n + \frac{1}{(n+1)!}x^{n+1}e^{c_x}\right) - 1}{x} \\
&= 1 + \frac{1}{2}x + \frac{1}{6}x^2 + \cdots + \frac{1}{n!}x^{n-1} + \frac{1}{(n+1)!}x^n e^{c_x},
\end{aligned}
$$

where c_x is between x and 0; the value of n will depend on our required accuracy. We would thus define f, for computational purposes, as

$$
f(x) = \begin{cases} 1 + \frac{1}{2}x + \frac{1}{6}x^2 + \cdots + \frac{1}{n!}x^{n-1} & |x| \text{ close to } 0 \\ x^{-1}(e^x - 1) & \text{otherwise.} \end{cases}
$$

If we wanted more accuracy, we would use more terms in the Taylor expansion.

We close this section with a definition. In a floating-point computer system, there will exist many nonzero numbers x such that

$$1 + x = 1$$

within computer precision. For instance, in the 32-bit implementation outlined at the beginning of the section, it is clear that this will hold for any $x < 2^{-24}$, since $1 + 2^{-24}$ in binary will require 25 bits of storage, and we only have 24 to work with. We define the *machine rounding unit*, or *machine epsilon*, **u**, to be the largest such number.

Definition 1.2 (*Machine Epsilon*)   *The machine epsilon (alternatively, the machine rounding unit),* **u***, is the largest floating-point number x such that $x + 1$ cannot be*

TABLE 1.1 Illustration of Subtractive Cancellation, Using $f(x) = e^x - 1/x$.

k	$x_k = 2^{-k}$	Single precision	Double precision
1	0.50000000000000E+00	0.12974424362183E+01	0.12974425414003E+01
2	0.25000000000000E+00	0.11361017227173E+01	0.11361016667510E+01
3	0.12500000000000E+00	0.10651874542236E+01	0.10651876245346E+01
4	0.62500000000000E−01	0.10319118499756E+01	0.10319113426858E+01
5	0.31250000000000E−01	0.10157890319824E+01	0.10157890399713E+01
6	0.15625000000000E−01	0.10078506469727E+01	0.10078533495479E+01
7	0.78125000000000E−02	0.10039215087891E+01	0.10039164424253E+01
8	0.39062500000000E−02	0.10019531250000E+01	0.10019556706170E+01
9	0.19531250000000E−02	0.10009765625000E+01	0.10009771985934E+01
10	0.97656250000000E−03	0.10004882812500E+01	0.10004884402344E+01
11	0.48828125000000E−03	0.10002441406250E+01	0.10002441803663E+01
12	0.24414062500000E−03	0.10000000000000E+01	0.10001220802469E+01
13	0.12207031250000E−03	0.10000000000000E+01	0.10000610376392E+01
14	0.61035156250000E−04	0.10000000000000E+01	0.10000305182002E+01
15	0.30517578125000E−04	0.10000000000000E+01	0.10000152589419E+01
16	0.15258789062500E−04	0.10000000000000E+01	0.10000076294382E+01
17	0.76293945312500E−05	0.10000000000000E+01	0.10000038146973E+01
18	0.38146972656250E−05	0.10000000000000E+01	0.10000019073486E+01
19	0.19073486328125E−05	0.10000000000000E+01	0.10000009536743E+01
20	0.95367431640625E−06	0.10000000000000E+01	0.10000004768372E+01
21	0.47683715820312E−06	0.10000000000000E+01	0.10000002384186E+01
22	0.23841857910156E−06	0.10000000000000E+01	0.10000001192093E+01
23	0.11920928955078E−06	0.10000000000000E+01	0.10000000596046E+01
24	0.59604644775391E−07	0.00000000000000E+00	0.10000000298023E+01
25	0.29802322387695E−07	0.00000000000000E+00	0.10000000149012E+01
26	0.14901161193848E−07	0.00000000000000E+00	0.10000000000000E+01
27	0.74505805969238E−08	0.00000000000000E+00	0.10000000000000E+01
28	0.37252902984619E−08	0.00000000000000E+00	0.10000000000000E+01
29	0.18626451492310E−08	0.00000000000000E+00	0.10000000000000E+01
30	0.93132257461548E−09	0.00000000000000E+00	0.10000000000000E+01

distinguished from 1 *on the computer:*

$$\mathbf{u} = \max\{x \mid 1 + x = 1, in\ computer\ arithmetic\}.$$

It is possible to compute the machine epsilon based on knowledge of the floating-point number system on the computer in question.

EXAMPLE 1.7

Suppose we want to compute the machine epsilon for a simple three-digit decimal computer that rounds, i.e., a machine that does decimal arithmetic, keeps only three digits, and rounds its results. Thus we can write any number x that is stored on the computer as $x = d_1 d_2 d_3 \times 10^t$, where t is the exponent and $d_1 d_2 d_3$ represent the decimal digits of the fraction. Now, based on the definition of machine epsilon, we know that $x_1 = 1.00 \times 10^{-2}$ is too large, because

$$1 + x_1 = 1.00 + 0.01 = 1.01 \neq 1.00.$$

On the other hand, we know that $x_2 = 1.00 \times 10^{-3}$ is certainly small enough (perhaps too small) because we have

$$1 + x_2 = 1.00 + 0.001 = 1.001 \quad \Rightarrow \quad 1.00 = 1.00.$$

Thus, the computer cannot distinguish between 1 and $1 + x_2$. But $\mathbf{u}$ is the *largest* such number, so we have to look a bit further. It's not a matter of using the right formula; it is really a matter of experimentation and trial and error. We have

$$1 + 0.002 = 1.002 \quad \Rightarrow \quad 1.00 = 1,$$

so we know that $x_3 = 2.00 \times 10^{-3}$ is small enough; the same argument would apply to $x_4 = 3.00 \times 10^{-3}$ and $x_5 = 4.00 \times 10^{-3}$. But

$$1 + 0.005 = 1.005 \quad \Rightarrow \quad 1.01 \neq 1$$

(because the machine rounds its computations); thus $x_6 = 5.00 \times 10^{-3}$ is too large. But it is just barely too large, as any smaller number would have resulted in the sum being rounded down to 1.00. Thus we want the next smaller number *within the floating-point number system*; i.e., we want $\mathbf{u} = 4.99 \times 10^{-3}$. This gives us

$$1.00 + 0.00499 = 1.00499 \quad \Rightarrow \quad 1.00 = 1,$$

and it is clear that any larger number that we can represent in our floating-point system would result in the sum being rounded up to $1.01 \neq 1$.

The student ought to be able to show, now, that if the machine chops instead of rounds, $\mathbf{u} = 9.99 \times 10^{-3}$. ∎

There are a number of interesting consequences of the existence of the machine epsilon. For example, consider the numbers

$$a = 1, \qquad b = \mathbf{u}, \qquad c = \mathbf{u}.$$

If we want to add up these numbers, we quickly learn that the way we organize the calculation matters. For example,

$$(a + b) + c = (1 + \mathbf{u}) + \mathbf{u} = 1 + \mathbf{u} = 1,$$

whereas

$$a + (b + c) = 1 + 2\mathbf{u} \neq 1.$$

From a technical point of view, this means that the associative law of arithmetic (for addition) does not hold in floating-point arithmetic systems. In other words, *the order in which we do operations sometimes matters*.

It is possible to estimate the machine epsilon by constructing a loop that adds increasingly small numbers to 1, and terminates this only when the result cannot be distinguished from 1.

The basic result for computer arithmetic is the following.

Theorem 1.8 *(Computer Arithmetic Error)   Let ∗ denote any of the basic binary operations (addition, subtraction, multiplication, division), and let $fl(x)$ denote the floating-point value of x. Then there exists a constant $C > 0$ such that, for all x and y,*

$$|x * y - fl(x * y)| \le C\mathbf{u}|x * y|.$$

Thus the computer value of $x * y$ is relatively accurate to within $O(\mathbf{u})$.

The point of this section is that we should be aware that computer arithmetic is not the 100% reliable process we might have thought it was. However, the errors that crop up do tend to appear only when extremely large or small numbers are involved (exceptions do occur, so we have to be careful), and are themselves very small. Moreover, rounding errors tend to cancel themselves out over the long run. Rounding error and related effects are things we have to watch out for and be aware of, but they are *usually* dominated by the mathematical error that is made in constructing the approximations. Exceptions to this rule of thumb do exist, however, as we shall see in Sections 2.2 and 4.11.1.

EXERCISES

1. In each example below, A is the exact value and A_h is an approximation to A. Find the absolute error and the relative error.

(a) $A = \pi$, $A_h = 22/7$

(b) $A = e$, $A_h = 2.71828$

(c) $A = \frac{1}{6}$, $A_h = 0.1667$

(d) $A = \frac{1}{6}$, $A_h = 0.1666$

2. Perform the indicated computations in each of three ways: (i) exactly, (ii) using three-digit decimal arithmetic with chopping, and (iii) using three-digit decimal arithmetic with rounding.   For both approximations, compute the absolute error and the relative error.

(a) $\frac{1}{6} + \frac{1}{10}$

(b) $\frac{1}{6} \times \frac{1}{10}$

(c) $\left(\frac{1}{7} + \frac{1}{10}\right) + \frac{1}{9}$

(d) $\frac{1}{7} + \left(\frac{1}{10} + \frac{1}{9}\right)$

3. For each function below explain why a naive construction will be susceptible to significant rounding error (for x near certain values), and explain how to avoid this error.

(a) $f(x) = (\sqrt{x+9} - 3)x^{-1}$

(b) $f(x) = x^{-1}(1 - \cos x)$

(c) $f(x) = (1-x)^{-1}(\ln x - \sin \pi x)$

(d) $f(x) = (\cos(\pi + x) - \cos \pi)x^{-1}$

(e) $f(x) = (e^{1+x} - e^{1-x})(2x)^{-1}$

4. For $f(x) = (e^x - 1)/x$, how many terms in a Taylor expansion are needed to get single-precision accuracy (7 decimal digits) for all $x \in [0, \frac{1}{2}]$? How many terms are needed for double-precision accuracy (14 decimal digits) over this same range?

5. Using single-precision arithmetic only, carry out each of the following computations, first using the form on the left side of the equals sign, and then using the form on the right side, and compare the two results. Comment on what you get in light of the material in this section.

(a) $(x+\epsilon)^3 - 1 = x^3 + 3x^2\epsilon + 3x\epsilon^2 + \epsilon^3 - 1$, $x = 1.0$, $\epsilon = 0.000001$

(b) $-b + \sqrt{b^2 - 2c} = 2c(-b - \sqrt{b^2 - 2c})^{-1}$, $b = 1,000$, $c = \pi$

6. Consider the sum

$$S = \sum_{k=0}^{m} e^{-14(1 - e^{-0.05k})},$$

where $m = 2 \times 10^5$. Again using only single precision, compute this in two ways: first by summing in the order indicated in the formula, and second by summing *backward*, i.e., starting with the $k = 200,000$ term and ending with the $k = 0$ term. Compare your results and comment on them.

7. Find three values a, b, and c such that

$$(a + b) + c \neq a + (b + c)$$

on the computer of your choice. Repeat for your pocket calculator.

8. Assume we are using three-digit decimal arithmetic. For $\epsilon = 0.0001$, $a_1 = 5$, compute

$$a_2 = a_0 + \left(\frac{1}{\epsilon}\right) a_1$$

for a_0 equal to each of 1, 2, and 3. Comment.

9. Let $\epsilon \leq \mathbf{u}$. Explain, in your own words, why the computation

$$a_2 = a_0 + \left(\frac{1}{\epsilon}\right) a_1$$

is potentially rife with rounding error. (Assume that a_0 and a_1 are of comparable size.) (Hint: See the previous problem.)

10. Using the computer and language of your choice, write a program to estimate the machine epsilon.

11. We can compute e^{-x} using Taylor polynomials in two ways, either using

$$e^{-x} \approx 1 - x + \frac{1}{2}x^2 - \frac{1}{6}x^3 + \cdots$$

or using

$$e^{-x} \approx \frac{1}{1 + x + \frac{1}{2}x^2 + \frac{1}{6}x^3 + \cdots}.$$

Discuss, in your own words, which approach is more accurate. In particular, which one is more (or less) susceptible to rounding error?

12. What is the machine epsilon for a computer that uses binary arithmetic, 24 bits for the fraction, and rounds? What if it chops?

13. What is the machine epsilon for a computer that uses *octal* (base 8) arithmetic, assuming it retains eight octal digits in the fraction?

1.4 A WORD ON COMPUTER LANGUAGES AND SOFTWARE

In the early 1970s the standard computer language for scientific computation was FORTRAN, with Algol perhaps the second choice. BASIC, in many incarnations, was also a possibility. By the late 1980s, Pascal had entered the fray and FORTRAN was considered passé in some quarters, especially since various easy-to-use integrated environment packages for Pascal were being marketed to personal computer users. By the early 1990s, Pascal was fading away, but we now had C, or C++, or even Java. And FORTRAN, despite the predictions of many and the desires of a few, was still with us. In addition, the increased power of personal computers meant that software packages such as MATLAB, Mathcad, Maple, or Mathematica might be used to do scientific computation. (MATLAB, in particular, has had a tremendous influence on scientific computation.)

In short, if you don't like the present state of affairs with regard to computing languages, wait around a little while; it will change.

Given the inevitable changes in computing languages, this text strives to avoid the entire issue by being as "language neutral" as possible. Most examples will be given in a generic pseudocode. The idea is to make the text as independent as possible of anyone's prejudices about languages, and also to pay attention to the notion that the student should be comfortable in as many different languages as possible; even though most programming *today* might be done in C or C++ or Java, it is still the case that there is a lot of computer code (called *legacy code*) still being used that was written in FORTRAN, Pascal, or Algol.

Finally, the more involved algorithms will be presented in more of an "outline" style, rather than a line-by-line code style.

A number of sources exist for good mathematical software. Traditionally, two of the best sources were the IMSL and NAG libraries, collections of FORTRAN routines for a wide variety of computational tasks. More specialized packages have also been developed, notably QUADPACK (numerical integration), LINPACK (linear algebra), EISPACK (eigenvalue methods), and LAPACK (an updated package that combines and actually replaces LINPACK and EISPACK). A repository of public-domain mathematics software is maintained at Netlib.[8] Recently, of course, we have seen the development of commercial packages such as MATLAB, Mathcad, Maple, and Mathematica.

Despite the wide availability of general-purpose mathematical software, it is still important for the student to learn how to write and (most important) debug their own codes. For this reason, many of the exercises in this text involve computer programming.

The text assumes that students are familiar with the use of elementary packages (such as Maple or Mathematica) for producing simple plots of functions. Whenever the exercises call for graphs or plots to be produced, it is assumed that such modern technology will be used. Students should also feel free (with their instructor's permission, of course) to use Maple or Mathematica to simplify some of the more involved manipulations in the exercises.

1.5 SIMPLE APPROXIMATIONS

We have already used Taylor series to construct a simple approximation to the exponential function. Here we will use a similar but slightly more involved example to reinforce the basic ideas and also to illustrate the usage of the asymptotic order notation introduced earlier.

The *error function* occurs often in probability theory and other areas of applied mathematics (the solution of heat conduction problems, for example). It is defined by an integral:

$$\operatorname{erf}(x) = \frac{2}{\sqrt{\pi}} \int_0^x e^{-t^2} dt.$$

It is not possible to evaluate this integral by means of the fundamental theorem of calculus; there is no elementary antiderivative for e^{-t^2}. (In Chapter 5 we will derive techniques that can be applied to directly approximate the integral.) Here we will use Taylor's Theorem to approximate the integrand as a polynomial, and exactly integrate that polynomial.

---

[8]Its site on the World Wide Web is http://www.netlib.org.

This is a fundamental idea in numerical methods: When confronted with a computation that cannot be done exactly, we often replace the relevant function with something simpler that approximates it, and carry out the computation exactly *on the simple approximation.*

We might be tempted to compute the Taylor approximation to e^{-t^2} by appealing directly to formula (1.1), but this will lead to a lot of unnecessary work. We know that

$$e^x = 1 + x + \frac{1}{2!}x^2 + \frac{1}{3!}x^3 + \cdots + \frac{1}{k!}x^k + \frac{x^{k+1}}{(k+1)!}e^c$$
$$= p_k(x) + R_k(x),$$

where

$$p_k(x) = \sum_{i=0}^{k} \frac{x^i}{i!}$$

and

$$R_k(x) = \frac{x^{k+1}}{(k+1)!}e^c.$$

Therefore, we can get the expansion and error for e^{-t^2} by a simple substitution:

$$e^{-t^2} = p_k(-t^2) + R_k(-t^2).$$

Thus we have

$$\text{erf}(x) = \frac{2}{\sqrt{\pi}} \int_0^x p_k(-t^2)dt + \frac{2}{\sqrt{\pi}} \int_0^x R_k(-t^2)\,dt$$
$$= \frac{2}{\sqrt{\pi}} \left(\sum_{i=0}^{k} \frac{(-1)^i x^{2i+1}}{(2i+1)i!} \right) + \frac{2}{\sqrt{\pi}} \int_0^x R_k(-t^2)\,dt.$$

For simplicity, define the polynomial that approximates the error function by q_k:

$$q_k(x) = \frac{2}{\sqrt{\pi}} \left(\sum_{i=0}^{k} \frac{(-1)^i x^{2i+1}}{(2i+1)i!} \right)$$
$$= \frac{2}{\sqrt{\pi}} \left(x - \frac{1}{3}x^3 + \frac{1}{10}x^5 - \frac{1}{42}x^7 + \cdots + \frac{(-1)^k x^{2k+1}}{(2k+1)k!} \right)$$

so that we have

$$\text{erf}(x) - q_k(x) = \frac{2}{\sqrt{\pi}} \int_0^x R_k(-t^2)\,dt.$$

We want to simplify this error and produce an error bound, so we will also set

$$E_k(x) = \frac{2}{\sqrt{\pi}} \int_0^x R_k(-t^2)\,dt.$$

Thus,

$$\text{erf}(x) - q_k(x) = E_k(x). \tag{1.10}$$

To simplify and bound the error, we note that

$$
\begin{aligned}
E_k(x) &= \frac{2}{\sqrt{\pi}} \int_0^x R_k(-t^2)\, dt \\
&= \frac{2}{\sqrt{\pi}} \int_0^x \frac{(-t^2)^{k+1}}{(k+1)!} e^c\, dt \\
&= \frac{2(-1)^{k+1}}{(k+1)!\sqrt{\pi}} \int_0^x t^{2k+2} e^c\, dt,
\end{aligned}
$$

where c depends on t, and $-t^2 \le c \le 0$. Now, since both functions in the integrand above are positive, we can apply the Integral Mean Value Theorem to get

$$\int_0^x t^{2k+2} e^c\, dt = e^\xi \int_0^x t^{2k+2}\, dt = e^\xi \frac{x^{2k+3}}{2k+3}$$

for some ξ between 0 and $-x^2$. Thus,

$$E_k(x) = \frac{2(-1)^{k+1} x^{2k+3}}{(2k+3)(k+1)!\sqrt{\pi}} e^\xi. \tag{1.11}$$

This error statement, although applicable only to a very elementary problem, nonetheless contains all of the common features that will appear in the error estimates for more realistic algorithms applied to more complicated problems. Thus it is worthwhile to discuss it before we proceed to simplify it further.

Let us write the error in a more structured form:

$$E_k(x) = \left(\frac{2}{\sqrt{\pi}} \right) \left(\frac{(-1)^{k+1} x^{2k+3}}{(2k+3)(k+1)!} \right) e^\xi = C\delta_k(x)M$$

where

$$C = \frac{2}{\sqrt{\pi}}, \qquad \delta_k(x) = \frac{(-1)^{k+1} x^{2k+3}}{(2k+3)(k+1)!}, \qquad M = e^\xi.$$

This divides the error into three distinct parts:

1. A raw numeric constant, C B

2. An expression depending on the computational parameters (k, x) of the problem; this is $\delta_k(x)$

3. A factor that depends on the function or its derivatives, evaluated at an indeterminant, unknown point; this is M

We are most interested in the second part of the error. The raw numeric constant is usually of less interest, since we cannot do anything to make it smaller. The function-dependent part of the error is of some concern to us in more complicated approximations; here we simply note that we are not able to actually compute that value in most cases, so we resort to upper bounds.

It is the parameter-dependent part of the error that determines convergence and accuracy, and how fast we achieve either of them, and that is, at least partly, under our control. In fact, we might use the asymptotic order notation to write (1.10) as

$$\text{erf}(x) = q_k(x) + O(\delta_k(x)).$$

Alternatively, we could use the approximate equality notation to write

$$\text{erf}(x) \approx q_k(x),$$

where it is understood (perhaps only implicitly) that this approximation is valid only for k large or for x small. We might want to simplify the error a little more by removing one of the variables from $\delta_k(x)$. For example, if we were interested only in values of x between 0 and 2, then we could easily establish that

$$|\delta_k(x)| \le \frac{2^{2k+3}}{(2k+3)(k+1)!}$$

for all $x \in [0,2]$. Moreover, the $2k+3$ factor in the denominator is not going to affect the error bound nearly as much as the factorial, so we can further simplify to get

$$|\delta_k(x)| \le \frac{8}{5} \frac{2^{2k}}{(k+1)!} = \frac{8}{5} \frac{4^k}{(k+1)!}$$

for all $k \ge 1$. Thus we can write

$$\text{erf}(x) = q_k(x) + O\left(\frac{4^k}{(k+1)!}\right), \ 0 \le x \le 2.$$

This extra simplification has indeed increased our estimate of the error, but only slightly, and not in a manner that ignores the most important factors in the convergence of the approximation—the factorial and the power. The benefit we get from this slightly increased error estimate is the ability to quickly and easily gauge the accuracy of an approximation using a specified number of terms. For example, we have that the error in a 20-term approximation is on the order of 2.2×10^{-8}, whereas a 10-term approximation is accurate only to within about 0.026.

EXERCISES

1. Consider the error (1.11) in approximating the error function. If we restrict ourselves to $k \le 3$, over what range of values of x is the approximation accurate to within 10^{-3}?

2. If we are interested only in $x \in [0,\frac{1}{2}]$, then how many terms in the error function approximation do we need to take to get an accuracy of 10^{-4}?

3. Repeat the preceding for $x \in [0,1]$.

4. Assume that $x \in [0,1]$ and write the error in the approximation to the error function using the asymptotic order notation.

5. For

$$f(x) = \int_0^x t^{-1} \sin t \, dt, \ x \in [-\pi/4, \pi/4],$$

construct a Taylor approximation that is accurate to within 10^{-4} over the indicated interval.

6. Repeat the preceding for

$$f(x) = \int_0^x e^{-t^2} dt, \, x \in \left[-\frac{1}{2}, \frac{1}{2} \right].$$

7. Construct a Taylor approximation for

$$f(x) = \int_0^x t^{-p} e^{-t^2} dt, \, 0 \le p < 1, \, x \in \left[-\frac{1}{2}, \frac{1}{2} \right]$$

that is accurate to within 10^{-3} for *all* values of p in the indicated range.

8. Does it make a difference in the previous problem if we restrict p to $p \in [0, \frac{1}{2}]$?

9. What is the error in the Taylor polynomial of degree 5 for $f(x) = 1/x$, using $x_0 = \frac{1}{3}$, for $x \in [\frac{1}{2}, 1]$?

10. How many terms must be taken in the preceding exercise to get an error of less than 10^{-2}? 10^{-4}?

11. What is the error in a Taylor polynomial of degree 4 for $f(x) = \sqrt{x}$ using $x_0 = \frac{9}{16}$, for all $x \in [\frac{1}{4}, 1]$?

1.6   APPLICATION: APPROXIMATING THE NATURAL LOGARITHM

In this section we will put together many of the basic ideas from previous sections to construct a reasonable approximation to the natural logarithm function. The primary tool will be Taylor's Theorem, and our goal will be to produce an approximation to the logarithm that is accurate to within $\epsilon = 10^{-16}$.

We first observe that if we can assume that $\tau = \ln 2$ is known to arbitrary precision, then we really only need to construct a logarithm approximation that is valid over a short interval. This follows because of the way that the computer stores numbers. Since any z is stored as $z = f \cdot 2^\beta$ for $\frac{1}{2} \le f \le 1$ and some β, we have

$$\ln z = \ln f + \beta \ln 2.$$

Since $f \in [\frac{1}{2}, 1]$, we will get the best results by choosing the center of the expansion to be $x_0 = \frac{3}{4}$. This is a little unusual, but not unprecedented. The Taylor expansion for the logarithm is then (you should verify this)

$$\ln x = \ln x_0 + \frac{x - x_0}{x_0} - \frac{1}{2} \left(\frac{x - x_0}{x_0} \right)^2 + \cdots + (-1)^{n-1} \frac{1}{n} \left(\frac{x - x_0}{x_0} \right)^n \qquad (1.12)$$
$$+ (-1)^n \int_{x_0}^x (x - t)^n t^{-n-1} \, dt.$$

Since the remainder here is a little more complicated than in the usual case, let's look at it carefully. We have

$$R_n(x) = (-1)^n \int_{x_0}^x (x - t)^n t^{-n-1} \, dt$$

so that

$$|R_n(x)| = \left| \int_{x_0}^x (x - t)^n t^{-n-1} \, dt \right|.$$

We now take upper bounds using the elementary fact that

$$\left| \int_a^b f(x) \, dx \right| \le |b - a| \max_{x \in [a,b]} |f(x)|.$$

Thus,

$$|R_n(x)| \le |x - x_0| \max_t \left| \frac{(x - t)^n}{t^{n+1}} \right|.$$

Now, we know that $x_0 = \frac{3}{4}$, that $x \in [\frac{1}{2}, 1]$, and that t is between x and x_0. It follows, then, that

$$|x - x_0| \le \frac{1}{4}$$

and that

$$\left| \frac{(x - t)^n}{t^{n+1}} \right| = |t|^{-1} \left| \frac{x}{t} - 1 \right|^n \le 2 \left| \frac{x - t}{t} \right|^n.$$

At this point we have to proceed carefully. We want to bound the function

$$g(t) = \frac{x - t}{t}$$

in absolute value. What do we know about t? Since t is between x and x_0, we have either $x_0 \le t \le x$ or $x \le t \le x_0$. We have $g'(t) = -x/t^2$, so there are no critical points (since $x \ne 0$). Therefore, the Extreme Value Theorem tells us that the maximum and minimum of g occurs at the endpoints, that is, for $t = x$ and for $t = x_0$. Since $g(x) = 0$, we clearly have $|g(t)| \le |g(x_0)|$. Thus, substituting into the previous inequality, we have

$$\left| \frac{(x - t)^n}{t^{n+1}} \right| \le 2 \left| \frac{x - x_0}{x_0} \right|^n \le 2 \left| \frac{\frac{1}{4}}{\frac{3}{4}} \right|^n.$$

Therefore,

$$|R_n(x)| \le \frac{1}{2} \left(\frac{1}{3} \right)^n,$$

from which we conclude that $n = 33$ is sufficient to guarantee that $|R_n(x)| \le 10^{-16}$ for all $x \in [\frac{1}{2}, 1]$. It thus requires a 33-degree polynomial to approximate the logarithm in this fashion, along with an accurate representation for $\ln 2$ and $\ln \frac{3}{4}$.

Exercise 5 asks you to implement this as a subprogram and check it against the intrinsic natural logarithm function on your computer.

EXERCISES

1. Write each of the following in the form $x = f \times 2^\beta$ for some $f \in [\frac{1}{2}, 1]$.
(a) $x = 13$
(b) $x = 25$
(c) $x = \frac{1}{3}$
(d) $x = \frac{1}{10}$

2. For each value in the previous problem, compute the logarithm approximation using the degree-4 Taylor polynomial from (1.12). What is the error compared with the logarithm on your calculator?

3. Repeat the preceding for the degree-6 Taylor approximation.

4. Repeat the preceding for the degree-10 Taylor approximation.

5. Implement (as a computer program) the logarithm approximation constructed in this section. Compare it with the intrinsic logarithm function over the interval $[\frac{1}{2}, 1]$. What is the maximum observed error?

6. Let's consider how we might improve on our logarithm approximation from this section.
(a) Compute the Taylor expansions, with remainder, for $\ln(1+x)$ and $\ln(1-x)$ (use the integral of the geometric series).

(b) Combine the two to get the Taylor expansion for

$$f(x) = \ln\left(\frac{1-x}{1+x}\right).$$

What is the remainder in this expansion?
(c) Given $z \in [\frac{1}{2}, 1]$ show how to compute x such that $z = (1-x)(1+x)^{-1}$. What interval contains x?
(d) Use this to construct an approximation to $\ln z$ that is accurate to within 10^{-16}.

7. Use the logarithm expansion from the previous problem, but limited to the degree-4 case, to compute approximations to the logarithm of each value in the first exercise of this section.

8. Repeat the above, using the degree-10 approximation.

9. Implement (as a computer program) the logarithm approximation constructed in Exercise 6. Compare it with the intrinsic logarithm function over the interval $[\frac{1}{2}, 1]$. What is the maximum observed error?

10. Try to use the ideas from this section to construct an approximation to the reciprocal function, $f(x) = x^{-1}$, that is accurate to within 10^{-16} over the interval $[\frac{1}{2}, 1]$.

LITERATURE REVIEW

There are a lot of textbooks on numerical analysis and numerical methods. Some, such as [8], [9], and [15], are considered classics. A list, by no means exhaustive, of numerical analysis or numerical methods texts is given below.

All of these books give decent treatments of the basic topics. Some are more mathematical than the others; some are designed for less-prepared students. The books [2],[4], and [18] are relatively current texts intended for a graduate student audience; [3], [5], [6], [7], [10], [13], [16], and [17] are current texts intended for an undergraduate audience. The presentation in this text has been heavily influenced by [4] and the earlier editions of [6].

An interesting and light-hearted collection of projects for a numerical methods course can be found in [12].

There are more specialized books that treat only the root-finding problem, or numerical integration, etc. These will be cited in the appropriate chapters.

REFERENCES

[1] ACTON, FORMAN S. *Numerical Methods That Work*. Mathematical Association of America, Washington, DC, 1990 (originally published by Harper & Row, 1970).

[2] ALLEN, MYRON B., & ISAACSON, ELI L. *Numerical Analysis for Applied Science*. John Wiley & Sons, New York, 1998.

[3] ASAITHAMBI, A.S. *Numerical Analysis, Theory and Practice*. Saunders College Publishing, Fort Worth, TX, 1995.

[4] ATKINSON, KENDALL. *An Introduction to Numerical Analysis*. John Wiley & Sons, New York, 1987 (2nd ed.).

[5] ATKINSON, KENDALL. *Elementary Numerical Analysis*. John Wiley & Sons, New York, 1993 (2nd ed.).

[6] BURDEN, RICHARD, & FAIRES, J. DOUGLAS. *Numerical Analysis*. Brooks/Cole, Pacific Grove, CA, 1997 (6th ed.).

[7] CHENEY, WARD, & KINCAID, DAVID. *Numerical Mathematics and Computing*. Brooks/Cole, Pacific Grove, CA, 1994 (3rd ed.).

[8] CONTE, S.D., & DE BOOR, CARL. *Elementary Numerical Analysis*. McGraw-Hill, New York, 1980 (3rd ed.).

[9] DAHLQUIST, GERMUND, & BJORCK, AKE. *Numerical Methods*. Prentice-Hall, Englewood Cliffs, NJ, 1974.

[10] FAIRES, J. DOUGLAS, & BURDEN, RICHARD. *Numerical Methods*. Brooks/Cole, Pacific Grove, CA, 1998 (2nd ed.).

[11] FORSYTHE, GEORGE, MALCOLM, MICHAEL, & MOLER, CLEVE. *Computer Methods for Numerical Computations*. Prentice-Hall, Englewood Cliffs, NJ, 1977.

[12] GRANDINE, THOMAS. *The Numerical Methods Programming Projects Book*. Oxford Science, Oxford, 1970.

[13] GREGORY, JOHN, & REDMOND, DON. *Introduction to Numerical Analysis*. Jones and Bartlett, Boston, 1994.

[14] HENRICI, PETER. *Essentials of Numerical Analysis*. John Wiley & Sons, New York, 1982.

[15] ISAACSON, EUGENE, & KELLER, HERBERT B. *Analysis of Numerical Methods*. Dover, New York, 1994 (originally published by John Wiley & Sons, 1966).

[16] KAHANER, DAVID, MOLER, CLEVE, & NASH, STEPHEN. *Numerical Methods and Software*. Prentice-Hall, Englewood Cliffs, NJ, 1989.

[17] MARON, MELVIN J., & LOPEZ, ROBERT J. *Numerical Analysis. A Practical Approach*. Wadsworth, Belmont, CA, 1991 (3rd ed.).

[18] STOER, J., & BULIRSCH, R. *Introduction to Numerical Analysis*. Springer-Verlag, New York, 1980.

[19] YOUNG, DAVID M., & GREGORY, ROBERT TODD. *A Survey of Numerical Mathematics in Two Volumes*. Dover, New York, 1988 (originally published by Addison-Wesley in 1972 and 1973).

CHAPTER 2

A SURVEY OF SIMPLE METHODS AND TOOLS

In this brief chapter we offer a few very simple examples of approximations, algorithms, and error estimates, partly because they will be useful in developing or implementing some of the methods presented in later chapters, and partly to ease our way into the subject material. What we want to do is acclimate the student into the broad area of numerical computations without restricting ourselves to one small corner of the subject. In addition, we show how some of the simplest approximation techniques (difference methods for derivative approximation, linear interpolation, solution of tridiagonal systems) can be used as the basis for computational schemes in more involved settings (Euler's method for the initial value problem, the trapezoid rule for numerical integration, the approximate solution of two-point boundary value problems). The goal here is to introduce the reader to the basic ideas of numerical methods and analysis by looking at simple techniques across a broad spectrum of problem areas.

2.1 HORNER'S RULE AND NESTED MULTIPLICATION

In the previous chapter we devoted some time to the construction of polynomial approximations to given functions. This section discusses the best way to efficiently evaluate those approximations.

The most efficient way to evaluate a polynomial is by *nested multiplication*. If we have

$$p_n(x) = a_0 + a_1 x + a_2 x^2 + \cdots + a_n x^n,$$

then we factor out each power of x as far as it will go, thus getting

$$p_n(x) = a_0 + x(a_1 + x(a_2 + \cdots + x(a_{n-1} + a_n x) \cdots)).$$

Computation with the second form of the polynomial will take $n + 1$ multiplications and n additions. Computation with the first form will take the same amount of work, *plus* the cost of forming the powers $x^2, x^3, \ldots, x^n$.

For example, we could write

$$q(x) = 1 + x + 3x^2 - 6x^3$$

as

$$q(x) = 1 + x(1 + x(3 - 6x)).$$

An algorithmic form of nested multiplication can be written very simply, as the pseudocode in Algorithm 2.1 illustrates.

Algorithm 2.1 *Horner's rule for polynomial evaluation*

```
!
!    Assumes that the polynomial coefficients
!    are stored in the array a(j),j=0..n.
!    px = value of polynomial upon completion
!         of the code.
!

     px = a(n)
     for k = n-1 downto 0
           px = a(k) + px*x
     endfor
```

Algorithm 2.2 *Horner's rule for polynomial derivative evaluation*

```
!
!    Assumes that the polynomial coefficients
!    are stored in the array a(j),j=0..n.
!    dp = value of derivative upon completion
!

     dp = n*a(n)
     for k=n-1 down to 1
           dp = k*a(k) + dp*x
     endfor
```

This is known as *Horner's rule*.[1] It can be easily modified to give the first derivative as well. Returning to the above notation and examples, we have

$$p'_n(x) = a_1 + 2a_2x + 3a_3x^2 + \cdots + na_nx^{n-1},$$

so that

$$p'_n(x) = a_1 + x(2a_2 + x(3a_3 + \cdots + x((n-1)a_{n-1} + na_nx)\cdots)).$$

In the specific case of q, we have

$$q'(x) = 1 + (3 \times 2)x + (-6 \times 3)x^2,$$

which we write as

$$q'(x) = 1 + x(6 - 18x).$$

Pseudocode for the derivative (Algorithm 2.2) is a modest change to the original algorithm. (The student should check that both programs perform as is asserted here.)

If the intermediate values in the computation of $p(x)$ are saved, then the subsequent computation of the derivative can be done more cheaply. Suppose we define the

---

[1] William George Horner (1786–1837) was born in Bristol, England, and spent much of his life as a schoolmaster in Bristol and, after 1809, in Bath. His paper on solving algebraic equations, which contains the algorithm we know as Horner's rule, was published in 1819, although the Italian Ruffini had previously published a very similar idea.

Algorithm 2.3 *A more efficient implementation of Horner's rule*

```
!
!     Assumes that the polynomial coefficients
!     are stored in the array a(j),j=0..n.
!     b(n) = value of polynomial upon completion
!          of the code.
!
      b(0) = a(n)
      for k = 1 to n
            b(k) = x*b(k-1) + a(n-k)
      endfor
!
!     c(n-1) = value of derivative upon completion
!          of the code
!
      c(0) = b(0)
      for k = 1 to n-1
            c(k) = x*c(k-1) + b(k)
      endfor
```

Programming Hint: Be careful when writing Horner's rule for a polynomial that has some coefficents that are negative. It is best to simply store the coefficients as negative numbers rather than try to (correctly) distribute the negative signs throughout the nested multiplication. Thus we write

$$p(x) = 1 - 2x - x^2 + x^3$$

as

$$p(x) = 1 + x((-2) + x((-1) + x)).$$

To do otherwise is to invite a lot of headache and potential error.

---

values b_k, $k = 1, 2, \ldots, n$, according to

$$b_k = xb_{k-1} + a_{n-k}, \quad b_0 = a_n,$$

so that $b_n = p(x)$. Now define the values c_k according to

$$c_k = xc_{k-1} + b_k, \quad c_0 = b_0.$$

Then (note that the b_k and c_k are functions of x)

$$b_0' = 0, \quad b_k' = b_{k-1} + xb_{k-1}',$$

and, in particular,

$$b_1' = b_0.$$

Therefore,

$$\begin{aligned} c_1 &= xc_0 + b_1 = xb_0 + b_1 = xb_1' + b_1 = b_2', \\ c_2 &= xc_1 + b_2 = xb_2' + b_2 = b_3', \end{aligned}$$

and so forth. In general, as can be established by an inductive proof, $c_k = b_{k+1}'$. Therefore, since $b_n = p(x)$, it follows that $c_{n-1} = p'(x)$.

So a more efficient algorithm might be Algorithm 2.3.

EXERCISES

1. Write each of the following polynomials in nested form.
(a) $x^2 + 5x + 6$
(b) $x^3 + 3x + 2$
(c) $x^6 + 2x^4 + 4x^2 + 1$
(d) $5x^6 + x^5 + 3x^4 + 3x^3 + x^2 + 1$

2. Write each of the following polynomials in nested form, but this time take advantage of the fact that they involve only even powers of x to minimize the computations.
(a) $1 - \frac{1}{2}x^2 + \frac{1}{24}x^4$
(b) $1 + x^2 + \frac{1}{2}x^4 + \frac{1}{6}x^6$

3. Write each of the following polynomials in nested form.
(a) $1 - x^2 + \frac{1}{2}x^4 - \frac{1}{6}x^6$
(b) $1 - x + x^2 - x^3$
(c) $1 - x + \frac{1}{2}x^2 - \frac{1}{3}x^3 - \frac{1}{4}x^5$

4. Write a computer code that takes a polynomial, defined by its coefficients, and evaluates that polynomial and its first derivative using Horner's rule. Test this code by applying it to each of the polynomials in Exercise 1.

5. Repeat the preceding, using the polynomials in Exercise 2 as the test set.

6. Repeat the preceding, using the polynomials in Exercise 3 as the test set.

7. Consider the polynomial

$$p(x) = 1 + (x-1) + \frac{1}{6}(x-1)(x-2) + \frac{1}{7}(x-1)(x-2)(x-4).$$

This can be written in "nested-like" form by factoring out each binomial term as far as it will go:

$$p(x) = 1 + (x-1)\left(1 + (x-2)\left(\frac{1}{6} + \frac{1}{7}(x-4)\right)\right).$$

Write each of the following polynomials in this kind of nested form.
(a) $p(x) = 1 + \frac{1}{3}x - \frac{1}{60}x(x-3)$
(b) $p(x) = 3 + \frac{1}{5}(x-8) - \frac{1}{60}(x-8)(x-3)$
(c) $p(x) = -1 + \frac{6}{7}(x-1/2) - \frac{5}{21}(x-1/2)(x-4) + \frac{1}{7}(x-1/2)(x-4)(x-2)$.

8. Write a computer code that computes polynomial values using the kind of nested form used in the previous exercise, and test it on each of the polynomials in that exercise.

9. Write a computer code to carry out Horner's rule on a polynomial defined by its coefficients. Test it out by using the polynomials in the previous exercises. Verify that the same values are obtained when Horner's rule is used as when a naive evaluation is done.

10. Write out the Taylor polynomial of degree 5 for approximating the exponential function, using $x_0 = 0$, by means of the Horner form. Repeat for the degree-5 Taylor approximation to the sine function. (Be sure to take advantage of the fact that the Taylor expansion to the sine uses only odd powers.)

11. For each function in Exercise 11 of Section 1.1, write the polynomial approximation in Horner form, and use this as the basis for a computer program that approximates the function. Compare the accuracy you actually achieve (based on the built-in intrinsic functions on your computer) to that which was theoretically established. Be sure to check that the required accuracy is achieved over the entire interval in question.

12. Repeat the preceding, except this time compare the accuracy of the *derivative* approximation constructed by taking the derivative of the approximating polynomial. Be sure to use the derivative form of Horner's rule to evaluate the polynomial.

2.2 DIFFERENCE APPROXIMATIONS TO THE DERIVATIVE

One of the simplest uses of Taylor's Theorem as a means of constructing approximations involves the use of difference quotients to approximate the derivative of a known function f. Intuitively, this is obvious from the definition of the derivative:

$$f'(x) = \lim_{h \to 0} \frac{f(x+h) - f(x)}{h} \quad \Rightarrow \quad f'(x) \approx \frac{f(x+h) - f(x)}{h}.$$

The challenge for us is to make this vague statement more precise (just how accurate is this approximation, in terms of the parameter h?) and to see if anything better (more accurate) can be found.

We determine the accuracy of the approximation by a very simple computation involving Taylor's Theorem:

$$f'(x) - \frac{f(x+h) - f(x)}{h} \quad = \quad f'(x) - \frac{hf'(x) + \frac{1}{2}h^2 f''(\xi_{x,h})}{h} = -\frac{1}{2}hf''(\xi_{x,h}),$$

so that we have

$$f'(x) - \frac{f(x+h) - f(x)}{h} = -\frac{1}{2}hf''(\xi_{x,h}) = O(h). \tag{2.1}$$

Thus the error is roughly proportional to the parameter h.

Can we do better? Yes, and the improvement does not take a lot of work. Consider the two Taylor expansions

$$f(x+h) = f(x) + hf'(x) + \frac{1}{2}h^2 f''(x) + \frac{1}{6}h^3 f'''(\xi_1) \tag{2.2}$$

and

$$f(x-h) = f(x) - hf'(x) + \frac{1}{2}h^2 f''(x) - \frac{1}{6}h^3 f'''(\xi_2). \tag{2.3}$$

Now subtract (2.3) from (2.2) to get

$$f(x+h) - f(x-h) = 2hf'(x) + \frac{1}{6}h^3 f'''(\xi_1) + \frac{1}{6}h^3 f'''(\xi_2).$$

We can solve this for $f'(x)$ to get

$$f'(x) = \frac{f(x+h) - f(x-h)}{2h} - \frac{1}{6}h^2 \frac{f'''(\xi_1) + f'''(\xi_2)}{2} \tag{2.4}$$

or, using the Discrete Average Value Theorem,

$$f'(x) = \frac{f(x+h) - f(x-h)}{2h} - \frac{1}{6}h^2 f'''(\xi_{x,h}), \tag{2.5}$$

where $\xi_{x,h}$ depends on both x and h. It is worthwhile to compare the two approximations and error estimates. We note first that the two approximations cost about the same to compute: both require two function evaluations and only a handful of other arithmetic operations. Since the function is arbitrary, its cost of computation might be quite large, and so we view that as the most significant part of the cost. However, the error estimates are quite different. The first approximation has an error that depends on f'' and is proportional to h; the second approximation depends on f''' and is proportional to h^2. Clearly, we will want to use h small to get the best accuracy, and thus the second estimate will tend to be better, since $h^2 < h$ will tend to be more important than the different sizes of f'' and f'''.

EXAMPLE 2.1

Consider, for example, the task of approximating the derivative of $f(x) = e^x$ at $x = 1$. The exact value, of course, is $f'(x) = e^x \Rightarrow f'(1) = e$. Using $h = 1/8$ and the "one-sided difference" (2.1), we get

$$f'(1) \approx \frac{e^{1.125} - e}{0.125} = 2.895480164,$$

while the same value of h with the "centered difference" (2.5) yields

$$f'(1) \approx \frac{e^{1.125} - e^{0.875}}{0.25} = 2.72536622.$$

The error in the first approximation is $-0.177...$, but the error in the second approximation is only $-7.084... \times 10^{-3}$. ∎

To further illustrate these differences in accuracy, let's continue computing with the same example, but take more and smaller values of h. Let

$$D_1(h) = \frac{f(1+h) - f(1)}{h}$$

and

$$D_2(h) = \frac{f(1+h) - f(1-h)}{2h},$$

with the corresponding errors

$$E_1(h) = f'(1) - D_1(h)$$

and

$$E_2(h) = f'(1) - D_2(h).$$

Table 2.1 gives the results of the computation using a sequence of decreasing values of h.

A couple of comments might be in order here. Note that the error E_2 went down by about a factor of 4 each time we cut h in half, whereas the error E_1 only went down by a factor of 2. This, of course, follows directly from the error estimates for both methods. The importance here is that we learn that the error in the second approximation goes down at a faster rate than the error in the first one.

However, if we continue the sequence of computations, something disturbing happens. Look at Table 2.2, in which we keep going with smaller and smaller values of h for $D_2(h)$. The last column gives the ratio of errors; the error estimate asserts that this should be nearly 4, yet in the last two entries it is clearly not even close to 4, and in fact the error actually begins to increase. What is going on here?

Our error estimates, made earlier, completely ignored the issue of rounding error. This is not unusual; most of our error estimates will be made in this fashion, concentrating on the mathematical error instead of the rounding error. But we know that the rounding error can intrude on a computation, and this is our first example of that.

We can illustrate what is going on here by a little bit of abstraction. Let $\tilde{f}(x)$ denote the function computation as actually done on the computer, i.e., the one polluted with rounding error. Define $\epsilon(x) = f(x) - \tilde{f}(x)$ as the error between the function as computed in infinite precision and as actually computed on the machine. This error will be small, but it won't be zero. The approximate derivative that we compute is constructed with $\tilde{f}$, not f, so we define

$$\tilde{D}_2(h) = \frac{\tilde{f}(x+h) - \tilde{f}(x-h)}{2h},$$

and we now want to bound the error $f'(x) - \tilde{D}_2(h)$. Note that this is the error between the quantity we want to compute ($f'(1)$) and the quantity we can actually compute

TABLE 2.1 Example of Derivative Approximation to $f(x) = e^x$ at $x = 1$

h^{-1}	$D_1(h)$	$E_1(h) = f'(1) - D_1(h)$	$D_2(h)$	$E_2(h) = f'(1) - D_2(h)$
2	3.526814461	−0.8085327148E+00	2.832967758	−0.1146860123E+00
4	3.088244438	−0.3699626923E+00	2.746685505	−0.2840375900E−01
8	2.895481110	−0.1771993637E+00	2.725366592	−0.7084846497E−02
16	2.805027008	−0.8674526215E−01	2.720052719	−0.1770973206E−02
32	2.761199951	−0.4291820526E−01	2.718723297	−0.4415512085E−03
64	2.739639282	−0.2135753632E−01	2.718391418	−0.1096725464E−03
128	2.728942871	−0.1066112518E−01	2.718307495	−0.2574920654E−04

TABLE 2.2 Illustration of Rounding Error in Derivative Approximations, Using $f(x) = e^x$, $x = 1$

h^{-1}	$D_2(h)$	$E_2(h) = f'(1) - D_2(h)$	$E_2(h/2)/E_2(h)$
2	2.832967758	−0.1146860123E+00	NA
4	2.746685505	−0.2840375900E−01	4.038
8	2.725366592	−0.7084846497E−02	4.009
16	2.720052719	−0.1770973206E−02	4.001
32	2.718723297	−0.4415512085E−03	4.011
64	2.718391418	−0.1096725464E−03	4.026
128	2.718307495	−0.2574920654E−04	4.259
256	2.718292236	−0.1049041748E−04	2.455
512	2.718261719	0.2002716064E−04	−0.524

$(\tilde{D}_2(h))$. We have

$$f'(x) - \tilde{D}_2(h) = f'(x) - \frac{\tilde{f}(x+h) - \tilde{f}(x-h)}{2h},$$

which we write as

$$f'(x) - \tilde{D}_2(h) = f'(x) - \frac{f(x+h) - f(x-h)}{2h} \quad + \frac{f(x+h) - f(x-h)}{2h} - \frac{\tilde{f}(x+h) - \tilde{f}(x-h)}{2h}$$

$$= -\frac{1}{6}h^2 f'''(\xi_{x,h}) \quad + \frac{f(x+h) - f(x-h) - \tilde{f}(x+h) + \tilde{f}(x-h)}{2h}$$

$$= \underbrace{-\frac{1}{6}h^2 f'''(\xi_{x,h})}_{\text{Error due to approximation}} \quad + \underbrace{\frac{\epsilon(x+h) - \epsilon(x-h)}{2h}}_{\text{Error due to rounding}}.$$

Generally, the numerator in the rounding error fraction will *not* go to zero as $h \to 0$, since we will always have some (small) amount of rounding error present. (Bear in mind that the error due to rounding can be assumed to include the error in doing the subtraction $f(x+h) - f(x-h)$, which, as we saw in Section 1.3, will be prone to large amounts of error as h gets smaller.) But since we are dividing the rounding error by $2h$, we do expect that this term will begin to grow as h gets small, and this in fact is what is observed in Table 2.2.

Figure 2.1 shows a plot of $\log_{10}|f'(1) - D_2(h)|$ versus $1/h$ for h ranging from $\frac{1}{2}$ all the way to $1/2048$. Note that for h much smaller than 0.01 (approximately),

FIGURE 2.1 Illustration of rounding error in approximating derivatives.

TABLE 2.3 **Illustration of Rounding Error in Derivative Approximations, Using $f(x) = e^x$, $x = 1$ (Double Precision)**

h^{-1}	$D_2(h)$	$E_2(h) = f'(1) - D_2(h)$	$E_2(h/2)/E_2(h)$
2	2.832967800	−0.1146859712E+00	NA
4	2.746685882	−0.2840405324E−01	4.038
8	2.725366220	−0.7084391345E−02	4.009
16	2.720051889	−0.1770060412E−02	4.002
32	2.718724279	−0.4424502865E−03	4.001
64	2.718392437	−0.1106085209E−03	4.000
128	2.718309480	−0.2765187706E−04	4.000
256	2.718288741	−0.6912953448E−05	4.000
512	2.718283557	−0.1728237360E−05	4.000
1024	2.718282261	−0.4320591627E−06	4.000
2048	2.718281936	−0.1080143859E−06	4.000
4096	2.718281855	−0.2700387380E−07	4.000
8192	2.718281835	−0.6751245785E−08	4.000
16,384	2.718281830	−0.1688998275E−08	3.997
32,768	2.718281829	−0.4193436709E−09	4.028

the plot ceases to show a decrease in the error, and instead the error oscillates wildly about, creating a "cloud" rather than a smooth curve. Note also that this cloud is trending slightly upward, meaning that the error is acually beginning to *increase*. Smaller values of h would only have continued to confirm this trend.

To confirm that rounding error is the culprit, we can repeat the calculation in a higher precision. The data for Tables 2.1 and 2.2 were produced using computer code in single precision (about 6–7 decimal digits of accuracy). If we repeat the computations for Table 2.2 in double precision (about 15 decimal digits), we get the results shown in Table 2.3. Note that we can go all the way down to $h = 2^{-15}$ without rounding error seriously affecting the computation, although it appears from the last two entries that, even at this level of precision, the effects of rounding error are beginning to become apparent. Figure 2.2 shows the same plot as in Figure 2.1, except this time we used the double-precision data. Note the absence of the cloud effect and the steady decrease of the error.

More involved combinations of $f(x \pm h)$ can be used to construct more accurate approximations to f' and higher-order derivatives. Exercise 12 asks you to derive an approximation to the second derivative that will be used in Section 2.7 to construct approximate solutions to certain differential equation problems.

It might seem silly to use these methods to compute approximate derivative values for the exponential function—after all, we can compute that derivative rather easily by basic calculus—but we need to keep in mind that we won't always have a neat and tidy formula for f that allows us to compute f' directly. The function f might be the result of a lengthy computation for which a simple formula is not really

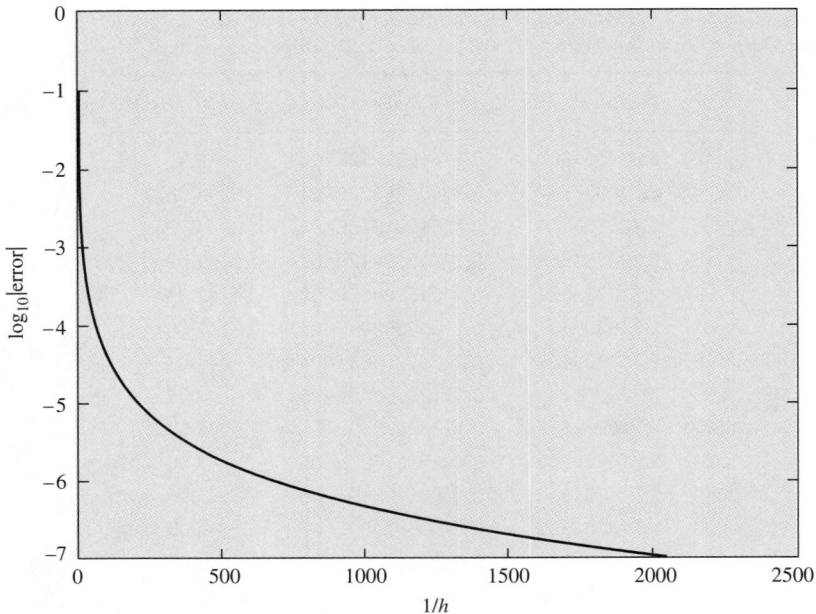

FIGURE 2.2   Double-precision version of Figure 2.1.

available. In addition, we can use formulas such as those developed in this section to compute (approximate) derivative values for functions defined only as a table of values. Finally, we can use the formulas derived in this section to replace the derivative in other equations in order to produce approximation schemes for other problems. The next section, on Euler's method, is just one example of this.

EXERCISES

1. Use the methods of this section to show that

$$f'(x) = \frac{f(x) - f(x-h)}{h} + O(h).$$

2. Compute, by hand, approximations to $f'(1)$ for each of the following functions, using $h = 1/16$ and each of the derivative approximations contained in (2.1) and (2.5).
(a) $f(x) = \sqrt{x+1}$
(b) $f(x) = e^{-x}$
(c) $f(x) = \ln x$

(d) $f(x) = \arctan x$
(e) $f(x) = \sin \pi x$

3. Write a computer program that uses the same derivative approximations as in the previous exercise to approximate the first derivative at $x = 1$ for each of the following functions, using $h^{-1} = 4,8,16,32$. Verify that the predicted theoretical accuracy is obtained.
(a) $f(x) = \sqrt{x+1}$
(b) $f(x) = e^{-x}$
(c) $f(x) = \ln x$
(d) $f(x) = \arctan x$
(e) $f(x) = \sin \pi x$

TABLE 2.4 Table for Exercise 4

x	$f(x)$	$f'(x)$
1.00	1.0000000000	
1.10	0.9513507699	
1.20	0.9181687424	
1.30	0.8974706963	
1.40	0.8872638175	
1.50	0.8862269255	
1.60	0.8935153493	
1.70	0.9086387329	
1.80	0.9313837710	
1.90	0.9617658319	
2.00	1.0000000000	

4. Use the approximations from this section to fill in approximations to the missing values in Table 2.4.

5. Use the error estimate (2.5) for the centered difference approximation to the first derivative to prove that this approximation will be *exact* for any quadratic polynomial.

6. Find coefficients A, B, and C such that
(a) $f'(x) = Af(x) + Bf(x+h) + Cf(x+2h) + O(h^2)$
(b) $f'(x) = Af(x) + Bf(x-h) + Cf(x-2h) + O(h^2)$

(Hint: Use Taylor's Theorem.)

7. Fill in the data from Exercise 4 using methods that are $O(h^2)$ at each point. (Hint: See the previous exercise.)

8. Use Taylor's Theorem to show that the approximation

$$f'(x) \approx \frac{8f(x+h) - 8f(x-h) - f(x+2h) + f(x-2h)}{12h}$$

is $O(h^4)$.

9. Use the derivative approximation from Exercise 8 to approximate $f'(1)$ for the same functions as in Exercise 3. Verify that the expected rate of decrease is observed for the error.

10. Use the derivative approximation from Exercise 8 to fill in as much as possible of Table 2.4.

11. Let $f(x) = \arctan x$. Use the derivative approximation from Exercise 8 to approximate $f'(\frac{1}{4}\pi)$ using $h^{-1} = 2,4,8,\ldots$. Try to take h small enough that the rounding error effect begins to dominate the mathematical error. For what value of h does this begin to occur? (You may have to restrict yourself to working in single precision.)

12. Use Taylor expansions for $f(x \pm h)$ to derive an $O(h^2)$ accurate approximation to f'' using $f(x)$ and $f(x \pm h)$. Provide all the details of the error estimate. (Hint: Go out as far as the fourth derivative term, and then add the two expansions.)

13. Let $h > 0$ and $\eta > 0$ be given, where $\eta = \theta h$, for $0 < \theta < 1$. Let f be some smooth function. Use Taylor expansions for $f(x+h)$ and $f(x-\eta)$ in terms of f and its derivatives at x in order to construct an approximation to $f'(x)$ that depends on $f(x+h)$, $f(x)$, and $f(x-\eta)$, and that is $O(h^2)$ accurate. Check your work by verifying that for $\theta = 1 \Rightarrow \eta = h$ you get the same results as in the text.

14. Write a computer program to test the approximation to the second derivative from Exercise 12 by applying it to estimate $f''(1)$ for each of the following functions, using $h^{-1} = 4,8,16,32$. Verify that the predicted theoretical accuracy is obtained.
(a) $f(x) = e^{-x}$
(b) $f(x) = \ln x$
(c) $f(x) = \cos \pi x$
(d) $f(x) = \sqrt{1+x}$

15. Define the following function:

$$f(x) = \ln(e^{\sqrt{x^2+1}}\sin(\pi x) + \tan \pi x).$$

Compute values of f' over the range $[0,\frac{1}{4}]$ using two methods:
(a) The centered-difference formula from (2.5)
(b) Ordinary calculus to find the formula for f'

Comment.

16. Let $f(x) = e^x$, and consider the problem of approximating $f'(1)$, as in the text. Let $D_1(h)$ be the difference approximation in (2.1). Using the appropriate values in Table 2.1, compute the new approximations

$$\Delta_1(h) = 2D_1(h) - D_1(2h),$$

and compare these values to the exact derivative value. Are they more or less accurate than the corresponding values of D_1? Try to deduce, from your calculations, how the error depends on h.

17. Repeat the preceding idea for $f(x) = \arctan(x), x = 1$ (but this time you will have to compute the original $D_1(h)$ values).

18. By keeping more terms in the Taylor expansion for $f(x+h)$, show that the error in the derivative approximation (2.1) can be written as

$$f'(x) - \left(\frac{f(x+h) - f(x)}{h} \right) = -\frac{1}{2}hf''(x)$$
$$-\frac{1}{6}h^2 f'''(x) - \cdots. \quad (2.6)$$

Use this to construct a derivative approximation involving $f(x), f(x+h)$, and $f(x+2h)$ that is $O(h^2)$ accurate. (Hint: Use (2.6) to write down the error in the approximation

$$f'(x) \approx \frac{f(x+2h) - f(x)}{2h},$$

and combine the two error expansions so that the terms that are $O(h)$ are eliminated.)

19. Apply the method derived above to the list of functions in Exercise 3, and confirm that the method is as accurate in practice as is claimed.

20. Let $f(x) = e^x$, and consider the problem of approximating $f'(1)$, as in the text. Let $D_2(h)$ be the difference approximation in (2.5). Using the appropriate values in Table 2.1, compute the new approximations

$$\Delta_2(h) = (4D_2(h) - D_2(2h))/3,$$

and compare these values to the exact derivative value. Are they more or less accurate than the corresponding values of D_2? Try to deduce, from your calculations, how the error depends on h.

21. Repeat the preceding idea for $f(x) = \arctan(x), x = 1$ (but this time you will have to compute the original $D_2(h)$ values).

22. The same ideas as in Exercise 18 can be applied to the centered-difference approximation (2.5). Show that in this case the error satisfies

$$f'(x) - \frac{f(x+h) - f(x-h)}{2h} = -\frac{1}{6}h^2 f'''(x)$$
$$-\frac{1}{120}h^4 f'''''(x) - \cdots. \quad (2.7)$$

Use this to construct a derivative approximation involving $f(x \pm h)$, and $f(x \pm 2h)$ that is $O(h^4)$ accurate.

23. Apply the method derived above to the list of functions in Exercise 3, and confirm that the method is as accurate in practice as is claimed.

2.3 APPLICATION: EULER'S METHOD FOR INITIAL VALUE PROBLEMS

One immediate application of difference methods for approximating derivatives is the approximate solution of initial value problems for ordinary differential equations. The usual general form of such a problem is

$$y' = f(t,y), \quad y(t_0) = y_0, \quad (2.8)$$

where f is a known function of t and y, and t_0, y_0 are given values. The object in solving this problem is to find y as a function of t; in the usual sophomore-level course in ordinary differential equations, the student learns a number of techniques for analytically solving (2.8), based on assuming any one of a number of special forms for f. Here we will use one of our derivative approximations to construct a method for approximately solving (2.8).

We use (2.1) to replace the derivative in (2.8):

$$\frac{y(t+h) - y(t)}{h} = f(t,y(t)) + \frac{1}{2}hy''(t_h),$$

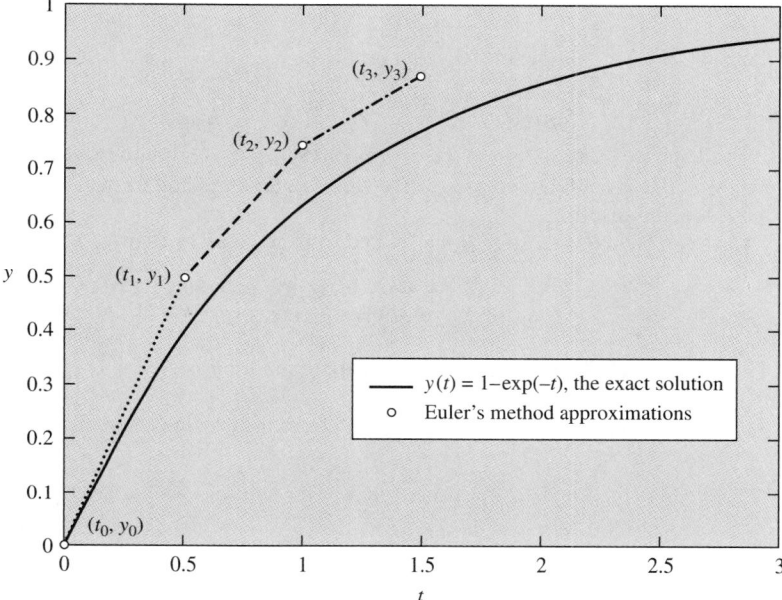

FIGURE 2.3   Geometric derivation of Euler's Method.

which can be simplified slightly to

$$y(t+h) = y(t) + hf(t,y(t)) + \frac{1}{2}h^2 y''(t_h).$$ (2.9)

This suggests the following numerical method:

1. Define a sequence of t values (called a *grid*) according to $t_n = t_0 + nh$, where h is a set parameter (called the *mesh spacing* or *grid size*; we will encounter this kind of thing often in later topics).

2. Compute the values y_n from y_0 and the t grid values, according to

$$y_{n+1} = y_n + hf(t_n, y_n).$$ (2.10)

Note that this follows from (2.9) by dropping the error term and adjusting the notation slightly.

Equation (2.10) defines what is known as *Euler's method* for solving (approximately) initial value problems for ordinary differential equations. Figure 2.3 shows what is happening geometrically.

An example is in order at this point.

EXAMPLE 2.2

Consider the very simple initial value problem

$$y' = -y + \sin t, \quad y(0) = 1.$$

This has exact solution $y(t) = \frac{3}{2}e^{-t} + \frac{1}{2}(\sin t - \cos t)$, found by using the kinds of methods taught in the usual ODE course. If we apply Euler's method to this, using $h = \frac{1}{4}$, then we get the following results.

Step 1: We have $h = \frac{1}{4}$, so $t_1 = h = \frac{1}{4}$, and y_0 is given as 1. Then

$$y_1 = y_0 + hf(t_0, y_0) = 1 + \frac{1}{4}(-1 + \sin 0) = \frac{3}{4}.$$

Thus $y(1/4) \approx 0.75$, and the error in this approximation is $e_1 = y(1/4) - y_1 = 0.8074469434 - 0.75 = 0.0574469434$.

Step 2: We have $t_2 = 2h = \frac{1}{2}$ and $y_1 = 0.75$ from the previous step. Then

$$y_2 = y_1 + hf(t_1, y_1) = 3/4 + \frac{1}{4}\left(-3/4 + \sin\frac{1}{4}\right) = 0.6243509898.$$

Thus $y(1/2) \approx 0.6243509898$, and the error in this approximation is $e_2 = y(1/2) - y_2 = 0.7107174779 - 0.6243509898 = 0.0863664881$. ∎

If we instead use $h = 1/8$ and continue the computation out to $t = 1$, then we get the results in Table 2.5.

If we cut the mesh size in half, again, to $h = 1/16$, then we get the results in Table 2.6. Note that for $h = 1/8$, the maximum error is given by 4.425×10^{-2}, whereas for $h = 1/16$ it is given by 2.140×10^{-2}. This suggests (but does not prove) that Euler's method is $O(h)$ accurate, something we will prove in Chapter 6, where we undertake a more wide-ranging study of numerical methods for ordinary differential equations. Figure 2.4 shows the exact solution (solid line), the approximate solution computed

TABLE 2.5 **Euler's Method Applied to $y' + y = \sin t$, $h = 1/8$.**

t_k	y_k	Error $= y(t_k) - y_k$
0.00000	0.1000000000E+01	0.0000000000E+00
0.12500	0.8750000000E+00	0.1498388695E–01
0.25000	0.7812093417E+00	0.2623760171E–01
0.37500	0.7144836689E+00	0.3433270290E–01
0.50000	0.6709572764E+00	0.3976020153E–01
0.62500	0.6470158092E+00	0.4294341032E–01
0.75000	0.6392759921E+00	0.4424878254E–01
0.87500	0.6445713381E+00	0.4399501342E–01
1.00000	0.6599428586E+00	0.4246064258E–01

FIGURE 2.4   Exact and approximate solutions to $y' + y = \sin\ t$; $*$ denotes the approximate solution for $h = 1/8$, and $+$ denotes the approximate solution for $h = 1/16$.

TABLE 2.6   **Euler's Method Applied to $y' + y = \sin t$, $h = 1/16$.**

t_k	y_k	Error $= y(t_k) - y_k$
0.00000	0.1000000000E+01	0.0000000000E+00
0.06250	0.9375000000E+00	0.3825497791E−02
0.12500	0.8828099574E+00	0.7173929590E−02
0.18750	0.8354265059E+00	0.1008216323E−01
0.25000	0.7948625553E+00	0.1258438808E−01
0.31250	0.7606463930E+00	0.1471233364E−01
0.37500	0.7323209006E+00	0.1649547113E−01
0.43750	0.7094428774E+00	0.1796119911E−01
0.50000	0.6915824637E+00	0.1913501427E−01
0.56250	0.6783226558E+00	0.2004066841E−01
0.62500	0.6692589069E+00	0.2070031255E−01
0.68750	0.6639988048E+00	0.2113462907E−01
0.75000	0.6621618220E+00	0.2136295267E−01
0.81250	0.6633791306E+00	0.2140338100E−01
0.87500	0.6672934759E+00	0.2127287562E−01
0.93750	0.6735591026E+00	0.2098735402E−01
1.00000	0.6818417279E+00	0.2056177329E−01

Algorithm 2.4: Euler's method

```
input h, N, y0, t0
external f
y = y0
t = t0
for k=1 to N do
     yn = y + h*f(t,y)
     y = yn
     t = t + h
endfor
```

Programming Hints: (1) Note that this code assumes that there is some externally written subprogram or function that computes the values of $f(t,y)$. (2) When writing a code for the exercises, be very sure that you compare y_n to the exact solution at $t = t_n$. It is very easy to accidentally write the program so that you are comparing y_n to $y(t_{n+1})$.

with $h = 1/8$ (denoted by the * signs), and the approximate solution computed with $h = 1/16$ (denoted by the + signs).

Writing a computer code for Euler's method is not very hard. If we assume that h, the mesh size, is given, along with N, the number of steps to take, then the code will look something like Algorithm 2.4.

EXERCISES

1. Use Euler's method with $h = 0.25$ to compute approximate solution values for the initial value problem
$$y' = \sin(t+y), \quad y(0) = 1.$$

You should get $y_4 = 1.851566895$ (be sure your calculator is set in radians).

2. Repeat the preceding with $h = 0.20$. What value do you now get for $y_5 \approx y(1)$?

3. Repeat the preceding with $h = 0.125$. What value do you now get for $y_8 \approx y(1)$?

4. Use Euler's method with $h = 0.25$ to compute approximate solution values for
$$y' = e^{t-y}, \quad y(0) = -1.$$

What approximate value do you get for $y(1) = 0.7353256638$?

5. Repeat the preceding with $h = 0.20$. What value do you now get for $y_5 \approx y(1)$?

6. Repeat the preceding with $h = 0.125$. What value do you now get for $y_8 \approx y(1)$?

7. Use Euler's method with $h = 0.0625$ to compute approximate solution values over the interval $0 \le t \le 1$ for the initial value problem
$$y' = t - y, \quad y(0) = 2,$$
which has exact solution $y(t) = 3e^{-t} + t - 1$. Plot your approximate solution as a function of t, and plot the error as a function of t.

8. Repeat the preceding for the equation
$$y' = e^{t-y}, \quad y(0) = -1,$$
which has exact solution $y = \ln(e^t - 1 + e^{-1})$.

9. Repeat the preceding for the equation
$$y' + y = \sin t, \quad y(0) = -1,$$
which has exact solution $y = (\sin t + \cos t - e^{-t})/2$.

10. Use Euler's method to compute approximate solutions to each of the following initial value problems, using $h^{-1} = 2, 4, 8, 16$. Compute the maximum error over the interval $[0,1]$ for each value of h. Plot your approximate solutions for the $h = 1/16$ case. (Hint: Verify that your code works by using it to reproduce the results given for the examples in the text.)

(a) $y' + 4y = 1$, $y(0) = 1$; $y(t) = \frac{1}{4}(3e^{-4t} + 1)$

(b) $y' = -y \ln y$, $y(0) = 3$; $y(t) = e^{(\ln 3)e^{-t}}$

(c) $y' = t - y$, $y(0) = 2$; $y(t) = 3e^{-t} + t - 1$

11. Consider the approximate values in Tables 2.5 and 2.6. Let y_k^8 denote the approximate values for $h = 1/8$, and y_k^{16} denote the approximate values for $h = 1/16$. Note that

$$y_k^8 \approx y(k/8)$$

and

$$y_{2k}^{16} \approx y(2k/16) = y(k/8) \approx y_k^8$$

Thus y_k^8 and y_{2k}^{16} are both approximations to the same value. Compute the set of new approximations

$$u_k = 2y_{2k}^{16} - y_k^8$$

and compare these to the corresponding exact solution values. Are they better or worse as approximations?

12. Apply the basic idea from the previous problem to the approximation of solutions to

$$y' = e^{t-y}, \quad y(0) = -1,$$

which has exact solution $y = \ln(e^t - 1 + e^{-1})$.

13. Assume that the function f satisfies

$$|f(t,y) - f(t,z)| \leq K|y - z|$$

for some constant K. Use this and (2.9)–(2.10) to show that the error $|y(t_{n+1}) - y_{n+1}|$ satisfies the recursion

$$|y(t_{n+1}) - y_{n+1}| \leq (1 + Kh)|y(t_n) - y_n| + \frac{1}{2}h^2 Y_2,$$

where

$$Y_2 = \max_t |y''(t)|.$$

2.4 LINEAR INTERPOLATION

Before hand calculators became a commonplace tool, high school algebra students were taught how to estimate intermediate values for tables of log and trig functions, for example, by a process known as *interpolation*. Even though this particular application is no longer as important as it once was, interpolation is still an important tool in computational mathematics. For example, almost all graphs produced by computers are actually the result of something called *piecewise linear polynomial interpolation*, in which the machine draws a very large number of very small straight lines to represent the curve.

Given a set of data points x_k, typically called *nodes*, we say that the function p interpolates the function f at these nodes if $p(x_k) = f(x_k)$, for all k. Sometimes we only have one function, p, and a set of function values y_k; in this case we say that p interpolates the data if $p(x_k) = y_k$, for all k. In the most common kind of interpolation, the function p (the *interpolant*) is a polynomial. Usually we are most interested in the extent to which $p \approx f$ or the extent to which p represents the data.

The interpolation problem gets a full treatment in Chapter 4; here we want to consider only the special case of *linear* interpolation, by which we mean using a straight line to approximate a given function. Since it takes only two points to determine a straight line, we assume that we are given two nodes x_0 and x_1, and a function f. Since we want to find the equation of a straight line that passes through the two points $(x_0, f(x_0))$ and $(x_1, f(x_1))$, it is not hard to show that

$$p_1(x) = \frac{x_1 - x}{x_1 - x_0} f(x_0) + \frac{x - x_0}{x_1 - x_0} f(x_1) \tag{2.11}$$

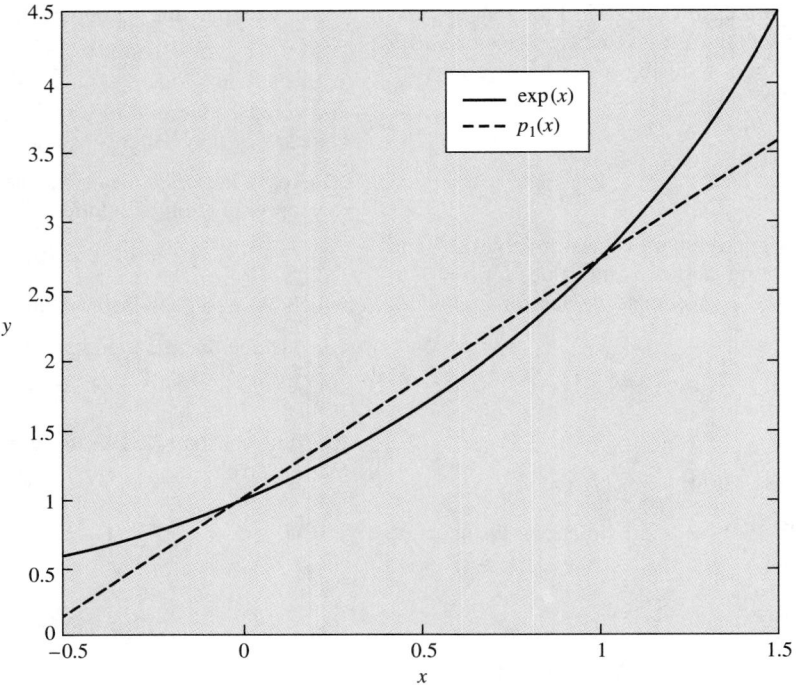

FIGURE 2.5   Linear interpolation to the exponential function on $[0,1]$.

is what we are looking for. The student should confirm that the interpolatory conditions are satisfied, i.e., that $p_1(x_i) = f(x_i)$, $i = 0,1$.

Figure 2.5 shows an example of linear interpolation to the exponential function using the two points $x_0 = 0$ and $x_1 = 1$. Clearly, the approximation is not extremely accurate, but if the nodes were closer together, then it might be more accurate, at least between the nodes.

Just how accurate is linear interpolation? We can investigate this using not much more than Rolle's Theorem.[2] Define

$$E(x) = f(x) - p_1(x), \quad w(x) = (x - x_0)(x - x_1), \quad G(x) = E(x) - \frac{w(x)}{w(t)}E(t),$$

where t is some fixed value in (x_0, x_1), i.e., $x_0 < t < x_1$. (The point t can actually be anywhere, but this restriction is the usual case.) Note that

$$G(x_0) = 0, \quad G(x_1) = 0, \quad G(t) = 0.$$

---

[2]Recall that Rolle's Theorem is a special case of the Mean Value Theorem in which $f(a) = f(b)$; thus there exists ξ such that $f'(\xi) = 0$. Michel Rolle (1652–1719), born in France, was largely self-taught and supported himself as an assistant to several attorneys, and then as a scribe when he moved to Paris in 1675. Rolle's Theorem first appeared in 1692.

Then Rolle's Theorem states that there exists a point η_0 between x_0 and t such that $G'(\eta_0) = 0$, and similarly, a point η_1 between x_1 and t such that $G'(\eta_1) = 0$. But now we can apply Rolle's Theorem to G' and assert that there exists a point ξ between η_0 and η_1 such that $G''(\xi) = 0$. But

$$G''(x) = f''(x) - \frac{2}{w(t)}E(t),$$

so that

$$G''(\xi) = 0 \Rightarrow f''(\xi) - \frac{2}{w(t)}E(t) = 0;$$

and we finally have

$$f(t) - p_1(t) = \frac{1}{2}(t - x_0)(t - x_1)f''(\xi) \tag{2.12}$$

for any point $t \in [x_0, x_1]$. The example plot given in Figure 2.5 shows clearly that the error in the approximation will grow rapidly outside the interval $[x_0, x_1]$, so we will confine our efforts at getting an upper bound to the error on that interval. Taking absolute values, and assuming the worst case for the second derivative term (since we can't do anything else with it),

$$|f(x) - p_1(x)| \leq \frac{1}{2}|(x - x_0)(x - x_1)| \max_{x_0 \leq t \leq x_1} |f''(t)|$$
$$\leq \frac{1}{2}\left(\max_{x_0 \leq t \leq x_1} |(t - x_0)(t - x_1)|\right)\left(\max_{x_0 \leq t \leq x_1} |f''(t)|\right),$$

so the upper bound on the error depends on our obtaining the maximum of the function

$$g(x) = |(x - x_0)(x - x_1)| = (x_1 - x)(x - x_0).$$

Since $g(x_0) = g(x_1) = 0$, the Extreme Value Theorem says that the maximum value of g on the interval $[x_0, x_1]$ will be found at a critical point. Thus we apply some ordinary calculus to get

$$g'(x) = x_1 - x - x + x_0 = (x_0 + x_1) - 2x,$$

which implies that the critical point is

$$x_c = \frac{1}{2}(x_0 + x_1),$$

i.e., the midpoint of the interval. The maximum value of the function is then given by

$$g(x_c) = \frac{1}{4}(x_1 - x_0)^2,$$

giving us, finally, that our error is bounded according to

$$|f(x) - p_1(x)| \leq \frac{1}{8}(x_1 - x_0)^2 \left(\max_{x_0 \leq x \leq x_1} |f''(x)| \right). \tag{2.13}$$

We can summarize this as a theorem.

Theorem 2.1 (Linear Interpolation Error) *Let $f \in C^2([x_0,x_1])$ and let $p_1(x)$ be the linear polynomial that interpolates f at x_0 and x_1. Then, for all $x \in [x_0,x_1]$,*

$$|f(x) - p_1(x)| \leq \frac{1}{8}(x_1 - x_0)^2 \max_{x_0 \leq x \leq x_1} |f''(x)|.$$

Let's now use this to bound the error in an old-time application: estimating intermediate values from tables of data.

EXAMPLE 2.3

Recall the error function

$$\text{erf}(x) = \frac{2}{\sqrt{\pi}} \int_0^x e^{-t^2} dt.$$

Table 2.7 gives values of $\text{erf}(x)$ in increments of 0.1, over the interval $[0,1]$. Suppose we need to know $\text{erf}(0.14)$; then we can interpolate using the values at $x_0 = 0.1$ and $x_1 = 0.2$. The polynomial becomes

$$p_1(x) = \frac{0.2 - x}{0.1}(0.11246291601828) + \frac{x - 0.1}{0.1}(0.22270258921048),$$

TABLE 2.7 **Table of $\text{erf}(x)$ Values**

x	$\text{erf}(x)$
0.0	0.00000000000000
0.1	0.11246291601828
0.2	0.22270258921048
0.3	0.32862675945913
0.4	0.42839235504667
0.5	0.52049987781305
0.6	0.60385609084793
0.7	0.67780119383742
0.8	0.74210096470766
0.9	0.79690821242283
1.0	0.84270079294971

so that

$$p_1(0.14) = \frac{0.06}{0.1}(0.11246291601828) + \frac{0.04}{0.1}(0.22270258921048)$$
$$= 0.15655878529516 \approx \text{erf}(0.14).$$

We can bound the error in this approximation using Theorem 2.1. We have

$$\text{erf}'(x) = \frac{2}{\sqrt{\pi}}e^{-x^2} \Rightarrow \text{erf}''(x) = -\frac{4x}{\sqrt{\pi}}e^{-x^2}.$$

A crude upper bound on the interval $[0.1, 0.2]$ is

$$|\text{erf}''(x)| \le \frac{4 \times 0.2}{\sqrt{\pi}}e^{-0.1^2} = 0.4468606427;$$

thus the error is bounded above by

$$|\text{erf}(0.14) - p_1(0.14)| \le \frac{1}{8}(0.1)^2 \times 0.4468606427 = 5.585758034 \times 10^{-4}.$$

Similarly, we get

$$\text{erf}(0.72) \approx 0.69066114801147$$

with an error that is bounded by $1.38254972 \times 10^{-3}$. ∎

A common and related technique for approximating a curve is to use *piecewise linear* interpolation, in which we break up the interval of interest into subintervals and then use a different linear interpolant on each subinterval. The effect is to approximate the curve by a set of connected straight lines.

EXAMPLE 2.4

Consider the problem of constructing a piecewise linear approximation to $f(x) = \log_2(x)$ using the nodes $\frac{1}{4}, \frac{1}{2}, 1$. We construct the separate linear polynomials over each pair of adjacent nodes:

$$Q_1(x) = \left(\frac{\frac{1}{2} - x}{\frac{1}{2} - \frac{1}{4}}\right)\log_2\left(\frac{1}{4}\right) + \left(\frac{x - \frac{1}{4}}{\frac{1}{2} - \frac{1}{4}}\right)\log_2\left(\frac{1}{2}\right) = 4x - 3$$

$$Q_2(x) = \left(\frac{1 - x}{1 - \frac{1}{2}}\right)\log_2\left(\frac{1}{2}\right) + \left(\frac{x - \frac{1}{2}}{1 - \frac{1}{2}}\right)\log_2(1) = 2x - 2.$$

Thus, the piecewise polynomial function is defined as

$$q(x) = \begin{cases} 4x - 3 & \frac{1}{4} \le x \le \frac{1}{2} \\ 2x - 2 & \frac{1}{2} \le x \le 1. \end{cases}$$

Figure 2.6 shows a plot of $\log_2(x)$ and the piecewise polynomial approximation.

The error is estimated by looking at the error in each individual polynomial approximation. We have

$$|\log_2(x) - Q_1(x)| \le \frac{1}{8}\left(\frac{1}{2} - \frac{1}{4}\right)^2 \max_{t \in [\frac{1}{4}, \frac{1}{2}]} |\log_2(e)t^{-2}| = 0.1803368801...,$$

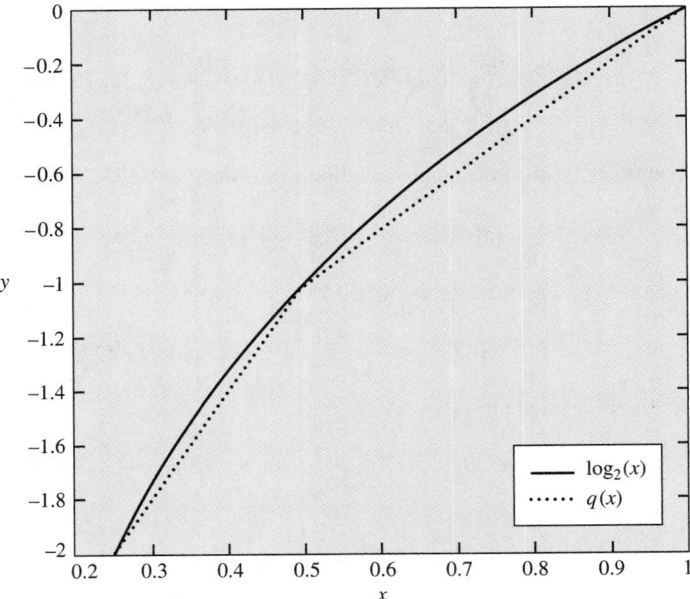

FIGURE 2.6 Piecewise linear approximation to $y = \log_2(x)$.

and this holds for all $x \in [\frac{1}{4}, \frac{1}{2}]$; furthermore,

$$\left| \log_2(x) - Q_2(x) \right| \leq \frac{1}{8} \left(1 - \frac{1}{2} \right)^2 \max_{t \in [\frac{1}{2}, 1]} \left| \log_2(e) t^{-2} \right| = 0.1803368801...,$$

and this holds for all $x \in [\frac{1}{2}, 1]$. Thus

$$\left| \log_2(x) - q(x) \right| \leq 0.1803368801...$$

for all $x \in [\frac{1}{4}, 1]$. (Usually the error in each piece will *not* be the same. This is an artifact of the example.) ∎

    The more general case of piecewise polynomial interpolation with higher-degree polynomials will be taken up in Section 4.7.

    Linear interpolation is a powerful tool that we will return to again and again to construct approximations for more involved computations, and the accuracy of those approximations will depend heavily on the accuracy of the underlying linear interpolation. In fact, we will use the estimate in this section to establish the accuracy and convergence properties of the secant method for finding roots of equations (Section 3.10.3), the trapezoid rule for numerically approximating integrals (Section 2.5), and several other methods for more sophisticated problems.

EXERCISES

1. Use linear interpolation to find approximations to the following values of the error function, using Table 2.7. For each case, give an upper bound on the error in the approximation.

(a) erf(0.07)

(b) erf(0.12)

(c) erf(0.34)

(d) erf(0.56)

(e) erf(0.89)

2. The gamma function, denoted by $\Gamma(x)$, occurs in a number of applications, most notably probability theory and the solution of certain differential equations. It is basically the generalization of the factorial function to noninteger values, in that $\Gamma(n+1) = n!$. Table 2.8 gives values of $\Gamma(x)$ for x between 1 and 2. Use linear interpolation to approximate values of $\Gamma(x)$ as given below.

(a) $\Gamma(1.005) = 0.9971385354$

(b) $\Gamma(1.290) = 0.8990415863$

(c) $\Gamma(1.635) = 0.8979334930$

(d) $\Gamma(1.930) = 0.9723969178$

3. Theorem 2.1 requires an upper bound on the second derivative of the function being interpolated, and this is

TABLE 2.8 Table of $\Gamma(x)$ values

x	$\Gamma(x)$
1.00	1.0000000000
1.10	0.9513507699
1.20	0.9181687424
1.30	0.8974706963
1.40	0.8872638175
1.50	0.8862269255
1.60	0.8935153493
1.70	0.9086387329
1.80	0.9313837710
1.90	0.9617658319
2.00	1.0000000000

not always available as a practical matter. However, if a table of values is available, we can use a difference approximation to the second derivative to *estimate* the upper bound on the derivative and hence the error. Recall, from Exercise 12 of Section 2.2 that

$$f''(x) \approx \frac{f(x-h) - 2f(x) + f(x+h)}{h^2}.$$

Assume the function values are given at the equally spaced grid points $x_k = a + kh$ for some grid spacing h. Using the estimate

$$\max_{x_k \le x \le x_{k+1}} |f''(x)| \approx$$
$$\max \left\{ \frac{f(x_{k-1}) - 2f(x_k) + f(x_{k+1})}{h^2}, \right.$$
$$\left. \frac{f(x_k) - 2f(x_{k+1}) + f(x_{k+2})}{h^2} \right\}$$

to approximate the derivative upper bound, estimate the error made in using linear interpolation to approximate each of the following values of $\Gamma(x)$, based on Table 2.8.

(a) $\Gamma(1.290) = 0.8990415863$

(b) $\Gamma(1.314) = 0.8954464400$

(c) $\Gamma(1.456) = 0.8856168100$

(d) $\Gamma(1.579) = 0.8913230181$

(e) $\Gamma(1.713) = 0.9111663772$

4. Construct a linear interpolating polynomial to the function $f(x) = x^{-1}$ using $x_0 = \frac{1}{2}$ and $x_1 = 1$ as the nodes. What is the upper bound on the error over the interval $[\frac{1}{2}, 1]$, according to the error estimate?

5. Repeat the preceding for $f(x) = \sqrt{x}$, using the interval $[\frac{1}{4}, 1]$.

6. Repeat the preceding for $f(x) = x^{1/3}$, using the interval $[\frac{1}{8}, 1]$.

7. If we want to use linear interpolation to the sine function and obtain an accuracy of 10^{-6}, how close together do the entries in the table have to be? What if we change the error criterion to 10^{-3}?

8. Repeat the preceding for $f(x) = \tan x$, for $x \in [-\pi/4, \pi/4]$.

9. If we try to approximate the logarithm by using a table of logarithm entries, together with linear interpolation, to construct the approximation to $\ln(1+x)$ over the interval $[-\frac{1}{2},0]$, how many points are required to get an approximation that is accurate to within 10^{-14}?

10. Construct a piecewise linear approximation to

$f(x) = \sqrt{x}$ over the interval $[\frac{1}{4},1]$ using the nodes $\frac{1}{4}, \frac{9}{16}, 1$. What is the maximum error in this approximation?

11. Repeat the above for $f(x) = x^{1/3}$ over the interval $[\frac{1}{8},1]$, using the nodes $\frac{1}{8}, \frac{27}{64}, 1$.

2.5 APPLICATION: THE TRAPEZOID RULE

One of the most important applications of linear interpolation is the construction of the trapezoid rule for approximating definite integrals. Define the integration of interest as $I(f)$, thus:

$$I(f) = \int_a^b f(x)\, dx.$$

Let $p_1(x)$ be the linear polynomial that interpolates f at $x = a$ and $x = b$:

$$p_1(x) = \frac{x-a}{b-a}f(b) + \frac{b-x}{b-a}f(a).$$

Then the basic trapezoid rule is defined by exactly integrating $p_1(x)$:

$$T_1(f) = I(p_1) = \int_a^b p_1(x)\, dx = \frac{1}{2}(b-a)(f(b)+f(a)). \qquad (2.14)$$

Note here that we are constructing an approximation by replacing the "exact function" (f) by a simpler function that approximates it (p_1), and doing the desired calculation (integration) *exactly* on the simpler function.

Figure 2.7 shows an illustration of this approximation. The integrand is $\frac{1}{2} + \sin \pi x$ over the interval $[\frac{1}{2},1]$. How accurate is this? Clearly the error is $I(f) - T_1(f)$, and the illustration suggests this might be large. We can analyze the error because the construction of T_1 as the exact integral of p_1 allows us to use the interpolation error theory from the previous section:

$$I(f) - T_1(f) = I(f) - I(p_1)$$
$$= \int_a^b f(x)\, dx - \int_a^b p_1(x)\, dx$$
$$= \int_a^b (f(x) - p_1(x))\, dx$$
$$= \frac{1}{2}\int_a^b (x-a)(x-b)f''(\xi_x)\, dx,$$

FIGURE 2.7   Illustration of the basic trapezoid rule.

where $\xi_x \in [a,b]$ depends on x. Since $(x-a)(x-b)$ does not change sign on $[a,b]$, we can now apply the Integral Mean Value Theorem to get an error estimate:

$$\int_a^b (x-a)(x-b)f''(\xi_x)\,dx = f''(\eta) \int_a^b (x-a)(x-b)\,dx$$
$$= -\frac{1}{6}(b-a)^3 f''(\eta),$$

so that we have

$$I(f) - T_1(f) = -\frac{1}{12}(b-a)^3 f''(\eta), \tag{2.15}$$

where $\eta \in [a,b]$. Formally, then, we have proved the following theorem.

Theorem 2.2 (Trapezoid Rule Error Estimate, Single Subinterval)   Let $f \in C^2([a,b])$ and let p_1 interpolate f at a and b. Define $T_1(f) = I(p_1)$. Then there exists $\eta \in [a,b]$ such that

$$I(f) - T_1(f) = -\frac{1}{12}(b-a)^3 f''(\eta). \tag{2.16}$$

While perfectly valid, this result does not appear to be especially useful, since the error will be small only if the length of the integration interval, $b-a$, is small; but it is only a first step. What we need to do now is use more points in the approximation.

To this end we subdivide the interval $[a,b]$ into n subintervals using mesh points $x_i, 0 \leq i \leq n$:

$$a = x_0 < x_1 < x_2 < \cdots < x_{n-1} < x_n = b. \tag{2.17}$$

Over each subinterval $[x_{i-1}, x_i]$ we apply the basic trapezoid rule (2.14) to get the *composite* trapezoid rule, which is more commonly referred to as "the *n*-subinterval trapezoid rule" (or just "the trapezoid rule" when there is no possibility of confusion):

$$
\begin{aligned}
I(f) &= \sum_{i=1}^{n} \int_{x_{i-1}}^{x_i} f(x) \, dx, \\
&\approx \sum_{i=1}^{n} \frac{1}{2}(x_i - x_{i-1})(f(x_{i-1}) + f(x_i)) \\
&= T_n(f).
\end{aligned}
$$

If we use a *uniform* grid, in which the mesh points are equally spaced so that $x_i - x_{i-1} = h$, for all i, then this simplifies a great deal:

$$T_n(f) = \frac{h}{2}(f(x_0) + 2f(x_1) + 2f(x_2) + \cdots + 2f(x_{n-1}) + f(x_n)). \tag{2.18}$$

Figure 2.8 shows this applied to the same integral of $f(x) = \frac{1}{2} + \sin \pi x$ as was used in Figure 2.7. Note that even with very few additional points in the grid, the apparent error is substantially reduced from the previous case. What we have done here, essentially, is exactly integrate a piecewise linear interpolation of f.

FIGURE 2.8   Illustration of the trapezoid rule.

The accuracy of the n-subinterval rule can be easily established from the previous result.

Theorem 2.3 (Trapezoid Rule Error Estimate, Uniform Grid)   *Let $f \in C^2([a,b])$ and let $T_n(f)$ be the n-subinterval trapezoid rule approximation to $I(f)$, using a uniform grid. Then there exists $\xi_h \in [a,b]$, depending on h, such that*

$$I(f) - T_n(f) = -\frac{b-a}{12}h^2 f''(\xi_h). \tag{2.19}$$

Proof   We simply apply the previous error result to each subinterval:

$$
\begin{aligned}
I(f) - T_n(f) &= \sum_{i=1}^{n} \left(\int_{x_{i-1}}^{x_i} f(x)\,dx - \frac{1}{2}(x_i - x_{i-1})(f(x_i) + f(x_{i-1})) \right) \\
&= -\sum_{i=1}^{n} \frac{1}{12}h^3 f''(\xi_{i,h}),\ \xi_{i,h} \in [x_{i-1}, x_i] \\
&= -\frac{1}{12}h^3 \sum_{i=1}^{n} f''(\xi_{i,h}) \\
&= -\frac{1}{12}h^2 \frac{b-a}{n} \sum_{i=1}^{n} f''(\xi_{i,h}), \\
&= -\frac{b-a}{12}h^2 \left(\frac{1}{n} \sum_{i=1}^{n} f''(\xi_{i,h}) \right).
\end{aligned}
$$

We can now apply the Discrete Average Value Theorem to simplify the remaining sum: There exists $\xi_h \in [a,b]$ such that

$$\frac{1}{n} \sum_{i=1}^{n} f''(\xi_{i,h}) = f''(\xi_h), \tag{2.20}$$

where we note that $\xi_h \in [a,b]$ does depend on the value of h. This completes the proof of the theorem.   □

This theorem tells us quite a bit about the n-subinterval trapezoid rule. It first of all tells us that the numerical approximation will converge to the exact value, so long as $f \in C^2([a,b])$:

$$\lim_{n\to\infty} |I(f) - T_n(f)| = \lim_{h\to 0} \frac{b-a}{12}h^2 |f''(\xi_h)| \leq \left(\lim_{h\to 0} \frac{b-a}{12}h^2 \right) \left(\max_{x\in[a,b]} |f''| \right) = 0,$$

so that

$$\lim_{n\to\infty} T_n(f) = I(f).$$

Second, it tells us how fast this convergence occurs, since the error goes like h^2. Perhaps the best way to see this is to note that doubling the number of subintervals results in the error going down by roughly a factor of 4. We have to say "roughly" because the value of ξ_h does depend on h and so will shift about a bit as we change the mesh size.

It is worth emphasizing at this point that the value of the error estimate is not that we are able to a priori predict the error—although we sometimes can use the error theory to *estimate* the error—but rather that the error estimate tells us how rapidly the error will decrease as a function of the mesh size, h. This will be useful in Chapter 5 when we seek to *improve* the trapezoid rule.

An example will illustrate.

EXAMPLE 2.5

Let $f(x) = e^x$ and $[a,b] = [0,1]$ so that $I(f) = e - 1 = 1.71828....$ Then the trapezoid rule using a single subinterval is given by

$$T_1(f) = \frac{1}{2} \times 1 \times (e^0 + e^1) = (1 + e)/2 = 1.859140914,$$

whereas the trapezoid rule using two subintervals is given by

$$T_2(f) = \frac{1}{2} \times \frac{1}{2} \times (e^0 + 2e^{1/2} + e^1) = \frac{1}{4}(1 + 2\sqrt{e} + e) = 1.753931093.$$

We note that the second value is considerably more accurate, as expected.

If we now continue the computation using a sequence of equally spaced grids with $h^{-1} = 2, 4, 8, \ldots, 2048$, we get the results in Table 2.9. Note that the last column, which gives the ratio of the previous error to the current error, does show that the error decreases by roughly a factor of 4 as we double the number of points. ∎

TABLE 2.9 Trapezoid example, $f(x) = e^x$, $[a;b] = [0;1]$

n	$T_n(f)$	$I(f) - T_n(f)$	Error ratio
2	1.753931092	−0.356493E−01	N/A
4	1.727221905	−0.894008E−02	3.9876
8	1.720518592	−0.223676E−02	3.9969
16	1.718841129	−0.559300E−03	3.9992
32	1.718421660	−0.139832E−03	3.9998
64	1.718316787	−0.349584E−04	4.0000
128	1.718290568	−0.873962E−05	4.0000
256	1.718284013	−0.218491E−05	4.0000
512	1.718282375	−0.546227E−06	4.0000
1024	1.718281965	−0.136557E−06	4.0000
2048	1.718281863	−0.341392E−07	4.0000

Programming Hint   We should note that doubling the number of subintervals and observing that the error goes down by roughly a factor of 4 is a useful device for debugging trapezoid rule programs. If your program does not produce errors that behave this way, then there are only three possible conclusions to draw: (1) Your test function is not C^2 on the entire interval of integration, (2) your code has an error, or (3) your example is one of a special class of functions (periodic functions, integrated over an integer multiple of the period) for which the trapezoid rule is "super-accurate," in which case the error starts out very small and does not always decrease in a regular fashion (see Section 5.8.1). You should not accept results that do not display the correct rate of error decrease unless you can demonstrate that either the first or third of these possibilities holds for your example.

EXAMPLE 2.6

The error theory can sometimes be used to predict how small the mesh must be to achieve a specified desired error. For example, consider the integral

$$I(f) = \int_0^1 e^{-x^2} dx.$$

How small does h have to be to guarantee that $|I(f) - T_n(f)| \leq 10^{-3}$?
     This is a relatively simple exercise. We know that

$$|I(f) - T_n(f)| = \frac{1}{12}h^2|f''(\xi)| \leq \frac{1}{12}h^2 \max_{x\in[0,1]} |f''(x)|.$$

Ordinary calculus shows us that

$$f''(x) = e^{-x^2}\left(4x^2 - 2\right),$$

so that, for all $x \in [0,1]$,

$$|f''(x)| = |e^{-x^2}\left(4x^2 - 2\right)| \leq e^0\left(4 + 2\right) = 6.$$

Therefore,

$$|I(f) - T_n(f)| \leq 10^{-3} \;\Leftarrow\; \frac{1}{12}h^2(6) \leq 10^{-3},$$

which implies that we need $h \leq \sqrt{0.002} = 0.044721....$ The hard part of this computation is in getting an upper bound on the second derivative factor in the error estimate.   ∎

     The assumption of a uniform grid is useful but not required. If the mesh is not uniform, then the result of Theorem 2.3 is changed slightly:

Theorem 2.4 (Trapezoid Rule Error Estimate, Nonuniform Grid)   Let $f \in C^2([a,b])$ and let $T_n(f)$ be the n-subinterval trapezoid rule approximation to $I(f)$, using the nonuniform grid defined by

$$a = x_0 < x_1 < x_2 < \cdots < x_{n-1} < x_n = b,$$

Algorithm 2.5 Simple Trapezoid Rule Pseudocode

```
input a, b, n; external f:
     sum = 0.5*(f(a) + f(b))
     h = (b - a)/n
     for i=1 to n-1 do
          x = a + i*h
          sum = sum + f(x)
     endfor
     trapezoid = h*sum
end code
```

Programming Hints: (1) When writing a program for *any* integration method, test it first on something for which you know the exact value of the integral, and verify that your approximations are converging to the exact value at the proper *rate* as *h* decreases. (This assumes that the integrand has enough continuous derivatives, of course.) (2) If your code works on one example but not on others, it is a good bet that the problem is with the coding of the other examples. Make sure your "exact values" are indeed correct, for instance.

We close this section with a brief discussion of the *stability* of the trapezoid rule. Consider the integral with $h_i = x_i - x_{i-1}$ and $h = \max_i h_i$. Then

$$|I(f) - T_n(f)| \leq \frac{b-a}{12} h^2 \max_{x \in [a,b]} |f''|.$$

The student should perhaps think about how to prove this result.

We close this section with a piece of pseudocode that shows how to implement the trapezoid rule (see Algorithm 2.5). Note that this code explicitly makes use of the assumed uniformity of the mesh spacing.

We close this section with a brief discussion of the *stability* of the trapezoid rule. Consider the integral

$$I(f) = \int_a^b f(x)\, dx$$

and its trapezoid rule approximation, $T_n(f)$. What happens if we slightly perturb f to produce a new function, say $g(x) = f(x) + \epsilon(x)$, where $\epsilon(x)$ is always "small"? How much change do we produce in the trapezoid rule values? In other words, can we bound the difference $|T_n(f) - T_n(g)|$ in terms of $|f(x) - g(x)| = |\epsilon(x)|$? We have

$$T_n(f) = h\sum_{i=0}^{n}{}'' f(x_i), \quad T_n(g) = h\sum_{i=0}^{n}{}'' g(x_i),$$

where the double prime on the summation symbol means that the first and last terms are multiplied by $\frac{1}{2}$. Thus,

$$T_n(f) - T_n(g) = h\sum_{i=0}^{n}{}'' (f(x_i) - g(x_i)) = h\sum_{i=0}^{n}{}'' \epsilon(x_i),$$

so that

$$|T_n(f) - T_n(g)| \leq \left(\max_{x \in [a,b]} |\epsilon(x)| \right) h \sum_{i=0}^{n} {}''1 = (b-a) \max_{x \in [a,b]} |\epsilon(x)|.$$

Therefore, if $\max_{x \in [a,b]} |\epsilon(x)|$ is "small," then so is $|T_n(f) - T_n(g)|$; and in fact if $\max_{x \in [a,b]} |\epsilon(x)| \to 0$, then $T_n(g) \to T_n(f)$. Thus, we conclude that the trapezoid rule is a stable numerical method. Almost all methods for numerically approximating integrals are stable, in fact.

Chapter 5 contains more material on the approximation of definite integrals.

EXERCISES

1. Use the trapezoid rule with $h = \frac{1}{4}$ to approximate the integral

$$I = \int_0^1 x^3 \, dx = \frac{1}{4}.$$

You should get that $T_4 = 17/64$. How small does h have to be to get error less than 10^{-3}? 10^{-6}?

2. Use the trapezoid rule and $h = \pi/4$ to approximate the integral

$$I = \int_0^{\pi/2} \sin x \, dx = 1.$$

How small does h have to be to get an error less than 10^{-3}? 10^{-6}?

3. Repeat the preceding with $h = \pi/6$.

4. Apply the trapezoid rule with $h = \frac{1}{8}$ to approximate the integral

$$I = \int_0^1 x(1 - x^2) \, dx = \frac{1}{4}.$$

Feel free to use a computer program or a calculator, as you wish. How small does h have to be to get an error less than 10^{-3}? 10^{-6}?

5. Apply the trapezoid rule with $h = \frac{1}{8}$ to approximate the integral

$$I = \int_0^1 \frac{1}{\sqrt{1+x^4}} \, dx = 0.92703733865069.$$

How small does h have to be to get an error less than 10^{-3}? 10^{-6}?

6. Apply the trapezoid rule with $h = \frac{1}{8}$ to approximate the integral

$$I = \int_0^1 \ln(1+x) \, dx = 2 \ln 2 - 1.$$

How small does h have to be to get an error less than 10^{-3}? 10^{-6}?

7. Apply the trapezoid rule with $h = \frac{1}{8}$, to approximate the integral

$$I = \int_0^1 \frac{1}{1+x^3} \, dx = \frac{1}{3} \ln 2 + \frac{1}{9} \sqrt{3} \pi.$$

How small does h have to be to get an error less than 10^{-3}? 10^{-6}?

8. Apply the trapezoid rule with $h = \frac{1}{8}$ to approximate the integral

$$I = \int_1^2 e^{-x^2} \, dx = 0.1352572580.$$

How small does h have to be to get an error less than 10^{-3}? 10^{-6}?

9. Let I_8 denote the value you obtained in the previous problem. Repeat the computation, this time using $h = \frac{1}{4}$, and call this approximate value I_4. Then compute

$$I_R = (4I_8 - I_4)/3$$

and compare this with the given exact value of the integral.

10. Repeat the preceding for the integral

$$I = \int_0^1 \frac{1}{1+x^3} \, dx = \frac{1}{3} \ln 2 + \frac{1}{9} \sqrt{3} \pi.$$

11. For each integral below, write a program to do the trapezoid rule using the sequence of mesh sizes $h = \frac{1}{2}(b-a), \frac{1}{4}(b-a), \frac{1}{8}(b-a), \ldots, \frac{1}{128}(b-a)$, where $b-a$ is the length of the given interval. Verify that the expected rate of decrease of the error is obtained.

(a) $f(x) = 1/(1+25x^2)$, $[0,1]$, $I(f) = \frac{1}{5}\arctan(5)$

(b) $f(x) = x^2 e^{-x}$, $[0,2]$ $I(f) = 2 - 10e^{-2} = 0.646647168$

(c) $f(x) = \ln x$, $[1,3]$, $I(f) = 3\ln 3 - 2 = 1.295836867$

(d) $f(x) = \sqrt{1-x^2}$, $[-1,1]$, $I(f) = \pi/2$

(e) $f(x) = x^{5/2}$, $[0,1]$, $I(f) = 2/7$

(f) $f(x) = e^{-x}\sin(4x)$, $[0,\pi]$, $I(f) = \frac{4}{17}(1 - e^{-\pi}) = 0.2251261368$

12. For each integral in Exercise 11, how small does h have to be to get accuracy, according to the error theory, of at least 10^{-3}? 10^{-6}?

13. Apply the trapezoid rule to the integral

$$I = \int_0^1 \sqrt{x}\, dx = \frac{2}{3}$$

using a sequence of uniform grids with $h = \frac{1}{2}, \frac{1}{4}, \ldots$. Do we get the expected rate of convergence? Explain.

14. The length of a curve $y = g(x)$, for x between a and b, is given by the integral

$$L(g) = \int_a^b \sqrt{1 + [g'(x)]^2}\, dx.$$

Use the trapezoid rule with $h = \pi/4$ and $h = \pi/16$ to find the length of one "arch" of the sine curve.

15. Use the trapezoid rule to find the length of the logarithm curve between $a = 1$ and $b = e$, using $n = 4$ and $n = 16$.

16. What should h be to guarantee an accuracy of 10^{-8} using the trapezoid rule for each of the following integrals?

(a) $I(f) = \int_0^1 e^{-x^2} dx$

(b) $I(f) = \int_{-5}^5 \frac{1}{1+x^2}\, dx$

(c) $I(f) = \int_1^3 \ln x\, dx$

(d) $I(f) = \int_0^1 \cos(\pi x/2)\, dx$

17. Since the natural logarithm is defined as an integral,

$$\ln x = \int_1^x \frac{1}{t}\, dt \qquad (2.21)$$

it is possible to use the trapezoid rule (or any other numerical integration rule) to construct approximations to $\ln x$.

(a) Show that using the trapezoid rule on the integral (2.21) results in the series approximation (for $x \in [1,2]$)

$$\ln x \approx \frac{x^2 - 1}{2nx} + \sum_{k=1}^{n-1} \frac{x-1}{n + k(x-1)}.$$

(Hint: What are a and b in the integral defining the logarithm?)

(b) How many terms are needed in this approximation to get an error of less than 10^{-8} for all $x \in [1,2]$? How many terms are needed for an error of less than 10^{-15} over the same interval?

(c) Implement this series for a predicted accuracy of 10^{-8} and compare it with the intrinsic natural logarithm function on your computer, over the interval $[1,2]$. Is the expected accuracy achieved?

(d) If we were interested only in the interval $[1,3/2]$, how many terms would be needed for the accuracy specified in (b)?

(e) Is it possible to reduce the computation of $\ln x$ for all $x > 0$ to the computation of $\ln z$ for $z \in [1,3/2]$? Explain.

18. How small must h be to compute the error function,

$$\text{erf}(x) = \frac{2}{\sqrt{\pi}} \int_0^x e^{-t^2} dt,$$

using the trapezoid rule, to within 10^{-8} accuracy for all $x \in [0,1]$?

19. Use the data in Table 2.7 to compute

$$I = \int_1^2 \Gamma(x)\, dx$$

using $h = 0.2$ and $h = 0.1$.

20. Use the data in Table 2.7 to compute

$$I = \int_0^1 \text{erf}(x)\, dx$$

using $h = 0.2$ and $h = 0.1$.

21. Show that the trapezoid rule is *exact* for all linear polynomials.

22. Prove Theorem 2.4.

23. Extend the discussion on stability to include changes in the interval of integration instead of changes in the function. State and prove a theorem that bounds the change in the trapezoid rule approximation to

$$I(f) = \int_a^b f(x)\, dx$$

when the upper limit of integration changes from b to $b + \epsilon$, but f remains the same.

24. Consider a function $\tilde{f}(x)$ that is the floating-point representation of $f(x)$; thus $f - \tilde{f}$ is the rounding error in computing f. If we assume that $|f(x) - \tilde{f}(x)| \le \epsilon$ for all x, show that

$$|T_n(f) - T_n(\tilde{f})| \le \epsilon(b - a).$$

What does this say about the effects of rounding error on the trapezoid rule?

2.6 SOLUTION OF TRIDIAGONAL LINEAR SYSTEMS

The solution of linear systems of equations is one of the most important areas of computational mathematics. A complete treatment is often given in a separate course, and a number of texts exist that are devoted entirely to the subject of numerical linear algebra. We present some of that material in Chapter 7; in this section we discuss one particular problem of special interest: the solution of tridiagonal systems of linear equations.

Recall that a system of linear equations can be written in the form

$$
\begin{aligned}
a_{11}x_1 + a_{12}x_2 + \cdots + a_{1n}x_n &= f_1 \\
a_{21}x_1 + a_{22}x_2 + \cdots + a_{2n}x_n &= f_2 \\
&\vdots \\
a_{n1}x_1 + a_{n2}x_2 + \cdots + a_{nn}x_n &= f_n,
\end{aligned}
$$

where the a_{ij} and f_i are known and the x_i are the unknowns we wish to find. This can be put into matrix-vector form as

$$Ax = f, \tag{2.22}$$

where A is the $n \times n$ matrix having a_{ij} as its entries, and x and f are real-valued vectors with components x_i and f_i, respectively.

A common special case of a linear system occurs when A is *tridiagonal*, i.e., there are only three "diagonals" in A that contain nonzero elements: the main diagonal, the first diagonal above the main diagonal (the superdiagonal), and the first diagonal below the main diagonal (the subdiagonal). Thus A has the special form

$$
A = \begin{bmatrix}
a_{11} & a_{12} & 0 & \cdots & & 0 \\
a_{21} & a_{22} & a_{23} & \ddots & & 0 \\
0 & a_{32} & a_{33} & \ddots & & 0 \\
\vdots & \ddots & \ddots & \ddots & & a_{n-1,n} \\
0 & \cdots & 0 & a_{n,n-1} & & a_{nn}
\end{bmatrix}.
$$

For example,

$$A = \begin{bmatrix} 4 & 2 & 0 & 0 & 0 \\ 1 & 5 & 6 & 0 & 0 \\ 0 & 3 & 9 & 5 & 0 \\ 0 & 0 & 0 & 1 & 2 \\ 0 & 0 & 0 & 2 & 4 \end{bmatrix}$$

is tridiagonal. This makes the solution of the system, under certain assumptions, quite easy. Before proceeding with the algorithm, let's make a notational simplification. Instead of writing the system in the traditional matrix notation and indexing style, we will use l_i, d_i, u_i to denote the lower-diagonal, diagonal, and upper-diagonal elements:

$$\begin{aligned} l_i &= a_{i,i-1}, & 2 \le i \le n \\ d_i &= a_{ii}, & 1 \le i \le n \\ u_i &= a_{i,i+1}, & 1 \le i \le n-1, \end{aligned}$$

where we adopt the convention that $l_1 = 0$ and $u_n = 0$. Under this notation, the augmented matrix corresponding to the system is

$$[A \mid f] = \begin{bmatrix} d_1 & u_1 & 0 & \dots & 0 & \bigm| & f_1 \\ l_2 & d_2 & u_2 & \ddots & 0 & \bigm| & f_2 \\ 0 & l_3 & d_3 & \ddots & 0 & \bigm| & \vdots \\ \vdots & \ddots & \ddots & \ddots & u_{n-1} & \bigm| & f_{n-1} \\ 0 & \dots & 0 & l_n & d_n & \bigm| & f_n \end{bmatrix}, \tag{2.23}$$

and we can store the entire problem using just the four vectors for l, d, u, and f, instead of the entire $n \times n$ matrix, which is mostly zeroes anyway.

Recall from linear algebra that the standard means of solving a linear system (called *Gaussian elimination*) is to eliminate all the components of A below the main diagonal; i.e., we reduce A to *triangular* form. In our case this is easy because we have to eliminate only a single term each time, since there is only a single element below the main diagonal in each column. Thus, we would multiply the first equation by l_2/d_1 and subtract this from the second equation to get

$$[A \mid f] \sim \begin{bmatrix} d_1 & u_1 & 0 & \dots & 0 & \bigm| & f_1 \\ 0 & d_2 - u_1(l_2/d_1) & u_2 & \ddots & 0 & \bigm| & f_2 - f_1(l_2/d_1) \\ 0 & l_3 & d_3 & \ddots & 0 & \bigm| & \vdots \\ \vdots & \ddots & \ddots & \ddots & u_{n-1} & \bigm| & f_{n-1} \\ 0 & \dots & 0 & l_n & d_n & \bigm| & f_n \end{bmatrix}$$

and then continue on with each successive row. Note that carrying out this step requires that $d_1 \neq 0$, and continuing to the next step will require that $d_2 - u_1(l_2/d_1) \neq 0$. Assuming the corresponding statement to be the case for each remaining row, we can reduce (2.23) to the equivalent system

$$[T \mid g] = \left[\begin{array}{ccccc|c} \delta_1 & u_1 & 0 & \ldots & 0 & g_1 \\ 0 & \delta_2 & u_2 & \ddots & 0 & g_2 \\ 0 & 0 & \delta_3 & \ddots & 0 & \vdots \\ \vdots & \ddots & \ddots & \ddots & u_{n-1} & g_{n-1} \\ 0 & \ldots & 0 & 0 & \delta_n & g_n \end{array} \right],$$

where

$$\delta_1 = d_1, \qquad \delta_2 = d_2 - u_1(l_2/\delta_1), \qquad \delta_3 = d_3 - u_2(l_3/\delta_2),$$

and, generally,

$$\delta_k = d_k - u_{k-1}(l_k/\delta_{k-1}), \, 2 \leq k \leq n.$$

Similarly,

$$g_1 = f_1, \quad g_2 = f_2 - g_1(l_2/\delta_1), \qquad g_3 = f_3 - g_2(l_3/\delta_2),$$

so that the general form is

$$g_k = f_k - g_{k-1}(l_k/\delta_{k-1}), \, 2 \leq k \leq n.$$

The matrix $[T \mid g]$ is *row equivalent* to the original augmented matrix $[A \mid f]$, meaning that we can progress from one to the other using elementary row operations; thus the two augmented matrices represent systems with exactly the same solution sets. Moreover, the solution is now easy to obtain, since we can solve the last equation $\delta_n x_n = g_n$ to get $x_n = g_n/\delta_n$, and then use this value in the previous equation to get x_{n-1}, and so on, to get each solution component. Again, carrying out this stage of the computation requires the assumption that each $\delta_k \neq 0$, $1 \leq k \leq n$.

The first stage of the computation (reducing the tridiagonal matrix A to the *triangular* one T) is generally called the elimination step, and the second stage is generally called the backward solution (or *back-solve*) step. A pseudocode for this process is given in Algorithm 2.6. Note that we do not store the entire matrix, but only the three vectors needed to define the elements in the nonzero diagonals; in addition, we did not use different variable names for the d_i, δ_i, etc., but rather *overwrote* the original variable with the new values. This saves storage when working with large problems.

At this point we look at a couple of examples.

Algorithm 2.6 *Pseudocode for the solution of tridiagonal systems*

```
!
!     Elimination Stage
!
          for i=2 to n
               d(i)  =  d(i)  -  u(i-1)*l(i)/d(i-1)
               f(i)  =  f(i)  -  f(i-1)*l(i)/d(i-1)
          endfor
!
!     Backsolve stage (bottom row is a special case)
!
          x(n)  =  f(n)/d(n)
          for i=n-1 downto 1
               x(i)  =  (f(i)  -  u(i)*x(i+1))/d(i)
          endfor
```

Programming Hints: Always test your codes on simple examples for which you know the solution, and for which you can easily do the computation by hand. In the case of a tridiagonal system, start with a small (3×3, or 4×4) system; if your computed solution is not almost exactly the same as the true solution, then your code has an error someplace. Do the algorithm by hand and then print out what the computer has for the final values of d, u, etc., so you can see where your code deviates from the correct computation. When you don't know the exact solution, compute the *residual*, defined by $r = b - Ax$. If your computed solution is correct, the residual should be almost exactly zero in each of the components. (*Note:* If the example matrix is "nearly singular"—a concept that will be more precisely defined in Chapter 7—then it is possible that the rounding error effects will be quite large, even for small 3×3 examples. To avoid this potential problem, make sure your examples have diagonal elements that are large compared with the absolute sum of the off-diagonal elements.)

EXAMPLE 2.7

Consider the system

$$
\begin{bmatrix}
4 & 1 & 0 & 0 \\
1 & 4 & 1 & 0 \\
0 & 1 & 4 & 1 \\
0 & 0 & 1 & 4
\end{bmatrix}
\begin{bmatrix}
x_1 \\
x_2 \\
x_3 \\
x_4
\end{bmatrix}
=
\begin{bmatrix}
6 \\
12 \\
18 \\
19
\end{bmatrix}.
\tag{2.24}
$$

If we do the elimination stage of the algorithm, we get the new system (written now as a single augmented matrix)

$$
[T \mid g] =
\begin{bmatrix}
4 & 1 & 0 & 0 & \Big| & 6 \\
0 & \frac{15}{4} & 1 & 0 & \Big| & \frac{21}{2} \\
0 & 0 & \frac{56}{15} & 1 & \Big| & \frac{76}{5} \\
0 & 0 & 0 & \frac{209}{56} & \Big| & \frac{209}{14}
\end{bmatrix},
$$

from which we get the solution

$$x_1 = 1, \qquad x_2 = 2, \qquad x_3 = 3, \qquad x_4 = 4.$$

∎

On the other hand, if we look at the system

$$\begin{bmatrix} 1 & 4 & 0 & 0 \\ 1 & 4 & 1 & 0 \\ 0 & 1 & 4 & 1 \\ 0 & 0 & 1 & 4 \end{bmatrix} \begin{bmatrix} x_1 \\ x_2 \\ x_3 \\ x_4 \end{bmatrix} = \begin{bmatrix} 9 \\ 12 \\ 18 \\ 19 \end{bmatrix}, \tag{2.25}$$

the elimination stage produces the following augmented matrix after a single pass through the first loop:

$$[T \mid g] = \begin{bmatrix} 1 & 4 & 0 & 0 & \vline & 9 \\ 0 & 0 & 1 & 0 & \vline & 3 \\ 0 & 1 & 4 & 1 & \vline & 18 \\ 0 & 0 & 1 & 4 & \vline & 19 \end{bmatrix}.$$

Clearly, we cannot continue the process, for we would have to divide by zero in the next step. We should point out, however, that this does *not* mean that the system in question does not have a solution; it simply means that the algorithm fails to work on this example. The reader should check that

$$x_1 = 1, \qquad x_2 = 2, \qquad x_3 = 3, \qquad x_4 = 4$$

is indeed the solution to the system.

A complete discussion of this problem and similar issues in numerical linear algebra is deferred until Chapter 7; for now we content ourselves with stating and proving a common condition that is sufficient to guarantee that the tridiagonal solution algorithm presented here will work. We require a definition first.

Definition 2.1 (Diagonal Dominance for Tridiagonal Matrices)   A tridiagonal matrix is called diagonally dominant *if*

$$d_i > |l_i| + |u_i| > 0, \quad 1 \le i \le n. \tag{2.26}$$

For example, the matrix

$$A_1 = \begin{bmatrix} 6 & 1 & 0 & 0 \\ 2 & 6 & 3 & 0 \\ 0 & 6 & 9 & 0 \\ 0 & 0 & 3 & 4 \end{bmatrix}$$

is diagonally dominant, but

$$A_2 = \begin{bmatrix} 6 & 1 & 0 & 0 \\ 2 & 5 & 3 & 0 \\ 0 & 6 & 9 & 0 \\ 0 & 0 & 3 & 4 \end{bmatrix}$$

is not. (The inequality does not hold in the second row.)

If the matrix is diagonally dominant, then we can easily show that we do not need to worry about the diagonal elements becoming zero during the elimination process.

Theorem 2.5   *If the tridiagonal matrix A is diagonally dominant, then Algorithm 2.6 will succeed in producing the correct solution to the original linear system, within the limitations of rounding error.*

Proof   The diagonal dominance condition directly shows that $d_1 = \delta_1 \neq 0$, so all that remains is to show that each $\delta_k = d_k - u_{k-1}l_k/\delta_{k-1} \neq 0$, for $2 \leq k \leq n$. Assume momentarily that $l_2 \neq 0$. We have

$$\begin{aligned} \delta_2 &= d_2 - u_1 l_2/d_1 \\ &\geq d_2 - |u_1 l_2/d_1| \\ &\geq |u_2| + |l_2| - |l_2|\theta_1 \\ &\geq (|u_2| + |l_2|)(1 - \theta_1), \end{aligned}$$

for $\theta_1 = |u_1|/|d_1| < 1$. Therefore, $\delta_2 > 0$ since $l_2 \neq 0$. If, on the other hand, $l_2 = 0$, then we have $\delta_2 = d_2 - 0 = d_2 > 0$. We can repeat the same argument for each index, and that completes the proof.   □

Note that in our example (2.24) the matrix was indeed diagonally dominant, whereas in (2.25) it was not; the condition (2.26) fails in the first row.

Note also that we did not discuss any error issues in this section. The reason is actually very simple: There is no approximation being made, so there is no mathematical error to worry about. In the absence of rounding error, this algorithm will produce the exact solution of the tridiagonal linear system, *if* it is able to run to completion (i.e., without divisions by zero).

EXERCISES

1. Use the tridiagonal algorithm to compute the solution to the following system of equations:

$$\begin{bmatrix} 4 & 2 & 0 & 0 \\ 1 & 4 & 1 & 0 \\ 0 & 1 & 4 & 1 \\ 0 & 0 & 2 & 4 \end{bmatrix} \begin{bmatrix} x_1 \\ x_2 \\ x_3 \\ x_4 \end{bmatrix} = \begin{bmatrix} \pi/9 \\ \sqrt{3}/2 \\ \sqrt{3}/2 \\ -\pi/9 \end{bmatrix}.$$

2. Write a computer code to solve the previous exercise.

3. Use the algorithm of this section to solve the following system of equations:

$$\begin{bmatrix} 6 & 1 & 0 & 0 \\ 2 & 4 & 1 & 0 \\ 0 & 1 & 4 & 2 \\ 0 & 0 & 1 & 6 \end{bmatrix} \begin{bmatrix} x_1 \\ x_2 \\ x_3 \\ x_4 \end{bmatrix} = \begin{bmatrix} 8 \\ 13 \\ 22 \\ 27 \end{bmatrix}.$$

You should get the solution $x = (1,2,3,4)^T$.

4. Write a computer code to solve the previous exercise.

5. The diagonal dominance condition is an example of a *sufficient* but not *necessary* condition. That is, the algorithm will often work for systems that are *not* diagonally dominant. Show that the following system is not diagonally dominant, but use the tridiagonal algorithm in this section to compute the solution to it:

$$
\begin{bmatrix}
1 & \frac{1}{2} & 0 & 0 \\
\frac{1}{2} & \frac{1}{3} & \frac{1}{4} & 0 \\
0 & \frac{1}{4} & \frac{1}{5} & \frac{1}{6} \\
0 & 0 & \frac{1}{6} & \frac{1}{7}
\end{bmatrix}
\begin{bmatrix}
x_1 \\ x_2 \\ x_3 \\ x_4
\end{bmatrix}
=
\begin{bmatrix}
2 \\ 23/12 \\ 53/30 \\ 15/14
\end{bmatrix}.
$$

You should get the solution $x = (1,2,3,4)^T$.

6. Use the tridiagonal algorithm in this section to compute the solution to the following system of equations:

$$
\begin{bmatrix}
1 & \frac{1}{2} & 0 & 0 \\
\frac{1}{2} & \frac{1}{3} & \frac{1}{4} & 0 \\
0 & \frac{1}{4} & \frac{1}{5} & \frac{1}{6} \\
0 & 0 & \frac{1}{6} & \frac{1}{7}
\end{bmatrix}
\begin{bmatrix}
x_1 \\ x_2 \\ x_3 \\ x_4
\end{bmatrix}
=
\begin{bmatrix}
2 \\ 2 \\ 53/30 \\ 15/14
\end{bmatrix}.
$$

Note that this is a very small change from the previous problem, since the only difference is that b_2 has changed by only $1/12$. How much has the answer changed?

7. Write a computer code to do the previous two exercises.

8. Verify that the following system is diagonally dominant and use the algorithm of this section to find the solution.

$$
\begin{bmatrix}
\frac{1}{2} & \frac{10}{21} & 0 & 0 \\
\frac{1}{4} & \frac{1}{3} & \frac{1}{13} & 0 \\
0 & \frac{1}{5} & \frac{1}{4} & \frac{1}{21} \\
0 & 0 & \frac{1}{6} & \frac{1}{5}
\end{bmatrix}
\begin{bmatrix}
x_1 \\ x_2 \\ x_3 \\ x_4
\end{bmatrix}
=
\begin{bmatrix}
\frac{61}{42} \\ \frac{179}{156} \\ \frac{563}{420} \\ \frac{13}{10}
\end{bmatrix}.
$$

9. Use the algorithm of this section to find the solution to this system.

$$
\begin{bmatrix}
\frac{1}{2} & \frac{10}{21} & 0 & 0 \\
\frac{1}{4} & \frac{1}{3} & \frac{1}{13} & 0 \\
0 & \frac{1}{5} & \frac{1}{4} & \frac{1}{21} \\
0 & 0 & \frac{1}{6} & \frac{1}{5}
\end{bmatrix}
\begin{bmatrix}
x_1 \\ x_2 \\ x_3 \\ x_4
\end{bmatrix}
=
\begin{bmatrix}
\frac{61}{42} \\ \frac{180}{156} \\ \frac{563}{420} \\ \frac{13}{10}
\end{bmatrix}.
$$

Note that the right side here is different from that in the previous exercise by only a small amount in the b_2 component. Comment on your results here as compared with those in the previous exercise.

10. Write a computer code to do the previous two exercises.

11. Write a code that carries out the tridiagonal solution algorithm, and test it on the following system of equations:

$$Tx = b$$

where T is 10×10 with

$$
t_{ij} =
\begin{cases}
1 & |i - j| = 1 \\
j + 1 & i = j \\
0 & \text{otherwise}
\end{cases}
$$

and $b_i = 1$ for all i. Check your results by computing the residual $r = b - Tx$. What is the largest component (in absolute value) of r?

12. Extend the tridiagonal algorithm to a *pentadiagonal* matrix, i.e., one with five nonzero diagonals. Write a program to carry out this solution algorithm, and apply it to the system

$$
\begin{bmatrix}
4 & 2 & 1 & 0 \\
1 & 4 & 1 & 1 \\
1 & 1 & 4 & 1 \\
0 & 1 & 2 & 4
\end{bmatrix}
\begin{bmatrix}
x_1 \\ x_2 \\ x_3 \\ x_4
\end{bmatrix}
=
\begin{bmatrix}
1 \\ 1 \\ 1 \\ 1
\end{bmatrix}.
$$

Check your results by again computing the residual vector.

2.7 APPLICATION: SIMPLE TWO-POINT BOUNDARY VALUE PROBLEMS

Consider the problem of finding a function $u(x)$ such that

$$-u'' + u = f(x), \quad x \in [0,1] \tag{2.27}$$

$$u(0) = u(1) = 0. \tag{2.28}$$

Here f is a known function, and we seek u that satisfies the differential equation and the conditions at $x = 0$ and $x = 1$. This is an example of a *two-point boundary value problem* (BVP), something we treat in more detail in Section 6.10; we have enough tools now to compute solutions, although we cannot completely analyze the error. Problems of this sort occur in many areas of applied mathematics and engineering, most notably perhaps in structural mechanics.

Divide the interval $[0,1]$ into n equal subintervals $[x_{k-1}, x_k]$, according to

$$0 = x_0 < x_1 < x_2 < \cdots < x_{n-1} < x_n = 1,$$

with $x_k - x_{k-1} = h$ (therefore $x_k = kh$), and let $U_h(x)$ denote the approximation to $u(x)$. We can use Taylor expansions similar to (2.2) and (2.3) (just take more terms) to derive an approximation to the second derivative, by adding them (instead of subtracting them). After a little manipulation (see Exercise 12 in Section 2.2), we get

$$u''(x) - \frac{u(x-h) - 2u(x) + u(x+h)}{h^2} = \frac{1}{12}h^2 u^{(4)}(\eta_x), \qquad (2.29)$$

for some $\eta_x \in [x-h, x+h]$. Therefore, the differential equation $-u'' + u = f(x)$ implies that

$$\frac{-u(x-h) + 2u(x) - u(x+h)}{h^2} + u(x) = f(x) + \frac{1}{12}h^2 u^{(4)}(\eta_x).$$

To define our approximation, we drop the remainder term and replace u with U_h, which yields

$$\frac{-U_h(x-h) + 2U_h(x) - U_h(x+h)}{h^2} + U_h(x) = f(x). \qquad (2.30)$$

This holds for all $x \in [0,1]$. To get a practical computational problem, we will impose (2.30) only on our grid points; i.e., we seek the $n-1$ values $U_k = U_h(x_k)$, $1 \le k \le n-1$, that satisfy (recall that the boundary conditions will force $U_0 = U_n = 0$)

$$-U_{k-1} + (2 + h^2)U_k - U_{k+1} = h^2 f(x_k), \quad 1 \le k \le n-1.$$

This is a tridiagonal system of linear equations. Written out in matrix-vector form, we have

$$
\begin{bmatrix}
2+h^2 & -1 & 0 & \cdots & \cdots & 0 \\
-1 & 2+h^2 & -1 & 0 & \cdots & 0 \\
0 & -1 & 2+h^2 & -1 & 0 & \vdots \\
\vdots & & \ddots & \ddots & \ddots & 0 \\
\vdots & & & \ddots & \ddots & -1 \\
0 & \cdots & \cdots & 0 & -1 & 2+h^2
\end{bmatrix}
\begin{bmatrix}
U_1 \\
U_2 \\
\vdots \\
\vdots \\
U_{n-2} \\
U_{n-1}
\end{bmatrix}
=
\begin{bmatrix}
h^2 f(x_1) \\
h^2 f(x_2) \\
\vdots \\
\vdots \\
h^2 f(x_{n-2}) \\
h^2 f(x_{n-1})
\end{bmatrix}.
$$

Moreover, it is diagonally dominant, so we can apply the algorithm developed in the previous section to produce solutions.

TABLE 2.10   **Two-Point Boundary Value Problem Results**

h^{-1}	$\max|u(x_k) - U_k|$
4	0.529469582786E–03
8	0.133142281084E–03
16	0.333344381367E–04
32	0.833667068484E–05
64	0.208435910093E–05
128	0.521101741688E–06
256	0.130276180538E–06
512	0.325690907577E–07
1024	0.814227994750E–08

EXAMPLE 2.8

If we take the specific case of $h = \frac{1}{4}$ and $f(x) = 1$, then we have a 3×3 system to solve, i.e.,

$$
\begin{bmatrix}
2.0625 & -1 & 0 \\
-1 & 2.0625 & -1 \\
0 & -1 & 2.0625
\end{bmatrix}
\begin{bmatrix}
U_1 \\
U_2 \\
U_3
\end{bmatrix}
=
\begin{bmatrix}
0.0625 \\
0.0625 \\
0.0625
\end{bmatrix}.
$$

The solution is easily found, using the algorithm from the previous section, to be

$$U_1 = 0.08492201039861, \qquad U_2 = 0.11265164644714, \qquad U_3 = 0.08492201039861.$$

Thus we have $u(1/4) \approx 0.08492201039861$, $u(1/2) \approx 0.11265164644714$, and $u(3/4)$ ≈ 0.08492201039861.

Table 2.10 shows the errors we get between $u(x_k)$ and U_k as h decreases from $1/4$ to $1/1024$, for this same choice of $f(x)$. The exact solution is given by

$$u(x) = 1 - \frac{e-1}{e^2-1}e^x - \frac{e^2-e}{e^2-1}e^{-x}.$$

Note that the error is decreasing by roughly a factor of 4 each time we double the number of points in the grid. Based on what we saw with the difference approximations to the derivatives, as well as the trapezoid rule, we suspect this means that the method is $O(h^2)$ accurate— something we will be able to prove in Section 6.10.1, assuming that the solution is smooth enough. Figure 2.9 shows a plot of the approximate solution and exact solution for the $h = \frac{1}{8}$ case. Note that the approximate solution is virtually exact, even for this coarse mesh. (This, of course, will not always be the case.)   ∎

Programming Hint   Although our grid on $[0,1]$ has n subintervals, note that the matrix system is $(n-1) \times (n-1)$, *not $n \times n$*. This is because the values at both $x = 0$ and $x = 1$ are known and so are not part of the system.

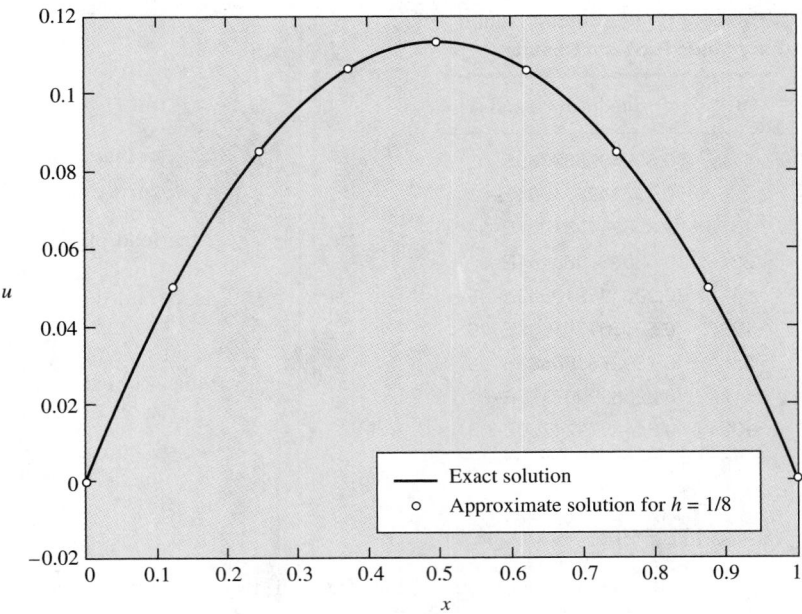

FIGURE 2.9   Approximate solution to the two-point BVP $-u'' + u = 1$, for $h = 1/8$.

EXERCISES

1. Solve, by hand, the two-point BVP (2.27)–(2.28) when $f(x) = x$, using $h = \frac{1}{4}$. Write out the linear system explicitly prior to solution. You should get the following 3×3 system:

$$\begin{bmatrix} 2.0625 & -1 & 0 \\ -1 & 2.0625 & -1 \\ 0 & -1 & 2.0625 \end{bmatrix} \begin{bmatrix} U_1 \\ U_2 \\ U_3 \end{bmatrix} = \begin{bmatrix} 0.015625 \\ 0.03125 \\ 0.046875 \end{bmatrix}.$$

2. Repeat the preceding, this time using $h = \frac{1}{5}$. What is the system now?

3. Repeat it again, this time using $h = \frac{1}{8}$. What is the system now? (For this exercise, you probably will want to use a computer code to actually solve the system.)

4. Write a program that solves the two-point BVP (2.27)–(2.28), where f is as given below.
(a) $f(x) = (\pi^2 + 1) \sin \pi x, u(x) = \sin \pi x$

(b) $f(x) = 4e^{-x} - 4xe^{-x}, u(x) = x(1-x)e^{-x}$
(c) $f(x) = \pi(\pi \sin \pi x + 2\cos \pi x)e^{-x}, u(x) = e^{-x}\sin \pi x$
(d) $f(x) = 3 - \dfrac{1}{x} - (x^2 - x - 2)\log x, u(x) = x(1-x)\log x$

The exact solutions are as given. Using $h^{-1} = 4, 8, 16, \ldots$, do we get the same kind of accuracy as in Table 2.10? Explain why or why not.

5. Try to apply the ideas of this section to approximating the solution of the two-point boundary value problem

$$-u'' + u' + u = 1, \quad x \in [0,1]$$
$$u(0) = 0, u(1) = 0.$$

Can we get a tridiagonal system that Algorithm 2.6 can be applied to? (Hint: Consider some of the approximations from Section 2.2; use the most accurate ones that can be easily used.)

6. Solve the two-point boundary value problem

$$-u'' + 64u' + u = 1, \quad x \in [0,1]$$
$$u(0) = 0, u(1) = 0,$$

using a range of mesh sizes, starting with $h = 1/4, 1/8$, and going as far as $h = 1/256$. Comment on your results.

7. Generalize the solution of the two-point boundary value problem to the case where $u(0) = g_0 \neq 0$ and $u(1) = g_1 \neq 0$. Apply this to the solution of the problem

$$-u'' + u = 0, \quad x \in [0,1]$$
$$u(0) = 2, \quad u(1) = 1,$$

which has exact solution

$$u(x) = \left(\frac{e-2}{e^2 - 1} \right) e^x + \left(\frac{2e - 1}{e^2 - 1} \right) e^{-x}.$$

Solve this for a range of values of the mesh. Do we get the expected $O(h^2)$ accuracy?

8. Consider the problem of determining the deflection of a thin beam, supported at both ends, due to a uniform load being placed along the beam. In one simple model, the deflection $u(x)$ as a function of position x along the beam satisfies the boundary value problem

$$-u'' + pu = qx(L - x), \quad 0 < x < L$$
$$u(0) = u(L) = 0.$$

Here p is a constant that depends on the material properties of the beam, L is the length of the beam, and q depends on the material properties of the beam as well as the size of the load placed on the beam. For a six-foot-long beam, with $p = 7 \times 10^{-6}$ and $q = 4 \times 10^{-7}$, what is the maximum deflection of the beam? Use a fine enough grid that you can be confident of the accuracy of your results. (Note that this problem is slightly more general than our example (2.27)–(2.28); you will have to adapt our method to this more general case.)

9. Repeat Exercise 8, except this time use a three-foot-long beam. How much does the maximum deflection change, and is it larger or smaller?

10. Repeat Exercise 8 again, but this time use a 12-foot-long beam.

11. Try to apply the ideas of this section to the solution of the *nonlinear* boundary value problem defined by

$$cc - u'' + e^u = 0, \quad 0 < x < 1;$$
$$u(0) = u(1) = 0.$$

Write out the systems of equations for the specific case of $h = \frac{1}{4}$. What goes wrong? Why can't we proceed with the approximate solution?

CHAPTER 3

ROOT FINDING

Consider the towns of Ziegler and Felts, located as shown in the following diagram. Pennsyltucky Petroleum wants to build a pipeline connecting the two towns. Because of differences in terrain, the cost to build the pipeline will be C_1 million dollars per mile north of the line WE, and C_2 million dollars per mile south of the line. How do we build the pipeline to minimize the cost of construction? This amounts to finding out where to place the point P at which the pipeline crosses the line WE.

Ordinary calculus, together with considerations of plane geometry and some trigonometry, can be used to show that the cost is minimized when P is located so that $C_1 \sin\theta_1 = C_2 \sin\theta_2$. In terms of x, the distance from W to P, we get the equation

$$C_2^2(L-x)^2(a^2+x^2) = C_1^2 x^2(b^2+(L-x)^2),$$

which cannot be solved by elementary methods. Here L is the length of the line WE. For the specific case $a=3$, $b=1$, $L=4$, $C_1=1$, $C_2=2$, we get the equation

$$0 = 3x^4 - 24x^3 + 83x^2 - 288x + 576.$$

Using the methods discussed in this chapter, we can find that $x = 3.5848$, which tells Pennsyltucky Petroleum to build the pipeline from Ziegler to a point P, where P is 3.5848 miles east of W, and then from P to Felts. (This is loosely based on Problem 47 from Section 3-5 of the second edition of *Calculus and Analytic Geometry* by C.H. Edwards and D.E. Penny.)

A problem that most students should be familiar with from ordinary algebra is that of finding the root of an equation $f(x) = 0$, i.e., the value of the argument that makes f zero. More precisely, if the function is defined as $y = f(x)$, we seek the value α such that

$$f(\alpha) = 0.$$

The precise terminology is that α is a *zero* of the function f, or a *root* of the equation $f(x) = 0$.[1] Note that we have not specified what kind of function f is yet. The obvious case is when f is an ordinary real-valued function of a single real variable x, but we can also consider the problem when f is a vector-valued function of a vector-valued variable, in which case we would have is a system of equations; this more complicated case will be discussed in Chapter 7.

In this chapter we will consider only the simple case, where f is a scalar real-valued function of a single real-valued variable. We will discuss three basic methods for finding the point α: the bisection method, Newton's method, and the secant method. We then will consider a broad class of ideas coming under the heading of *fixed-point theory*, which will enable us to broaden and extend our understanding of iterations in general, whether applied to root-finding problems or not. Finally, we will discuss some variants of Newton's method and other advanced topics.

3.1 THE BISECTION METHOD

Bisection is a marvelously simple idea that is based on little more than the continuity of the function f. Suppose we know that $f(a)f(b) < 0$. This means that f is negative at one point and positive at the other. If we assume that f is continuous, then it follows (by the Intermediate Value Theorem) that there must be some value between a and b at which f is zero. In other words, we know that there is a root α between a and b. (Note: There may be more than one root in the interval.)

Now let's try to use these ideas to find α. Let c be the midpoint of the interval $[a,b]$, i.e.,

$$c = \frac{1}{2}(a+b),$$

and consider the product $f(a)f(c)$. There are three possibilities:

1. $f(a)f(c) < 0$. This means that a root (there might be more than one) is between a and c, i.e., $\alpha \in [a,c]$.

2. $f(a)f(c) = 0$. If we assume that we already know $f(a) \neq 0$, this means that $f(c) = 0$; thus $\alpha = c$ and we have found a root.

3. $f(a)f(c) > 0$. This means that a root must lie in the other half of the interval, i.e., $\alpha \in [c,b]$.

At first glance this is helpful only if we get the second case and land smack-dab on top of a root, which does not seem very likely. However, a second look reveals that if (1) or (3) hold, we now have a root localized to an interval ($[a,c]$ or $[c,b]$) that is *half* the length of the original interval $[a,b]$. If we repeat the process, the interval of uncertainty is again decreased by half, and so on, until we have the root localized to within any tolerance we desire.

---

[1]The distinction is often muddled in practice, however. For example, it is commonplace to speak of the "roots of a polynomial," when one should in fact speak only of the "zeroes of a polynomial." In this text we have used the terms interchangeably, as denoting where a function is equal to zero.

FIGURE 3.1   Bisection applied to $y = 2 - e^x$.

EXAMPLE 3.1

If $f(x) = 2 - e^x$ and we take the original interval to be $[a,b] = [0,1]$, then the first several steps of the computation are as follows:

$$f(a) = 1, f(b) = -0.7183 \quad \Rightarrow \quad c = (0+1)/2 = 1/2; f(c) = 0.3513 > 0$$
$$\Rightarrow \quad [a,b] \leftarrow [1/2,1]. f(a) = 0.3513, f(b) = -0.7183$$
$$\Rightarrow \quad c = (1/2+1)/2 = 3/4; f(c) = -0.1170 < 0$$
$$\Rightarrow \quad [a,b] \leftarrow [1/2,3/4] f(a) = 0.3513, f(b) = -0.1170$$
$$\Rightarrow \quad c = (1/2+3/4)/2 = 5/8; f(c) = 0.1318 > 0$$
$$\Rightarrow \quad [a,b] \leftarrow [5/8,3/4].$$

Thus we have reduced the "interval of uncertainty" from $[0,1]$, which has length 1, to $[5/8,3/4]$, which has length $\frac{1}{8} = 0.125$. If we were to continue the process, we would eventually have the root localized to within an interval of length as small as we want, since each step cuts the interval of uncertainty in half. Figure 3.1 shows a graphical view of this example, with the marks on the horizontal axis showing how the interval around the root changes with each iteration. ■

A formal statement is given in Algorithm 3.1.

Each step is decreasing an upper bound on the (absolute) error by a factor of 2, a fact we can rigorously establish.

Algorithm 3.1 *Bisection Method (outline form)*

Given an initial interval $[a_0, b_0] = [a, b]$, set $k = 0$ and proceed as follows:

1. Compute $c_{k+1} = a_k + \frac{1}{2}(b_k - a_k)$.

2. If $f(c_{k+1})f(a_k) < 0$, then set $a_{k+1} = a_k$, $b_{k+1} = c_{k+1}$.

3. If $f(c_{k+1})f(b_k) < 0$, then set $b_{k+1} = b_k$, $a_{k+1} = c_{k+1}$.

4. Update k and go to Step 1.

Theorem 3.1 (Bisection Convergence and Error)   Let $[a_0, b_0] = [a, b]$ be the initial interval, with $f(a)f(b) < 0$. Define the approximate root as $x_n = c_n = (b_{n-1} + a_{n-1})/2$. Then there exists a root $\alpha \in [a, b]$ such that

$$|\alpha - x_n| \leq \left(\frac{1}{2}\right)^n (b - a). \tag{3.1}$$

Moreover, to achieve accuracy of

$$|\alpha - x_n| \leq \epsilon,$$

it suffices to take

$$n \geq \frac{\log(b - a) - \log \epsilon}{\log 2}. \tag{3.2}$$

Proof   We first observe, trivially, that

$$b_n - a_n = \frac{1}{2}(b_{n-1} - a_{n-1}),$$

which implies that

$$b_n - a_n = \left(\frac{1}{2}\right)^n (b_0 - a_0).$$

Therefore,

$$|\alpha - x_n| \leq \frac{1}{2}(b_{n-1} - a_{n-1})$$

implies

$$|\alpha - x_n| \leq \frac{1}{2}(b_{n-1} - a_{n-1})$$

$$= \frac{1}{2}\left(\frac{1}{2}\right)^{n-1}(b_0 - a_0)$$

$$= \left(\frac{1}{2}\right)^n (b_0 - a_0),$$

and we have proved the estimate. To establish the bound on the number of iterations, n, we simply observe that

$$\left(\frac{1}{2}\right)^n (b-a) \le \epsilon \tag{3.3}$$

together with (3.1) implies

$$|\alpha - x_n| \le \epsilon,$$

and we solve (3.3) for n to get (3.2)   □

The bisection process can be easily reduced to an algorithm (see Algorithm 3.2).

Table 3.1 shows the result of running this algorithm on the function $f(x) = 2 - e^x$, for which $\alpha = \log 2 = 0.6931471806$. Note the use of the quantity $\log_{10}(b_n - x_n)$; this is an upper bound on the base 10 logarithm of the error, and thus it gives an estimate of the number of correct digits in the approximation, which is a useful measure of the accuracy of the approximation. (This estimate is valid only if the error is less than 1.)

TABLE 3.1   **Bisection Method Applied to $f(x) = 2 - e^x$**

n	x_n	$b_{n-1} - a_{n-1}$	$\log_{10}(b_n - x_n)$	a_n	b_n
1	0.500000000000	1.000000000000	−0.3010	0.500000000000	1.000000000000
2	0.750000000000	0.500000000000	−0.6021	0.500000000000	0.750000000000
3	0.625000000000	0.250000000000	−0.9031	0.625000000000	0.750000000000
4	0.687500000000	0.125000000000	−1.2041	0.687500000000	0.750000000000
5	0.718750000000	0.062500000000	−1.5051	0.687500000000	0.718750000000
6	0.703125000000	0.031250000000	−1.8062	0.687500000000	0.703125000000
7	0.695312500000	0.015625000000	−2.1072	0.687500000000	0.695312500000
8	0.691406250000	0.007812500000	−2.4082	0.691406250000	0.695312500000
9	0.693359375000	0.003906250000	−2.7093	0.691406250000	0.693359375000
10	0.692382812500	0.001953125000	−3.0103	0.692382812500	0.693359375000
11	0.692871093750	0.000976562500	−3.3113	0.692871093750	0.693359375000
12	0.693115234375	0.000488281250	−3.6124	0.693115234375	0.693359375000
13	0.693237304688	0.000244140625	−3.9134	0.693115234375	0.693237304688
14	0.693176269531	0.000122070312	−4.2144	0.693115234375	0.693176269531
15	0.693145751953	0.000061035156	−4.5154	0.693145751953	0.693176269531
16	0.693161010742	0.000030517578	−4.8165	0.693145751953	0.693161010742
17	0.693153381348	0.000015258789	−5.1175	0.693145751953	0.693153381348
18	0.693149566650	0.000007629395	−5.4185	0.693145751953	0.693149566650
19	0.693147659302	0.000003814697	−5.7196	0.693145751953	0.693147659302
20	0.693146705627	0.000001907349	−6.0206	0.693146705627	0.693147659302

Algorithm 3.2 *Bisection Method (pseudocode)*

```
input a,b,eps
external f
fa = f(a)
fb = f(b)
if f(a)*f(b) > 0 then stop
n  = fix((log(b-a) - log(eps))/log(2)) + 1
for i=1 to n do
     c   = a + 0.5*(b - a)
     fc = f(c)
     if fa*fc < 0 then
          b = c
          fb = fc
     else
          if fa*fc > 0 then
               a = c
               fa = fc
          else
               alpha = c
               return
          endif
     endif
endfor
```

Programming Hints: Note that we have used the formula

$$c = a + 0.5(b - a) \tag{3.4}$$

to compute c, rather than the more obvious choice

$$c = (a + b)/2. \tag{3.5}$$

The reason is that for very large values of a and b, (3.5) can lead to a computational overflow, whereas (3.4) will not. Note also that we were very careful in the algorithm to reuse the function values from iteration to iteration. A careless implementation of bisection will have two function calls each time through the loop, but it is possible to get by with only one function call. This is important because the function evaluation is outside our control and might be the most expensive part of the computation. An example of a function that could be quite expensive to compute is given in the discussion of so-called shooting methods for nonlinear boundary value problems, in Section 6.10.2. Finally, although we will not usually bother with showing output statements in any of the algorithms in the text, it is a good idea to have the code print out the entire sequence of iterates, along with the corresponding function values, rather than just the final answer. This provides useful information for debugging and validating the code.

Bisection is an example of what is known as a *global* method; i.e., it always converges no matter how far you start from the actual root (assuming, of course, that you have the root "bracketed," i.e., $f(a)f(b) < 0$). Most of the other methods that we will study are only *locally* convergent, meaning that we must start the iteration with a "good enough" approximation to the root. One disadvantage of bisection is that it cannot be used to find roots when the function is tangent to the axis and does not pass through the axis; think about using bisection to find the root of $f(x) = x^2$, for example. Another disadvantage is that it converges slowly compared with other methods—note that Theorem 3.1 implies that a decrease of the initial error by a factor of $2^{10} \approx 1000$ requires 10 iterations—however, it can be used in conjunction with a more rapidly converging local method to make a rather effective root-finding algorithm. (See Section 3.10.5.)

EXERCISES

1. Do three iterations (by hand) of the bisection method, applied to $f(x) = x^3 - 2$, using $a = 0$ and $b = 2$.

2. For each of the following functions, do a calculation by hand (i.e., with a calculator) to find the root to an accuracy of 0.1. This will take at most five iterations, and fewer for several of them.

(a) $f(x) = x^6 - x - 1, [a,b] = [0,2]$
(b) $f(x) = x^3 - 3, [a,b] = [0,3]$
(c) $f(x) = x - e^{-x^2}, [a,b] = [0,1]$
(d) $f(x) = 3 - 2^x, [a,b] = [0,2]$
(e) $f(x) = \ln x + x, [a,b] = [\frac{1}{10}, 1]$

3. Write a program that uses the bisection method to find the root of a given function on a given interval, and apply this program to find the roots of the following functions on the indicated intervals. Use the relationship (3.2) to determine a priori the number of steps necessary for the root to be accurate to within 10^{-6}.

(a) $f(x) = x^3 - 2x - 5, [a,b] = [0,3]$
(b) $f(x) = e^x - 2, [a,b] = [0,1]$
(c) $f(x) = x^3 - 2, [a,b] = [0,2]$
(d) $f(x) = x - e^{-x}, [a,b] = [0,1]$
(e) $f(x) = x^6 - x - 1, [a,b] = [0,2]$
(f) $f(x) = 1 - 2xe^{-x/2}, [a,b] = [0,1]$
(g) $f(x) = 5 - x^{-1}, [a,b] = [0.1, 0.25]$
(h) $f(x) = x^2 - \sin x, [a,b] = [0.1, \pi]$
(i) $f(x) = 2 + x^{-1} \ln x, [a,b] = [0.1, 1]$.

4. Return to the pipeline-building problem from the beginning of the chapter. Let $C = C_1/C_2$ be the ratio of costs for the two regions, and find the value of x for the case $a = 3, b = 1, L = 4$ as a function of the cost ratio.

5. Consider the pipeline-building problem now for the case $a = 5, b = 1, L = 4$. Again, construct a table of va-

lues for x as a function of the cost ratio $C = C_1/C_2$, using the bisection method to solve the polynomial equation.

6. If you borrow L dollars at an annual interest rate of r, for a period of m years, then the size of the monthly payment, M, is given by the *annuity equation*

$$L = \frac{12M}{r} \left(1 - \left(1 + \frac{r}{12} \right)^{-12m} \right).$$

The author needs to borrow $150,000 to buy the new house that he wants, and he can afford to pay only $600 per month. Assuming a 30-year mortgage, use the bisection method to determine what interest rate he can afford to pay. (Should the author perhaps find some rich relatives to help him out here?)

7. What interest rate can the author afford if he has to borrow only $100,000?

8. Consider the problem of modeling the position of the liquid-solid boundary in a substance that is melting due to the application of heat at one end. In a simplified model,[2] if the initial position of the interface is taken to be $x = 0$, then the interface moves according to

$$x = 2\beta\sqrt{t}$$

where β satisfies the nonlinear equation

$$\frac{(T_M - T_0)k}{\lambda\sqrt{\pi}} e^{-\beta^2/k} = \beta \, \text{erf}(\beta/\sqrt{k}).$$

Here T_M is the melting temperature (absolute scale), $T_0 < T_M$ is the applied temperature, k and λ are parameters dependent on the material properties of the substance involved, and $\text{erf}(z)$ is the *error function* defined

---

[2] See L.I. Rubinstein, *The Stefan Problem*, American Mathematical Society, Providence, RI, 1971.

by

$$\mathrm{erf}(z) = \frac{2}{\sqrt{\pi}} \int_0^z e^{-t^2} dt.$$

The error function can be accurately approximated by

$$E(z) = 1 - (a_1\xi + a_2\xi^2 + a_3\xi^3)e^{-z^2},$$

where $\xi = 1/(1 + pz)$ and

$$p = 0.47047, \quad a_1 = 0.3480242$$

$$a_2 = -0.0958798, \quad a_3 = 0.747856.$$

(a) Show that finding β is equivalent to finding the root α of the one-parameter family of functions defined by

$$f(z) = \theta e^{-z^2} - z\,\mathrm{erf}(z). \qquad (3.6)$$

What is θ? How is α related to β?
(b) Find the value of α corresponding to $\theta = 0.001$, 0.1,10,1000. Use the bisection method, and $E(z)$ to approximate the error function.

9. A variation on the bisection method is known as *regula falsi*, or *the method of false position*. Given an interval $[a,b]$, with $f(a)f(b) < 0$, the new point c is defined by finding where the straight line connecting $(a,f(a))$

and $(b,f(b))$ crosses the axis. Show that this yields

$$c = b - f(b)(b - a)/(f(b) - f(a)).$$

10. Do three iterations (by hand) of regula falsi (see Exercise 9), applied to $f(x) = x^3 - 2$, using $a = 0$ and $b = 2$. Compare with your results for Exercise 1.

11. Modify the bisection algorithm to perform regula falsi (see Exercise 9), and use the new method to find the same roots as in Exercise 3. Stop the program when the difference between consecutive iterates is less than 10^{-6}, i.e., when $|x_k - x_k| \le 10^{-6}$.

12. Repeat Exercise 8(b), using your regula falsi program.

13. The bisection method will always cut the interval of uncertainty in half, but regula falsi might cut the interval by less, or might cut it by more. Do both bisection and regula falsi on the function $f(x) = e^{-4x} - \frac{1}{10}$, using the initial interval $[0,5]$. Which one gets to the root the fastest?

14. Apply both bisection and regula falsi to the following functions on the indicated intervals. Comment on your results in light of how the methods are supposed to behave.
(a) $f(x) = 1/(x - 1)$, $[a,b] = [0,3]$
(b) $f(x) = 1/(x^2 + 1)$, $[a,b] = [0,5]$

3.2 NEWTON'S METHOD: DERIVATION AND EXAMPLES

Newton's method[3] is the classic algorithm for finding roots of functions. It is often introduced in the calculus sequence as an application of the derivative of a function. Historically, it appears to have been first used by Newton in 1669, although the ideas

---

[3] Isaac Newton (1642–1727) is generally regarded as one of the greatest mathematicians of all time. He entered Trinity College, Cambridge, in 1661 and graduated with a BA degree in 1665. In 1668 he received a master's degree and was appointed Lucasian Professor of Mathematics, one of the most prestigious positions in English academia at the time. In his later years Newton served in Parliament and was warden of the Mint. In 1703 he was elected president of the Royal Society of London, of which he had been a member since 1672. Two years later he was knighted by Queen Anne.

Newton is given co-credit, along with the German Wilhelm Gottfried von Leibniz, for the discovery and development of calculus—work that Newton did in the period 1664–1666 but did not publish until years later, thus laying the groundwork for an ugly argument with Leibniz over who should get credit for the discovery. In 1687, at the urging of the astronomer Edmund Halley, Newton published his ground-breaking compilation of mathematics and science, *Principia Mathematica*, which is apparently the first place that the root-finding method that bears his name appears, although he probably had used it as early as 1669.

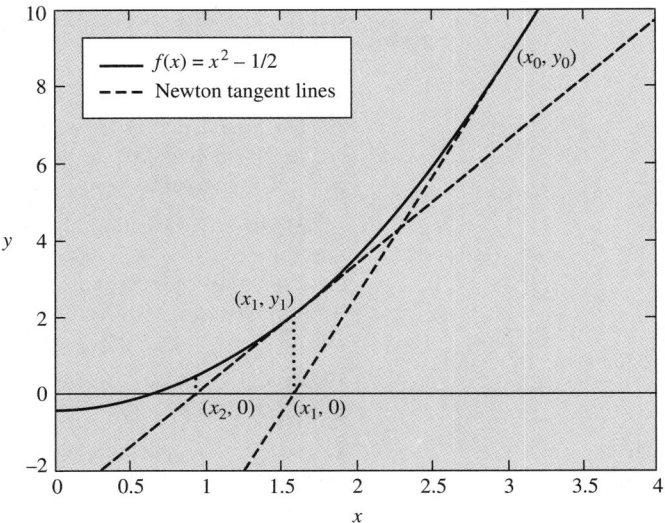

FIGURE 3.2   Newton's method for $y = x^2 - 1/2$.

were known to others beforehand.[4] In fact, the ancient Babylonians had a method for approximating square roots that is essentially Newton's method.[5]

There are two good derivations of Newton's method, one that is geometric and one that is analytic. We will discuss both, beginning with the geometric.

Consider Figure 3.2. We wish to find a root, α, of $y = f(x)$, given an "initial guess" of x_0. How do we improve upon this initial guess to get a better and better approximation? The fundamental idea in Newton's method is to use the tangent line approximation to the function f at the point $(x_0, f(x_0))$. The point-slope formula for the equation of the straight line gives us

$$\frac{y - y_0}{x - x_0} = f'(x_0);$$

thus we have the straight line with equation

$$y = \ell_0(x) = f(x_0) + f'(x_0)(x - x_0).$$

To find where this crosses the axis, we set $y = 0$ and solve for x:

$$x = x_0 - \frac{f(x_0)}{f'(x_0)}.$$

---

[4] See the book *A History of Numerical Analysis from the 16th through the 19th Century* (Springer-Verlag, New York, 1977, p. 64ff), by Herman Goldstine, for a discussion of this and other historical aspects of numerical methods. The method is sometimes called Newton-Raphson, in honor of Joseph Raphson, who published the idea before Newton did.

[5] Carl Boyer, *A History of Mathematics*, John Wiley & Sons, New York, 1968, p. 449.

Call this new approximate value x_1, and note that (at least in the case of Figure 3.2) it is much closer to the root α. Now continue the process with another straight line to get

$$x_2 = x_1 - \frac{f(x_1)}{f'(x_1)},$$

or, generally,

$$x_{n+1} = x_n - \frac{f(x_n)}{f'(x_n)}. \tag{3.7}$$

This is Newton's method. It is based on a very simple idea, one that is fundamental and repeated time and again in the derivation of numerical methods (in fact, we have already seen it in the derivation of the trapezoid rule in Section 2.5):

> *Replace a general function by a simpler function, and do the required computation* exactly *on the simpler function.*

Thus, in this instance, we replaced the general function f with the simple function (a straight line) ℓ_0 and found the *exact* root of ℓ_0.

The second derivation of Newton's method depends on our analytical workhorse, Taylor's Theorem. Given a value $x_n \approx \alpha$, we expand f in a Taylor series about x_n:

$$f(x) = f(x_n) + (x - x_n)f'(x_n) + \frac{1}{2}(x - x_n)^2 f''(\xi_n),$$

where ξ_n is between x and x_n. To get a useful algorithm out of this, we set $f(x) = 0$ and solve for x in terms of $f(x_n), f'(x_n)$, and the remainder:

$$x = x_n - \frac{f(x_n)}{f'(x_n)} - \frac{1}{2}(x - x_n)^2 \frac{f''(\xi_n)}{f'(x_n)}. \tag{3.8}$$

We now define x_{n+1} to be the quantity that results from (3.8) when we drop the remainder term:

$$x_{n+1} = x_n - \frac{f(x_n)}{f'(x_n)},$$

which defines the next point. This, of course, is Newton's method again. This derivation makes use of another fundamental and oft-repeated idea in numerical methods and analysis:

> *Given an expression in terms of something simple plus a remainder, generate a numerical approximation by dropping the remainder term (and adjusting the notation, perhaps).*

Thus, in (3.8) we dropped the remainder term to define x_{n+1}.

The geometric derivation suggests that Newton's method ought to work very well, although the formula (3.7) does hint that there might be problems if $f'(x_n) \approx 0$ were ever true. Before proceeding with any analysis, let's look at an example.

TABLE 3.2   **Newton's Method for $f(x) = 2 - e^x$.**

n	x_n	$\alpha - x_n$	$\log_{10}(\alpha - x_n)$
0	0.000000000000	0.693147180560	−0.1592
1	1.000000000000	0.306852819440	−0.5131
2	0.735758882343	0.042611701783	−1.3705
3	0.694042299919	0.000895119359	−3.0481
4	0.693147581060	0.000000400500	−6.3974
5	0.693147180560	0.000000000000	−13.0961

EXAMPLE 3.2

Consider applying Newton's method to the function $f(x) = 2 - e^x$, which is the same as we used for bisection in the previous section. We choose $x_0 = 0$ and then compute as follows:

$$x_1 = x_0 - \frac{2 - e^{x_0}}{-e^{x_0}} = -\frac{2 - 1}{-1} = 1$$

$$x_2 = x_1 - \frac{2 - e^{x_1}}{-e^{x_1}} = 1 - \frac{2 - e}{-e} = 0.7357588823$$

$$x_3 = x_2 - \frac{2 - e^{x_2}}{-e^{x_2}} = 0.7357588823 - \frac{2 - e^{0.7357588823}}{-e^{0.7357588823}} = 0.6940422999.$$

Table 3.2 gives the results of continuing the computation for several more steps. Note that the convergence here was much more rapid than for bisection: In six iterations bisection had achieved only about a digit and a half of accuracy, while Newton's method has over *13* accurate digits. ■

Note, by the way, that the number of correct digits in the Newton calculation is doubling, roughly, every iteration. If this were typical behavior, it would indeed suggest that Newton's method is very fast.

Newton's method is *not* a global method, however. There are examples for which convergence will be poor, or even for which convergence doesn't occur. Usually this can be cured by obtaining a better initial guess, but sometimes we have to take x_0 *very* close to α in order to obtain convergence.

Consider Figure 3.3; this is the graph of $f(x) = \frac{4}{3}e^{2-x/2}(1 + x^{-1}\log x)$. Note that there is a single root in the vicinity of $x = 0.5$. But note that the shape of the graph of f will cause problems for Newton's method, unless the initial point is carefully chosen. If $x_0 \in [0.8, 1.2]$ (approximately), then the tangent line generated by Newton's method will predict that $x_1 < 0$, which is outside the domain of definition of f. If $x_0 > 1.2$ (approximately), then Newton's method will generate a sequence of values $x_{n+1} > x_n$, growing without bound as $n \to \infty$. (Note that, in a sense, Newton's method is working in this case as $\lim_{x \to \infty} f(x) = 0$; i.e., there is a root at infinity. But this is not the root we want to find.)

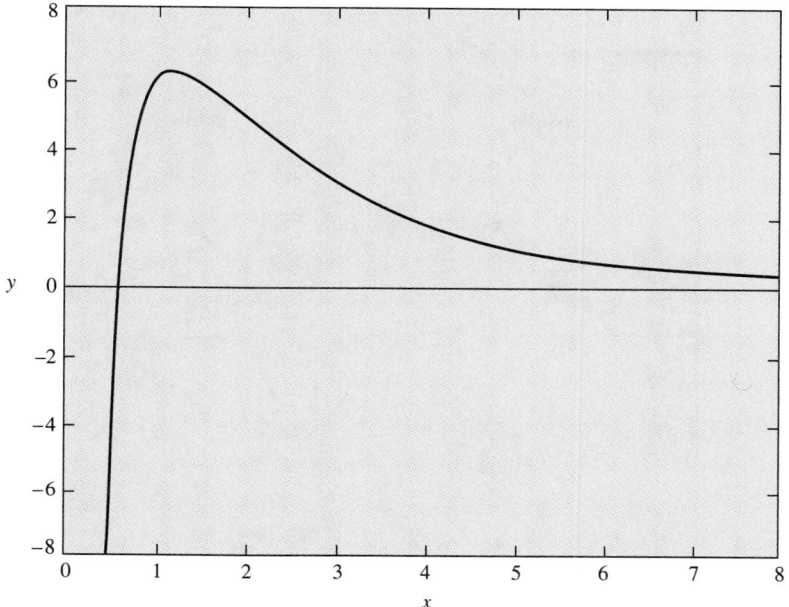

FIGURE 3.3 A function for which Newton's method will not work well, unless x_0 is carefully chosen.

A simpler example can be found by considering the function

$$f(x) = \frac{20x - 1}{19x},$$

which has a single root at $\alpha = 0.05$. The graph of f is given in Figure 3.4, and from this we can see that convergence will not occur except for a narrow range of values near the root, since Newton's method will predict $x_1 < 0$ for x_0 much larger than α. Table 3.3 shows the value of x_1 produced for each of a sequence of values of x_0. Note that we do not get $x_1 > 0$ until $x_0 = \frac{1}{16}$, which is very close to the actual root.

It is even possible to construct a function f such that, for some initial guesses, Newton's method will cycle indefinitely. Consider $f(x) = \arctan(x)$. This has a single root at $x = 0$, of course. However, for $x_0 = 1.39174520027\ldots$, it is not hard to show that $x_1 = -1.39174520027\ldots$ and $x_2 = 1.39174520027\ldots = x_0$. In other words, Newton's method will just hop back and forth between these two values. (The student might want to consider how to compute this special value of x. See Exercise 9.)

So, where are we with the convergence of Newton's method? Example 3.2 showed rapid convergence, but we quickly followed that up with two examples of what is clearly less than marvelous performance. Under what conditions can we expect Newton's method to converge? Is there any sort of theory that we can rely upon to give us confidence in this method? Note that in the absence of any such theory, we cannot use Newton's method with any expectation or confidence that the results it reports will have any meaning.

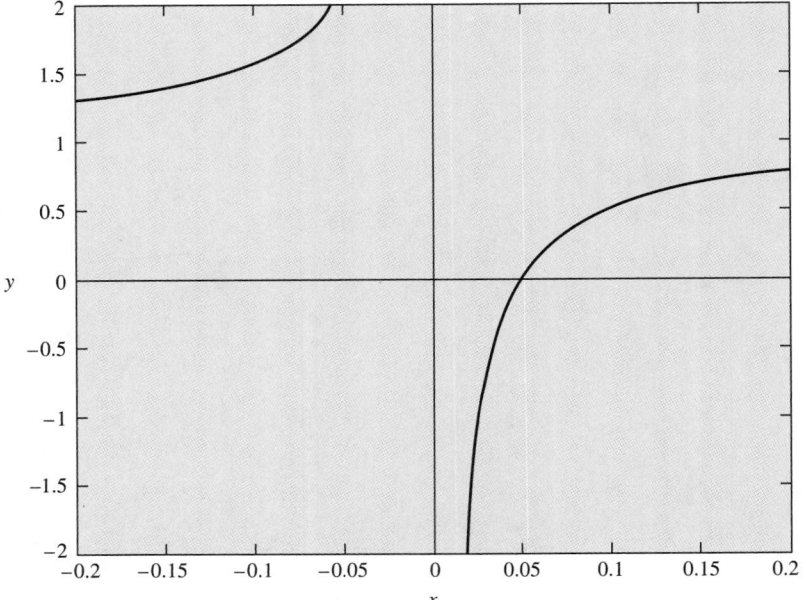

FIGURE 3.4 A second function for which Newton's method will not work well, unless x_0 is very close to α.

TABLE 3.3 Newton's Method (First Step Only)
Applied to $f(x) = \frac{20x-1}{19x}$.

x_0	x_1
1.00000000	−18.00000000
0.50000000	−4.00000000
0.25000000	−0.75000000
0.12500000	−0.06250000
0.06250000	0.04687500

The answer is actually very simple. We will defer a detailed discussion of it until Section 3.6, but the conclusion can be stated fairly succinctly:

If f, f', and f'' are all continuous near the root, and if f' does not equal zero at the root, then Newton's method will converge whenever the initial guess is sufficiently close to the root. Moreover, this convergence will be very rapid, with the number of correct digits roughly doubling each iteration.

In this context, "sufficiently close" implies that if we keep taking x_0 closer and closer to the root, we will eventually find an x_0 such that the iteration converges.

EXERCISES

1. Write down Newton's method as applied to the function $f(x) = x^2 - 2$. Simplify the computation as much as possible. What has been accomplished if we find the root of this function?

2. Write down Newton's method as applied to the function $f(x) = x^3 - 2$. Simplify the computation as much as possible. What has been accomplished if we find the root of this function?

3. Generalize the preceding two exercises by writing down Newton's method as applied to $f(x) = x^n - a$.

4. Write down Newton's method as applied to the function $f(x) = a - x^{-1}$. Simplify the resulting computation as much as possible. What has been accomplished if we find the root of this function?

5. Do three iterations of Newton's method (by hand) for each of the following functions:
(a) $f(x) = x^3 - 2x - 5$, $x_0 = 2$
(b) $f(x) = e^x - 2$, $x_0 = 1$
(c) $f(x) = x^3 - 2$, $x_0 = 1$
(d) $f(x) = x - e^{-x}$, $x_0 = 1$
(e) $f(x) = x^6 - x - 1$, $x_0 = 1$
(f) $f(x) = 1 - 2xe^{-x/2}$, $x_0 = 0$
(g) $f(x) = 5 - x^{-1}$, $x_0 = \frac{1}{2}$
(h) $f(x) = x^2 - \sin x$, $x_0 = \frac{1}{2}$
(i) $f(x) = x + \tan x$, $x_0 = 3$
(j) $f(x) = 2 - x^{-1} \ln x$, $x_0 = \frac{1}{3}$.

6. Do three iterations of Newton's method for $f(x) = 3 - e^x$, using $x_0 = 1$. Repeat, using $x_0 = 2, 4, 8, 16$. Comment on your results.

7. Draw the graph of a single function f that satisfies all of the following:
(a) f is defined and differentiable for all x.
(b) There is a unique root $\alpha > 0$.
(c) Newton's method will converge for any $x_0 > \alpha$.
(d) Newton's method will diverge for all $x_0 < 0$.

8. Draw the graph of a single function f that satisfies all of the following:
(a) f is defined and differentiable for all x.
(b) There is a single root $\alpha \in [a,b]$, for some interval $[a,b]$.
(c) There is a second root, $\beta < a$.
(d) Newton's method will converge to α for all $x_0 \in [a,b]$.
(e) Newton's method will converge to β for all $x_0 \leq \beta$.

9. Write down Newton's method for finding the root of the arctangent function. From this, formulate an equation that must be satisfied by the value $x = \beta$, in order to have the Newton iteration cycle back and forth between β and $-\beta$. (Hint: If $x_n = \beta$, and Newton's method is supposed to give us $x_{n+1} = -\beta$, what is an equation satisfied by β?)

10. Compute the value of β from the previous problem. (Hint: Use bisection to solve the equation for β.)

11. Use Newton's method on the computer of your choice to compute the root $\alpha = 0$ of the arctangent function. Use the value of β from the previous problem as your x_0, and comment on your results. Repeat using $x_0 = \beta/2$ and $x_0 = \beta - \epsilon$, where ϵ is $O(\mathbf{u})$.

3.3   HOW TO STOP NEWTON'S METHOD

With the bisection method we were able to compute, ahead of time, the number of iterations needed to achieve a given desired accuracy. We will not be able to do this for Newton's method. (More correctly, the computation that might be used is too difficult to be practical; see Section 3.6.) So the question arises: How do we stop the iteration?

Ideally, we would want to stop when the error. $\alpha - x_n$ is sufficiently small. Of course, we cannot use this rationale since we don't know the value of α! However, so long as f' is not zero near the root, we can relate this error to a computable quantity.

The Mean Value Theorem tells us that

$$f(\alpha) - f(x_n) = f'(c_n)(\alpha - x_n),$$

where c_n is some value between α and x_n. Therefore, we can solve for $\alpha - x_n$ and write

$$
\begin{aligned}
\alpha - x_n &= \frac{f(\alpha) - f(x_n)}{f'(c_n)} \\
&= -\frac{f(x_n)}{f'(c_n)} \\
&= -\frac{f(x_{n-1})}{f'(x_{n-1})} \frac{f(x_n)}{f(x_{n-1})} \frac{f'(x_{n-1})}{f'(c_n)} \\
&= (x_n - x_{n-1}) \left(\frac{f(x_n)}{f(x_{n-1})} \frac{f'(x_{n-1})}{f'(c_n)} \right).
\end{aligned}
$$

Note that in the last step we used the Newton iteration to replace $-f(x_{n-1})/f'(x_{n-1})$ with $x_n - x_{n-1}$. We have, then, that

$$\alpha - x_n = C_n(x_n - x_{n-1}) \tag{3.9}$$

for $C_n = f(x_n)/f(x_{n-1}) \times f'(x_{n-1})/f'(c_n)$. Thus the error $\alpha - x_n$ is a simple multiple of the *computable*[6] quantity $x_n - x_{n-1}$. Morever, if we assume that convergence is occurring, and that $f'(\alpha) \neq 0$, then the constant of proportionality satisfies

$$\lim_{n \to \infty} |C_n| = 1. \tag{3.10}$$

(The student should consider why this is true.) It follows, then, that

$$\lim_{n \to \infty} \frac{|\alpha - x_n|}{|x_n - x_{n-1}|} = 1.$$

As a consequence, then, we can use the computable quantity $|x_n - x_{n-1}|$ to measure convergence. Usually one stops the iteration when a modest multiple of this is small—for example,

$$5|x_n - x_{n-1}| \leq \epsilon,$$

where $\epsilon > 0$ is a user-defined tolerance.

A word of warning is in order, however. The results in this section are based on some very casual analysis, and while it can all be made precise and rigorous, the fact remains that it is possible to have $|x_{n+1} - x_n|$ small and yet have x_{n+1} not very close to α. This can happen, for example, if $f'(x_n)$ is very large compared with $f(x_n)$. For this reason it is commonplace to add a term to the error check to make sure that the function value itself is small, i.e., we stop when

$$|f(x_n)| + |x_n - x_{n-1}| \leq \epsilon/5.$$

---

[6]This is another important rule in numerical methods and analysis: Use computable *estimates* of quantities that cannot be easily computed.

EXERCISES

1. Under the assumption that $f'(\alpha) \neq 0$ and $x_n \to \alpha$, prove (3.10); be sure to provide all the details. (Hint: Expand f and f' in Taylor series about $x = \alpha$.)

2. We could also stop the iteration when $|f(x_n)|$ was sufficiently small. Use the Mean Value Theorem plus the fact that $f(\alpha) = 0$ to show that, if f' is continuous and nonzero near α, then there are constants c_1 and c_2 such that

$$c_1|f(x_n)| \leq |\alpha - x_n| \leq c_2|f(x_n)|.$$

Comment on this result.

3. Write a computer program that uses Newton's method to find the root of a given function, and apply this program to find the root of the following functions, using x_0 as given. Stop the iteration when the error as estimated by $|x_{n+1} - x_n|$ is less than 10^{-6}. Compare with your results for bisection.

(a) $f(x) = x^3 - 2x - 5$, $x_0 = 2$
(b) $f(x) = e^x - 2$, $x_0 = 1$
(c) $f(x) = x^3 - 2$, $x_0 = 1$
(d) $f(x) = x - e^{-x}$, $x_0 = 1$
(e) $f(x) = x^6 - x - 1$, $x_0 = 1$
(f) $f(x) = 1 - 2xe^{-x/2}$, $x_0 = 0$
(g) $f(x) = 5 - x^{-1}$, $x_0 = \frac{1}{2}$
(h) $f(x) = x^2 - \sin x$, $x_0 = \frac{1}{2}$
(i) $f(x) = x + \tan x$, $x_0 = 3$
(j) $f(x) = 2 - x^{-1}\ln x$, $x_0 = \frac{1}{3}$.

4. Figure 3.5 shows the geometry of a planetary orbit[7] around the sun. The position of the sun is given by S, and the position of the planet is given by P. Let x denote the angle defined by P_0OA, measured in radians. The dotted line is a circle concentric to the ellipse and having a radius equal to the major axis of the ellipse. Let T be the total orbital period of the planet, and let t be the time required for the planet to go from A to P. Then Kepler's equation from orbital mechanics, relating x and t, is

$$x - \epsilon \sin x = \frac{2\pi t}{T}.$$

Here ϵ is the eccentricity of the elliptical orbit (the extent to which it deviates from a circle). For an orbit of

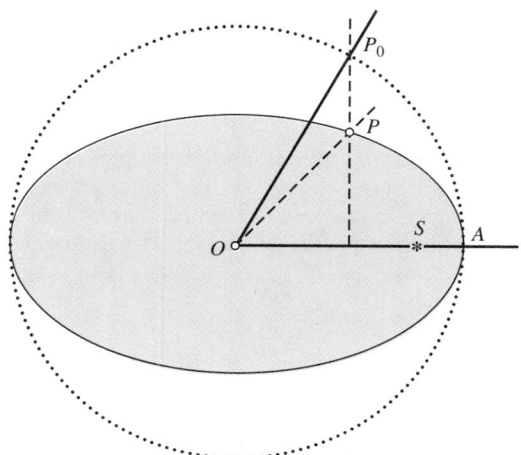

FIGURE 3.5 Orbital geometry for Exercises 4 and 5.

eccentricity $\epsilon = 0.01$ (roughly equivalent to that of the earth), what is the value of x corresponding to $t = T/4$? What is the value of x corresponding to $t = T/8$? Use Newton's method to solve the required equation.

5. Consider now a highly eccentric orbit, such as that of a comet, for which $\epsilon = 0.9$ might be appropriate. What is the value of x corresponding to $t = T/4$? What is the value of x corresponding to $t = T/8$?

6. Consider the problem of putting water pipes far enough underground to avoid frozen pipes should the external temperature suddenly drop. Let T_0 be the initial temperature of the ground, and assume that the external air temperature suddenly drops to a new value $T_1 < T_0$. Then a simple model of how the underground temperature responds to the change in external temperature tells us that

$$\frac{u(x,t) - T_1}{T_0 - T_1} = \operatorname{erf}\left(\frac{x}{2\sqrt{at}}\right).$$

Here $u(x,t)$ is the temperature at a depth x feet and time t seconds after the temperature change, and a is the thermal conductivity of the soil.

(a) Suppose that $a = 5.4 \times 10^{-6}$ ft^2=s and that the temperature below ground is a constant 40°F. How deep

---

[7]For the background material for this and the next problem, the author is indebted to the interesting new calculus text by Alexander J. Hahn, *Basic Calculus: From Archimedes to Newton to Its Role in Science*, published in 1998 by Springer-Verlag.

must the pipe be buried to guarantee that the temperature does not reach 32°F for 30 days after the surface temperature drops to 0°F ? If your computing environment does not have an intrinsic error function, use the approximation presented in Exercise 8 of Section 3.1.
(b) Repeat the preceding with $a = 9.7 \times 10^{-6}$.

7. In the previous exercise, produce a plot of temperature u at depths of 6 feet and 12 feet, as a function of time (days), in response to a temperature shift of $-40°$ for the following initial temperatures (in degrees Farenheit):
(a) $T_0 = 40$
(b) $T_0 = 50$

(c) $T_0 = 60$
Use both values of a.

8. Use Newton's method to solve Exercise 6 from Section 3.1.

9. Repeat the above, assuming that the mortgage is only 15 years in length.

10. Use Newton's method to solve the pipeline problem from the beginning of the chapter, for the case $a = 3$, $b = 1, L = 4$; construct a table of values for x as a function of the cost ratio $C = C_1/C_2$.

11. Repeat the preceding for the case $a = 5$, $b = 1$, $L = 6$.

3.4 APPLICATION: DIVISION USING NEWTON'S METHOD

In this section we will investigate an old algorithm that was actually used in the early days of computing to approximate the division operation. The purpose is to illustrate the use of Newton's method and the analysis of the resulting iteration—in particular, to demonstrate how we can predict the performance of the iteration for a wide range of parameter values.

Consider the function

$$f(x) = a - \frac{1}{x}.$$

Clearly, $f(\alpha) = 0 \Rightarrow \alpha = a^{-1}$, so that we can "divide" by a by finding the root of f and then multiplying. If we use Newton's method to find the root, we get the iteration

$$x_{n+1} = x_n - \left(\frac{a - x_n^{-1}}{x_n^{-2}} \right) = x_n(2 - ax_n).$$

Thus, we can approximate the reciprocal of a without having to do division.

The questions of most interest to us are: When does this iteration converge and how fast? What initial guesses x_0 will work for us?

We begin by using the information about the way the computer stores numbers. We have

$$a = b \times 2^{t-p}$$

for $b \in [\frac{1}{2}, 1]$, so it suffices to be able to compute the reciprocal for numbers between $\frac{1}{2}$ and 1. We assume, then, without loss of generality, that $a \in [\frac{1}{2}, 1]$.

To analyze the behavior of the iteration, we use the *residual* defined by

$$r_n = 1 - ax_n.$$

We then have that the error satisfies $\alpha - x_n = e_n = r_n/a = 1/a - x_n$ and, moreover,

$$r_{n+1} = 1 - ax_{n+1} = 1 - a[x_n(2 - ax_n)] = 1 - a[x_n(1 + r_n)] = 1 - (1 - r_n)(1 + r_n) = r_n^2,$$

from which it follows that

$$r_n = r_0^{2^n}.$$

If $|r_0| < 1$, then we will get $r_n \to 0$, and very rapidly. In fact, we have that the absolute error satisfies

$$|e_{n+1}| = |ae_n^2| = ae_n^2$$

and the relative error satisfies (recall that $\alpha = 1/a$)

$$\left|\frac{e_{n+1}}{\alpha}\right| = \left(\frac{e_n}{\alpha}\right)^2,$$

so that we have

$$\left|\frac{e_n}{\alpha}\right| = \left(\frac{e_0}{\alpha}\right)^{2^n} = (r_0)^{2^n}.$$

Since the error e_n goes to zero if and only if the residual r_n does, we have convergence so long as $|r_0| < 1$, which is equivalent to saying

$$0 < x_0 < \frac{2}{a} = 2\alpha.$$

How do we get an x_0 that satisifes this?

Recall linear interpolation from the previous chapter. We can construct the linear interpolant to the function $y = x^{-1}$ and use this to *estimate* a^{-1} and then use this value as x_0. The linear interpolant using the endpoints of $[\frac{1}{2}, 1]$ as the nodes is easily found to be

$$p_1(x) = 3 - 2x,$$

so that

$$x_0 = p_1(a) = 3 - 2a.$$

The linear interpolation error formula (2.13) shows us that

$$\left|\frac{1}{a} - p_1(a)\right| \leq \frac{1}{2}. \tag{3.11}$$

Hence, $|x_0 - \alpha| \leq \frac{1}{2}$, which is equivalent to saying that $|r_0| = a|e_0| = a/2 \leq \frac{1}{2}$; thus we will get convergence using this initial value. In terms of the relative error, we have

$$\left|\frac{e_0}{\alpha}\right| \leq \frac{\frac{1}{2}}{\alpha} \leq \frac{1}{2}.$$

A mere six iterations (about 21 total operations) yield relative accuracy on the order of 10^{-20}.

The estimate (3.11) is actually very conservative, because of the way that the bound on the second derivative of $1/x$ was treated. Because of the simplicity of the functions involved, we can actually apply some ordinary calculus to determine that the initial error is *much* smaller than predicted above.

Define the error in the initial guess as

$$g(x) = \frac{1}{x} - (3 - 2x);$$

we want to find the maximum of this (in absolute value) over the interval $[\frac{1}{2}, 1]$. Taking the derivative gives us

$$g'(x) = \frac{-1}{x^2} + 2,$$

which implies that critical points exist at $x = \pm\frac{1}{2}\sqrt{2}$. The critical point at $x = -\frac{1}{2}\sqrt{2}$ is outside our interval of interest, so we ignore it and compute

$$g(\sqrt{2}/2) = \frac{2}{\sqrt{2}} - 3 + \sqrt{2} = 2\sqrt{2} - 3 = -0.1716\ldots.$$

In addition, we have that $g(1/2) = g(1) = 0$. Therefore, the *smallest* value of $g(x)$ over $[\frac{1}{2}, 1]$ is (to four digits) -0.1716. But it therefore follows that the *largest* value of $|g(x)|$ over the same interval is $0.1716 < 0.172$. Using this as the upper bound on e_0 shows that we only need five iterations to get relative accuracy on the order of 10^{-20}.

EXAMPLE 3.3

Let's try this out on $a = 0.8$, so that we are trying to approximate $(0.8)^{-1} = 1.25$. The initial guess is found from the linear interpolant as

$$x_0 = 3 - 2 \times 0.8 = 1.4.$$

Newton's method then gives us

$$
\begin{aligned}
x_1 &= x_0(2 - (0.8)x_0) = (1.4)(2 - (0.8)(1.4)) = 1.232 \\
x_2 &= x_1(2 - (0.8)x_1) = (1.232)(2 - (0.8)(1.232)) = 1.2497408 \\
x_3 &= x_2(2 - (0.8)x_2) = (1.2497408)(2 - (0.8)(1.2497408)) = 1.249999946 \\
x_4 &= x_3(2 - (0.8)x_3) = (1.249999946)(2 - (0.8)(1.249999946)) = 1.25
\end{aligned}
$$

Thus we converge in only four iterations (a total of about 14 operations). ∎

EXAMPLE 3.4

Suppose we had a *quadratic* approximation to $1/t$ that we used to generate x_0. For example, consider the quadratic

$$p_2(t) = \frac{1}{3}\left(8t^2 - 18t + 13\right).$$

Note that $p_2(\frac{1}{2}) = 2$, $p_2(3/4) = 4/3$, and $p_2(1) = 1$; therefore we see that p_2 interpolates to $1/t$ for $t = \frac{1}{2}$, $t = \frac{3}{4}$, and $t = 1$. (We will see in the next chapter how to construct p_2.) Since we don't have an error theory (yet) for this kind of interpolation, how can we estimate the error between $p_2(a)$ and $1/a$?

Again, we can use ordinary calculus. Define the error as

$$E(t) = t^{-1} - p_2(t),$$

so that

$$E'(t) = -t^{-2} - \frac{1}{3}(16t - 18).$$

Then $E'(t) = 0$ if and only if

$$0 = -t^{-2} - \frac{1}{3}(16t - 18)$$

or

$$16t^3 - 18t^2 + 3 = 0.$$

This doesn't factor, but we can use a computer algebra package to plot it and see that it has roots near $t = 1$, $t = \frac{1}{2}$, and $t = -\frac{1}{2}$. We can then use Newton's method or bisection to find those roots, getting (to three places)

$$t_1 = 0.886, \quad t_2 = 0.595, \quad t_3 = -0.356$$

as our three critical points. Since t_3 is not in the interval $[1/2,1]$ we are not interested in it anymore. We then evaluate $E(t)$ at the remaining critical points plus the endpoints, getting

$$E(t_1) = 0.018012294, \quad E(t_2) = -0.026728214,$$
$$E(1/2) = 0, \quad E(1) = 0.$$

Therefore (note the use of the Extreme Value Theorem here),

$$|E(t)| \leq 0.026728214$$

is the upper bound on the error for $t \in [1/2,1]$. Hence,

$$e_0 = |1/a - p_2(a)| \leq 0.026728214.$$

This is substantially smaller than we got with the linear polynomial. We get around 10^{-25} accuracy in only four iterations, and the total number of operations is now 16. ∎

The analysis we have done here shows how to reliably choose x_0 to ensure convergence for all values of a that are of interest, for this particular application. However, for most cases we will need more powerful tools, which we will derive in the next section.

EXERCISES

1. Test the method derived in this section by using it to approximate $1/0.75 = 1.333\ldots$, with x_0 as suggested in the text. Don't write a computer program; just use a hand calculator.

2. Repeat the preceding for $a = 2/3$.

3. Repeat the preceding for $a = 0.6$.

4. Based on the material in this section, write each of the following numbers in the form

$$a = b \times 2^k,$$

where $b \in [\frac{1}{2}, 1]$, and then use Newton's method to find the reciprocal of b and hence of a.
(a) $a = 3$
(b) $a = \pi$
(c) $a = 5$
(d) $a = 7$
(e) $a = 6$

Be sure to use the initial value as generated in this section, and do only as many iterations as necessary for 10^{-16} accuracy. Compare your values with the intrinsic reciprocal function on your computer or calculator. Note that this can be done on computer or hand calculator.

5. Test the method derived in this section by writing a computer program to implement it. Test the program by having it it compute $1/0.75 = 1.333\ldots$, with x_0 as suggested in the text, as well as $1/(2/3) = 1.5$.

6. How close an initial guess is needed for the method to converge to within a relative error of 10^{-14} in only *three* iterations?

7. Consider the quadratic polynomial

$$q_2(x) = 4.328427157 - 6.058874583x + 2.745166048x^2.$$

What is the error when we use this to generate the initial guess? How many steps are required to get 10^{-20} accuracy? How many operations?

8. Repeat Exercise 7, this time with the polynomial

$$q_3(x) = 5.742640783 - 12.11948252x + 11.14228488x^2 - 3.767968453x^3.$$

9. Modify your program from Exercise 5 to use each of p_2, q_2, and q_3 to compute $1/a$ for $a = 0.75$ and $a = 0.6$. Compare with the values produced by ordinary division on your computer. (Remember to use Horner's rule to evaluate the polynomials!)

3.5   THE NEWTON ERROR FORMULA

The derivation of Newton's method from Taylor's Theorem strongly suggested, but did not prove, that the error at each step of the iteration is related to the square of the previous error. We can in fact make that a precise statement, and it becomes a key tool in deriving a precise theory of convergence for Newton's method.

Theorem 3.2 (Newton Error Formula)   Let $f \in C^2(I)$ be given, for some interval $I \subset \mathbf{R}$, with $f(\alpha) = 0$ for some $\alpha \in I$. For a given $x_n \in I$, define

$$x_{n+1} = x_n - \frac{f(x_n)}{f'(x_n)}.$$

Then there exists a point ξ_n between α and x_n such that

$$(\alpha - x_{n+1}) = -\frac{1}{2}(\alpha - x_n)^2 \frac{f''(\xi_n)}{f'(x_n)}. \tag{3.12}$$

Proof   We begin by expanding f in a Taylor series about $x = x_n$:

$$f(x) = f(x_n) + (x - x_n)f'(x_n) + \frac{1}{2}(x - x_n)^2 f''(\xi_n);$$

here ξ_n is between x and α. Now set $x = \alpha$ to get

$$0 = f(x_n) + (\alpha - x_n)f'(x_n) + \frac{1}{2}(\alpha - x_n)^2 f''(\xi_n).$$

Divide both sides by $f'(x_n)$ and rearrange, which yields

$$(x_n - \alpha) - \frac{f(x_n)}{f'(x_n)} = \frac{1}{2}(\alpha - x_n)^2 \frac{f''(\xi_n)}{f'(x_n)}.$$

The left side simplifies to

$$(x_n - \alpha) - \frac{f(x_n)}{f'(x_n)} = \left(x_n - \frac{f(x_n)}{f'(x_n)} \right) - \alpha = x_{n+1} - \alpha,$$

from which (3.12) follows by a simple substitution, and we are done.   □

Note that this result shows that the error at one step goes like the *square* of the error at the previous step. Thus, once the error becomes small enough, it begins to decrease rapidly. In fact, if we assume that convergence is occurring (with $f'(\alpha) \neq 0$), so that

$$\lim_{n \to \infty} x_n = \alpha,$$

then we can say that $f'(x_n) \approx f'(\alpha)$ and $f''(\xi_n) \approx f''(\alpha)$, so that

$$(\alpha - x_{n+1}) \approx C(\alpha - x_n)^2, \tag{3.13}$$

where

$$C = -\frac{1}{2}\frac{f''(\alpha)}{f'(\alpha)}.$$

The approximate equality (3.13) suggests, to some extent, how close x_0 has to be to α for convergence to occur. We will address this more precisely in Section 3.6.

In addition, if convergence is occurring, we can easily show that

$$\lim_{n \to \infty} \frac{\alpha - x_{n+1}}{(\alpha - x_n)^2} = -\frac{f''(\alpha)}{2f'(\alpha)}. \tag{3.14}$$

This follows from the squeeze principle of calculus (together with the continuity of f' and f''), which shows that the convergence of the iteration ($x_n \to \alpha$) forces $\xi_n \to \alpha$.

We will now explain why it is often said that a converging Newton iteration will double the number of accurate decimal digits each step.

EXAMPLE 3.5

Let x_n be a sequence of values generated by applying Newton's method to a smooth function f, and assume that the sequence is converging to the root α. Then it follows from (3.12) that

$$|\alpha - x_{n+1}| = \frac{1}{2}(\alpha - x_n)^2 \frac{|f''(\xi_n)|}{|f'(x_n)|},$$

so that, taking common logarithms of both sides, we have

$$\log_{10}|\alpha - x_{n+1}| = 2\log_{10}|\alpha - x_n| + \log_{10}\left(\frac{|f''(\xi_n)|}{2|f'(x_n)|}\right) = 2\log_{10}|\alpha - x_n| + b_n. \qquad (3.15)$$

As discussed previously, the quantity

$$d_n = \log_{10}|\alpha - x_n|$$

can be interpreted as the number of correct decimal digits in the approximation (so long as the error is less than 1), and (3.15) shows that (ignoring the additive b_n term) this doubles each iteration. ∎

We close this section with a definition that allows us to characterize how rapidly a sequence of values is converging to a limit. Bear in mind that, while we are indeed applying this to the root-finding problem, the definition can be used for any sequence of numbers that are approaching a limit.

Definition 3.1 (Order of Convergence for Sequences)   Let x_n be a sequence of values converging to α, such that

$$\lim_{n \to \infty} \frac{\alpha - x_{n+1}}{(\alpha - x_n)^p} = C \qquad (3.16)$$

for some nonzero (finite) constant C and some p. Then p is called the order of convergence *for the sequence.*

The requirement that C be nonzero and finite actually forces p to be a single, unique value. If p is too large, then the denominator shrinks too fast and the ratio grows without bound to $\pm\infty$; if p is too small, then the numerator vanishes too quickly and the ratio shrinks to zero as $n \to \infty$. One special case that is of interest to us is when $p = 2$, as in Newton's method; this we call *quadratic convergence*. The case $p = 1$ is called linear convergence, and requires that $|C| < 1$ (see Exercise 9).

Another interesting special case is when the sequence satisfies

$$\lim_{n \to \infty} \frac{\alpha - x_{n+1}}{\alpha - x_n} = 0 \qquad (3.17)$$

but

$$\lim_{n \to \infty} \frac{\alpha - x_{n+1}}{(\alpha - x_n)^2} = \pm\infty, \qquad (3.18)$$

i.e., the sequence does not converge quadratically. In this case, the sequence is called *superlinearly convergent*, because (3.17) suggests that it converges faster than linear but (3.18) indicates it is not quadratic.

Generally speaking, having p larger means convergence is more rapid, since (3.16) implies

$$\alpha - x_{n+1} \approx C(\alpha - x_n)^p. \tag{3.19}$$

The Newton error estimate (3.12) strongly suggests that Newton's method is quadratically convergent, something we will prove to be the case in Section 3.6. We can illustrate its use in analyzing a particular example here; a more thorough discussion (for another example) is in Section 3.7.

EXAMPLE 3.6

Consider the problem of finding the root of $f(x) = e^{-ax} - x$. We want to be able to compute this root for a in the range $0 < a \leq 1$, and right now our only tools are bisection and Newton's method. If we apply the Newton error formula to this problem, we get

$$\alpha - x_{n+1} = -\frac{1}{2}(\alpha - x_n)^2 \frac{a^2 e^{-a\xi_n}}{-ae^{-ax_n} - 1} = \frac{1}{2}(\alpha - x_n)^2 \frac{a^2 e^{-a\xi_n}}{ae^{-ax_n} + 1},$$

where ξ_n is between x_n and α. Note that this implies (since the right side is positive) that $\alpha - x_{n+1} \geq 0$; thus, for $n \geq 1$, we have $x_n \leq \xi_n \leq \alpha$. Moreover, for any $x_0 > 0$, the Newton iteration itself always stays positive: $x_n > 0$. This enables us to write

$$\left| \frac{a^2 e^{-a\xi_n}}{ae^{-ax_n} + 1} \right| \leq a \left| \frac{ae^{-ax_n}}{ae^{-ax_n} + 1} \right| \leq a \leq 1.$$

Thus,

$$|\alpha - x_{n+1}| \leq \frac{1}{2}(\alpha - x_n)^2.$$

If the initial error is less than 1, then we can quickly show (just by going through the recursion) that

$$|\alpha - x_5| \leq 5 \times 10^{-10}.$$

Hence, five iterations are enough to compute our root, no matter what the value of a is (so long as $a \in (0,1]$). How do we find an initial guess such that $|\alpha - x_0| \leq 1$? We note that $f(0) = 1$ and $f(2) < 0$ for any $a > 0$; hence the root must lie in the interval $[0,2]$. Therefore, $x_0 = 1$ will always satisfy $|\alpha - x_0| \leq 1$. ∎

Note that we have performed this analysis without doing any numerical computations, and that it applies for all $a \in (0,1]$. Even though we don't know the value of the root, or the value of the one parameter in the problem, we are able to determine an initial guess such that the iteration will converge in five steps. We won't always be able to do this complete an analysis of a problem, but this example does illustrate the issues that are involved and some of the tools that are required.

EXERCISES

1. If f is such that $|f''(x)| \leq 3$ for all x and $|f'(x)| \geq 1$ for all x, and if the initial error in Newton's method is less than $\frac{1}{2}$, what is an upper bound on the error at each of the first three steps? (Hint: Use the Newton error formula from this section.)

2. If f is now such that $|f''(x)| \leq 4$ for all x but $|f'(x)| \geq 2$ for all x, and if the initial error in Newton's method is less than $\frac{1}{3}$, what is an upper bound on the error at each of the first three steps?

3. Consider the data in Table 3.4. Supposedly this comes from applying Newton's method to a smooth function whose derivative does not vanish at the root. Use the limit result (3.14) to determine whether or not the program is working. (Hint: Use $\alpha \approx x_7$.)

4. Repeat the previous exercise, using the data in Table 3.5.

5. Apply Newton's method to find the root of $f(x) = 2 - e^x$, for which the exact solution is $\alpha = \ln 2$. Perform the following experiments:
(a) Compute the ratio

$$R_n = \frac{\alpha - x_{n+1}}{(\alpha - x_n)^2}$$

and observe whether or not it converges to the correct value as $n \to \infty$.
(b) Compute the modified ratio

$$R_n(p) = \frac{\alpha - x_{n+1}}{(\alpha - x_n)^p}$$

for various $p \neq 2$ but near 2. What happens? Comment, in light of the definition of order of convergence.
(c) Consider applying Newton's method to find the root of the function $f(x) = 4x - \cos x$. Assume we want ac-

curacy to within 10^{-8}. Use the Newton error estimate (3.12) to show that the iteration will converge for all $x_0 \in [-2, 2]$. How many iterations will be needed for the iteration to converge? Compare with the corresponding results for bisection, using an initial interval of $[-2, 2]$.

6. Verify by actual computation that your results in the previous exercise are correct; i.e., apply Newton's method to $f(x) = 4x - \cos x$. Do you converge to the specified accuracy in the correct number of iterations, and does this convergence appear to be occurring for all choices of x_0?

7. Consider now the function $f(x) = 7x - \cos(2\pi x)$. Show that a root exists on the interval $[0, 1]$, and then use the Newton error estimate to determine how close x_0 has to be to the root to guarantee convergence.

8. Investigate your results in the previous problem by applying Newton's method to $f(x) = 7x - \cos(2\pi x)$, using several choices of x_0 within the interval $[0, 1]$. Comment on your results in light of the theory of the method. (Note: This can be done by a very modest modification of your existing Newton program.)

9. Show that if x_n is a sequence converging linearly to a value α, then the constant C in (3.16) must satisfy $|C| < 1$. (Hint: Assume $|C| > 1$ and prove a contradiction to the convergence assumption.)

10. Explain, in your own words, why the assumptions
$$f'(x_n) \approx f'(\alpha)$$
and
$$f''(\xi_n) \approx f''(\alpha)$$
are valid if we have $x_n \to \alpha$.

TABLE 3.5   Data for Exercise 4

n	x_n
0	0.000000000000
1	0.626665778573
2	0.889216318667
3	0.970768039664
4	0.992585155212
5	0.998139415613
6	0.999534421170
7	0.999883578197
8	0.999970892852
9	0.999992723105
10	0.999998180783

TABLE 3.4   Data Table for Exercise 3.

n	x_n
0	10.0
1	5.25
2	3.1011904761905
3	2.3567372726442
4	2.2391572227372
5	2.2360701085329
6	2.2360679775008
7	2.2360679774998

3.6 NEWTON'S METHOD: THEORY AND CONVERGENCE

The Newton error estimate (3.12) gives us a very precise idea of how the error changes from iteration to iteration, but it is not specific enough by itself to imply convergence or establish any kind of accuracy in practice, save for specific examples like those in the exercises at the end of the previous section. To get a general theory that is widely applicable to more general problems, we need to do a little more work, and that is the purpose of this section.

We first prove a theorem that makes very stringent assumptions on the behavior of the function f. We then show that those assumptions are not entirely necessary to prove *local* convergence, i.e., convergence under the assumption of a sufficiently close initial guess.

Theorem 3.3 *Assume that f is defined and twice continuously differentiable for all x, with $f(\alpha) = 0$ for some α. Define the ratio*[8]

$$M = \frac{\max_{x \in \mathbf{R}} |f''(x)|}{2 \min_{x \in \mathbf{R}} |f'(x)|} \qquad (3.20)$$

and assume that $M < \infty$. Then, for any x_0 such that

$$M|\alpha - x_0| < 1, \qquad (3.21)$$

the Newton iteration converges. Moreover,

$$|\alpha - x_n| \le M^{-1} (M|\alpha - x_0|)^{2^n}.$$

Proof   The proof is based almost completely on the Newton error formula (3.12). We have

$$
\begin{aligned}
|\alpha - x_{n+1}| &= \frac{1}{2}(\alpha - x_n)^2 \frac{|f''(\xi_n)|}{|f'(x_n)|} \\
&\le (\alpha - x_n)^2 M.
\end{aligned}
$$

To simplify the notation, let $e_n = |\alpha - x_n|$. Then we have

$$e_{n+1} \le e_n^2 M,$$

so that

$$
\begin{aligned}
e_1 &\le e_0^2 M \\
e_2 &\le e_1^2 M \le (e_0^2 M)^2 M = e_0^4 M^3 = M^{-1}(Me_0)^4 \\
e_3 &\le e_2^2 M \le (e_1^2 M)^2 M = e_0^8 M^7 = M^{-1}(Me_0)^8,
\end{aligned}
$$

---

[8]Recall that $\mathbf{R} = (-\infty, \infty)$.

and an inductive argument easily shows that

$$e_n \leq M^{-1} (Me_0)^{2^n}. \tag{3.22}$$

It is clear that the right side goes to zero as $n \to \infty$ if and only if $Me_0 = M|\alpha - x_0| < 1$, which completes the proof. $\qquad \square$

A couple of observations might be in order here. First, note that the convergence, when it occurs, is *very* rapid, since the exponent in (3.22) grows exponentially. For example, if $Me_0 = 0.5$, we have

$$e_5 \leq M^{-1}(0.5)^{32} = M^{-1} \times 2.328 \ldots \times 10^{-10}.$$

Second, it is not very practical to use condition (3.21) to predict a priori what value of x_0 to use. The point of the theorem is not to help us choose x_0, but to explain to us how it is that the iteration converges, to show us what issues are important in obtaining convergence, and to assure us that if we can find a value of x_0 that is close enough to the root, then we will get convergence (and rapid convergence).

But recall that this theorem assumes that the function f is defined and twice continuously differentiable everywhere, in addition to the assumption of the bound (3.20). These are very restrictive assumptions, and we would like to remove them, if possible. The key issue is to understand that assuming continuity of f and its first two derivatives *near* the root is actually enough to achieve essentially the same result as in the previous theorem. Precisely establishing this is a somewhat involved task, however.

Theorem 3.4   *Let $f \in C^2(I)$, where $\alpha \in I \subset \mathbf{R}$ is a root and I is an open interval. Assume that $f'(\alpha) \neq 0$, and let the values x_n be defined by applying Newton's method to f. Then, for x_0 sufficiently close to α, we have*

$$\lim_{n \to \infty} x_n = \alpha \tag{3.23}$$

and

$$\lim_{n \to \infty} \frac{\alpha - x_{n+1}}{(\alpha - x_n)^2} = -\frac{f''(\alpha)}{2f'(\alpha)}. \tag{3.24}$$

Proof   Since $f'(\alpha) \neq 0$ and f' is continuous, we can find a closed interval around α (perhaps a very small interval) such that $f'(x) \neq 0$ throughout this interval. Call this interval J, and note that we can assume, without loss of generality, that $J \subset I$ and $J = \{x \mid \alpha - \epsilon \leq x \leq \alpha + \epsilon\}$, where $\epsilon > 0$ is small enough for J to be contained in I and to satisfy $f'(x) \neq 0$ in J.

Now, since J is closed, and in addition f' does not vanish on J and f'' is continuous on J, the ratio

$$M = \frac{\max_{x \in J} |f''(x)|}{2 \min_{x \in J} |f'(x)|}$$

is bounded. (This follows from the Extreme Value Theorem.) Take $x_0 \in J$ and use Newton's method to define x_1:

$$x_1 = x_0 - \frac{f(x_0)}{f'(x_0)}.$$

We need to take x_0 sufficiently close to α that $x_1 \in J$. The Newton error formula (3.12) tells us that

$$\alpha - x_1 = -\frac{1}{2}(\alpha - x_0)^2 \frac{f''(\xi_0)}{f'(x_0)},$$

where ξ_0 is between α and x_0 and therefore $\xi_0 \in J$. Thus

$$|\alpha - x_1| = \frac{1}{2}(\alpha - x_0)^2 \frac{|f''(\xi_0)|}{|f'(x_0)|} \leq (\alpha - x_0)^2 \frac{\max_{x \in J} |f''(x)|}{2 \min_{x \in J} |f'(x)|} = (\alpha - x_0)^2 M.$$

If we choose x_0 so that $M|\alpha - x_0| < 1$, we have

$$|\alpha - x_1| \leq (\alpha - x_0)^2 M = |\alpha - x_0| (|\alpha - x_0|M) < |\alpha - x_0|,$$

which forces $x_1 \in J$ (since it is closer to α than x_0 is). The same argument can now be used to show that x_2 (defined from x_1 by Newton's method) is also in J and, recursively, the entire sequence of Newton iterates is in J. We can now essentially follow the same argument used in the previous theorem.[9] We have

$$|\alpha - x_{n+1}| = \frac{1}{2}(\alpha - x_n)^2 \frac{|f''(\xi_n)|}{|f'(x_n)|}$$

$$\leq (\alpha - x_n)^2 M,$$

where the fact that all the iterates are in J is crucial because it allows us to bound the derivative ratio on the right side. From this we quickly get that, for $e_n = |\alpha - x_n|$, again,

$$e_n \leq M^{-1} (Me_0)^{2^n},$$

so that for $Me_0 = M|\alpha - x_0| < 1$ (which defines "sufficiently close to α") convergence occurs and (3.23) is proven. To get (3.24), we have

$$\alpha - x_{n+1} = -\frac{1}{2}(\alpha - x_n)^2 \frac{f''(\xi_n)}{f'(x_n)},$$

from which we get

$$\frac{\alpha - x_{n+1}}{(\alpha - x_n)^2} = -\frac{1}{2} \frac{f''(\xi_n)}{f'(x_n)}.$$

---

[9]Note that all of the hard work with the technical issues of continuity and intervals was done so that we could get to this point and use an already verified argument, i.e., the argument from Theorem 3.3.

Now take the limit as $n \to \infty$ of both sides. Since $x_n \to \alpha$ and ξ_n is between α and x_n, it follows (from the "squeeze theorem" of calculus) that $\xi_n \to \alpha$ as well. Continuity then shows that

$$\lim_{n \to \infty} \frac{\alpha - x_{n+1}}{(\alpha - x_n)^2} = -\lim_{n \to \infty} \frac{1}{2} \frac{f''(\xi_n)}{f'(x_n)},$$

$$= -\frac{f''(\lim_{n \to \infty} \xi_n)}{2 f'(\lim_{n \to \infty} x_n)},$$

$$= -\frac{f''(\alpha)}{2 f'(\alpha)},$$

which completes the proof. $\square$

EXERCISES

1. Consider a function f that satisfies the following properties:
(a) There exists a unique root $\alpha \in [0,1]$.
(b) For all real x, $f'(x) \geq 2$ and $0 \leq f''(x) \leq 3$.

Show that for $x_0 = \frac{1}{2}$, Newton's method will converge to within 10^{-6} of the actual root in four iterations. How long would bisection take to achieve this accuracy?

2. Consider a function f that satisfies the following properties:
(a) There exists a unique root $\alpha \in [2,3]$.
(b) For all real x, $f'(x) \geq 3$ and $0 \leq f''(x) \leq 5$.

Using $x_0 = 5/2$, will Newton's method converge, and if so, how many iterations are required to get 10^{-4} accuracy?

3. Repeat the preceding, this time aiming for 10^{-20} accuracy.

4. Consider a function f that satisfies the following properties:
(a) There exists a unique root $\alpha \in [-1,3]$.
(b) For all real x, $f'(x) \geq 4$ and $-6 \leq f''(x) \leq 3$.

Using $x_0 = 1$, will Newton's method converge, and if so, how many iterations are required to get 10^{-4} accuracy?

5. Repeat the preceding, this time aiming for 10^{-20} accuracy.

6. Consider a function that satisfies the following properties:
(a) f is defined and twice continuously differentiable for all x.
(b) f has a unique root $\alpha \in [-1,1]$.
(c) $|f'(x)| \geq 2$ for all x.
(d) $|f''(x)| \leq 5$ for all x.

Can we conclude that Newton's method will converge for all $x_0 \in [-1,1]$? If so, how many iterations are required to get 10^{-6} accuracy? If not, how many steps of bisection must we do to get the initial interval small enough so that Newton's method will converge (for *any* choice of x_0)? How many total function evaluations (bisection plus Newton) are required to get 10^{-6} accuracy?

7. Repeat the preceding for a function satisfying the following properties:
(a) f is defined and twice continuously differentiable for all x.
(b) f has a unique root $\alpha \in [-1,2]$.
(c) $f'(x) \leq -3$ for all x.
(d) $|f''(x)| \leq 4$ for all x.

8. Consider the function $f(x) = x - a\sin x - b$, where a and b are positive parameters, with $a < 1$. Will the initial guess $x_0 = b$ always lead to convergence? If not, what additional condition on a or the initial guess needs to be made?

9. Write a program using Newton's method to find the root of $f(x) = 1 - e^{-x^2}$; the exact value is $\alpha = 0$. Compute the ratio

$$R_n = \frac{\alpha - x_{n+1}}{(\alpha - x_n)^2}$$

as the iteration progresses and comment (carefully!) on your results in light of the material in this section.

10. A *monotone* function is one whose derivative never changes sign; the function is either always increasing or always decreasing. Show that a monotone function can have at most one root.

11. If f is a smooth monotone function with a root $x = \alpha$, will Newton's method always converge to this root, for any choice of x_0? (Provide either a counterexample or a valid proof.)

3.7 APPLICATION: COMPUTATION OF THE SQUARE ROOT

(Much of this presentation is based on an exercise found in Kendall Atkinson's text, *An Introduction to Numerical Analysis*, published by John Wiley & Sons.)

An interesting application of Newton's method is to the computation of $\sqrt{a}$. Consider now the function $f(x) = x^2 - a$. Clearly, the roots of this function are $\alpha = \pm\sqrt{a}$, so if we can find the (positive) root of f, we have computed $\sqrt{a}$.

Recall that the floating-point representation of a (assuming a binary base) is

$$a = \tilde{b} \times 2^{t-p},$$

where $\tilde{b} \in [\frac{1}{2}, 1]$ is the fraction and $t - p$ is the binary exponent. If $t - p$ is odd, we shift one factor of 2 from the fraction to the exponent; this allows us to write

$$a = b \times 2^{2k},$$

where $b \in [\frac{1}{4}, 1]$. Thus

$$\sqrt{a} = \sqrt{b} \times 2^k,$$

and all we have to do is compute the square root of a number in the interval $[\frac{1}{4}, 1]$. Can we find an initial guess such that Newton's method will always converge for b on this interval? If so, how rapidly will it converge? The Newton error formula (3.12) applied to $f(x) = x^2 - b$ tells us that

$$\sqrt{b} - x_{n+1} = -(\sqrt{b} - x_n)^2 \left(\frac{2}{4x_n}\right)$$
$$= -(\sqrt{b} - x_n)^2 \left(\frac{1}{2x_n}\right), \tag{3.25}$$

so that the relative error satisfies

$$\left| \frac{\sqrt{b} - x_{n+1}}{\sqrt{b}} \right| = \left(\frac{\sqrt{b} - x_n}{\sqrt{b}} \right)^2 \left| \frac{\sqrt{b}}{2x_n} \right|. \tag{3.26}$$

Now Newton's method applied to f is

$$x_{n+1} = x_n - \frac{x_n^2 - b}{2x_n} = \frac{1}{2}\left(x_n + \frac{b}{x_n}\right),$$

so that if $x_0 > 0$, then $x_n > 0$ for all n. But then (3.25) implies that

$$\sqrt{b} - x_{n+1} \leq 0$$

for all $n \geq 0$. Thus $\alpha = \sqrt{b} \leq x_n$ for all $n \geq 1$. (We could choose x_0 less than $\sqrt{b}$, but all subsequent iterates would be greater than $\sqrt{b}$.) It therefore follows, from (3.26), that since $|\sqrt{b}/x_n| \leq 1$,

$$\left|\frac{\sqrt{b} - x_{n+1}}{\sqrt{b}}\right| \leq \frac{1}{2}\left(\frac{\sqrt{b} - x_n}{\sqrt{b}}\right)^2. \tag{3.27}$$

We can solve the recursion quickly to get

$$\left|\frac{\sqrt{b} - x_n}{\sqrt{b}}\right| \leq 2\left(\frac{\sqrt{b} - x_0}{2\sqrt{b}}\right)^{2^n}. \tag{3.28}$$

We now turn our attention to finding an initial guess. The fact that we are going to be applying the Newton iteration to finding $\sqrt{b}$ means that we need only look at the interval $[\frac{1}{4}, 1]$. We could just take $x_0 = \frac{5}{8}$, which is the midpoint of the interval and implies that $|\alpha - x_0| \leq \frac{3}{8}$, but we can do better by using *linear interpolation* to the square root function. We take our nodes to be $\frac{1}{4}$ and 1 and apply the linear interpolation formula (2.11) to $g(x) = \sqrt{x}$. We then have that

$$p_1(x) = \frac{x - 1/4}{3/4}\sqrt{1} + \frac{1 - x}{3/4}\sqrt{1/4} = \frac{2x + 1}{3}$$

and we take $x_0 = p_1(b)$; therefore $x_0 = \frac{1}{3}(2b + 1)$. Moreover, we can use the error estimate (2.13) to show that

$$|\sqrt{b} - x_0| \leq \frac{1}{8}\left(1 - \frac{1}{4}\right)^2 \max_{\frac{1}{4} \leq t \leq 1}\left|\frac{1}{4}t^{-3/2}\right|$$
$$= \frac{9}{64},$$

so that the relative error in $\sqrt{b}$ (which equals the relative error in $\sqrt{a}$) becomes

$$\left|\frac{\sqrt{b} - x_n}{\sqrt{b}}\right| \leq 2\left(\frac{9}{128\sqrt{b}}\right)^{2^n} \leq 2\left(\frac{9}{64}\right)^{2^n}. \tag{3.29}$$

Since $\frac{9}{64} < 0.15$, this will decrease very rapidly. In fact, we have that for $n = 4$, the relative error is less than 4.7×10^{-14}. Thus, four steps of Newton's method—about 15 total operations, counting the cost of forming x_0—suffice to compute the square root of a number to a very high (relative) accuracy.

EXERCISES

1. Based on the material in this section, write each of the following numbers in the form

$$a = b \times 2^k,$$

where $b \in [\frac{1}{4}, 1]$ and k is even, and then use Newton's method to find the square root of b and hence of a.
(a) $a = 3$
(b) $a = \pi$
(c) $a = 5$
(d) $a = 7$
(e) $a = 6$

Be sure to use the initial value as generated in this section and do only as many iterations as are necessary for 10^{-16} accuracy. Compare your values to the intrinsic square root function on your computer or calculator. Note this can be done on computer or hand calculator.

2. In the example of this section, how many iterations are required for the *absolute* error to be less than 2^{-48}?

3. How accurate is the result in only three iterations, using the initial guess used here?

4. What does the initial error have to be in order for the relative error, after only *two* iterations, to be less than 10^{-14}?

5. Extend the derivation and analysis of this section to the problem of computing *cube* roots by solving for the roots of $f(x) = x^3 - a$. Be sure to cover all the major

points that were covered in the text. Is the use of linear interpolation to get x_0 accurate enough to guarantee convergence of the iteration?

6. Consider using the polynomial

$$p_2(x) = \frac{9}{35} + \frac{22}{21}x - \frac{32}{105}x^2$$

to generate the initial guess. What is the maximum error in $|\sqrt{x} - p_2(x)|$ over the interval $[\frac{1}{4}, 1]$? How many iterations (and hence how many operations) are needed to get accuracy of 10^{-16}? (Feel free to compute the maximum value experimentally, but you must justify the accuracy of your value in this case with some kind of argument.)

7. Repeat the preceding using

$$q_2(x) = 0.2645796916$$
$$+ 1.0302824818x - 0.2983646911x^2.$$

8. Repeat again, using

$$q_1(x) = 0.647941993x + 0.3667102689.$$

9. If we use a piecewise linear interpolation to construct x_0, using the nodes $\frac{1}{4}$, $\frac{9}{16}$, and 1, what is the initial error and how many iterations are now needed to get 10^{-16} accuracy? How many operations are involved in this computation?

3.8 THE SECANT METHOD: DERIVATION AND EXAMPLES

An obvious drawback of Newton's method is that it requires a formula for the derivative of f. For classroom examples this is not an issue, but in the "real world" it can be. Suppose, for example, that f is not defined by a simple and tidy formula like $f(x) = 2 - e^x$, but is instead "defined" by a separate subprogram that uses around 2000 lines of involved computer code. Even if we could theoretically write a formula for f from which f' could be constructed, is this a practical task to set for ourselves?

One obvious way to deal with this problem is to use an approximation to the derivative in the Newton formula. For example, in Section 2.2 we saw that

$$f'(x) \approx \frac{f(x+h) - f(x)}{h},$$

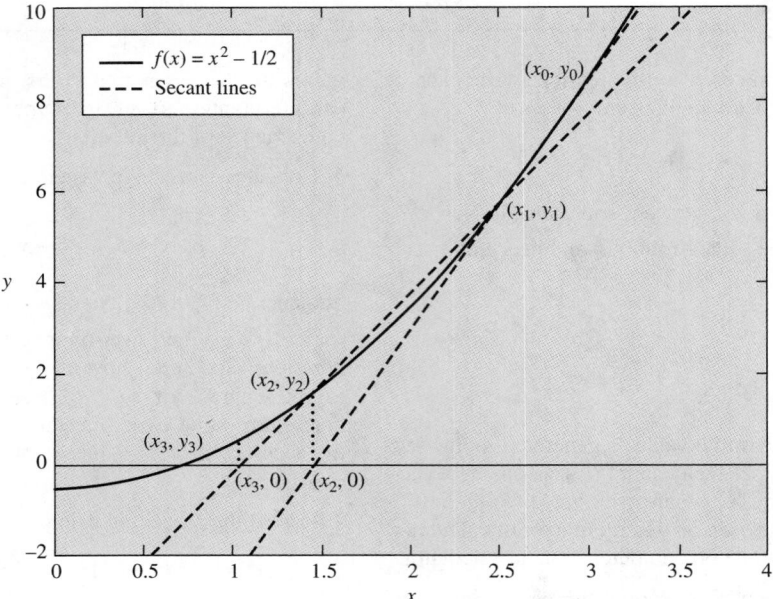

FIGURE 3.6 Illustration of the secant method.

so we could use this in (3.7) to get the new iteration

$$x_{n+1} = x_n - f(x_n) \left[\frac{h}{f(x_n + h) - f(x_n)} \right].$$

We will, in fact, analyze the convergence of this method in Section 3.10.2. In this section we want to pursue a similar but slightly different idea.

Newton's method is derived, geometrically, by drawing a tangent line from the current approximate root down to the axis. The derivative is required because we use the tangent line. If, instead, we used a *secant line* (i.e., one passing through two points on the curve instead of just one), then no derivative would be required.

Let x_0 and x_1 be given—thus we have two initial guesses—and consider Figure 3.6. Construct the line that passes through $(x_0, f(x_0))$ and $(x_1, f(x_1))$ and use its root to define the next iterate, x_2. The line is defined by the formula

$$\frac{y - f(x_1)}{x - x_1} = \frac{f(x_1) - f(x_0)}{x_1 - x_0},$$

so that the next iterate is given by (setting $y = 0$, and solving for $x = x_2$ in the above)

$$x_2 = x_1 - f(x_1) \left[\frac{x_1 - x_0}{f(x_1) - f(x_0)} \right].$$

More generally, then, we have

$$x_{n+1} = x_n - f(x_n) \left[\frac{x_n - x_{n-1}}{f(x_n) - f(x_{n-1})} \right]. \tag{3.30}$$

This is the *secant method*. Note that it is consistent with Newton's method, where we use the approximation

$$f'(x_n) \approx \frac{f(x_n) - f(x_{n-1})}{x_n - x_{n-1}}.$$

The error in this derivative approximation is proportional to $x_n - x_{n-1}$. Thus, if we assume that the iteration is converging (so that $x_n - x_{n-1} \to 0$), then the secant method becomes more and more like Newton's method. Hence we expect rapid convergence for x_0 near α.

The secant method has a number of advantages over Newton's method. Not only does it not require the derivative, but it can be coded in such a way (see below) as to require only a single function evaluation per iteration. Newton requires two: one for the function and one for the derivative. Thus the secant method is about half as costly per step as Newton's.

But the important question is: How well does it perform?

EXAMPLE 3.7

If we apply the secant method to the same example used heretofore, $f(x) = 2 - e^x$, using $x_0 = 0$ and $x_1 = 1$, then we get the following results for the first few iterations:

$$\begin{aligned}
x_2 &= x_1 - f(x_1)(x_1 - x_0)/(f(x_1) - f(x_0)) \\
&= 1 - (2 - e)(1 - 0)/(1 - e) = 0.5819767068 \\
x_3 &= x_2 - f(x_2)(x_2 - x_1)/(f(x_2) - f(x_1)) \\
&= 0.5819767068 - (2 - e^{0.5819767068})(0.5819767068 - 1)/(e - e^{0.5819767068}) \\
&= 0.6766927037.
\end{aligned}$$

As we continue the iteration, we get the results shown in Table 3.6. Note that we converged to almost the same root as for Newton's method, but this time in six iterations as opposed to five for Newton's. However, recall that Newton's method is more costly per iteration: It achieved its accuracy with a total of 10 calls to the function f and the derivative f'; the secant method did it with only six calls, all to f. The secant method (for this example) was actually *more* efficient in terms of the number of total function calls. ■

The secant method suffers from many of the ills that Newton's method does, however, since the underlying geometric idea is pretty much the same. The fundamental convergence theory is almost exactly the same, in essence: If the initial guesses are both sufficiently close to the root, then the method will converge. We will give a fairly

TABLE 3.6   The Secant Method Applied to $f(x) = 2 - e^x$

n	x_n	$\alpha - x_n$	$\log_{10}(\alpha - x_n)$
0	0.000000000000	0.693147180560	−0.1592
1	1.000000000000	0.306852819440	−0.5131
2	0.581976706869	0.111170473691	−0.9540
3	0.676692703760	0.016454476800	−1.7837
4	0.694081399681	0.000934219121	−3.0296
5	0.693139474645	0.000007705915	−5.1132
6	0.693147176961	0.000000003599	−8.4438
7	0.693147180560	0.000000000000	−13.8577

complete discussion of the convergence of the secant method in Section 3.10.3. At this point we simply note, without proof, that an error formula for the secant method can be derived that says

$$\alpha - x_{n+1} = -\frac{1}{2}(\alpha - x_n)(\alpha - x_{n-1})\frac{f''(\xi_n)}{f'(\eta_n)} \tag{3.31}$$

where $\min\{\alpha, x_n, x_{n-1}\} \leq \xi_n, \eta_n \leq \max\{\alpha, x_n, x_{n-1}\}$. Note that this error formula is very much akin to the one for Newton's method. Note also that it shows

$$|\alpha - x_{n+1}| \approx C|\alpha - x_n||\alpha - x_{n-1}|$$

for $x_n \approx \alpha$, $x_{n+1} \approx \alpha$, and $x_{n-1} \approx \alpha$. Thus the error goes like the product of the two previous errors, and it follows easily that the secant method is superlinear when it converges.

In addition, we can easily establish that a relationship similar to (3.9) holds for the secant iterates as well as the Newton iterates (the definition of the constant C_n is slightly different); thus we can use the difference of consecutive approximations as a stopping criterion (See Algorithm 3.3).

We close this section with an informal statement of the convergence result for the secant method.

If f, f', and f'' are all continuous near the root, and if f' does not equal zero at the root, then the secant method will converge whenever the initial guess is sufficiently close to the root. Moreover, this convergence will be superlinear, in the sense that

$$\lim_{n\to\infty} \frac{\alpha - x_{n+1}}{\alpha - x_n} = 0.$$

Note that, fundamentally, this is almost the same as for Newton's method.

Algorithm 3.3: *Secant Method*

```
input x0, x1, tol, n
external f
        f0 = f(x0)
        f1 = f(x1)
        for i=1 to n do
                x = x1 - f1*(x1 - x0)/(f1 - f0)
                fx = f(x)
                x0 = x1
                x1 = x
                f0 = f1
                f1 = fx
                if abs(x1 - x0) < tol then
                        root = x1
                          stop
                endif
        endfor
```

EXERCISES

1. Do three steps of the secant method for $f(x) = x^3 - 2$, using $x_0 = 0$, $x_1 = 1$.

2. Repeat the preceding using $x_0 = 1$, $x_1 = 0$. Comment.

3. Apply the secant method to the same functions as in Exercise 3 of Section 3.1, using x_0, x_1 equal to the endpoints of the given interval. Stop the iteration when the error as estimated by $|x_n - x_{n-1}|$ is less than 10^{-6}. Compare with your results for Newton's method and bisection in the previous problems.

4. For the secant method, prove that

$$\alpha - x_{n+1} = C_n(x_{n+1} - x_n)$$

where $C_n \to 1$ as $n \to \infty$, so long as the iteration converges. (Hint: Follow what we did in Section 3.3 for Newton's method.)

5. Assume (3.31) and prove that if the secant method converges, then it is superlinear.

6. Assume (3.31) and consider a function f such that
(a) There is a unique root on the interval [0,4].
(b) $|f''(x)| \le 2$ for all $x \in [0,4]$.
(c) $f'(x) \ge 5$ for all $x \in [0,4]$.

Can we prove that the secant iteration will converge for any $x_0, x_1 \in [0,4]$? If so, how many iterations are required to get an error that is less than 10^{-8}? If con-vergence is not guaranteed for all $x \in [0,4]$, how many steps of bisection are needed before convergence *will* be guaranteed for the secant method?

7. Repeat the preceding problem under the following assumptions:
(a) There is a unique root on the interval [0,9].
(b) $|f''(x)| \le 6$ for all $x \in [0,9]$.
(c) $f'(x) \ge 2$ for all $x \in [0,9]$.

Can we prove that the secant iteration will converge for any $x_0, x_1 \in [0,9]$? If so, how many iterations are re-quired to get an error that is less than 10^{-8}? If con-vergence is not guaranteed for all $x \in [0,9]$, how many steps of bisection are needed before convergence *will* be guaranteed for the secant method?

8. (This exercise and the next one are based on an ex-ample in Martin Braun's *Differential Equations and Their Applications*, Springer-Verlag, New York, 1993 (4th ed.).) It has been suggested that nuclear waste ma-terial be placed in specially designed drums, which will then be dumped in the ocean. So long as the drums remain intact, contamination of the surrounding sea is minimal. Solving the equations of motion for the bar-rels as they fall through the water yields the following relationship between the impact velocity and the depth of water:

$$D = \frac{1}{k^2 g} \left[W(W - B) \ln \left(1 + \frac{kv}{W - B} \right) - W k v \right].$$

Here D is the depth of the water, W is the weight of the barrels, B is the buoyancy of the barrels, g is the gravitational constant, k is the drag coefficient for the barrels, and v is the impact velocity. Experiments show that the drums will break upon impact with the sea bottom if they hit at a (*downward*) velocity greater than 40 feet per second. The buoyancy of the barrels can be determined from their volume and is given as 470 pounds. The drag coefficient is experimentally determined to be $k = 0.08$ pounds per foot per second. The gravitational constant is $g = 32$ feet per second per second. If the barrels weigh 527 pounds, use the secant method to determine if we can safely dump the barrels in 300 feet of water.

9. If the barrels are fitted with "aquatic parachutes" to slow their descent, then the drag coefficient is effectively raised to $k = 0.20$ pounds per foot per second. Can we now drop the barrels in 300 feet of water?

10. Repeat Exercise 8 of Section 3.1, except this time find α for the set of θ values defined by

$$\theta_k = 10^{k/4}$$

for k ranging from -24 all the way to 24. Construct a plot of α versus $\log_{10} \theta$.

11. Use the secant method to solve the pipeline problem from the beginning of the chapter, for the case $a = 3$, $b = 1, L = 4$; construct a table of values for x as a function of the cost ratio $C = C_1/C_2$.

12. Repeat the preceding for the case $a = 5$, $b = 1$, $L = 6$.

13. Comment, in your own words, on the differences between the secant method and regula falsi (see Exercise 9 of Section 3.1).

3.9   FIXED-POINT ITERATION

So far we have looked at three methods for approximating the roots of a given function $f(x)$: bisection, Newton's method, and the secant method. The first two can be considered instances of simple iteration, in which we recursively substitute values into a function or process to obtain the next value. (The secant method can be put into this framework as well, but only imperfectly, since each secant value is determined by the *two* previous values.)

In this section we will study simple iteration for its own sake, generally divorced from any considerations or connections to root-finding problems. Our goal, however, is to use the added understanding of simple iteration to enhance our understanding of and ability to solve root-finding problems.

Consider Newton's method as applied to $f(x) = x^2 - a$:

$$x_{n+1} = \frac{1}{2} \left(x_n + \frac{a}{x_n} \right). \tag{3.32}$$

As $n \to \infty$, we know that $x_n \to \alpha = \sqrt{a}$. (In this case, convergence occurs for any $x_0 > 0$.) We can write (3.32) more abstractly as

$$x_{n+1} = g(x_n) \tag{3.33}$$

for $g(x) = \frac{1}{2}(x + ax^{-1})$. Note that

$$f(\alpha) = 0 \iff \alpha = g(\alpha).$$

The fact that $\alpha = g(\alpha)$ is interesting. This defines α, which we already know as the root of f (the point where the graph of $y = f(x)$ crosses the x axis), to be a point where the graph of the new function $y = g(x)$ crosses the line $y = x$. Because $\alpha = g(\alpha)$ shows that $g(\alpha)$ "stays" at α, this kind of point is called a *fixed point* of the function g, and an iteration of the form (3.33) is called a *fixed-point iteration* for g.

We have already seen that we can establish a connection between a root-finding problem and a fixed-point problem using Newton's method. But this is clearly not the only way to do so. For example, each of the following iterations can be derived from the equation $x^2 - a = 0$; thus, for each one, $\alpha = g(\alpha) \iff f(\alpha) = 0 \iff \alpha = \sqrt{a}$. To verify, simply let $x_n = x_{n+1} = x$, and solve for x.

1. $x_{n+1} = x_n + \frac{1}{2}(x_n^2 - a)$; $g(x) = x + \frac{1}{2}(x^2 - a)$
2. $x_{n+1} = a/x_n$; $g(x) = a/x$
3. $x_{n+1} = a + x - x^2$; $g(x) = a + x - x^2$

So, for a given function g, a number of questions can be raised:

1. Under what conditions does a fixed point exist?

2. Under what conditions does the iteration (3.33) converge?

3. If the iteration converges, how fast does it converge?

In the remainder of this section we will outline the theory and practice of fixed-point iteration, mostly from the perspective of applying it to the root-finding problem. Our objective is to use the fixed-point theory to inform us about certain aspects of root-finding problems.

Do not confuse the notion of a root of a function with that of a fixed point for the same function. Generally, for a given function f the roots and fixed points (if any exist) are not the same. Figure 3.7 might be instructive in this regard; it shows the graph of

FIGURE 3.7   Fixed point versus root. The root is where the curve crosses the x axis; the fixed point is where the curve crosses the line $y = x$.

FIGURE 3.8 Newton's method.

$y = f(x) = x^2 - \frac{1}{2}$ and $y = x$. The roots of f are $\alpha = \pm\sqrt{1/2}$, but the fixed point of f is where it intersects the graph of $y = x$, and this occurs at $x = \frac{1}{2}(1 + \sqrt{3}) \approx 1.366$. When we use a fixed-point iteration to find a root of a function f, the root is a fixed point of a *different* function, g, not f. The fixed point of f, if one exists, is usually not related at all to the roots of f. Further clarification in this regard might come from comparing Figures 3.8 and 3.9. Figure 3.8 shows two iterations of Newton's method, graphically, as done on $f(x) = \frac{1}{2}x^2 - \frac{1}{2}$. The dashed lines are the tangent lines being used to find the next iterate. Figure 3.9 shows fixed-point iteration being performed on $g(x) = \frac{1}{2}e^{-x}$. Here the dashed lines show how x_0 is mapped onto $y = x$ and then projected onto the curve $y = g(x)$ to get the next iterate. Geometrically, we are looking for different points on the curve.

Before proceeding with the formal development of the important results, let's do some informal investigation of a fixed-point iteration. We have

$$\alpha = g(\alpha),$$

and

$$x_{n+1} = g(x_n), \tag{3.34}$$

so that

$$\alpha - x_{n+1} = g(\alpha) - g(x_n) = g'(\xi_n)(\alpha - x_n), \tag{3.35}$$

where we have of course assumed that g is differentiable, in order to use the Mean Value Theorem. If $|g'(x)| < 1$ near the fixed point, then we ought to be able to show

FIGURE 3.9 Fixed-point iteration.

that

$$|\alpha - x_n| < c^n |\alpha - x_0|$$

for some positive value $c < 1$, which would of course imply convergence. Once we have convergence established, a limit result quickly follows from (3.35), establishing that the iteration converges linearly, since (3.35) implies that

$$\frac{\alpha - x_{n+1}}{\alpha - x_n} = g'(\xi_n) \to g'(\alpha).$$

We can investigate the iteration (3.34) and the results we have rather casually derived by experimenting with a function g whose derivative is less than 1.

EXAMPLE 3.8

Let's take $g(x) = \frac{1}{2}e^{-x}$ as our example. Will the iteration

$$x_{n+1} = \frac{1}{2}e^{-x_n}$$

converge?

TABLE 3.7   **Fixed-Point Iteration Example, $g(x) = \frac{1}{2}e^{-x}$**

n	x_n, when $x_0 = 0$	x_n, when $x_0 = 5$	x_n, when $x_0 = -5$
0	0	5.0000000000000	-5.0000000000000
1	0.50000000000000	3.3689734995427E-03	74.206579551288
2	0.30326532985632	0.49831834756204	2.9611607506452E-33
3	0.36920157498737	0.30377574578843	0.50000000000000
4	0.34564302521408	0.36901317670616	0.30326532985632
5	0.35388254815500	0.34570814990043	0.36920157498737
6	0.35097870435331	0.35385950241548	0.34564302521408
7	0.35199937290228	0.35098679301031	0.35388254815500
8	0.35164028150092	0.35199652571160	0.35097870435331
9	0.35176657517651	0.35164128268928	0.35199937290228
10	0.35172215208802	0.35176622299208	0.35164028150092
11	0.35173777701935	0.35172227595910	0.35176657517651
12	0.35173228118368	0.35173773344922	0.35172215208802
13	0.35173421425181	0.35173229650870	0.35173777701935
14	0.35173353432626	0.35173420886147	0.35173228118368
15	0.35173377347896	0.35173353622223	0.35173421425181

Let's take $x_0 = 0$; this is a somewhat arbitrary choice, but let's see where it leads us. We have

$$x_1 = g(x_0) = \frac{1}{2}e^0 = \frac{1}{2}$$

$$x_2 = g(x_1) = \frac{1}{2}e^{-1/2} = 0.3032653299$$

$$x_3 = g(x_2) = \frac{1}{2}e^{-0.3032653299} = 0.369201575$$

$$x_4 = g(x_3) = \frac{1}{2}e^{-0.369201575} = 0.3456430253.$$

If we continue on, we get the values in Table 3.7, along with the results for two other values of x_0. The computation was stopped at $n = 15$; clearly, all three iterations are converging to the fixed point $\alpha \approx 0.3517$, but it is worth noting that none of the sequences is converging to its limit as fast as with either Newton's method or the secant method. Figure 3.10 shows the graph of $y = g(x)$ and $y = x$ to illustrate where the fixed point is located.   ∎

The important results in this section can be covered in three theorems, which we will now state and prove. While this makes for a lengthy and perhaps abstract theoretical interlude, it is necessary to understand the theory underlying fixed-point iteration in order to understand (and appreciate) what it gives us in terms of practical algorithms for root finding.

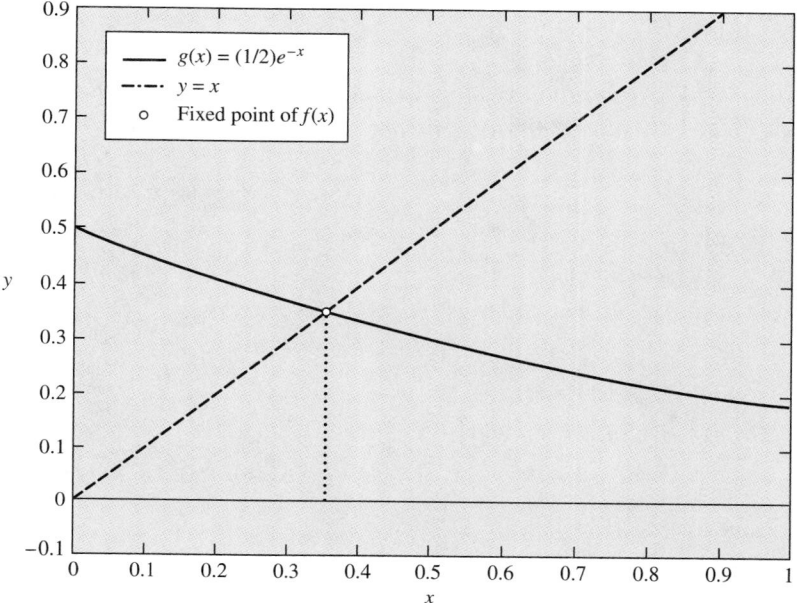

FIGURE 3.10 Fixed-point example.

The first theorem summarizes the conditions under which a given function g will have a fixed point and the conditions under which a fixed-point iteration will converge for any x_0 on a given interval, and it also provides us with error estimates for this convergence.

Theorem 3.5 (Fixed-Point Existence and Convergence Theory)   *Let* $g \in C([a,b])$ *with* $a \le g(x) \le b$ *for all* $x \in [a,b]$; *then*

  1. g has at least one fixed point $\alpha \in [a,b]$.

  2. If there exists a value $\gamma < 1$ such that

$$|g(x) - g(y)| \le \gamma |x - y| \tag{3.36}$$

  for all x and y in $[a,b]$, then

  (a) α is unique.

  (b) The iteration $x_{n+1} = g(x_n)$ converges to α for any initial guess $x_0 \in [a,b]$.

  (c) We have the error estimate

$$|\alpha - x_n| \le \frac{\gamma^n}{1 - \gamma} |x_1 - x_0|. \tag{3.37}$$

  3. If g is continuously differentiable on $[a,b]$ with

$$\max_{x \in [a,b]} |g'(x)| = \gamma < 1, \tag{3.38}$$

then

(a) α is unique.

(b) The iteration $x_{n+1} = g(x_n)$ converges to α for any initial guess $x_0 \in [a,b]$.

(c) We have the error estimate

$$|\alpha - x_n| \le \frac{\gamma^n}{1-\gamma}|x_1 - x_0|.$$

(d) The limit

$$\lim_{n \to \infty} \frac{\alpha - x_{n+1}}{\alpha - x_n} = g'(\alpha)$$

holds.

Proof   Define $h(x) = g(x) - x$. Then

$$h(b) = g(b) - b \le 0$$

and

$$h(a) = g(a) - a \ge 0$$

Therefore, the Intermediate Value Theorem implies that h has a root α on the interval $[a,b]$; thus $h(\alpha) = 0$, which implies $\alpha = g(\alpha)$. This proves (1).

Suppose now that (3.36) holds, and that a second fixed point, β, exists on $[a,b]$. Then we have

$$\alpha = g(\alpha) \tag{3.39}$$

and

$$\beta = g(\beta),$$

so that (from (3.36))

$$|\alpha - \beta| = |g(\alpha) - g(\beta)| \le \gamma|\alpha - \beta|,$$

which implies

$$|\alpha - \beta|(1 - \gamma) \le 0.$$

Since $0 < \gamma < 1$, the only way this can be true is for $|\alpha - \beta| \le 0$, and the only way for this to be true is for $\alpha = \beta$, which implies that the fixed point α is unique. (Why?) This proves (2a).

Consider now the iteration (3.33) and the definition of fixed point (3.39). If we subtract and take absolute values, we get

$$|\alpha - x_{n+1}| = |g(\alpha) - g(x_n)| \le \gamma|\alpha - x_n|.$$

Now write $e_n = |\alpha - x_n|$ so that the preceding becomes $e_{n+1} \le \gamma e_n$. The recursion can be solved readily to get $e_n \le \gamma^n e_0$, from which it follows that $e_n \to 0$ as $n \to \infty$; hence the iteration converges. This proves (2b).

Finally, we note that,

$$|\alpha - x_0| = |\alpha - g(x_0) + x_1 - x_0| \le |g(\alpha) - g(x_0)| + |x_1 - x_0| \le \gamma|\alpha - x_0| + |x_1 - x_0|,$$

from which it follows that

$$|\alpha - x_0| \le \frac{1}{1-\gamma}|x_1 - x_0|,$$

so that

$$e_n \le \gamma^n e_0 \le \frac{\gamma^n}{1-\gamma}|x_1 - x_0|,$$

which proves (2c).

We now note that (3.38) implies (3.36), so that (3a)–(3c) follow just as (2a)–(2c) did. It remains only to prove (3d).

We have, from (3.39) and (3.33), that

$$\alpha - x_{n+1} = g(\alpha) - g(x_n) = g'(\xi_n)(\alpha - x_n)$$

so that

$$\frac{\alpha - x_{n+1}}{\alpha - x_n} = g'(\xi_n) \rightarrow g'(\alpha)$$

since $\xi_n \rightarrow \alpha$ is forced by the convergence of x_n to α.   □

Note: It is important to keep in mind here the condition $a \le g(x) \le b$, which is one of our hypotheses. This assumption guarantees that the iterates computed according to $x_{n+1} = g(x_n)$ stay within the interval $[a,b]$, which is the only place where we know that (3.36) or (3.38) holds.

We can illustrate this theorem by using it to analyze the example we looked at previously.

EXAMPLE 3.9

Consider the iteration

$$x_{n+1} = g(x_n),$$

for $g(x) = \frac{1}{2}e^{-x}$. Since $0 \le g(x) \le \frac{1}{2}$ for all $x > 0$, we have that $g(x) \in [0,\frac{1}{2}]$ for all $x \in [0,\frac{1}{2}]$. Thus we can take our interval $[a,b]$ as $[0,\frac{1}{2}]$. Since g is continuous on this interval, we know that a fixed point must exist there. Further, since g is continuously differentiable on $[0,\frac{1}{2}]$ and

$$|g'(x)| \le \frac{1}{2} < 1$$

for all $x \in [0,\frac{1}{2}]$, we have that this fixed point is unique. Moreover, the iteration converges linearly and the error estimate (3.37) applies, with $\gamma = \frac{1}{2}$.   ■

The second theorem is a local convergence result; i.e., if we only know information about g near the fixed point, we can still deduce that convergence occurs for a sufficiently close initial guess.

Theorem 3.6 (Local Convergence for Fixed-Point Iterations)   *Let g be continuously differentiable in an open interval of a fixed point α with $|g'(\alpha)| < 1$; then, for all*

x_0 *sufficiently close to* α, *the iteration* $x_{n+1} = g(x_n)$ *converges,*

$$\lim_{n \to \infty} \frac{\alpha - x_{n+1}}{\alpha - x_n} = g'(\alpha),$$

and

$$|\alpha - x_n| \le \frac{\gamma^n}{1 - \gamma} |x_1 - x_0|,$$

for some $\gamma < 1$.

Proof   Since g is continuously differentiable in an open interval of the fixed point, with $|g'(\alpha)| < 1$, we can (as we did in the proof of the Newton convergence theorem) find a closed interval J, centered on the fixed point α, such that $|g'(x)| \le \gamma < 1$ for all $x \in J$. It follows from the definition of the iteration that

$$|\alpha - x_1| = |g(\alpha) - g(x_0)| \le \gamma |\alpha - x_0|;$$

therefore, x_1 is closer to the fixed point than x_0 is, and so are all the rest of the iterates. Thus $g(x) \in J$ for all $x \in J$, and we can apply Theorem 3.5 to complete the proof.   □

The third theorem provides us with a local criterion under which a fixed-point iteration will have a higher-than-linear order of convergence.

Theorem 3.7   *Consider the fixed-point iteration*

$$x_{n+1} = g(x_n), \tag{3.40}$$

where g is p times continuously differentiable, and $\alpha = g(\alpha)$. If

$$g'(\alpha) = g''(\alpha) = \cdots = g^{(p-1)}(\alpha) = 0$$

but

$$g^{(p)}(\alpha) \ne 0,$$

then the iteration (3.40) converges with order p for x_0 sufficiently close to α.

Proof   The fact that $g'(\alpha) = 0 < 1$ means that the iteration will converge for x_0 sufficiently close to α; this follows from Theorem 3.6. All we have to do is establish the higher convergence rate. We have, by Taylor's Theorem,

$$g(x_n) = g(\alpha) + (x_n - \alpha)g'(\alpha) + \cdots + \frac{(x_n - \alpha)^{p-1}}{(p-1)!} g^{(p-1)}(\alpha) + \frac{(x_n - \alpha)^p}{p!} g^{(p)}(\xi_n),$$

where ξ_n is between x_n and α. Now, all the derivative terms except that in the remainder vanish, so we quickly have that

$$g(x_n) - g(\alpha) = \frac{(x_n - \alpha)^p}{p!} g^{(p)}(\xi_n),$$

so that the iteration implies

$$\frac{x_{n+1} - \alpha}{(x_n - \alpha)^p} = \frac{1}{p!} g^{(p)}(\xi_n),$$

from which the convergence with order p follows.   □

To illustrate the use of this theorem, we consider Newton's method,

$$x_{n+1} = x_n - \frac{f(x_n)}{f'(x_n)}$$

which can be viewed as a fixed-point iteration with

$$g(x) = x - \frac{f(x)}{f'(x)}.$$

Note that

$$g'(x) = 1 - \left[\frac{[f'(x)]^2 - f(x)f''(x)}{[f'(x)]^2}\right],$$

so

$$g'(\alpha) = 1 - \left[\frac{[f'(\alpha)]^2 - f(\alpha)f''(\alpha)}{[f'(\alpha)]^2}\right] = 1 - 1 = 0,$$

from which we conclude that Newton's method has (local) order of convergence of at least 2. This assumes, of course, that $f'(\alpha) \neq 0$. See Section 3.10.4 for what happens when this assumption fails to hold.

With the theory that underlies fixed-point iterations behind us, we can apply this theory to obtain a greater understanding of how certain root-finding methods work, and how to improve them. Much of the next section is devoted to this.

EXERCISES

1. Do three steps of each of the following fixed-point iterations, using the indicated x_0.
(a) $x_{n+1} = \cos x_n$, $x_0 = 0$ (be sure to set your calculator in radians)
(b) $x_{n+1} = e^{-x_n}$, $x_0 = 0$
(c) $x_{n+1} = \ln(1 + x_n)$, $x_0 = 1/2$
(d) $x_{n+1} = \frac{1}{2}(x_n + 3/x_n)$, $x_0 = 3$

2. Let $Y = 1/2$ be fixed, and take $h = \frac{1}{8}$. Do three steps of the fixed-point iteration

$$y_{n+1} = Y + \frac{1}{2}h(-Y \ln Y - y_n \ln y_n)$$

using $y_0 = Y$.

3. Let $Y_0 = 1/2$ and $Y_1 = 0.54332169878500$ be fixed, and take $h = \frac{1}{8}$. Do three steps of the fixed-point iteration

$$y_{n+1} = \frac{4}{3}Y_1 - \frac{1}{3}Y_0 - 2hy_n \ln y_n$$

using $y_0 = Y_1$.

4. Consider the fixed-point iteration $x_{n+1} = 1 + e^{-x_n}$. Show that this iteration converges for any $x_0 \in [1,2]$. How many iterations does the theory predict it will take to achieve 10^{-5} accuracy?

5. For each of the following functions find an interval $[a,b]$ such that $g([a,b]) \subset [a,b]$. Draw a graph of $y = g(x)$ and $y = x$ over this interval, and confirm that a fixed point exists there. Estimate (by eye) the value of the fixed point, and use this as a starting value for a fixed-point iteration. Does the iteration converge? Explain.

(a) $g(x) = \frac{1}{2}(x + 2/x)$

(b) $g(x) = \cos x$

(c) $g(x) = 1 + e^{-x}$

(d) $g(x) = x + e^{-x} - \frac{1}{4}$

(e) $g(x) = \frac{1}{2}(x^2 + 1)$

6. Let $h(x) = 1 - x^2/4$. Show that this function has a root at $x = \alpha = 2$. Now, using $x_0 = 1/2$, do the iteration $x_{n+1} = h(x_n)$ to approximate the fixed point of h. Comment on your results.

7. Let $h(x) = 3 - e^{-x}$. Using $x_0 = 2$, perform as many iterations of Newton's method as are needed to accurately approximate the root of h. Then, using the same initial point, do the iteration $x_{n+1} = h(x_n)$ to approximate the fixed point of h. Comment on your results.

8. Consider the iteration defined by

$$x_{n+1} = x_n - f(x_n)\left[\frac{f(x_n)}{f(x_n + f(x_n)) - f(x_n)}\right].$$

This is also known as Steffenson's method. Show that it is (locally) quadratically convergent.

9. Apply Steffenson's method to find the root of $f(x) = 2 - e^x$, using $x_0 = 0$. Compare your convergence results with those in the text for Newton's method and the secant method.

10. Apply Steffenson's method to $f(x) = x^2 - a$ for the computation of $\sqrt{a}$. For the following values of a, how does the performance compare with Newton's method?

(a) $a = 2$

(b) $a = 3$

(c) $a = \pi$

3.10 SPECIAL TOPICS IN ROOT-FINDING METHODS

3.10.1 Extrapolation and Acceleration

One of the more important aspects of numerical analysis is that we can sometimes accelerate or improve the convergence of an algorithm with very little additional effort, simply by using the output of the algorithm to estimate some of the uncomputable quantities in the analysis of the algorithm. The best example of this is found in the chapter on numerical integration. Here we use this basic idea to speed up the convergence of a linearly convergent sequence. The process is known as *Aitken extrapolation.*[10] It can be used to accelerate a linear algorithm into a quadratic one, and thus it makes the fixed-point iteration ideas somewhat competitive with Newton's method. Moreover, we will use these ideas to recover the speed of convergence of Newton's method when the theory of Section 3.6 does not apply; this is done in Section 3.10.4.

We begin, as we almost always do with fixed-point iteration, by considering the error. If we assume that g is continuously differentiable, then the Mean Value Theorem says that there exists ξ_{n-1} between α and x_{n-1} such that

$$\alpha - x_n = g(\alpha) - g(x_{n-1}) = g'(\xi_{n-1})(\alpha - x_{n-1}).$$

---

[10] Alexander Craig Aitken (1895–1967) was born at Dunedin, New Zealand, of Scottish parents and educated at the University of Otago, located in Dunedin. In 1923 he obtained a scholarship for graduate study at the University of Edinburgh in Scotland, where he spent the rest of his life. The idea of accelerating the convergence of a linearly converging sequence, sometimes called the Aitken δ^2 process, is presented in a 1925 paper published in the *Proceedings of the Royal Society of Edinburgh.*

Now consider

$$\alpha - x_n = (\alpha - x_{n-1}) + (x_{n-1} - x_n) = \frac{1}{g'(\xi_{n-1})}(\alpha - x_n) + (x_{n-1} - x_n),$$

so that solving for α implies

$$\alpha = x_n + \frac{g'(\xi_{n-1})}{1 - g'(\xi_{n-1})}(x_n - x_{n-1}). \tag{3.41}$$

This formula is the starting point for the construction of our extrapolation algorithm. Note that it gives an expression for α in terms of x_n, x_{n-1}, and $g'(\xi_{n-1})$. The problem is that the last quantity is not computable; but it can be *estimated*, and that is the key.

Since we are assuming that $x_n \to \alpha$, we also know that $g'(\xi_n) \to g'(\alpha)$; thus

$$g'(\xi_{n-1}) \approx g'(\alpha).$$

On the other hand, consider the ratio

$$\begin{aligned}
\gamma_n &= \frac{x_{n-1} - x_n}{x_{n-2} - x_{n-1}} \\
&= \frac{(\alpha - x_{n-1}) - (\alpha - x_n)}{(\alpha - x_{n-2}) - (\alpha - x_{n-1})} \\
&= \frac{(\alpha - x_{n-1}) - g'(\xi_{n-1})(\alpha - x_{n-1})}{(\alpha - x_{n-1})/g'(\xi_{n-2}) - (\alpha - x_{n-1})} \\
&= \frac{1 - g'(\xi_{n-1})}{1/g'(\xi_{n-2}) - 1} \\
&= g'(\xi_{n-2})\frac{1 - g'(\xi_{n-1})}{1 - g'(\xi_{n-2})} \\
&\to g'(\alpha).
\end{aligned}$$

Thus, $\gamma_n \approx g'(\alpha)$ as well. The difference (and it is an important one) is that γ_n is *computable*. We thus can use

$$\gamma_n \approx g'(\xi_{n-1})$$

in (3.41) to get a completely computable estimate of α,

$$\alpha \approx x_n + \frac{\gamma_n}{1 - \gamma_n}(x_n - x_{n-1}),$$

which we use to define a new sequence of approximations:

$$\bar{x}_n = x_n + \frac{\gamma_n}{1 - \gamma_n}(x_n - x_{n-1}). \tag{3.42}$$

There are a couple of ways to use Aitken extrapolation in an algorithm. Perhaps the most obvious is to use the ordinary fixed-point iteration to produce the x_n sequence, with the $\bar{x}_n$ sequence being produced passively, as it were.

EXAMPLE 3.10

Let's return to our example iteration for which $g(x) = \frac{1}{2}e^{-x}$. Taking $x_0 = 0$, we compute the first several ordinary iterates as follows:

$$x_1 = g(x_0) = \frac{1}{2}e^0 = \frac{1}{2}$$

$$x_2 = g(x_1) = \frac{1}{2}e^{-1/2} = 0.3032653299$$

$$x_3 = g(x_2) = \frac{1}{2}e^{-0.3032653299} = 0.369201575.$$

Now, starting with x_2, we can begin to compute the accelerated values $\bar{x}_n$ according to

$$\bar{x}_n = x_n + \frac{\gamma_n}{1 - \gamma_n}(x_n - x_{n-1}),$$

where

$$\gamma_n = \frac{x_{n-1} - x_n}{x_{n-2} - x_{n-1}}.$$

Proceeding in this fashion, we get

$$\bar{x}_2 = 0.3032653299 + \left(\frac{-0.33515315365051}{1 + 0.33515315365051} \right)(0.3032653299 - 0.5)$$

$$= 0.35265011013472$$

$$\bar{x}_3 = 0.369201575 + \left(\frac{-0.35729286262005}{1 + 0.35729286262005} \right)(0.3692015759 - 0.3032653299)$$

$$= 0.35184456205221.$$

■

Pseudocode for this might go as in Algorithm 3.4; more computation with this example is given in Table 3.8.

While Algorithm 3.4 does result in faster convergence, it does not take as much advantage of the acceleration as is possible. An alternative version actually feeds the extrapolated value back into the computation; this is sometimes known as Steffenson's iteration, and is given in Algorithm 3.5 (although there is another root-finding method associated with this name, so confusion is possible).

EXAMPLE 3.11

Using the same function and initial value, we compute as follows:

$$x_1' = g(x_0) = \frac{1}{2}e^0 = \frac{1}{2}$$

$$x_2' = g(x_1') = \frac{1}{2}e^{-1/2} = 0.3032653299.$$

Algorithm 3.4

```
Aitken Extrapolation, Version 1
x1 = g(x0)
x2 = g(x1)
for k=1 to n do
        if (dabs(x1 - x0) > 1.d-20) then
                gamma = (x2 - x1)/(x1 - x0)
        else
                gamma = 0.0d0
        endif
        xbar = x2 + gamma*(x2 - x1)/(1 - gamma)
        if(abs(xbar - x2) < error) then
                alpha = xbar
                stop
        endif
        x = g(x2)
        x0 = x1
        x1 = x2
        x2 = x
enddo
```

Programming Note: Since division by zero—or a very small number—is possible in the algorithm's computation of γ, we put in a conditional test: Only if the denominator is greater than 10^{-20} do we carry through the division.

TABLE 3.8 Extrapolation Example, $g(x) = \frac{1}{2}e^{-x}$

n	Fixed-point	Aitken, version 1	Aitken, version 2	Newton
0	0	NA	0	0
1	0.50000000000000	NA	0.35881664959840	0.33333333333333
2	0.30326532985632	0.50000000000000	0.35173599679979	0.35168933155542
3	0.36920157498737	0.35881664959840	0.35173371124943	0.35173371099294
4	0.34564302521408	0.35265011013472	0.35173371124920	0.35173371124920
5	0.35388254815500	0.35184456205221	0.35173371124920	0.35173371124920
6	0.35097870435331	0.35174752132793	0.35173371124920	0.35173371124920
7	0.35199937290228	0.35173541541015	0.35173371124920	0.35173371124920
8	0.35164028150092	0.35173392226949	0.35173371124920	0.35173371124920
9	0.35176657517651	0.35173373734771	0.35173371124920	0.35173371124920
10	0.35172215208802	0.35173371447837	0.35173371124920	0.35173371124920
11	0.35173777701935	0.35173371164868	0.35173371124920	0.35173371124920
12	0.35173228118368	0.35173371129862	0.35173371124920	0.35173371124920
13	0.35173421425181	0.35173371125531	0.35173371124920	0.35173371124920
14	0.35173353432626	0.35173371124995	0.35173371124920	0.35173371124920
15	0.35173377347896	0.35173371124929	0.35173371124920	0.35173371124920

Now we do the acceleration using x_0, x_1', and x_2':

$$\gamma = \frac{x_1' - x_2'}{x_0 - x_1'} = -0.3934693402;$$

thus the next value of the iteration is

$$x_1 = 0.3032653299 + \left(\frac{-0.3934693402}{1 + 0.3934693402} \right)(0.3032653299 - 0.5) = 0.3588166496.$$

We then compute (note that, in this notation, x_2' gets redefined)

$$x_2' = g(x_1) = \frac{1}{2}e^{-0.3588166496} = 0.3492512051$$

$$x_3' = g(x_2') = \frac{1}{2}e^{-0.3492512051} = 0.3526079771.$$

Now use Aitken acceleration with x_1, x_2', and x_3':

$$\gamma = \frac{x_2' - x_3'}{x_1 - x_2'} = -0.3509269224;$$

thus the next iterate is

$$x_2 = 0.3526079771 + \left(\frac{-0.3509269224}{1 + 0.3509269224} \right)(0.3526079771 - 0.3492512051)$$

$$= 0.3517359968.$$

∎

Algorithm 3.5 *Aitken extrapolation, version 2*

```
for k=1 to n do
        x1 = g(x0)
        x2 = g(x1)
        if (dabs(x1 - x0) > 1.d-20) then
                gamma = (x2 - x1)/(x1 - x0)
        else
                gamma = 0.0d0
        endif
        x0 = x2 + gamma*(x2 - x1)/(1 - gamma)
        if(abs(x0 - x2) < error) then
                alpha = x0
                stop
        endif
enddo
```

Pseudocode for this is given in Algorithm 3.5. Note again that we checked for division by small numbers before computing gamma.

Table 3.8 shows more computations with the original fixed-point iteration $x_{n+1} = \frac{1}{2}e^{-x_n}$, together with both Aitken extrapolation algorithms. As a further comparison, we have included the data for using Newton's method to find the root of the function $f(x) = x - \frac{1}{2}e^{-x}$, whose root is the same as the fixed point of g.

Note that the first Aitken algorithm does converge somewhat more quickly than the unaccelerated fixed-point iteration, but the second Aitken algorithm is *much* faster than both. (The table is somewhat deceptive in this regard. The second Aitken algorithm is using two evaluations of g each step, so it really is converging in eight iterations instead of four.) Note that Newton's method and the second Aitken algorithm both converged in four iterations.

It can be shown, under the correct hypotheses, that the second Aitken algorithm is *quadratically* convergent. In fact, this is one application of Theorem 3.7; see Exercise 4.

One problem that we have conveniently avoided discussing so far is that Aitken extrapolation is very susceptible to subtractive cancellation, based on the formula for γ_n as well as the difference $x_{n-1} - x_n$ that appears in formula (3.42). For this reason the computations in Aitken extrapolation should be done in as high a precision as possible. It is sometimes suggested that, after the extrapolated method has converged, a few ordinary iterations be done to remove any error caused by the subtractive cancellation.

3.10.2 Variants of Newton's Method

One application of the fixed-point theory developed in Section 3.9 is to the study of other, Newton-like iterations that can be proposed. In this section we will look at three such ideas. All three are locally convergent, although the first two are only linear and not quadratic; the third is an example of a cubic convergent method.

The Chord Method   Newton's method works by using the tangent line approximation to find the approximate root of the function $f(x)$. This, of course, requires the derivative $f'(x)$ and uses two function evaluations in each step of the iteration. The chord method uses the original value of the derivative $f'(x_0)$ on each iteration instead of just the first one:

$$x_{n+1} = x_n - \frac{f(x_n)}{f'(x_0)}.$$

By giving up the need to evaluate the derivative in each step of the iteration, we save some cost, but we should also expect to lose some convergence and accuracy. The question is, How much do we lose?

If we apply Theorem 3.6 to

$$g(x) = x - \frac{f(x)}{f'(x_0)},$$

we get

$$g'(x) = 1 - \frac{f'(x)}{f'(x_0)} \qquad \Rightarrow \qquad g'(\alpha) = 1 - \frac{f'(\alpha)}{f'(x_0)} = \frac{f''(\xi)}{f'(x_0)}(\alpha - x_0),$$

so that $|g'(\alpha)| < 1$—which is what controls the existence of local convergence, according to Theorem 3.6—depends on how close x_0 is to α. Moreover, we won't get quadratic convergence unless $x_0 = \alpha$ (we know this from Theorem 3.7), so we know we have lost the extra speed that Newton's method gives us; the chord method is only linear, and only locally convergent. This might prompt the question, Then why bother with it? After all, we can use the secant method, which is faster than linear and does not use the derivative. What purpose does the chord method serve?

A complete answer requires almost another whole course. The chord method is useful in solving nonlinear *systems* of equations, i.e., problems involving 2 or 3 or 20 (or more) equations in 2 or 3 or 20 (or more) unknowns. In this setting, which we briefly discuss in Section 7.8, the derivative of f is actually a matrix, and so every evaluation of f' is actually the evaluation of an $n \times n$ matrix. Having a method, even if it is only linearly convergent, that avoids the need to reevaluate (and invert, since division by f' in the scalar case becomes matrix inversion in the systems case) that matrix every step of the way is valuable.

One interesting variant of the chord method updates the point at which the derivative is evaluated, but not every iteration. Thus for the first, say, p iterations, we use

$$x_{n+1} = x_n - \frac{f(x_n)}{f'(x_0)},$$

but for the next p iterations we use

$$x_{n+1} = x_n - \frac{f(x_n)}{f'(x_p)},$$

and so on, periodically updating the derivative evaluation point. While this method will not converge as quickly as Newton's method, it will converge more quickly than the ordinary chord method, and is less costly than Newton's.

EXAMPLE 3.12

Let's use the chord method to find the root of $f(x) = 2 - e^x$, the same example we have used throughout this chapter. Taking $x_0 = 0$, we have $f'(x_0) = -1$ and so

$$x_1 = x_0 - \frac{f(x_0)}{f'(x_0)} = 0 - \frac{2-1}{-1} = 1$$

$$x_2 = x_1 - \frac{f(x_1)}{f'(x_0)} = 1 - \frac{2-e}{-1} = 0.2817181715$$

$$x_3 = x_2 - \frac{f(x_2)}{f'(x_0)} = 0.2817181715 - \frac{2 - e^{0.2817181715}}{-1} = 0.9563130411$$

$$x_4 = x_3 - \frac{f(x_3)}{f'(x_0)} = 0.9563130411 - \frac{2 - e^{0.9563130411}}{-1} = 0.3542280558.$$

If we continue the iteration, we find that it does converge, but much more slowly than did Newton's method. For instance, $x_{12} = 0.465017583563$, which is still off in the first decimal place, whereas Newton's method converged in only five iterations.

If we update the computation of f' every, say, three iterations, then we do get improved performance. The first three iterates are the same as in the pure chord method:

$$x_1 = x_0 - \frac{f(x_0)}{f'(x_0)} = 1$$

$$x_2 = x_1 - \frac{f(x_1)}{f'(x_0)} = 0.2817181715$$

$$x_3 = x_2 - \frac{f(x_2)}{f'(x_0)} = 0.9563130411.$$

But now we compute $f'(0.9563130411) = -2.602084985$, and so the next few iterates are

$$x_4 = x_3 - \frac{f(x_3)}{f'(x_3)} = 0.7249274450$$

$$x_5 = x_4 - \frac{f(x_4)}{f'(x_3)} = 0.7001083868$$

$$x_6 = x_5 - \frac{f(x_5)}{f'(x_3)} = 0.6947392372,$$

which is a much more accurate approximation (but still not as good as Newton's method). ∎

Other Approximations to the Derivative   In Section 3.8 we briefly mentioned the method

$$x_{k+1} = x_k - f(x_k) \frac{h}{f(x_k + h) - f(x_k)}, \tag{3.43}$$

which is based on using a finite difference approximation to the derivative in Newton's method.

We can now use Theorem 3.7 to analyze the convergence of this method. Here we have

$$g(x) = x - f(x) \frac{h}{f(x+h) - f(x)},$$

so that

$$g'(x) = 1 - f'(x) \left(\frac{h}{f(x+h) - f(x)} \right) - f(x) \left(\frac{h}{f(x+h) - f(x)} \right)'.$$

Since $f(\alpha) = 0$, we have

$$g'(\alpha) = 1 - f'(\alpha) \left(\frac{h}{f(\alpha+h) - f(\alpha)} \right).$$

But (2.1) implies that

$$\frac{f(\alpha+h) - f(\alpha)}{h} = f'(\alpha) + \frac{1}{2} h f''(\xi_{\alpha,h}),$$

so we have

$$g'(\alpha) = 1 - f'(\alpha)\left(\frac{1}{f'(\alpha) + \frac{1}{2}hf''(\xi_{\alpha,h})}\right),$$

which simplifies to

$$g'(\alpha) = \frac{\frac{1}{2}hf''(\xi_{\alpha,h})}{f'(\alpha) + \frac{1}{2}hf''(\xi_{\alpha,h})} = \left(\frac{\frac{1}{2}f''(\xi_{\alpha,h})}{f'(\alpha) + \frac{1}{2}hf''(\xi_{\alpha,h})}\right)h.$$

Thus, for this method, so long as $f'(\alpha) \neq 0$, $g'(\alpha) = O(h) \neq 0$, and therefore we have only linear convergence, although the rate constant for the linear convergence will typically be small (since h is small). But the secant method will converge faster, in general, and requires only a single function evaluation per step; the method (3.43) requires two for each iteration. Thus this is not a reasonable method to use; it is just as costly as Newton's method, more costly than the secant method, and slower than both.

Higher-Order Convergence: Halley's Method   Could we perhaps get a method better than Newton's if we were willing to use the second derivative of f? The answer is yes, and the method is known as Halley's method.[11]

Actually there are two methods. The most obvious one is to use the quadratic term in a Taylor series expansion of f to define a more accurate x_{n+1} from x_n. This means we define x_{n+1} as the root of the quadratic

$$0 = f(x_n) + (x - x_n)f'(x_n) + \frac{1}{2}(x - x_n)^2 f''(x_n).$$

This leads to the iteration

$$x_{n+1} = x_n - \frac{2f(x_n)}{f'(x_n) \pm \sqrt{[f'(x_n)]^2 - 2f(x_n)f''(x_n)}}, \tag{3.44}$$

where we choose the sign in the denominator to make x_{n+1} closer to x_n, i.e., to maximize the magnitude of the denominator.

The second method is a bit more subtle. Define the rational function

$$R(x) = \frac{a - x}{bx + c}$$

---

[11]Edmund Halley (1656–1742) was born near London and educated at Oxford. He is most famous for analyzing the orbit of the comet of 1682 and predicting that it would return in 76 years, which it did. To this day the comet is known as Halley's Comet.

Although an astronomer and not a mathematician, Halley had a great influence on the development of mathematics because he encouraged Newton to finally write up and publish his work applying the calculus to problems of planetary and celestial motion, which led to the writing of *Principia Mathematica*, originally published in 1687 with funds from Halley's own pocket. The methods for solving equations that bear his name date from a paper written for the Royal Society of London, in 1684. The derivation used here is different from that used by Halley.

and note that $R(x) = 0 \Rightarrow x = a$. Now, define a, b, and c so that

$$R(x_n) = f(x_n), \qquad R'(x_n) = f'(x_n), \qquad R''(x_n) = f''(x_n).$$

Thus we have defined R as the rational function that matches f and its first two derivatives at x_n (essentially, R is a "rational Taylor approximation" to f); we thus expect $R \approx f$ near x_n, and we can use the (easily computable) root of R to approximate the root of f. We have a system of three equations in the three unknowns a, b, and c, which we can solve to get

$$a = x - \frac{2f(x)f'(x)}{2[f'(x)]^2 - f(x)f''(x)}.$$

This is the only one that matters; we thus have the iteration

$$x_{n+1} = x_n - \frac{2f(x_n)f'(x_n)}{2[f'(x_n)]^2 - f(x_n)f''(x_n)}. \tag{3.45}$$

Both of these methods converge with order $p = 3$, which can be proved by application of Theorem 3.7. (See Exercise 9.) Table 3.9 shows the results of using each of these iterations to find the root of $f(x) = 2 - e^x$. While this convergence is indeed extremely fast, bear in mind that both versions of Halley's method require a second derivative function, and that each iteration involves *three* function evaluations, whereas Newton's method requires only two and the secant method only one.

3.10.3 The Secant Method: Theory and Convergence

In this section we will establish the error theory for the secant method. The result here is similar to that for Newton's method, but somewhat more difficult to establish.

Theorem 3.8   Let f be twice continuously differentiable in a neighborhood of a root α, and assume that $f'(x) \neq 0$ for all x in this neighborhood. Then, for x_0 and x_1 sufficiently close to α, the secant iteration (3.30) converges to α, with

$$\lim_{n \to \infty} \frac{\alpha - x_{n+1}}{\alpha - x_n} = 0 \tag{3.46}$$

TABLE 3.9   **Halley's Method Examples, $f(x) = 2 - e^x$**

n	x_n, via (3.43)	Error	x_n, via (3.42)	Error
0	0.00000000	0.693147180560E+00	0.00000000	0.693147180560E+00
1	0.66666667	0.264805138933E−01	0.73205081	0.389036270089E−01
2	0.69314563	0.154727507107E−05	0.69313707	0.101121327991E−04
3	0.69314718	0.000000000000Ee+00	0.69314718	0.222044604925E−15
4	0.69314718	0.000000000000E+00	0.69314718	0.000000000000E+00

and

$$\lim_{n\to\infty} \frac{\alpha - x_{n+1}}{(\alpha - x_n)^p} = \left(\frac{1}{2}\frac{f''(\alpha)}{f'(\alpha)}\right)^{p-1} \tag{3.47}$$

for $p = (1+\sqrt{5})/2 \approx 1.618\ldots$.

Proof   We begin by establishing an error formula for the secant method similar to the one for Newton's method.

Recall that the secant method is based on finding the exact root of a linear polynomial that interpolates the original function f. Let s be this interpolating polynomial, so that we have

$$s(x) = f(x_n) + (x - x_n)\left(\frac{f(x_n) - f(x_{n-1})}{x_n - x_{n-1}}\right), \tag{3.48}$$

where the next iterate is defined by solving $0 = s(x_{n+1})$. Then the linear interpolation error estimate (2.12) implies that

$$f(x) - s(x) = \frac{1}{2}(x - x_n)(x - x_{n-1})f''(\xi_n)$$

for all x; here ξ_n is in the interval defined by x_n, x_{n-1}, and α. Set $x = \alpha$ so that we have

$$f(\alpha) - s(\alpha) = \frac{1}{2}(\alpha - x_n)(\alpha - x_{n-1})f''(\xi_n).$$

But $f(\alpha) = 0$, so we have

$$-s(\alpha) = \frac{1}{2}(\alpha - x_n)(\alpha - x_{n-1})f''(\xi_n).$$

On the other hand, we know that $s(x_{n+1}) = 0$, so we can substitute this in to get

$$s(x_{n+1}) - s(\alpha) = \frac{1}{2}(\alpha - x_n)(\alpha - x_{n-1})f''(\xi_n).$$

The Mean Value Theorem then implies that a value η_n exists between α and x_{n+1} such that $s(x_{n+1}) - s(\alpha) = s'(\eta_n)(x_{n+1} - \alpha)$; therefore,

$$s'(\eta_n)(x_{n+1} - \alpha) = \frac{1}{2}(\alpha - x_n)(\alpha - x_{n-1})f''(\xi_n)$$

or

$$x_{n+1} - \alpha = \frac{1}{2}(\alpha - x_n)(\alpha - x_{n-1})\frac{f''(\xi_n)}{s'(\eta_n)}. \tag{3.49}$$

But we can go directly to (3.48) to show that (again appealing to the MVT) there is a θ_n between x_n and x_{n-1} such that

$$s'(\eta_n) = \frac{f(x_n) - f(x_{n-1})}{x_n - x_{n-1}} = f'(\theta_n).$$

Thus, we have

$$x_{n+1} - \alpha = \frac{1}{2}(\alpha - x_n)(\alpha - x_{n-1})\frac{f''(\xi_n)}{f'(\theta_n)}. \tag{3.50}$$

Note that with this relationship in hand, it quickly follows that, if the secant iterates x_n are converging to α, then

$$\lim_{n \to \infty} \frac{\alpha - x_{n+1}}{\alpha - x_n} = -\frac{1}{2} \lim_{n \to \infty} (\alpha - x_{n-1}) \frac{f''(\alpha)}{f'(\alpha)} = 0.$$

Thus we have (3.46). Finishing the proof then requires two steps: (1) Establish that convergence occurs, if the initial points are close enough. (2) Show that the limit (3.47) holds.

Convergence: This part of the proof follows the Newton proof very closely. Let I be the neighborhood of α in which f is twice continuously differentiable, etc. As in the Newton theorem, we can find a closed interval $J \subset I$ such that $J = \{x | \alpha - \epsilon \le x \le \alpha + \epsilon\}$ for some $\epsilon > 0$, and the ratio

$$M = \frac{\max_{x \in J} |f''(x)|}{2 \min_{x \in J} |f'(x)|}$$

is bounded. Take x_0 and x_1 in J so that

$$|\alpha - x_i| \le \delta < \min\{\epsilon, 1\}, \qquad i = 0, 1,$$

and $M\delta < 1$. Note that this is possible simply by taking both initial points close enough to α. Therefore, from (3.50),

$$|\alpha - x_2| \le M |\alpha - x_1| |\alpha - x_0| \le |\alpha - x_0|,$$

so that $x_2 \in J$ is assured. The same argument shows that each secant iterate $x_n \in J$; thus we have ($e_k = |\alpha - x_k|$):

$$e_2 \le e_1 e_0 M \le M\delta^2$$
$$e_3 \le e_2 e_1 M \le (e_1 e_0 M) e_1 M = e_1^2 e_0 M^2 < M^2 \delta^3$$
$$e_4 \le e_3 e_2 M \le (e_1^2 e_0 M^2)(e_1 e_0 M) M = e_1^3 e_0^2 M^4 < M^4 \delta^5,$$

and an inductive argument easily shows that

$$e_n \le M^{-1} (M\delta)^{q_n}, \tag{3.51}$$

where q_n is the sequence of *Fibonnaci numbers*,[12] defined by

$$q_{n+1} = q_n + q_{n-1}, \quad q_1 = q_0 = 1.$$

---

[12]Leonardo Pisano (1170–1250) is more widely known by his nickname, Fibonacci. He traveled widely in the Mediterranean region with his father, a diplomat in the service of Pisa. Fibonacci wrote a number of texts, the most famous of which is probably *Liber abbaci*, which appeared in 1202. This contains the famous problem which leads to the Fibonacci sequence: "A certain man put a pair of rabbits in a place surrounded on all sides by a wall. How many pairs of rabbits can be produced from that pair in a year if it is supposed that every month each pair begets a new pair which from the second month on becomes productive?"

Fibonacci also dabbled in geometry and number theory, but his most significant work may have been as simple as spreading the use of the Hindu-Arabic system of numeration.

Since $M\delta < 1$ and $q_n \to \infty$, it follows immediately that $e_n \to 0$ and therefore x_n converges to α.

Order of convergence: The order of convergence is defined as the value p such that

$$\lim_{n\to\infty} \frac{\alpha - x_{n+1}}{(\alpha - x_n)^p} = C \tag{3.52}$$

for $C \neq 0$ and finite. Shifting the limit variable yields the equivalent statement

$$\lim_{n\to\infty} \frac{\alpha - x_n}{(\alpha - x_{n-1})^p} = C \tag{3.53}$$

or, taking powers of both sides,

$$\lim_{n\to\infty} \frac{(\alpha - x_n)^p}{(\alpha - x_{n-1})^{p^2}} = C^p. \tag{3.54}$$

Multiplying (3.52) and (3.54) then yields that p must satisfy

$$\lim_{n\to\infty} \frac{\alpha - x_{n+1}}{(\alpha - x_{n-1})^{p^2}} = C^{1+p}. \tag{3.55}$$

Now, at the same time, we note that the convergence of the iterates plus the error estimate (3.50) implies

$$\lim_{n\to\infty} \frac{\alpha - x_{n+1}}{(\alpha - x_n)(\alpha - x_{n-1})} = -\frac{f''(\alpha)}{2f'(\alpha)}, \tag{3.56}$$

which, when combined with (3.53), yields

$$\lim_{n\to\infty} \frac{\alpha - x_{n+1}}{(\alpha - x_{n-1})^{1+p}} = -C\frac{f''(\alpha)}{2f'(\alpha)}. \tag{3.57}$$

Since the order of convergence is unique,[13] it follows from comparing (3.55) and (3.57) that

$$1 + p = p^2$$

and

$$C^{1+p} = -C\frac{f''(\alpha)}{2f'(\alpha)}.$$

Solving these two yields

$$p = \frac{1}{2}(1 + \sqrt{5})$$

---

[13] We really ought to prove this, but it is not within the intended scope of the book.

and

$$C = \left(-\frac{f''(\alpha)}{2f'(\alpha)}\right)^{1/p},$$

which completes the proof. $\qquad\square$

3.10.4 Multiple Roots

So far our study of root-finding methods has assumed that the derivative of the function does not vanish at the root: $f'(\alpha) \neq 0$. The rapid convergence of Newton's method and the secant method both depend on this assumption. In this section we investigate what happens if the derivative *does* vanish at the root. We will see that both the Newton and, secant methods will continue to converge, but not as rapidly as we expect. Furthermore, there are other problems with trying to find a root when the derivative vanishes.

We start by pointing out some consequences of having $f'(\alpha) = 0$. Geometrically, this means that the graph of the function is tangent to the axis at the root.

Lemma 3.1   *If f is k times continuously differentiable in a neighborhood of α, and*

$$f(\alpha) = f'(\alpha) = \cdots = f^{(k-1)}(\alpha) = 0,$$

i.e., if the function and first $k-1$ derivatives vanish at α, but $f^{(k)}(\alpha) \neq 0$, then we can write

$$f(x) = (x-\alpha)^k F(x) \qquad (3.58)$$

where $F(\alpha) \neq 0$. Similarly, if we can write f in the form (3.58), where $F(\alpha) \neq 0$, then it follows that the first $k-1$ derivatives vanish at α.

Proof   See Exercise 17. $\qquad\square$

If $k = 2$, so that both the function and the first derivative are zero at the root, then α is called a "double root." If more derivatives are zero, then the root is generally called a "multiple root."

EXAMPLE 3.13

Let $f(x) = \cos^2 x$, which has a root at $x = \frac{1}{2}\pi$. Since $f'(x) = -2\sin x \cos x$, it follows that the derivative also vanishes at $x = \frac{1}{2}\pi$, so f has a double root. We can write f in the form called for in the lemma by the simple device of writing

$$f(x) = \cos^2 x = \left(x - \frac{\pi}{2}\right)^2 F(x),$$

where

$$F(x) = \frac{\cos^2 x}{(x - \pi/2)^2}.$$

So long as $x \neq \pi/2$, $F(x)$ is well defined. What happens at $x = \pi/2$? We can use L'Hôpital's rule to determine that

$$\lim_{x \to \pi/2} F(x) = \lim_{x \to \pi/2} \frac{-2\sin x \cos x}{2}(x - \pi/2) = \lim_{x \to \pi/2} \frac{-2\cos^2 x + 2\sin^2 x}{2} = 1.$$

Thus we would define

$$F(x) = \begin{cases} \dfrac{\cos^2 x}{(x - \pi/2)^2} & x \neq \pi/2 \\ 1 & x = \pi/2. \end{cases}$$

■

We might want to continue our study by conducting some experiments. Consider the function $f(x) = 1 - xe^{1-x}$. Note that $\alpha = 1$ is a root, and that $f'(\alpha) = 0$ as well. If we apply Newton's method and the secant method to this example, with $x_0 = 0$ for Newton and secant, and $x_{-1} = -1$ for secant, we get the results shown in Table **??**.

The data suggests that both iterations are converging, but neither one is converging as rapidly as we might have expected, based on previous results. Can we explain this?

Clearly, the fact that $f'(\alpha) = 0$ will have an effect on both the Newton and secant methods. The error formulas (3.12) and (3.50) and limits (3.24) and (3.47) all require that $f'(\alpha) \neq 0$ in order to hold. Can we find anything more in the way of an explanation?

Theorem 3.7 tells us when a fixed-point iteration exhibits quadratic convergence, and we applied it in Section 3.9 to Newton's method. However, that exercise was done under the assumption that $f'(\alpha) \neq 0$. Let us repeat that exercise, this time under the assumption that f has a double root and therefore

$$f(x) = (x - \alpha)^2 F(x).$$

Then the iteration function for Newton's method is

$$g(x) = x - \frac{f(x)}{f'(x)} = x - \frac{(x - \alpha)^2 F(x)}{2(x - \alpha)F(x) + (x - \alpha)^2 F'(x)} = x - \frac{(x - \alpha)F(x)}{2F(x) + (x - \alpha)F'(x)}$$

and we can compute (many details skipped; see Exercise 21)

$$g'(\alpha) = 1 - \frac{1}{2} = \frac{1}{2} \neq 0. \tag{3.59}$$

Note that we no longer have $g'(\alpha) = 0$; therefore (according to Theorem 3.7), we no longer have quadratic convergence. The simple fact that the function is tangent to the axis at the root costs us our rapid convergence. (Note that we still have linear convergence, based on Theorem 3.6.)

Note also that (3.59) gives us a clue as to how to recover the lost quadratic convergence. If we change the Newton iteration to be

$$x_{n+1} = x_n - 2\frac{f(x_n)}{f'(x_n)}$$

TABLE 3.10 Double-Root Example, $f(x) = 1 - xe^{1-x}$. (Computation Stopped after 20 Iterations)

	Newton			Secant		
n	x_n	$\alpha - x_n$	$\log_{10}|\alpha - x_n|$	x_n	$\alpha - x_n$	$\log_{10}|\alpha - x_n|$
0	0.000000000000	1.000000000000	0.0000	0.000000000000	1.000000000000	0.0000
1	0.367879441171	0.632120558829	-0.1992	0.135353283237	0.864664716763	-0.0632
2	0.626665778573	0.373334221427	-0.4279	0.421192747824	0.578807252176	-0.2375
3	0.792118936917	0.207881063083	-0.6822	0.586457122517	0.413542877483	-0.3835
4	0.889216318667	0.110783681333	-0.9555	0.724548464295	0.275451535705	-0.5600
5	0.942618075030	0.057381924970	-1.2412	0.817997254576	0.182002745424	-0.7399
6	0.970768039664	0.029231960336	-1.5341	0.882864790983	0.117135209017	-0.9313
7	0.985242636648	0.014757363352	-1.8310	0.925476813313	0.074523186687	-1.1277
8	0.992585155212	0.007414844788	-2.1299	0.953105630395	0.046894369605	-1.3289
9	0.996283431246	0.003716568754	-2.4299	0.970673923815	0.029326076185	-1.5327
10	0.998139415613	0.001860584387	-2.7304	0.981742402042	0.018257597958	-1.7386
11	0.999069131113	0.000930868887	-3.0311	0.988663870449	0.011336129551	-1.9455
12	0.999534421170	0.000465578830	-3.3320	0.992973792390	0.007026207610	-2.1533
13	0.999767174462	0.000232825538	-3.6330	0.995649804081	0.004350195919	-2.3615
14	0.999883578197	0.000116421803	-3.9340	0.997308460142	0.002691539858	-2.5700
15	0.999941786839	0.000058213161	-4.2350	0.998335397262	0.001664602738	-2.7787
16	0.999970892852	0.000029107148	-4.5360	0.998970783260	0.001029216740	-2.9875
17	0.999985446281	0.000014553719	-4.8370	0.999363742391	0.000636257609	-3.1964
18	0.999992723105	0.000007276895	-5.1381	0.999606707485	0.000393292515	-3.4053
19	0.999996361541	0.000003638459	-5.4391	0.999756907516	0.000243092484	-3.6142
20	0.999998180783	0.000001819217	-5.7401	0.999849751283	0.000150248717	-3.8232

then an application of Theorem 3.7 shows that we now have $g'(\alpha) = 0$ and therefore we again have quadratic convergence for Newton's method. More generally, it is not hard to show that, for a k-fold root, we would use

$$x_{n+1} = x_n - k\frac{f(x_n)}{f'(x_n)} \tag{3.60}$$

as the (modified) Newton iteration.

The problem with this technique is that it requires that we know the degree of multiplicity of the root ahead of time, and this is generally not the case in practice. So an alternative is needed. Given f, with a multiple root of unknown order at $x = \alpha$, define

$$u(x) = \frac{f(x)}{f'(x)}. \tag{3.61}$$

It can be shown (see Exercise 19) that u has only a single root at $x = \alpha$. Thus Newton's method applied to find a root of u will avoid any problems of multiple roots. The drawback of this method is that applying Newton's method to u will require that we have a formula for the *second* derivative of f.

Since the secant method is not a single-point iteration, it does not fit into the theory of Section 3.9, and so we have no way to rigorously analyze its behavior in the presence of multiple roots. Clearly, the results of Table **??** suggest that some degradation of performance does occur; this is not surprising, since the secant method is based on approximating Newton's method. However, it is worth noting that we can replace $f(x)$ with $u(x) = f(x)/f'(x)$ in the secant method as well as in Newton's method. This removes the multiple root, and with the secant method we need only the additional function for the first derivative.

Finally, we might note that the linear convergence that Theorem 3.9 predicts for Newton's method in the presence of multiple roots can be dealt with by using Aitken extrapolation.

To illustrate some of these ideas, we return to our example function $f(x) = 1 - xe^{1-x}$. In Table 3.11 we show the results of applying several different algorithms to this example:

1. Newton's method as modified in (3.60); this is x_n in the table
2. Newton's method based on the change of function (3.61); this is y_n in the table
3. Newton's method with Aitken extrapolation; this is z_n in the table
4. The secant method based on the change of function (3.61); this is s_n in the table

Some comments might be in order here. Note that all four modified methods converge to near the root $\alpha = 1$ much faster than did the original Newton and secant iterations in Table **??**. Note, however, that there appears to be a limit to how accurate any of these iterations can get, and this accuracy is not as good as in the past. What is going on?

There is a problem inherent in the geometry of a multiple root that makes it difficult—if not outright impossible—to approximate the root to high accuracy. To illustrate the problem, let's look at a graph of the polynomial

$$p(x) = x^5 - (5/2)x^4 + (5/2)x^3 - (5/4)x^2 + (5/16)x - (1/32) = \left(x - \frac{1}{2}\right)^5$$

TABLE 3.10   **Multiple Root Computations for $f(x) = 1 - xe^{1-x}$**

	Modified Newton's method (3.60)		Newton's method for $u(x) = f(x)/f'(x)$	
n	x_n	$\log_{10}(\alpha - x_n)$	y_n	$\log_{10}(\alpha - y_n)$
0	0.000000000000E+00	0.000	0.000000000000E+00	0.000
1	0.735758882343E+00	−0.578	0.139221119118E+01	−0.406
2	0.978185253678E+00	−1.661	0.104801490853E+01	−1.319
3	0.999842233626E+00	−3.802	0.100076234718E+01	−3.118
4	0.999999991704E+00	−8.081	0.100000019370E+01	−6.713
5	0.999999991704E+00	−8.081	0.999999999938E+00	−10.208
6	0.999999991704E+00	−8.081	0.999999999876E+00	−9.907
7	0.999999991704E+00	−8.081	0.999999999752E+00	−9.606
8	0.999999991704E+00	−8.081	0.999999999752E+00	−9.606
9	0.999999991704E+00	−8.081	0.999999999752E+00	−9.606
10	0.999999991704E+00	−8.081	0.999999999752E+00	−9.606

	Newton's method with Aitken acceleration		Secant method for $u(x) = f(x)/f'(x)$	
n	z_n	$\log_{10}(\alpha - z_n)$	s_n	$\log_{10}(\alpha - s_n)$
0	0.000000000000E+00	0.000	0.000000000000E+00	0.000
1	0.124054847245E+01	−0.619	0.677393774677e+00	−0.491
2	0.100856591658E+01	−2.067	0.111944928199E+01	−0.923
3	0.100001218112E+01	−4.914	0.986957103091E+00	−1.885
4	0.100000000001E+01	−11.147	0.999485342581E+00	−3.288
5	0.100000000001E+01	−11.147	0.100000224007E+01	−5.650
6	0.100000000001E+01	−11.147	0.999999999674E+00	−9.487
7	0.100000000001E+01	−11.147	0.100000052208E+01	−6.282
8	0.100000000001E+01	−11.147	0.100000029546E+01	−6.530
9	0.100000000001E+01	−11.147	0.999999999965E+00	−10.456
10	0.100000000001E+01	−11.147	0.999999999965E+00	−10.456

in a very narrow region around the root $\alpha = \frac{1}{2}$. Figure 3.11 shows a plot of 8000 points from this curve on the interval $[0.45, 0.55]$, computed in floating-point arithmetic. Note that we do not have a smooth curve but rather a "cloud" of values, distributed *along* the x axis. Thus, there are many values of x near the exact root that yield very small values of $p(x)$, and all of the iterations we are using will yield $x_{n+1} - x_n$ very small if $f(x_n)$ is very small. So, once the iteration lands on one of the points that produces a small function value, it essentially stays there, even if it is not all that close to the exact root. Thus we get "premature convergence" to a value that is not a good approximation to the root. This is not a flaw in the root-finding method; it is a difficulty inherent in the problem of finding a root when the derivative is also zero at the root, using finite-precision arithmetic. We should note that this effect does depend on the way that the polynomial is computed; see Exercise 25.

FIGURE 3.11   Close look near the root of the polynomial
$p(x) = x^5 - (5/2)x^4 + (5/2)x^3 - (5/4)x^2 + (5/16)x - (1/32) = \left(x - \frac{1}{2}\right)^5$, computed in floating-point arithmetic.

What this tells us is that we need to be very careful how we organize the computations when working near a multiple root, and we need to work in as high a precision as possible when doing so, or simply accept that our accuracy is going to be less than we are used to for simple roots.

3.10.5   In Search of Fast Global Convergence: Hybrid Algorithms

We began this chapter with the bisection method, which is perhaps like the tortoise in Aesop's fable: slow but steady and reliable. We then looked at Newton's method and the secant method, which are perhaps both like the rabbit: fast but potentially unreliable. Does there exist a fast *and* reliable method for finding roots of a function?

There is no such "single algorithm"; however, we can create one by combining the features that make the bisection method reliable with the features that make the Newton and secant methods fast. The basic ideas are incorporated, in a somewhat more sophisticated fashion than we present here, into an algorithm known as Brent's algorithm.

Given a function f and an interval $[a,b]$, with $f(a)f(b) < 0$, so we know that a root exists in the interval, we begin by doing the secant iteration, using the endpoints of the interval as our initial guesses. So long as the secant iterates are within the bracketing interval, we continue with the secant iteration. At the same time, we use the result of each secant iteration to update the bracketing interval so that it shrinks as much as possible. Finally, if the secant iterate ever leaves the bracketing interval, then we ignore that result, and do bisection in order to get closer to the root before restarting the secant iteration.

Formally, we have the sequence of steps shown in Algorithm 3.6.

Note that while this algorithm does maintain a set of brackets around the root, it is using the secant iterates x_k for the secant step, *not* the bracket values. This is what makes this method different from regula falsi.

Algorithm 3.6 *Hybrid algorithm.*

Given an initial interval $[a,b] = [a_1,b_1]$, with $f(a_1)f(b_1) < 0$, set $k = 1$, $x_0 = a_1$, $x_1 = b_1$, and proceed as follows:

1. Set $c = x_k - f(x_k)(x_k - x_{k-1})/(f(x_k) - f(x_{k-1}))$

2. If $c < a_k$ or $c > b_k$, then (do bisection)

   (a) Set $c = a_k + \frac{1}{2}(b_k - a_k)$

   (b) If $f(a_k)f(c) < 0$, then

       i. $[a_{k+1},b_{k+1}] = [a_k,c]$

       ii. Set $x_{k+1} = c$, $x_k = a_k$

   (c) If $f(c)f(b_k) < 0$, then

       i. $[a_{k+1},b_{k+1}] = [c,b_k]$

       ii. Set $x_{k+1} = c$, $x_k = b_k$

3. Else (update brackets for secant step)

   (a) If $f(a_k)f(c) < 0$, then $[a_{k+1},b_{k+1}] = [a_k,c]$

   (b) If $f(c)f(b_k) < 0$, then $[a_{k+1},b_{k+1}] = [c,b_k]$

   (c) Set $x_{k+1} = c$

4. Set $k = k+1$ and go to step 1.

EXAMPLE 3.14

This is a sufficiently complicated algorithm that it is worth a close look, via an example. Let $f(x) = 2xe^{-15} - 2e^{-15x} + 1$. It is a simple matter to compute $f(0) = -1$ and $f(1) = 1$; thus a root exists on the interval $[0,1]$.

1: The secant prediction is $c = 0.5$, which is indeed in the bracketed interval $[0,1]$, so we accept the secant step and update as follows:

$$f(c) = 0.99889\ldots \Rightarrow [a,b] \leftarrow [0,c] = [0,0.5]$$

$$x_2 = 0.5$$

$$x_1 = 1.0.$$

2: The secant prediction this time is

$$c = x_2 - f(x_2)(x_2 - x_1)/(f(x_2) - f(x_1)) = -0.4511 \times 10^3,$$

which is clearly outside the interval $[0, 0.5]$, so we do a step of bisection, and update as follows:

$$c = 0.25$$
$$f(c) = 0.95296\dots \qquad \Rightarrow \qquad [a,b] \leftarrow [0,c] = [0, 0.25]$$
$$x_3 = 0.25$$
$$x_2 = 0.0.$$

Note that in the bisection step, x_2 is redefined to be at the other end of the interval as x_3; this ensures that we start the next iteration with the two most recent (hence most accurate) approximations that "bracket" the root.   ■

The exercises ask the student to continue this computation by hand.

Consider another example function

$$f(x) = \frac{20x - 1}{19x}.$$

The root is at $x = 0.05$, of course, and we must have $x_n \approx 0.07$ before the secant method will begin to converge. Table 3.11 shows the result of applying our algorithm to this function, with $[a,b] = [0.01, 1]$ and $x_0 = a$, $x_1 = b$. The last column shows which algorithm was used to get the current approximate root (S = secant, B = bisection). This is not quite the same as Brent's algorithm, but it is in the same spirit and omits some of the details.

TABLE 3.11   Hybrid Algorithm Applied to $f(x) = 20x - 1/19x$.

n	x_n	$f(x_n)$	$\alpha - x_n$	$\log_{10}|\alpha - x_n|$	a	b	Method
1	0.81000000	0.98765432	$-0.7600000E+00$	-0.119	0.010	0.810	S
2	0.41000000	0.92426187	$-0.3600000E+00$	-0.444	0.010	0.410	B
3	0.33800000	0.89691685	$-0.2880000E+00$	-0.541	0.010	0.338	S
4	0.17400000	0.75015124	$-0.1240000E+00$	-0.907	0.010	0.174	B
5	0.14920000	0.69987301	$-0.9920000E-01$	-1.003	0.010	0.149	S
6	0.07960000	0.39143084	$-0.2960000E-01$	-1.529	0.010	0.080	B
7	0.07368000	0.33830505	$-0.2368000E-01$	-1.626	0.010	0.074	S
8	0.03598144	-0.41011085	$0.1401856E-01$	-1.853	0.036	0.074	S
9	0.05663919	0.12338844	$-0.6639190E-02$	-2.178	0.036	0.057	S
10	0.05186144	0.03778160	$-0.1861438E-02$	-2.730	0.036	0.052	S
11	0.04975283	-0.00522940	$0.2471688E-03$	-3.607	0.050	0.052	S
12	0.05000920	0.00019369	$-0.9201785E-05$	-5.036	0.050	0.050	S
13	0.05000005	0.00000096	$-0.4548788E-07$	-7.342	0.050	0.050	S
14	0.05000000	0.00000000	$0.8371401E-11$	-11.077	0.050	0.050	S

EXERCISES

1. Consider the fixed-point iteration $x_{n+1} = 1 + e^{-x_n}$, with $x_0 = 0$. Do four steps (by hand) and apply both of the Aitken acceleration algorithms to speed up the convergence of the iteration.

2. Repeat the preceeding for the iteration $x_{n+1} = \frac{1}{2}\ln(1 + x_n)$, using $x_0 = 1/2$.

3. Consider the iteration $x_{n+1} = e^{-x}$. Show (by computational experiment) that it converges for $x_0 = 1/2$; then use both Aitken acceleration algorithms to speed up the convergence. What is the gain in terms of function calls to the iteration function?

4. Write the second Aitken iteration in the form

$$x_{k+1} = G(x_k)$$

(Hint: Take

$$x_1 = g(x_0), \quad x_2 = g(x_1) = g(g(x_0))$$

and use this to write the updated value of x0 in Algorithm 3.5 entirely in terms of x0, $g(x_0)$, and $g(g(x_0))$.)

5. Consider the iteration

$$x_{k+1} = G(x_k),$$

where

$$G(x) = g(g(x)) + \frac{H(x)}{1 - H(x)}(g(g(x)) - g(x))$$

for

$$H(x) = \frac{g(g(x)) - g(x)}{g(x) - x}.$$

Assume that $\alpha = g(\alpha)$ is a fixed point for g.
(a) Use L'Hôpital's Rule to show that

$$H(\alpha) = g'(\alpha).$$

(b) Use Theorem 3.7 to show that the fixed-point iteration for G is quadratic.

6. Apply the chord method to each of the functions in Exercise 3 of Section 3.1, using the midpoint of the interval as x_0. Compare your results with what you got for bisection, Newton's method, and/or the secant method.

7. Repeat the preceding, except this time apply both Aitken acceleration algorithms to improve the conver-

gence of the iteration. Compute the values of the ratio

$$R_n = \frac{\alpha - x_{n+1}}{(\alpha - x_n)^2},$$

where α is the best value for the root, as found by previous methods. Does the sequence of R_n values appear to be converging to a limit? What does this tell you?

8. Repeat Exercise 6, but this time update the value used to compute the derivative in the chord method every other iteration. Comment on your results.

9. Show that both Halley method iterations are of third order. (Hint: Write them as fixed-point iterations

$$x_{n+1} = g(x_n)$$

and write the iteration function as $g(x) = f(x)G(x)$. The fact that $f(\alpha) = 0$ will make the derivative computations easier.)

10. Apply the Halley iteration (3.45) to $f(x) = x^2 - a$, $a > 0$, to derive a cubic method for finding the square root. Test this by using it to find $\sqrt{5}$.

11. Repeat the preceding for $f(x) = x^N - a$. Test this out by finding the cubic, quartic, and quintic roots of 2, 3, 4, and 5.

12. Apply Halley's second method to $f(x) = a - e^x$ to show that

$$x_{n+1} = x_n + 2\left(\frac{ae^{-x_n} - 1}{ae^{-x_n} + 1}\right)$$

is a cubic convergent iteration for $\alpha = \log a$. Analyze the convergence of this iteration, under the assumption that $a \in [\frac{1}{2}, 1]$. Comment on this as a practical means of computing the logarithm of an arbitrary a.

13. Use the secant method with $x_0 = 0$ and $x_1 = 1$ to find the root of $f(x) = 2 - e^x$, which has exact solution $\alpha = \ln 2$. Compute the ratio

$$R = \frac{\alpha - x_{n+1}}{(\alpha - x_n)^p}$$

for $p = (1 + \sqrt{5})/2$. Do we get convergence to the appropriate value?

14. Repeat the previous experiment, except use values of p just a bit above and below the correct value. What happens?

15. Consider applying the secant method to find the root of the function $f(x) = 4x - \cos x$. Assume we want accuracy to within 10^{-8}. Use the secant error estimate (3.50) to show that the iteration will converge for all $x_0 \in [-2, 2]$. How many iterations will be needed for the iteration to converge?

16. Verify by actual computation that your results in the previous problem were correct, i.e., apply the secant method to $f(x) = 4x - \cos x$. Do your results converge to the specified accuracy in the correct number of iterations, and does this convergence appear to be occurring for all choices of x_0?

17. Prove Lemma 3.1. (Hint: Expand f in a Taylor series about α.)

18. Let $f(x) = 1 - xe^{1-x}$. Write this function in the form (3.58). What is $F(x)$? Use Taylor's Theorem or L'Hôpital's Rule to determine the value of $F(\alpha)$.

19. For $u(x)$ as defined in (3.61), where f has a k-fold root at $x = \alpha$, show that $u(\alpha) = 0$ but $u'(\alpha) \neq 0$.

20. Use the modified Newton method (3.60) to find the root $\alpha = 1$ for the function $f(x) = 1 - xe^{1-x}$. Is the quadratic convergence recovered?

21. Let f have a double root at $x = \alpha$. Show that the Newton iteration function

$$g(x) = x - f(x)/f'(x)$$

is such that $g'(\alpha) \neq 0$. Provide all details of the calculation.

22. Using a hand calculator, carry out six iterations of the hybrid method for the function

$$f(x) = 2xe^{-15} - 2e^{-15x} + 1,$$

using $[a_0, b_0] = [0, 1]$. You should be able to match the values generated for x_2 and x_3 in the text. In addition, $x_6 = 0.0414641$.

23. Write a computer code to implement the hybrid method described in Section 3.10.5. Apply it to each of the following functions, using the given interval as the starting interval. On your output, be sure to indicate which iteration method was used at each step.
(a) $f(x) = x^{1/19} - 19^{1/19}$, $[1, 100]$
(b) $f(x) = x^2 - (1 - x)^{20}$, $[0, 1]$
(c) $f(x) = 2xe^{-5} - 2e^{-5x} + 1$, $[0, 1]$
(d) $f(x) = 2xe^{-20} - 2e^{-20x} + 1$, $[0, 1]$

24. In Exercise 9 of Section 3.1 we introduced the regula falsi method. In this problem we demonstrate how the hybrid method (Algorithm 3.6) is different from regula falsi. Let

$$f(x) = e^{-4x} - \frac{1}{10}.$$

(a) Show that f has a root in the interval $[0, 5]$.
(b) Using this interval as the starting interval, compute five iterations of regula falsi; be sure to tabulate the new interval in addition to the new approximate root value predicted by the method at each step. A graph showing the location of each approximate root might be useful.
(c) Using the same interval as the starting interval, compute five iterations of the hybrid method given in this section. Again, note the new interval as well as the new approximate root value at each step. A graphical illustration might be instructive.
(d) Based on your results, comment on the difference between regula falsi and the hybrid method.

25. Write a computer program that evaluates the polynomial $p(x) = (x - 1)^5$ using the following three forms for the computation:

$$p(x) = (x - 1)^5$$
$$p(x) = x^5 - 5x^4 + 10x^3 - 10x^2 + 5x - 1$$
$$p(x) = -1 + x(5 + x(-10 + x(10 + x(-5 + x)))).$$

Use this to evaluate the polynomial at 400 equally spaced points on the interval $[0.998, 1.002]$. Comment on the results.

LITERATURE AND SOFTWARE DISCUSSION

The problem of numerically solving a single nonlinear equation is an old one in numerical analysis, with a fairly rich and complete literature. In addition to the treatments in almost any decent numerical analysis text, the monographs by Ostrowski [7], Traub [8], Householder [5], and Wait [9] are worth mentioning for more in-depth discussions of many of the issues.

The standard "state-of-the-art" algorithm is known as Brent's algorithm. The original reference is [4] (which leads some to call it Dekker's method, or the Brent-Dekker method), and there is a good discussion in Atkinson's text [2]. Brent gave an algorithm in [3]. It has been suggested in at least one place [1] that Brent's method might be improved. An implementation of Brent's method can be found in the MATLAB function `fzero`.

The Brent-Dekker algorithm is a hybrid method, combining bisection, the secant method, and inverse quadratic interpolation (see Chapter 4) to obtain a very fast and nearly global method.

The reader might wonder why we did not treat the problem of polynomial root finding. There are a number of specialized methods for polynomials, but the state-of-the-art method, due to Jenkins and Traub [6], is based on finding the *eigenvalues* of a *matrix* whose characteristic polynomial is the same as the given polynomial; thus it was judged inappropriate to cover that material here.

BIBLIOGRAPHY

[1] ALEFELD, G.E., POTRA, F.A., & SHI, Y. "Enclosing zeroes of continuous functions," *ACM Transactions on Mathematical Software*, vol. 21 (1995), no. 3, pp. 327–344.

[2] ATKINSON, KENDALL. *An Introduction to Numerical Analysis*. John Wiley & Sons, New York, 1987 (2nd ed).

[3] BRENT, RICHARD. *Algorithms for Minimization without Derivatives*. Prentice-Hall, Englewood Cliffs, NJ, 1973.

[4] DEKKER, T. "Finding a zero by means of successive linear interpolation." In *Constructive Aspects of the Fundamental Theorem of Algebra*. Dejon and Henrici eds. John Wiley & Sons, New York, 1969.

[5] HOUSEHOLDER, ALSTON. *The Numerical Treatment of a Single Nonlinear Equation*. McGraw-Hill, New York, 1970.

[6] JENKINS, M., & TRAUB, J. "A three-stage variable shift algorithm for polynomial zeros and its relation to generalized Rayleigh iteration." *Numerische Mathematik*d,. vol. 14 (1970), pp. 252–263.

[7] OSTROWSKI, ALEXANDER. *Solution of Equations and Systems of Equations*. Academic Press, New York, 1966.

[8] TRAUB, JOSEPH. *Iterative Methods for the Solution of Equations*. Prentice-Hall, Englewood Cliffs, NJ, 1964.

[9] WAIT, R.A. *The Numerical Solution of Algebraic Equations*. John Wiley & Sons, Chichester, 1979.

CHAPTER 4

INTERPOLATION AND APPROXIMATION

The Metallurgy Department at Pennsyltucky Piping has developed a new alloy, which they have modestly named "pennsyltuckium." As they are good scientists and engineers, one of the first things they have done is study the material properties of this new alloy. The following table shows values of the *thermal conductivity* of the alloy, as a function of temperature. (The thermal conductivity indicates the ability of the material to conduct heat. Insulating materials have a very low thermal conductivity; pure metals such as copper have a very high thermal conductivity.) Having worked with pennsyltuckium in the lab extensively, the engineers are confident that the thermal conductivity varies smoothly with the temperature over the range of temperatures in this table, which is the critical range of temperatures for the anticipated applications of pennsyltuckium. However, in designing the new components to be made of pennsyltuckium, it is necessary that the thermal conductivity be computable at *all* temperatures in this range; thus, we need to convert the table of discrete values to a smooth curve.

There are a number of ways to do this, depending on the kind of data available and the specific requirements placed on the resulting curve. The most general and powerful tools fall under the heading of *approximation theory*; simpler but less robust techniques fall under the heading of *interpolation*. In this chapter we will study mostly interpolation processes, although in Section 4.10.2 we do discuss some elements of more general approximation.

Temperature (K), u	200	300	400	500	600	700
Conductivity (W/cm K), k	0.94	0.803	0.694	0.613	0.547	0.487

One of the oldest problems in mathematics—and, at the same time, one of the most applied—is the problem of constructing an approximation to a given function f from among simpler functions, typically (but not always) polynomials. A slight variation of this problem is that of constructing a smooth function from a discrete set of data points.

In this chapter we will study both of these problems and develop several methods for solving them. We start with a more general treatment of an idea we first saw back in Chapter 2.

4.1 LAGRANGE INTERPOLATION

The basic interpolation problem can be posed in one of two ways:

1. Given a set of *nodes* $\{x_i, 0 \leq i \leq n\}$ and corresponding data values $\{y_i, 0 \leq i \leq n\}$, find the polynomial $p_n(x)$ of degree less than or equal to n, such that
$$p_n(x_i) = y_i, \qquad 0 \leq i \leq n.$$

2. Given a set of *nodes* $\{x_i, 0 \leq i \leq n\}$ and a continuous function $f(x)$, find the polynomial $p_n(x)$ of degree less than or equal to n, such that
$$p_n(x_i) = f(x_i), \qquad 0 \leq i \leq n.$$

Note that in the first problem we are trying to fit a polynomial to the data, and in the second we are trying to approximate a given function with the interpolating polynomial.

While the two problems are in fact different, we can always consider the first one a special case of the second (by taking $y_i = f(x_i)$, for each i), so we will present most of the material here in terms of the second version of the problem.

It is actually very easy to prove that both of these problems have a unique solution. Moreover, the proof shows us how to construct the polynomial.

Theorem 4.1 (Polynomial Interpolation Existence and Uniqueness)   Let the nodes $x_i \in I$, $0 \leq i \leq n$, be given. So long as each of the x_i are distinct, i.e., $x_i = x_j$ if and only if $i = j$, then there exists a unique polynomial p_n, of degree less than or equal to n, that satisfies either of

$$p_n(x_i) = y_i, \quad 0 \leq i \leq n \tag{4.1}$$

for a given set of data values $\{y_i\}$, or

$$p_n(x_i) = f(x_i), \quad 0 \leq i \leq n \tag{4.2}$$

for a given function $f \in C(I)$.

Proof   We begin by defining the family of functions[1]

$$L_i^{(n)}(x) = \prod_{\substack{k=0 \\ k \neq i}}^{n} \frac{x - x_k}{x_i - x_k}$$

and note that they are polynomials of degree n and have the interesting property that

$$L_i^{(n)}(x_j) = \delta_{ij} = \begin{cases} 1 & i = j \\ 0 & i \neq j. \end{cases} \tag{4.3}$$

---

[1]The reader is reminded that the symbol $\prod$ is used for a product in the same way that $\sum$ is used for a sum. Thus

$$\prod_{k=1}^{n} a_k = a_1 \times a_2 \times a_3 \times \cdots \times a_n.$$

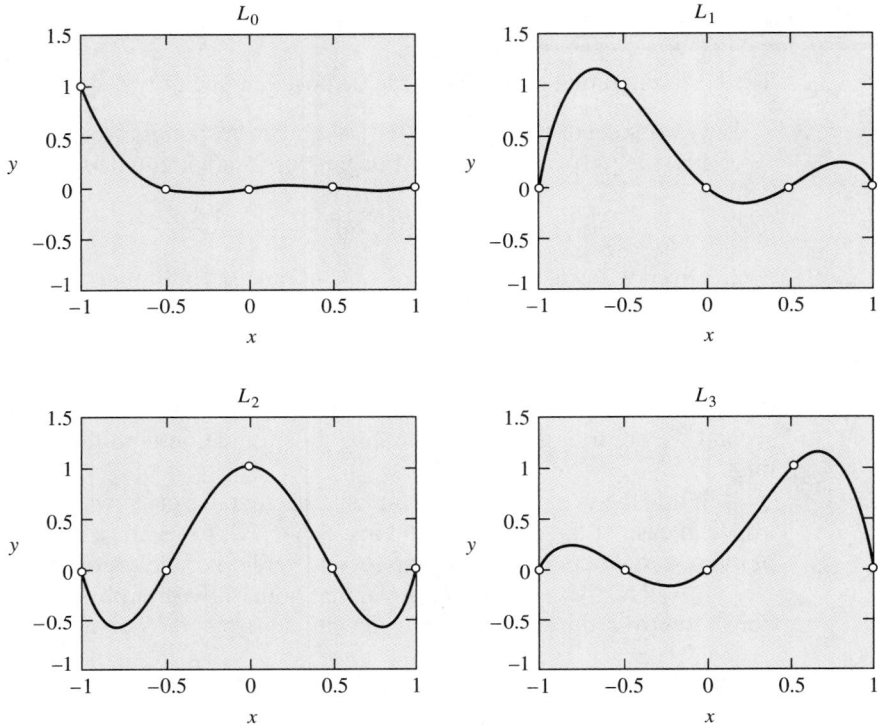

FIGURE 4.1 Lagrange polynomials for five equally spaced nodes on $[-1,1]$.

(The symbol δ_{ij} implicitly defined above is called the *Kronecker[2] delta*.) Figure 4.1 shows examples of these polynomials for five equally spaced nodes on $[-1,1]$. Based on (4.3), then, if we define the polynomial by

$$p_n(x) = \sum_{k=0}^{n} y_k L_k^{(n)}(x),$$

it follows that

$$p_n(x_i) = \sum_{k=0}^{n} y_k L_k^{(n)}(x_i) = y_i.$$

---

[2]Leopold Kronecker (1823–1891) was born in Liegnitz, Prussia, and educated in local schools and the University of Berlin, graduating with his doctorate in 1845.

Most of Kronecker's mathematical research was in what is now called modern algebra and the theory of equations. In 1845 he briefly left academia to enter the business world, but he remained mathematically active even during this interlude. The fortune he made while working as a banker enabled him to continue as a mathematician despite not having a regular professorship until 1883.

In his later years Kronecker advocated a somewhat extreme philosophy of mathematics in which only finite numbers and finite operations were valid.

The Kronecker delta notation came out of his work on determinants and matrices, done in 1869, under the title "Bemerkungen zur Determinanten-Theorie," published in the *Journal für die reine und angewandte Mathematik*.

Thus the interpolatory conditions are satisfied, and it remains only to prove the uniqueness of the polynomial. To this end, assume there exists a second interpolatory polynomial; call it q, and define

$$r(x) = p_n(x) - q(x).$$

Since both p_n and q are polynomials of degree less than or equal to n, so is their difference. However, we must note that

$$r(x_i) = p_n(x_i) - q(x_i) = y_i - y_i = 0$$

for each of the $n+1$ nodes. Thus we have a polynomial of degree less than or equal to n that has $n+1$ roots; the only such polynomial is the zero polynomial, i.e.,

$$r(x) \equiv 0 \quad \Rightarrow \quad p_n(x) \equiv q(x),$$

and thus p_n is unique. □

EXAMPLE 4.1

To illustrate this construction, let $f(x) = e^x$ and consider the problem of constructing an approximation to this function on the interval $[-1,1]$ using the nodes $\{-1, -\frac{1}{2}, 0, \frac{1}{2}, 1\}$. We have

$$
\begin{aligned}
p_4(x) &= \frac{(x+\frac{1}{2})(x-0)(x-\frac{1}{2})(x-1)}{(-1+\frac{1}{2})(-1-0)(-1-\frac{1}{2})(-1-1)} e^{-1} \\[6pt]
&+ \frac{(x+1)(x-0)(x-\frac{1}{2})(x-1)}{(-\frac{1}{2}+1)(-\frac{1}{2}-0)(-\frac{1}{2}-\frac{1}{2})(-\frac{1}{2}-1)} e^{1/2} \\[6pt]
&+ \frac{(x+1)(x+\frac{1}{2})(x-\frac{1}{2})(x-1)}{(0+1)(0+\frac{1}{2})(0-\frac{1}{2})(0-1)} e^0 \\[6pt]
&+ \frac{(x+1)(x+\frac{1}{2})(x-0)(x-1)}{(\frac{1}{2}+1)(\frac{1}{2}+\frac{1}{2})(\frac{1}{2}-0)(\frac{1}{2}-1)} e^{1/2} \\[6pt]
&+ \frac{(x+1)(x+\frac{1}{2})(x-0)(x-\frac{1}{2})}{(1+1)(1+\frac{1}{2})(1-0)(1-\frac{1}{2})} e^1,
\end{aligned}
$$

which simplifies to

$$p_4(x) = 1.0 + 0.997853749x + 0.499644939x^2 + 0.177347443x^3 + 0.043435696x^4.$$

Figure 4.2(a) shows the plot of f and $p_4(x)$. The approximation is sufficiently accurate that it is impossible to discern that there are two curves present, so it is probably better to look at Figure 4.2(b), which plots the error $e^x - p_4(x)$. Compare this with the error obtained by doing a fourth-degree Taylor polynomial, centered at the origin, which is given in Figure 4.2(c). Note that while the Taylor polynomial is *very* accurate near the middle of the interval, and much less accurate near the ends, the interpolating polynomial has less variation in its error, and a much smaller worst-case error over the same interval. ■

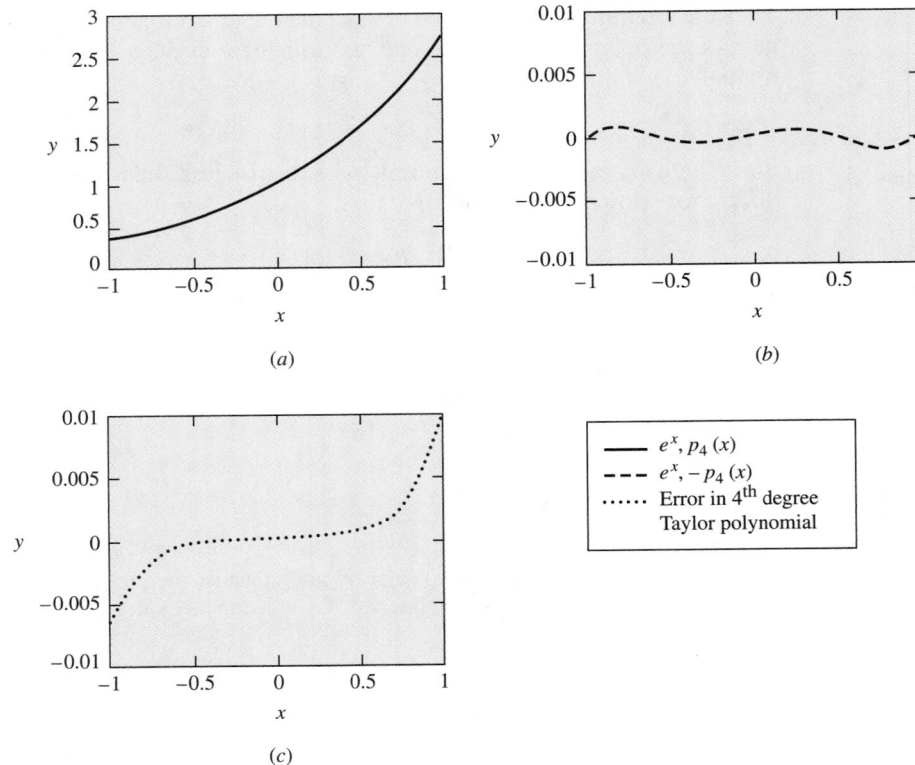

FIGURE 4.2   Interpolation to $f(x) = e^x$.

The construction presented in this section is called *Lagrange*[3] *interpolation*. Other approaches to the interpolation problem will be studied in other sections; this one is the most fundamental.

How good is interpolation at approximating a function? We will study this a little more carefully in Sections 4.3 and 4.11, but we can illustrate some of the issues with a few examples right now. We can already note that the fourth-degree approximation to the exponential function is very accurate, even compared with the fourth-degree Taylor polynomial. What about some other functions?

Consider now the function $f(x) = (1 + 25x^2)^{-1}$. If we use a fourth-degree inter-polating polynomial to approximate this function, the results are as shown in Figure

---

[3]Joseph-Louis Lagrange (1736–1813) was born in Turin, Italy, of French parents. At a very young age he was made professor at the Royal Artillery School in Turin. In 1766 he succeeded his friend and mentor Euler as Director of the Mathematics Section of the Berlin Academy. In 1787 he accepted a similar position at the Paris Academy of Sciences, where he was active in establishing two very influential French schools, the École Polytechnique, and the École Normale. In his later years he was granted a number of honors by the French Emperor, Napoleon Bonaparte.

His results on interpolation were a major part of his work while in Berlin, but the interpolation formula ascribed to him here was not published until 1795, after he had moved to France. Ironically, Lagrange claims to have gotten the idea from some of Newton's work.

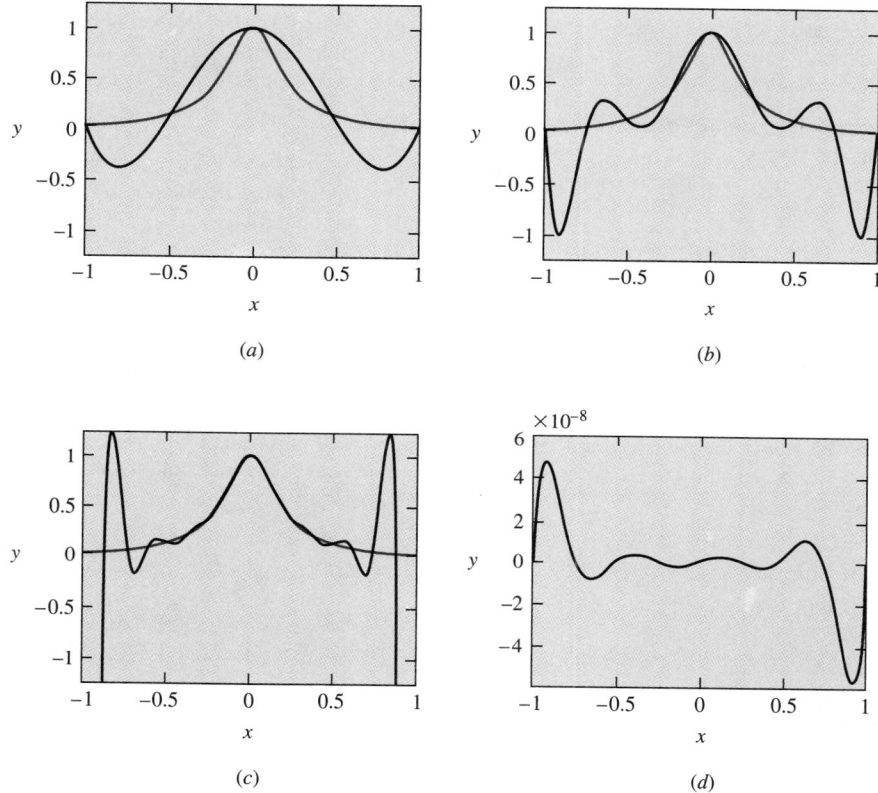

FIGURE 4.3 Polynomial interpolation to $f(x) = (1 + 25x^2)^{-1}$ with $n = 4$ (A), $n = 8$ (B), and $n = 16$ (C), and the error in interpolation for $n = 8$ to $f(x) = e^x$ (D).

4.3(a). These are not nearly as good as for the exponential function. If we increase the number of points (and therefore the degree of the interpolating polynomial), things improve on part of the interval, but not all of it. Figure 4.3(b) shows the plots for $n = 8$, and Figure 4.3(c) shows the $n = 16$ results. In contrast, Figure 4.3(d) plots the *error* for the degree 8 polynomial interpolation to the exponential function.[4]

Clearly, there are circumstances ($f(x) = e^x$) in which polynomial interpolation as approximation will work very well, and other circumstances ($f(x) = (1 + 25x^2)^{-1}$) in which it will not. More study is therefore required.

The attentive reader will have noted that we have not discussed an algorithm for computing Lagrange interpolating polynomials. The reason for this is quite simple, actually: The Lagrange form of the interpolating polynomial is not well suited for actual computations, and there is an alternative construction that is far superior to it. So we will defer all discussion of algorithms until the next section.

---

[4]The reader should note that all of these examples used equally spaced points to define the nodes. Other choices are possible, and there is a particular choice that is indeed better. See Section 4.11.3.

EXERCISES

1. Find the polynomial of degree 2 that interpolates at the data points $x_0 = 0$, $y_0 = 1$, $x_1 = 1$, $y_1 = 2$, and $x_2 = 4$, $y_2 = 2$. You should get $p_2(t) = -\frac{1}{4}t^2 + \frac{5}{4}t + 1$.

2. Find the polynomial of degree 2 that interpolates to $y = x^3$ at the nodes $x_0 = 0$, $x_1 = 1$, and $x_2 = 2$. Plot $y = x^3$ and the interpolating polynomial over the interval $[0,2]$.

3. Construct the quadratic polynomial that interpolates to $y = 1/x$ at the nodes $x_0 = 1/2$, $x_1 = 3/4$, and $x_2 = 1$.

4. Construct the quadratic polynomial that interpolates to $y = \sqrt{x}$ at the nodes $x_0 = 1/4$, $x_1 = 9/16$, and $x_2 = 1$.

5. For each function listed below, construct the Lagrange interpolating polynomial for the set of nodes specified. Plot both the function and the interpolating polynomial, and also the error, on the interval defined by the nodes.

(a) $f(x) = \sqrt{x}$, $x_i = 0,1,4$
(b) $f(x) = \ln x$, $x_i = 1, \frac{3}{2}, 2$
(c) $f(x) = \log_2 x$, $x_i = 1,2,4$
(d) $f(x) = \sin \pi x$, $x_i = -1,0,1$

6. Find the polynomial of degree 3 that interpolates $y = x^3$ at the nodes $x_0 = 0$, $x_1 = 1$, $x_2 = 2$, and $x_3 = 3$. (Simplify your interpolating polynomial as much as possible.) (Hint: This is easy if you think about the implications of the uniqueness of the interpolating polynomial.)

7. Construct the Lagrange interpolating polynomial to the function $f(x) = x^2 + 2x$, using the nodes $x_0 = 0$, $x_1 = 1$, $x_2 = -2$. Repeat, using the nodes $x_0 = 2$, $x_1 = 1$,

$x_2 = -1$. (For both sets of nodes, simplify your interpolating polynomial as much as possible.) Comment on your results, especially in light of the uniqueness part of Theorem 4.1, and then write down the polynomial for interpolating to $f(x) = x^3 + 2x^2 + 3x + 1$ at the nodes $x_0 = 0$, $x_1 = 1$, $x_2 = 2$, $x_3 = 3$, $x_4 = 4$, and $x_5 = 5$. (Hint: You should be able to do the last part without doing any computations.)

8. Let f be a polynomial of degree $\leq n$, and let p_n be a polynomial interpolant to f, at the $n+1$ distinct nodes $x_0, x_1, \ldots, x_n$. Prove that $p_n(x) = f(x)$ for all x, i.e., that interpolating to a polynomial will reproduce the polynomial, if you use enough nodes. (Hint: Consider the uniqueness part of Theorem 4.1.)

9. Let p be a polynomial of degree $\leq n$. Use the uniqueness of the interpolating polynomial to show that we can write

$$p(x) = \sum_{i=0}^{n} L_i^{(n)}(x) p(x_i)$$

for any distinct set of nodes $x_0, x_1, \ldots, x_n$. (Hint: See the previous exercise.)

10. Show that

$$\sum_{i=0}^{n} L_i^{(n)}(x) = 1$$

for any set of distinct nodes x_k, $0 \leq k \leq n$. (Hint: This does not require *any* computation with the sum, but rather a perceptive choice of polynomial p in the previous exercise.)

4.2 NEWTON INTERPOLATION AND DIVIDED DIFFERENCES

The Lagrange form of the interpolating polynomial gives us a very tidy construction, but it does not lend itself well to actual computation. This is partly because whenever we decide to add a point to the set of nodes, we have to completely recompute all of the $L_i^{(n)}$ functions; we cannot (easily) write p_{n+1} in terms of p_n, using the Lagrange construction.

An alternative form of the polynomial, known as the Newton form, avoids this problem and allows us to easily write p_{n+1} in terms of p_n. The basic result is contained in the following theorem.

Theorem 4.2 (Newton Interpolation Construction)   Let p_n be the polynomial that interpolates f at the nodes x_i, $i = 0,1,2,3,\ldots,n$. Let p_{n+1} be the polynomial that interpolates f at the nodes x_i, $i = 0,1,2,3,\ldots,n,n+1$. Then p_{n+1} is given by

$$p_{n+1}(x) = p_n(x) + a_{n+1}w_n(x), \tag{4.4}$$

where

$$w_n(x) = \prod_{i=0}^{n}(x - x_i)$$

$$a_{n+1} = \frac{f(x_{n+1}) - p_n(x_{n+1})}{w_n(x_{n+1})} \tag{4.5}$$

and

$$a_0 = f(x_0).$$

Proof   Since we know that the interpolation polynomial is unique, all we have to do is show that p_{n+1}, as given in (4.4), satisfies the interpolation conditions. For $k \leq n$, we have $w_n(x_k) = 0$, so

$$p_{n+1}(x_k) = p_n(x_k) + a_{n+1}w_n(x_k) = p_n(x_k) = f(x_k);$$

hence p_{n+1} interpolates all but the last point. To check x_{n+1}, we directly compute

$$\begin{aligned}
p_{n+1}(x_{n+1}) &= p_n(x_{n+1}) + a_{n+1}w_n(x_{n+1}) \\
&= p_n(x_{n+1}) + f(x_{n+1}) - p_n(x_{n+1}) \\
&= f(x_{n+1}).
\end{aligned}$$

Thus p_{n+1} interpolates at all the nodes; moreover, it clearly is a polynomial of degree less than or equal to $n+1$, and so we are done. $\square$

Corollary 4.1   For $\{a_k\}$ and w_n as defined in Theorem 4.2, we have

$$\begin{aligned}
p_n(x) &= a_0 + a_1(x - x_0) + a_2(x - x_0)(x - x_1) + \cdots + a_n(x - x_0)\cdots(x - x_{n-1}) \\
&= \sum_{k=0}^{n} a_k w_{k-1}(x)
\end{aligned}$$

where we have used $w_{-1}(x) \equiv 1$.

Proof   This is a straightforward inductive argument. See Exercise 5. $\square$

We can use this Newton form of the interpolating polynomial to construct and evaluate interpolating polynomials in an efficient fashion that is reminiscent of Horner's rule for nested multiplication (Section 2.1). We have

$$p_n(x) = a_0 + a_1(x - x_0) + a_2(x - x_0)(x - x_1) + \cdots + a_n(x - x_0)\cdots(x - x_{n-1}),$$

so that we can write

$$p_n(x) = a_0 + (x - x_0)(a_1 + (x - x_1)(a_2 + \cdots + (x - x_{n-1})a_n)\cdots)).$$

EXAMPLE 4.2

To consider an example, let us construct the quadratic polynomial that interpolates the sine function on the interval $[0, \pi]$, using equally spaced nodes. We have $f(x) = \sin x$, $x_i = 0, \frac{1}{2}\pi, \pi$, and $y_i = 0, 1, 0$. Then

$$a_0 = 0 \quad \Rightarrow \quad p_0(x) = 0$$

$$a_1 = \frac{\sin \frac{1}{2}\pi - p_0(\frac{1}{2}\pi)}{(\frac{1}{2}\pi - 0)} = \frac{2}{\pi} \quad \Rightarrow \quad p_1(x) = \frac{2}{\pi}x$$

$$a_2 = \frac{\sin \pi - p_1(\pi)}{(\pi - 0)(\pi - \frac{1}{2}\pi)} = -\frac{4}{\pi^2} \quad \Rightarrow \quad p_2(x) = \frac{2}{\pi}x - \frac{4}{\pi^2}x(x - \pi).$$

The reader should check that each interpolating polynomial does interpolate at the appropriate points. ∎

The coefficients a_k are called *divided differences*, and we can construct an algorithm (see Algorithm 4.1) for computing them, based on formula (4.5). It is also possible to slightly modify this algorithm to allow the user to *update* an existing set of divided-difference coefficients by adding new points to the set of nodes. (See Exercise 6.)

An alternative way of arriving at the divided-difference coefficients—and the original way it was done—is by means of a *divided-difference table*. Since this is an easy means by which hand calculations can be performed, it is worth looking at here as well.

We begin by defining some terminology and notation. Given a set of nodes and corresponding function values, the function values will be referred to as the *zeroth divided differences* and, in this context, we will write them as $f_0(x_k)$. The *first divided*

Algorithm 4.1: Newton interpolation (construction) via divided-difference coefficients

```
input n,x,y
a(0) = y(0)
for k=1 to n do
w = 1
p = 0
for j=0 to k-1 do
p = p + a(j)*w
w = w*(x(k) - x(j))
endfor
a(k) = (y(k) - p)/w
endfor
```

Algorithm 4.2: *Newton interpolation (evaluation) via divided-difference coefficients*

```
input n,xx,x,a
!
!       n = degree of polynomial
!       xx = point at which the polynomial is
!             to be evaluated
!       x = array of nodes
!       a = array of divided difference coefficients
!
px = a(n)
for k = n-1,0,-1
        xd = xx - x(k)
        px =  a(k) + px*xd
endfor
!
!       px = polynomial value at xx
!
```

Programming Hint: We know, from the exercises at the end of the previous section, that polynomial interpolation of sufficiently high degree to a polynomial will reproduce the original polynomial. Thus, one way for a user to check that an interpolation code is working is to see if it can reproduce a polynomial of degree n when asked to interpolate it at $n + 1$ nodes.

---

differences are then formed from the zeroth ones according to

$$f_1(x_k) = \frac{f_0(x_{k+1}) - f_0(x_k)}{x_{k+1} - x_k},$$

and then the *second divided differences* are given by

$$f_2(x_k) = \frac{f_1(x_{k+1}) - f_1(x_k)}{x_{k+2} - x_k},$$

and so on. The general formula is

$$f_j(x_k) = \frac{f_{j-1}(x_{k+1}) - f_{j-1}(x_k)}{x_{k+j} - x_k}.$$

Now using this terminology and notation,[5] we can build a set of tables that can then be readily used to construct the interpolating polynomials.

EXAMPLE 4.3

Take $f(x) = \log_2 x$ as the function to be interpolated, using the nodes $x_0 = 1$, $x_1 = 2$, $x_2 = 4$. We begin by arranging the nodes and corresponding functional data in two columns, as in Table

---

[5]We have broken with convention and tradition in our divided-difference notation here. Most texts use $f[x_k]$ for the zeroth divided difference; $f[x_k,x_{k+1}]$ for the first divided difference formed by the nodes x_k and x_{k+1}; $f[x_k,x_{k+1},x_{k+2}]$ for the second divided difference formed by x_k, x_{k+1}, and x_{k+2}; and so on. This becomes unwieldy in dealing with higher and higher differences, and while it is slightly more general, most texts (including this one) do very little to justify that extra generality.

TABLE 4.1 **Initial Table for Divided Differences**

k	x_k	$f_0(x_k)$
0	1	0
1	2	1
2	4	2

TABLE 4.2 **Table for Divided Differences after Computation of First Differences**

k	x_k	$f_0(x_k)$	$f_1(x_k)$
0	1	0	
			1
1	2	1	
			$\frac{1}{2}$
2	4	2	

4.1. Note that we write the functional data using the divided-difference notation.

$$1 = \frac{f(x_1) - f(x_0)}{x_1 - x_0} = \frac{1 - 0}{2 - 1} = f_1(x_0)$$

and

$$\frac{1}{2} = \frac{f(x_2) - f(x_1)}{x_2 - x_1} = \frac{2 - 1}{4 - 2} = f_1(x_1),$$

The third column of the table is formed by the first divided differences; thus so we get Table 4.2.

The fourth (and, in this case, final) column is formed by the second divided difference:

$$-\frac{1}{6} = \frac{1/2 - 1}{4 - 1} = f_2(x_0);$$

see Table 4.3. (If we had more data, subsequent columns would be formed in a similar fashion.) How do we use this result to construct the interpolating polynomial? The top diagonal row of table values gives us the divided-difference coefficients based on the nodes as numbered in ascending order. Thus

$$p_2(x) = 0 + (1)(x - 1) - \frac{1}{6}(x - 1)(x - 2) = -\frac{1}{6}(x - 1)(x - 8).$$

TABLE 4.3 Table for Divided Differences after Computation of Second Differences

k	x_k	$f_0(x_k)$	$f_1(x_k)$	$f_2(x_k)$
0	1	0		
			1	
1	2	1		$-\frac{1}{6}$
			$\frac{1}{2}$	
2	4	2		

(These are the same values that the pseudocode produces.) The bottom diagonal row gives the divided-difference coefficients based on numbering the nodes in reverse order. Thus we also have

$$p_2(x) = 2 + \frac{1}{2}(x-4) - \frac{1}{6}(x-4)(x-2) = -\frac{1}{6}(x-1)(x-8).$$

Finally, to add a node to the data set and construct the new polynomial is easy, based on these tables. Let's add the new node $x_3 = \frac{1}{2}$ to our current example. We have, initially, the arrangement in Table 4.4. (Note that the nodes do not have to be arranged in ascending order—or, in fact, in any particular order.)

Computing forward, we get the (new) final result in Table 4.5, from which we conclude that the interpolating polynomial is

$$p_3(x) = 0 + (1)(x-1) - \frac{1}{6}(x-1)(x-2) + \frac{1}{7}(x-1)(x-2)(x-4).$$

The reader should confirm that this does indeed interpolate correctly at the given points and therefore is the correct polynomial. ∎

TABLE 4.4 Table for Divided Differences with New Node Added

k	x_k	$f_0(x_k)$	$f_1(x_k)$	$f_2(x_k)$
0	1	0		
			1	
1	2	1		$-\frac{1}{6}$
			$\frac{1}{2}$	
2	4	2		
3	$\frac{1}{2}$	-1		

TABLE 4.5 **Table for Divided Differences after Computation of Third Differences**

k	x_k	$f_0(x_k)$	$f_1(x_k)$	$f_2(x_k)$	$f_3(x_k)$
0	1	0			
			1		
1	2	1		$-\frac{1}{6}$	
			$\frac{1}{2}$		$\frac{1}{7}$
2	4	2		$-\frac{5}{21}$	
			$\frac{6}{7}$		
3	$\frac{1}{2}$	-1			

Programming Hint   An interpolating polynominal if it is correctly constructed, has to reproduce the values at the nodes exactly, so this is an easy test to use in evaluating and debugging your code. If the values at the nodes are computed correctly, then the polynomial has to be correct; if the values at the nodes are not computed correctly, then there is something wrong, either with the construction of the polynomial or with the code that is evaluating the polynomial.

EXERCISES

1. Construct the polynomial of degree 3 that interpolates to the data $x_0 = 1$, $y_0 = 1$, $x_1 = 2$, $y_1 = 1/2$, $x_2 = 4$, $y_2 = 1/4$, and $x_3 = 3$, $y_3 = 1/3$. You should get $p(t) = (50 - 35t + 10t^2 - t^3)/24$.

2. For each function listed below, use divided-difference tables to construct the Newton interpolating polynomial for the set of nodes specified. Plot both the function and the interpolating polynomial, and also the error, on the interval defined by the nodes.
(a) $f(x) = \sqrt{x}$, $x_i = 0,1,4$
(b) $f(x) = \sin \pi x$, $x_i = 0, \frac{1}{4}, \frac{1}{2}, \frac{3}{4}, 1$
(c) $f(x) = \ln x$, $x_i = 1, \frac{3}{2}, 2$
(d) $f(x) = \log_2 x$, $x_i = 1,2,4$
(e) $f(x) = \sin \pi x$, $x_i = -1,0,1$

3. Let $f(x) = e^x$. Define $p_n(x)$ to be the Newton interpolating polynomial for $f(x)$, using $n+1$ equally spaced nodes on the interval $[-1,1]$. Thus we are taking increasingly higher-degree polynomial approximations to the exponential function. Write a program that computes $p_n(x)$ for $n = 2,4,8,16,32$ and that samples the error $f(x) - p_n(x)$ at 501 equally spaced points on $[-1,1]$. Record the maximum error as found by the sampling,

as a function of n; i.e., define E_n as

$$E_n = \max_{0 \le k \le 500} |f(t_k) - p_n(t_k)|,$$

where $t_k = -1 + 2k/500$, and plot E_n versus n.

4. In Section 3.7 we used linear interpolation to construct an initial guess for Newton's method as a means of approximating $\sqrt{a}$. Construct the quadratic polynomial that interpolates the square root function at the nodes $x_0 = \frac{1}{4}, x_1 = \frac{4}{9}, x_2 = 1$. Plot the error between p_2 and $\sqrt{x}$ over the interval $[\frac{1}{4},1]$ and try to estimate the worst error. What impact will this have on the use of Newton's method for finding $\sqrt{a}$?

5. Prove Corollary 4.1.

6. Write and test a computer code that takes a given set of nodes, function values, and corresponding divided-difference coefficients and computes new divided-difference coefficients for new nodes and function values. Test it by recursively computing interpolating polynomials that approximate $f(x) = e^x$ on the interval $[0,1]$.

7. In 1973, the horse Secretariat became the first (and so far, only) winner of the Kentucky Derby to finish the

TABLE 4.6   Data for Exercise 7

x	0.0	0.25	0.50	0.75	1.00	1.25
t	0.0	25.0	49.4	73.0	96.4	119.4

TABLE 4.7   Data for Exercise 9.

Temperature (K), u	300	400	500	600	700
Conductivity (watts/cm deg K), k	0.084	0.098	0.109	0.12	0.127

race in less than two minutes, running the $1\frac{1}{4}$-mile distance in 1:59.4. Remarkably, he ran each quarter mile *faster* than the previous one, as the data in Table 4.6 shows. Here t is the elapsed time (in seconds) since the race began, and x is the distance (in miles) that Secretariat has traveled.

(a) Find the cubic polynomial that interpolates this data at $x = 0, 1/2, 3/4, 5/4$.
(b) Use this polynomial to estimate Secretariat's speed at the finish of the race, by finding $p'_3(5/4)$.
(c) Find the quintic polynomial that interpolates the entire data set.
(d) Use the quintic polynomial to estimate Secretariat's speed at the finish line.

8. Use Newton interpolation to find a fifth-degree polynomial that interpolates the pennsyltuckium thermal conductivity data from the opening of this chapter. Plot both the interpolating polynomial and the nodal data. How well does the interpolating polynomial follow the overall trend of the data?

9. Table 4.7 gives the actual thermal conductivity data for the element mercury. Use Newton interpolation and the data for 300 K, 500 K, and 700 K to construct a quadratic interpolant for this data. How well does it predict the values at 400 K and 600 K?

10. The gamma function, denoted by $\Gamma(x)$, is an important special function in probability, combinatorics, and other areas of applied mathematics. Because it can be shown that $\Gamma(n+1) = n!$, the gamma function is considered a gen-

eralization of the factorial function to noninteger arguments. Table 4.8 gives the values of $\Gamma(x)$ on the interval $[1,2]$. Use these to construct the fifth-degree polynomial based on the nodes $x = 1, 1.2, 1.4, 1.6, 1.8, 2.0$, and then use this polynomial to estimate the values at $x = 1.1, 1.3, 1.5, 1.7, 1.9$. Plot your polynomial and compare it with the intrinsic gamma function on your computing system or calculator.

11. The error function, which we saw briefly in Chapters 1 and 2, is another important special function in applied mathematics, with applications to probability theory and the solution of heat conduction problems. The formal definition of the error function is as follows:

$$\text{erf}(x) = \frac{2}{\sqrt{\pi}} \int_0^x e^{-t^2} dt.$$

TABLE 4.8   Table of $\Gamma(x)$ Values.

x	$\Gamma(x)$
1.00	1.0000000000
1.10	0.9513507699
1.20	0.9181687424
1.30	0.8974706963
1.40	0.8872638175
1.50	0.8862269255
1.60	0.8935153493
1.70	0.9086387329
1.80	0.9313837710
1.90	0.9617658319
2.00	1.0000000000

TABLE 4.9 Table of erf(x) values for Exercise 11.

x	erf(x)
0.0	0.00000000000000
0.1	0.11246291601828
0.2	0.22270258921048
0.3	0.32862675945913
0.4	0.42839235504667
0.5	0.52049987781305
0.6	0.60385609084793
0.7	0.67780119383742
0.8	0.74210096470766
0.9	0.79690821242283
1.0	0.84270079294971

Table 4.9 gives values of erf(x) in increments of 0.1, over the interval $[0,1]$.
(a) Construct the quadratic interpolating polynomial to the error function using the data at the nodes $x_0 = 0$, $x_1 = 0.5$, and $x_2 = 1.0$. Plot the polynomial and the data in the table, and comment on the observed accuracy.
(b) Repeat the preceding, but this time construct the cubic interpolating polynomial using the nodes $x_0 = 0.0$, $x_2 = 0.3$, $x_2 = 0.7$, and $x_3 = 1.0$.

12. As steam is heated, the pressure it generates is increased. Over the temperature range $[220, 300]$ (degrees Fahrenheit) the pressure, in pounds per square inch, is given in Table 4.10.[6]
(a) Construct the quadratic interpolating polynomial to this data at the nodes $T_0 = 220$, $T_1 = 260$,

and $T_2 = 300$. Plot the polynomial and the data in the table, and comment on the observed accuracy.
(b) Repeat the preceding, but this time construct the quartic interpolating polynomial using the nodes $T_0 = 220$, $T_1 = 240$, $T_2 = 260$, $T_3 = 280$, and $T_4 = 300$.
(c) Which of the two polynomials do you think would be best to use to get values for $P(T)$ that are not in the table?

13. Similar data for gaseous ammonia is given in Table 4.11.
(a) Construct the quadratic interpolating polynomial to this data at the nodes $T_0 = 0$, $T_1 = 20$, and $T_2 = 40$. Plot the polynomial and the data in the table, and comment on the observed accuracy.
(b) Repeat the preceding, but this time construct the quartic interpolating polynomial using the nodes $T_0 = 0$, $T_1 = 10$, $T_2 = 20$, $T_3 = 30$, and $T_4 = 40$.
(c) Which of the two polynomials do you think would be best to use to get values for $P(T)$ that are not in the table?

14. In Exercise 8 of Section 3.1 and Exercise 10 of Section 3.8 we looked at the motion of a liquid-solid interface under a simplified model of the physics involved, in which the interface moved according to
$$x = 2\beta\sqrt{t}$$
for $\beta = \alpha/\sqrt{k}$. Here k is a material property and α is the root of $f(z) = \theta e^{-z^2} - z\,\text{erf}(z)$, where θ also depends on material properties. Figure 4.4 shows a plot of α versus $\log_{10}\theta$, based on finding the root of $f(z)$. Some of the data used to create this curve is given in Table 4.12.

TABLE 4.10   Temperature-Pressure Values for Steam; Exercise 12

T	220	230	240	250	260	270	280	290	300
P	17.188	20.78	24.97	29.82	35.42	41.85	49.18	57.53	66.98

---

[6]Taken from tables in *Introduction to Thermodynamics: Classical and Statistical*, by R.E. Sonntag and G.J. Van Wylen, John Wiley & Sons, New York, 1971.

(a) Use the data at the nodes $\{-6, -4, -2, 0, 2, 4, 6\}$ to construct an interpolating polynomial for this table. Plot the polynomial and compare it with the actual graph in Figure 4.4. Use the polynomial to compute values at $\log_{10}\theta = -5, -3, \ldots, 3, 5$. How do these compare with the actual values from the table?

(b) Compute the higher-degree Newton polynomial based on the entire table of data. Plot this polynomial and compare it with the one generated using only part of the data.

15. Write a computer program to construct the Newton interpolating polynomial to $f(x) = \sqrt{x}$ using equally spaced nodes on the interval $[0,1]$. Plot the error $f(x) - p_n(x)$ for $n = 4, 8, 16$ and comment on what you get.

TABLE 4.11 Data for Exercise 14.

$\log_{10}\theta$	α
-6.0000	0.944138E-03
-5.0000	0.298500E-02
-4.0000	0.941277E-02
-3.0000	0.297451E-01
-2.0000	0.938511E-01
-1.0000	0.289450E+00
0.0000	0.767736E+00
1.0000	0.141492E+01
2.0000	0.198151E+01
3.0000	0.245183E+01
4.0000	0.285669E+01
5.0000	0.321632E+01
6.0000	0.354269E+01

TABLE 4.12 Temperature-Pressure Values for Gaseous Ammonia; Exercise 13

T	0	5	10	15	20	25	30	35	40
P	30.42	34.27	38.51	43.14	48.21	53.73	59.74	66.26	73.32

FIGURE 4.4 Figure for Exercise 14.

4.3 INTERPOLATION ERROR

How good an approximation is the interpolating polynomial? In Section 2.4 an error estimate was fairly easily derived for linear interpolation, and here we will generalize that result to arbitrary polynomial interpolation. We expect, based on the examples $f(x) = e^x$ and $f(x) = (1 + 25x^2)^{-1}$, that using more points could make the approximation more accurate and yet could also lead to problems.

Theorem 4.3 (Polynomial Interpolation Error Theorem)   Let $f \in C^{n+1}([a,b])$ and let the nodes $x_k \in [a,b]$ for $0 \le k \le n$. Then, for each $x \in [a,b]$, there is a $\xi_x \in [a,b]$ such that

$$f(x) - p_n(x) = \frac{w_n(x)}{(n+1)!} f^{(n+1)}(\xi_x), \tag{4.6}$$

where
$$w_n(x) = \prod_{k=0}^{n} (x - x_k).$$

Proof   The proof is an involved argument using repeated applications of Rolle's Theorem, and is deferred to Appendix A.1.   □

This result is too general to be easily interpreted at this point in our study, so we will deal directly with some specific cases. Before doing so, we introduce some notation.

In measuring the error in certain approximations, it will be useful from now on to have some convenient means of measuring the *size* of a function. This is done via the concept of a *norm*. Briefly, a norm is any mapping from functions to the nonnegative real numbers (usually denoted by a double-bar notation, $\|f\|$) that satisfies certain basic axioms.

Definition 4.1 (Function Norm)   A function norm is any computation, denoted by the symbol $\|f\|$, that satisfies the following conditions:

1. $\|f\| > 0$ for any function f that is not identically zero.

2. $\|af\| = |a|\|f\|$ for any constant a.

3. $\|f + g\| \le \|f\| + \|g\|$ for any two functions f and g.

Note that the use of the double-bar notation ($\|f\|$) is similar to the single-bar notation for the absolute value; this is deliberate, since the notion of a norm plays much the same role for functions as the absolute value does for ordinary numbers: it helps us to measure size, or magnitude.

Some examples of common norms include

- The "infinity" norm or pointwise norm: If f is continuous on the closed interval $[a,b]$, then we define
$$\|f\|_{\infty,[a,b]} = \max_{x \in [a,b]} |f(x)|.$$

- The "2-norm": If f is such that its square can be integrated over the interval $[a,b]$, then we define

$$\|f\|_{2,[a,b]} = \left(\int_a^b [f(x)]^2 \, dx \right)^{1/2}.$$

Note the use of subscripts on the norm notation to distinguish different norms; the subscript defining the interval will often be dropped or omitted when there is no danger of confusion. Note also that different norms will give different numeric values for the same function, and will convey different kinds of information.

EXAMPLE 4.4

If $f(x) = e^{-x}$ on the interval $[-1,1]$, then

$$\|f\|_\infty = e = 2.71828...$$

and

$$\|f\|_2 = \left(\frac{e^2 - e^{-2}}{2} \right)^{1/2} = 1.90443....$$

∎

See Exercises 14 and 15 at the end of the section for some further exploration of what different norms mean.

We already know the interpolation error result for $n = 1$, i.e., linear interpolation. In Section 2.4 we saw that we could prove

$$|f(x) - p_1(x)| \le \frac{1}{8}(x_1 - x_0)^2 \max_{x_0 \le x \le x_1} |f''(x)|, \tag{4.7}$$

which holds for all $x \in [x_0, x_1]$. We can write this in norm notation as

$$\|f - p\|_{\infty, I} \le \frac{1}{8}(x_1 - x_0)^2 \|f''\|_{\infty, I}, \tag{4.8}$$

where $I = [x_0, x_1]$ is the interval defined by the nodes used in the interpolation. What happens for larger values of n?

If $n = 2$ (quadratic interpolation) we have, from (4.6),

$$f(x) - p_2(x) = \frac{1}{6}(x - x_0)(x - x_1)(x - x_2)f'''(\xi_x). \tag{4.9}$$

To get an upper bound like we did for linear interpolation, we assume that the nodes satisfy

$$x_1 - x_0 = x_2 - x_1 = h$$

for some $h > 0$. We thus can write

$$(x - x_0)(x - x_1)(x - x_2) = ((x - x_1) + h)(x - x_1)((x - x_1) - h) = t(t^2 - h^2)$$

for $t = x - x_1$. We now have

$$|f(x) - p_2(x)| \le \frac{1}{6} \left(\max_{-h \le t \le h} |t(t^2 - h^2)| \right) \max_{x_0 \le t \le x_2} |f'''(t)|, \tag{4.10}$$

which holds for all $x \in [x_0, x_2]$. Following the same kind of argument used in Section 2.4, we can show that

$$\max_{-h \le t \le h} |t(t^2 - h^2)| = \frac{2}{3\sqrt{3}} h^3,$$

so that the upper bound on the error becomes

$$|f(x) - p_2(x)| \leq \frac{1}{9\sqrt{3}} h^3 \max_{x_0 \leq t \leq x_2} |f'''(t)|. \tag{4.11}$$

This can be rendered into norm notation as

$$\|f - p_2\|_{\infty, I} \leq \frac{1}{9\sqrt{3}} h^3 \|f'''\|_{\infty, I}, \tag{4.12}$$

where I is again the interval defined by the smallest and largest nodes, and h is the spacing between nodes.

EXAMPLE 4.5

Let $f(x)$ be the error function, i.e.,

$$f(x) = \text{erf}(x) = \frac{2}{\sqrt{\pi}} \int_0^x e^{-t^2} dt.$$

Suppose we want to construct interpolating polynomial approximations to this function on the interval $[0,1]$ that are accurate to within 10^{-6}. We have

$$f''(x) = \frac{2}{\sqrt{\pi}} \left(e^{-x^2} \right)' = \frac{2}{\sqrt{\pi}} \left(-2xe^{-x^2} \right)$$

so that a crude upper bound on the second derivative is

$$\|f''\|_{\infty, [0,1]} \leq 2 \times 1 \times 2e^{-x^2}/\sqrt{\pi} \leq 2 \times 2e^0/\sqrt{\pi} \leq 2.5.$$

Therefore, the error in linear interpolation is bounded by $(h^2/8)(2.5) = h^2/3.2$; thus we need to have $h \leq \sqrt{3.2} \times 10^{-3}$ to get the error less than 10^{-6}. This tells us that the table of error function values must have $N = 1/h > 560$ points for linear interpolation between points to achieve the desired accuracy.

On the other hand, we have

$$f'''(x) = \frac{2}{\sqrt{\pi}} \left(-2e^{-x^2} + 4x^2 e^{-x^2} \right) = \frac{4}{\sqrt{\pi}} \left(2x^2 - 1 \right) e^{-x^2},$$

for which we have (again, crudely)

$$\|f'''\|_{\infty, [0,1]} \leq 4 \times (1) \times e^{-x^2}/\sqrt{\pi} \leq 4 \times e^0/\sqrt{\pi} \leq 5.$$

Hence the error in quadratic interpolation is bounded by $(h^3/9\sqrt{3})(5) \leq 0.4h^3$. To get the error less than the desired 10^{-6}, we then want to have

$$0.4h^3 \leq 10^{-6} \Rightarrow h \leq (0.4)^{-1/3} \times 10^{-2} \leq 1.4 \times 10^{-2}.$$

So we have to have $N = 1/h > 72$ points in the table to get the desired accuracy. ■

A similar analysis to that used to get (4.12) shows that the error in cubic interpolation satisfies

$$\|f - p_3\| \le \frac{1}{24}h^4 \|f^{(4)}\|_{\infty,I}. \tag{4.13}$$

See Exercise 4.

We could, of course, keep going on with specific cases, but since we will eventually learn (in Section 4.11.1) that interpolation with high-degree polynomials is not always a good idea, we will stop at this point.

EXERCISES

1. What is the error in quadratic interpolation to $f(x) = \sqrt{x}$, using equally spaced nodes on the interval $[\frac{1}{4},1]$?

2. Repeat the preceding for $f(x) = x^{-1}$ on $[\frac{1}{2},1]$.

3. Repeat the previous two exercises, using cubic interpolation.

4. Show that the error in third-degree polynomial interpolation satisfies

$$\|f - p_3\|_\infty \le \frac{1}{24}h^4 \|f^{(4)}\|_\infty,$$

if the nodes x_0, x_1, x_2, x_3 are equally spaced, with $x_i - x_{i-1} = h$. (Hint: Use the change of variable $t = x - x_1 - \frac{1}{2}h$.)

5. Show that the error in polynomial interpolation using six equally spaced points (quintic interpolation) satisfies

$$\|f - p_5\|_\infty \le Ch^6 \|f^{(6)}\|_\infty,$$

where $C \approx 0.0235$. (Hint: See the previous exercise.)

6. Generalize the derivation of the error bound (4.12) for quadratic interpolation to the case where the nodes are *not* equally spaced. Take $x_1 - x_0 = h$ and $x_2 - x_1 = \theta h$ for some $\theta > 0$.

7. Apply your result for the previous exercise to the error in quadratic interpolation to $f(x) = \sqrt{x}$ using the nodes $x_0 = \frac{1}{4}, x_1 = \frac{9}{16}$, and $x_2 = 1$.

8. If we want to use a table of exponential values to interpolate the exponential function on the interval $[-1,1]$, how many points are needed to guarantee 10^{-6} accuracy with linear interpolation? Quadratic interpolation?

9. If we want to use a table of values to interpolate the error function on the interval $[0,5]$, how many points are

needed to get 10^{-6} accuracy using linear interpolation? Quadratic interpolation? Would it make sense to use one grid spacing on, say, $[0,1]$ and another one on $[1,5]$? Explain.

10. If we want to use a table of values to interpolate the sine function on the interval $[0,\pi]$, how many points are needed for 10^{-6} accuracy with linear interpolation? Quadratic interpolation? Cubic interpolation?

11. Let's return to the computation of the natural logarithm. Consider a computer that stores numbers in the form $z = f \cdot 2^\beta$, where $\frac{1}{2} \le f \le 1$. We want to consider using this, in conjunction with interpolation ideas, to compute the natural logarithm function.

(a) Using piecewise linear interpolation over a grid of equally spaced points, how many table entries would be required to accurately approximate $\ln z$ to within 10^{-14}?

(b) Repeat the preceding, using piecewise quadratic interpolation.

(c) Repeat it again, using piecewise cubic interpolation.

Explain, in a brief essay, the importance here of restricting the domain of z to the interval $[\frac{1}{2},1]$.

12. Assume that, for any real c,

$$\lim_{n\to\infty} \frac{c^n}{n!} = 0.$$

Use this to prove that, if p_n is the polynomial interpolant to $f(x) = \sin x$ on the interval $[a,b]$, using *any* distribution of distinct nodes x_k, $0 \le k \le n$, then $\|f - p_n\|_\infty \to 0$ as $n \to \infty$. (Hint: Can we justify $|w_n(x)| \le c^{n+1}$ for some c?)

13. Can you repeat the preceding for $f(x) = e^x$? Why or why not?

14. Define the norms

$$\|f\|_\infty = \max_{x\in[0,1]} |f(x)|$$

and

$$\|f\|_2 = \left(\int_0^1 [f(x)]^2 \, dx \right)^{1/2}.$$

Compute $\|f\|_\infty$ and $\|f\|_2$ for the following list of functions:

(a) $\sin n\pi x$, n integer
(b) e^{-ax}, $a > 0$
(c) $\sqrt{x} e^{-ax^2}$, $a > 0$
(d) $1/\sqrt{1+x}$

15. Define

$$f_n(x) = \begin{cases} 1 - nx & 0 \le x \le 1/n \\ 0 & 1/n \le x \le 1. \end{cases}$$

Show that

$$\lim_{n \to \infty} \|f_n(x)\|_\infty = 1$$

but

$$\lim_{n \to \infty} \|f_n(x)\|_2 = 0.$$

(Hint: Draw a graph of $f_n(x)$.)

16. Show that

$$\|f\|_{1,[a,b]} = \int_a^b |f(x)| \, dx$$

defines a function norm.

4.4 APPLICATION: MULLER'S METHOD AND INVERSE QUADRATIC INTERPOLATION

We can use the idea of interpolation to develop more sophisticated root-finding methods.

The secant method can be viewed as finding the root of a linear interpolating polynomial. Our error results tell us that a quadratic (second-degree) polynomial should be a better approximation to the function; hence its root ought to be a better approximation to the root of the original function. Can we take this outline of an idea and use it to construct a practical method for finding roots?

The answer is yes, and the method is known as *Muller's method*. Given three points, x_0, x_1, and x_2, we find the quadratic polynomial $p(x)$ such that $p(x_i) = f(x_i)$, $i = 0,1,2$; and then define x_3 as the root of p that is closest to x_2. The actual implementation of Muller's method does present a few problems that need to be resolved, however.

The best way to write the interpolating polynomial is in the Newton form, as[7]

$$p(x) = f(x_2) + A(x - x_2) + B(x - x_2)(x - x_1), \tag{4.14}$$

where A and B are the divided differences (see Exercise 7)

$$A = \frac{f(x_2) - f(x_1)}{x_2 - x_1} \tag{4.15}$$

and

$$B = \frac{1}{x_2 - x_0} \left(\frac{f(x_2) - f(x_1)}{x_2 - x_1} - \frac{f(x_1) - f(x_0)}{x_1 - x_0} \right). \tag{4.16}$$

---

[7]Note that we are here treating x_2 as the "first" node, instead of x_0. This is possible because the order of the nodes is irrelevant in the computation, although this does affect the formulas somewhat.

Note that these coefficients will have to be recomputed each iteration.

Finding the root of (4.14) via the quadratic formula requires that some care be taken. Since we want to find the root nearest to x_2, we rewrite p as a quadratic in $x - x_2$:

$$
\begin{aligned}
p(x) &= f(x_2) + A(x - x_2) + B(x - x_2)(x - x_1) \\
&= f(x_2) + A(x - x_2) + B(x - x_2)(x - x_2) + B(x - x_2)(x_2 - x_1) \\
&= f(x_2) + (A + B(x_2 - x_1))(x - x_2) + B(x - x_2)(x - x_2) \\
&= f(x_2) + C(x - x_2) + B(x - x_2)^2,
\end{aligned}
$$

where $C = A + B(x_2 - x_1)$. The quadratic formula then says that the root of p is defined by

$$
x - x_2 = \frac{1}{2B}\left(-C \pm \sqrt{C^2 - 4f(x_2)B}\right)
$$

where we choose the sign to minimize the right side of the equation. To avoid the pitfalls of subtractive cancellation, we rationalize the numerator to get

$$
x - x_2 = -\frac{2f(x_2)}{C \pm \sqrt{C^2 - 4f(x_2)B}}
$$

where the sign is chosen to maximize the denominator. Thus, the next point in the iteration is given by

$$
x_3 = x_2 - \frac{2f(x_2)}{C + \operatorname{sgn}(C)\sqrt{C^2 - 4f(x_2)B}}
$$

where sgn is the "sign" function defined by

$$
\operatorname{sgn}(x) = \begin{cases} -1 & x < 0 \\ 1 & x \geq 0 \end{cases}.
$$

Thus, the general case is given by

$$
x_{n+1} = x_n - \frac{2f(x_n)}{C_n + \operatorname{sgn}(C_n)\sqrt{C_n^2 - 4f(x_n)B_n}},
$$

where

$$
B_n = \frac{1}{x_n - x_{n-2}}\left(\frac{f(x_n) - f(x_{n-1})}{x_n - x_{n-1}} - \frac{f(x_{n-1}) - f(x_{n-2})}{x_{n-1} - x_{n-2}}\right)
$$

and

$$
C_n = \frac{f(x_n) - f(x_{n-1})}{x_n - x_{n-1}} + B_n(x_n - x_{n-1}).
$$

TABLE 4.13 Muller's Method, $f(x) = 2 - e^x$

n	x_n	$f(x_n)$	$\alpha - x_n$	$\log_{10}(\alpha - x_n)$
0	0.000000000000	0.100000E+01	0.693147E+00	−0.1592
1	0.500000000000	0.351279E+00	0.193147E+00	−0.7141
2	1.000000000000	−0.718282E+00	0.306853E+00	−0.5131
3	0.687259367753	0.117410E−01	0.588781E−02	−2.2300
4	0.693087847691	0.118662E−03	0.593329E−04	−4.2267
5	0.693147161269	0.385812E−07	0.192906E−07	−7.7147
6	0.693147180560	−0.222045E−14	0.122125E−14	−14.9132
7	0.693147180560	0.000000E+00	0.111022E−15	−15.9546

Table 4.13 shows the result of applying Muller's method to the same example we used thoughout Chapter 3, $f(x) = 2 - e^x$, so that the exact root is $\alpha = \ln 2$. Compare the number of iterations used here to that of the secant method (Table 3.6); note that we can code Muller's method to use only a single new function evaluation each step.

An alternative to Muller's method that avoids the difficulties associated with the quadratic formula is *inverse quadratic interpolation*. In this case we find the quadratic polynomial $q(y)$ such that

$$q(f(x_i)) = x_i, \ i = 0,1,2.$$

Direct computation with the Newton form shows that

$$q(y) = x_2 + a(y - y_2) + b(y - y_2)(y - y_1),$$

where $y_k = f(x_k)$,

$$a = \frac{x_2 - x_1}{y_2 - y_1} \tag{4.17}$$

and

$$b = \frac{1}{y_2 - y_0} \left(\frac{x_2 - x_1}{y_2 - y_1} - \frac{x_1 - x_0}{y_1 - y_0} \right). \tag{4.18}$$

Thus q is quadratic in y and still passes through the points $(x_i, f(x_i))$; we have essentially reversed the notion of ordinate and abscissa in our coordinate system. The approximate root is the value of $q(y) = x$ corresponding to $y = 0$; thus it is obtained by the simple expedient of evaluating q at 0:

$$x_3 = q(0) = x_2 - ay_2 + by_2y_1.$$

In the general case, we have

$$x_{n+1} = x_n - a_n f(x_n) + b_n f(x_n) f(x_{n-1}),$$

TABLE 4.14 Inverse Quadratic Interpolation, $f(x) = 2 - e^x$

n	x_n	$f(x_n)$	$\alpha - x_n$	$\log_{10}(\alpha - x_n)$
0	0.000000000000	0.100000E+01	0.693147E+00	−0.1592
1	0.500000000000	0.351279E+00	0.193147E+00	−0.7141
2	1.000000000000	−0.718282E+00	0.306853E+00	−0.5131
3	0.708748678748	−0.314477E−01	0.156015E−01	−1.8068
4	0.692849949759	0.594373E−03	0.297231E−03	−3.5269
5	0.693146743432	0.874255E−06	0.437127E−06	−6.3594
6	0.693147180561	−0.134603E−11	0.673128E−12	−12.1719
7	0.693147180560	0.000000E+00	0.111022E−15	−15.9546

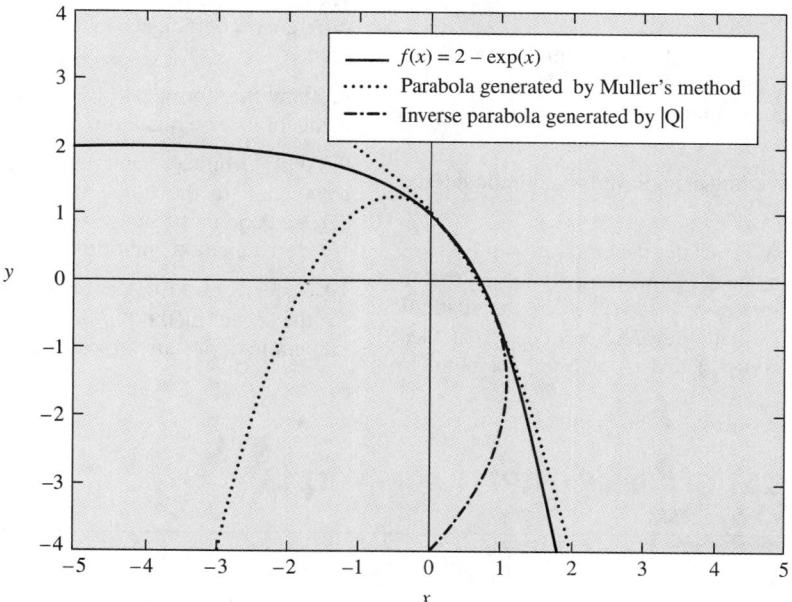

FIGURE 4.5 Muller's method and inverse quadratic interpolation.

where a_n and b_n are computed from appropriately generalized versions of (4.17) and (4.18). Table 4.14 shows the result of applying inverse quadratic interpolation to the example $f(x) = 2 - e^x$.

Figure 4.5 shows the graph of $f(x) = 2 - e^x$, along with the graphs of the parabola generated by Muller's method and the inverse parabola generated by inverse quadratic interpolation, using $x_0 = 0$, $x_1 = 1$, and $x_2 = 1/2$ as the initial values in each case.

The convergence theory for both Muller's method and inverse quadratic interpolation is beyond the intended scope of this text, so we will content ourselves with reporting that both methods are locally convergent and superlinear, with $p \approx 1.84$. Note that this order of convergence is not much greater than for the secant method.

One great utility of Muller's method is that it is able to find complex roots of real-valued functions, because of the square root in the computation. (The same is true of Halley's method (3.44).)

Inverse quadratic interpolation is used as part of the Brent-Dekker root-finding algorithm, which is a commonly implemented automatic root-finding program.

EXERCISES

1. Do three steps of Muller's method for $f(x) = 2 - e^x$, using the initial points $0, 1/2, 1$. Make sure you reproduce the values in the table in the text.

2. Repeat the preceding for inverse quadratic interpolation.

3. Let $f(x) = x^6 - x - 1$; show that this has a root on the interval $[1,3]$, and do three steps of Muller's method using the ends of the interval plus the midpoint as the initial values.

4. Repeat the preceding using inverse quadratic interpolation.

5. Let $f(x) = e^x$ and consider the nodes $x_0 = -1, x_1 = 0$, and $x_2 = 1$. Let p_2 be the quadratic polynomial that interpolates f at these nodes, and let q_2 be the quadratic polynomial (in y) that *inverse*-interpolates f at these nodes. Construct $p_2(x)$ and $q_2(y)$ and plot both, to-

gether with $f(x)$.

6. Repeat the preceding using $f(x) = \sin \pi x$ and the nodes $x_0 = 0$, $x_1 = \frac{1}{4}$, and $x_2 = \frac{1}{2}$.

7. Show that formulas (4.15) and (4.16) for the divided-difference coefficients in Muller's method (4.14) are correct.

8. Show that formulas (4.17) and (4.18) for the coefficients in inverse quadratic interpolation are correct.

9. Apply Muller's method and inverse quadratic interpolation to the functions in Exercise 3 of Section 3.1, and compare your results with those obtained with Newton's method and/or the secant method.

10. Refer back to the discussion of the error estimate for the secant method in Section 3.10.3. Adapt this argument to derive an error estimate for Muller's method.

4.5 APPLICATION: MORE APPROXIMATIONS TO THE DERIVATIVE

We can use polynomial interpolation to derive more formulas for approximating the derivative. Essentially, since interpolation error theory shows that polynomials can approximate more complicated functions reasonably well, then we infer that the derivative of a polynomial interpolant might approximate the derivative of the complicated function reasonably well.

The interpolation error theorem gives us

$$f(x) - p_n(x) = \frac{1}{(n+1)!} w_n(x) f^{(n+1)}(\xi_x),$$

and the fact that this is an equality is important, for we can now differentiate both sides to get

$$f'(x) - p_n'(x) = \frac{1}{(n+1)!} \frac{d}{dx} \left[w_n(x) f^{(n+1)}(\xi_x) \right].$$

We have to be careful evaluating the derivative of the bracketed term; recall that ξ_x depends on x, so that

$$\frac{d}{dx}\left[w_n(x)f^{(n+1)}(\xi_x)\right] = w_n'(x)f^{(n+1)}(\xi_x) + w_n(x)\frac{d}{dx}\left[f^{(n+1)}(\xi_x)\right]. \qquad (4.19)$$

The problem in using this formula is that we do not know how ξ_x depends on x, and thus we cannot estimate or predict the size of the derivative in the second term. However, if we evaluate the approximation at one of the nodes, x_i, then we will have $w_n(x_i) = 0$ and the second term on the right side of (4.19) will vanish, leaving us with:

$$f'(x_i) - p_n'(x_i) = \frac{1}{(n+1)!}w_n'(x_i)f^{(n+1)}(\xi_i), \qquad (4.20)$$

where we use $\xi_i = \xi_{x_i}$ for simplicity of notation.[8]

We can illustrate this by using a quadratic polynomial to approximate f and thus derive three formulas for approximating the first derivative. Let the nodes be denoted by x_0, x_1, and x_2, and assume they are equally spaced, so we have $x_1 - x_0 = h$, $x_2 - x_1 = h$, etc. Then the interpolating polynomial in Lagrange form is

$$p_2(x) = f(x_0)L_0^{(2)}(x) + f(x_1)L_1^{(2)}(x) + f(x_2)L_2^{(2)}(x).$$

The approximate derivative is then

$$f'(x_i) \approx p_2'(x_i) = f(x_0)(L_0^{(2)})'(x_i) + f(x_1)(L_1^{(2)})'(x_i) + f(x_2)(L_2^{(2)})'(x_i),$$

where $i = 0, 1, 2$. The error is given as in (4.20); thus

$$f'(x_i) - p_2'(x_i) = \frac{1}{6}w_2'(x_i)f'''(\xi_i)$$

and we need only evaluate $(L_j^{(2)})'(x_i)$ and $w_2'(x_i)$ for $i, j = 0, 1, 2$, to complete the construction of the approximations. We get

$$(L_0^{(2)})'(x_0) = -3/2h \qquad (L_1^{(2)})'(x_0) = 2/h \qquad (L_2^{(2)})'(x_0) = -1/2h$$

$$(L_0^{(2)})'(x_1) = -1/2h \qquad (L_1^{(2)})'(x_1) = 0 \qquad (L_2^{(2)})'(x_1) = 1/2h$$

$$(L_0^{(2)})'(x_2) = 1/2h \qquad (L_1^{(2)})'(x_2) = -2/h \qquad (L_2^{(2)})'(x_2) = 3/2h$$

and

$$w_2'(x_0) = 2h^2 \qquad w_2'(x_1) = -h^2 \qquad w_2'(x_2) = 2h^2$$

---

[8] Note that we have assumed here that ξ_x is a differentiable function of x. This is indeed the case, but actually proving it is a nontrivial exercise that we omit for the sake of a brief and concise presentation.

so that the approximations are

$$f'(x_0) \approx \frac{1}{2h}\left(-f(x_2)+4f(x_1)-3f(x_0)\right) \qquad (4.21)$$

$$f'(x_1) \approx \frac{1}{2h}\left(f(x_2)-f(x_0)\right) \qquad (4.22)$$

$$f'(x_2) \approx \frac{1}{2h}\left(3f(x_2)-4f(x_1)+f(x_0)\right) \qquad (4.23)$$

with errors

$$f'(x_0) - \frac{1}{2h}\left(-f(x_2)+4f(x_1)-3f(x_0)\right) = \frac{1}{3}h^2 f'''(\xi_0) \qquad (4.24)$$

$$f'(x_1) - \frac{1}{2h}\left(f(x_2)-f(x_0)\right) = -\frac{1}{6}h^2 f'''(\xi_1) \qquad (4.25)$$

$$f'(x_2) - \frac{1}{2h}\left(3f(x_2)-4f(x_1)+f(x_0)\right) = \frac{1}{3}h^2 f'''(\xi_2). \qquad (4.26)$$

We recognize the approximation (4.22) as the same central difference formula derived back in Section 2.2; the other two formulas are new. All three are second-order accurate, in that the error is $O(h^2)$, but (4.21) and (4.23) require three function evaluations whereas (4.22) requires only two. In some sense this makes (4.22) a "better" approximation to use, but in Sections 4.8, 5.2, and 6.6.2 we will see situations where (4.21) and (4.23) are more useful.

Consider, for example, a function defined only on a grid of equally spaced points. Thus we have $y_k = f(x_k)$ for only the nodes x_k. Suppose further that we need an accurate approximation to $f'(x_0)$ (the derivative at the left endpoint) or $f'(x_n)$ (the derivative at the right endpoint). We can't use the centered difference approximation (2.5) at the endpoints, but we can use (4.24) and (4.26) to get

$$f'(x) = \left(\frac{-f(x+2h)+4f(x+h)-3f(x)}{2h}\right) + O(h^2)$$

and

$$f'(x) = \left(\frac{3f(x)-4f(x-h)+f(x-2h)}{2h}\right) + O(h^2).$$

This will be useful when we need derivative approximations at the ends of intervals for constructing spline approximations (Section 4.8) or for improving the trapezoid rule (Section 5.2).

EXERCISES

1. Apply the derivative approximations (4.21) and (4.23) to the approximation of $f'(x)$ for $f(x) = x^3 + 1$ for $x = 1$ and $h = \frac{1}{8}$.

2. Apply the derivative approximations (4.21) and (4.23) to the same set of functions as in Exercise 3 of

Section 2.2, using a decreasing sequence of mesh values, $h^{-1} = 2,4,8,\ldots$. Do we achieve the expected rate of accuracy as h decreases?

3. Derive a version of (4.24) under the assumption that $x_1 - x_0 = h$, but $x_2 - x_1 = \eta = \theta h$ for some real, positive

θ. Be sure to include the error estimate as part of your work, and confirm that when $\theta = 1$ you get the same results as in the text.

4. Repeat the preceding for (4.25).

5. Repeat the preceding for (4.26).

6. Use the derivative approximations from this section to construct a table of values for the derivative of the gamma function, based on the data in Table 4.8.

7. Try to extend the ideas of this section to construct an approximation to $f''(x_k)$. Is it possible? What happens?

4.6 HERMITE INTERPOLATION

So far we have studied interpolation that involves only knowledge of function values. What happens if we include information about the derivative as well?

To be specific, we want to find a polynomial $p(x)$ that interpolates $f(x)$ in the sense that[9]

$$p(x_i) = f(x_i), \quad p'(x_i) = f'(x_i), \quad 1 \le i \le n.$$

This is known as the Hermite[10] interpolation problem. Can we do this? The answer is yes, as seen in the following theorem.

Theorem 4.4 (Hermite Interpolation Theorem)   Given the n nodes $x_i, 1 \le i \le n$, and a differentiable function $f(x)$, if the nodes are distinct, then there exists a unique polynomial H_n of degree less than or equal to $2n - 1$, such that

$$H_n(x_i) = f(x_i), \qquad H_n'(x_i) = f'(x_i), \qquad 1 \le i \le n. \tag{4.27}$$

Proof   This is basically the same as the proof of Theorem 4.1. We define the two families of polynomials

$$h_k(x) = [1 - 2[L_k^{(n)}]'(x_k)(x - x_k)][L_k^{(n)}(x)]^2$$

and

$$\tilde{h}_k(x) = (x - x_k)[L_k^{(n)}(x)]^2;$$

now observe that

$$h_k(x_j) = \delta_{kj}, \qquad h_k'(x_j) = 0,$$

and

$$\tilde{h}_k(x_j) = 0, \qquad \tilde{h}_k'(x_j) = \delta_{kj}.$$

---

[9]Note the change in indexing convention: We are now indexing from 1 to n, instead of from 0 to n.

[10]Charles Hermite (1822–1901) was born in Dieuze, France, on Christmas Eve. A poor test taker, Hermite was not able to pass the exams for his bachelor's degree until the age of nearly 26. He held professional positions at a number of French schools, most notably the École Polytechnique, the École Normale, and the Sorbonne.

Perhaps his best-known mathematical result is the first proof that e is a transcendental number (published in 1873). His name is attached to a number of mathematical ideas and concepts, including the Hermite differential equation, Hermite polynomials, and Hermitian matrices. The idea of interpolating to the derivative as well as the function values was part of a paper he published in 1878, *Sur la formule d'interpolation de Lagrange*, which actually considers interpolation not only of the first derivative, but of higher derivatives as well. Hermite interpolation is sometimes called "osculatory interpolation."

TABLE 4.15 **Table for Divided Differences for Hermite Interpolation: Initial Setup**

k	x_k	$f_0(x_k)$	$f_1(x_k)$
1	a	$f(a)$	
			$f'(a)$
1	a	$f(a)$	
2	b	$f(b)$	
			$f'(b)$
2	b	$f(b)$	

Therefore, the polynomial

$$H_n(x) = \sum_{k=1}^{n} \left(f(x_k)h_k(x) + f'(x_k)\tilde{h}_k(x) \right)$$

satisfies the interpolating conditions (4.27) and is clearly of degree less than or equal to $2n - 1$. Uniqueness follows by the same argument used before. □

Like the Lagrange form in ordinary interpolation, this form of the Hermite polynomial is not conducive to efficient computation. A variation of the Newton approach, including the use of divided-difference tables, can be constructed; however, since the most common usage of Hermite interpolation is the cubic case, we will forgo the more general development and show only the cubic construction.

We begin by setting up a divided-difference table (Table 4.15), much as we did in Section 4.2. Note, however, that we enter the functional data twice and use the derivative data for part of the first-differences column.

We then complete the table just as we would an ordinary divided-difference table, getting the results shown in Table 4.16. The elements denoted by letters are defined as follows:

$$A = \frac{f(b) - f(a)}{b - a}$$

$$B = \frac{A - f'(a)}{b - a}$$

$$C = \frac{f'(b) - A}{b - a}$$

$$D = \frac{C - B}{b - a}.$$

The polynomial is then given by

$$H_2(x) = f(a) + f'(a)(x - a) + B(x - a)^2 + D(x - a)^2(x - b). \tag{4.28}$$

TABLE 4.16   Table for Divided Differences for Hermite Interpolation: Final Form

k	x_k	$f_0(x_k)$	$f_1(x_k)$	$f_2(x_k)$	$f_3(x_k)$
1	a	$f(a)$			
			$f'(a)$		
1	a	$f(a)$		B	
			A		D
2	b	$f(b)$		C	
			$f'(b)$		
2	b	$f(b)$			

It can easily be confirmed that this does interpolate to f and f' at both a and b (see Exercise 1).

As an illustration of Hermite interpolation, consider the problem of interpolating to the exponential function on the interval $[-1,1]$ using data only at the endpoints. We have $x_1 = -1$, $x_2 = 1$, and the divided-difference table is shown as Table 4.17, so that the polynomial is

$$H_2(x) = 2.718281828 + 2.718281828(x-1) + 0.771540317(x-1)^2$$
$$+ 0.1839397203(x-1)^2(x+1),$$

or (after some simplification),

$$H_2(x) = 0.1839397206x^3 + 0.5876005967x^2 + 0.9912614728x + 0.9554800379.$$

Figure 4.6 shows a graph of both the exponential and H_2; Figure 4.7 shows the error $e^x - H_2(x)$.

TABLE 4.17   Table of Divided Differences for Cubic Hermite Interpolation to $f(x) = e^x$

k	x_k	$f_0(x_k)$	$f_1(x_k)$	$f_2(x_k)$	$f_3(x_k)$
1	1	2.718281828			
			2.718281828		
1	1	2.718281828		0.771540317	
			1.175201194		0.1839397203
2	−1	0.3678794412		0.4036608764	
			0.3678794412		
2	−1	0.3678794412			

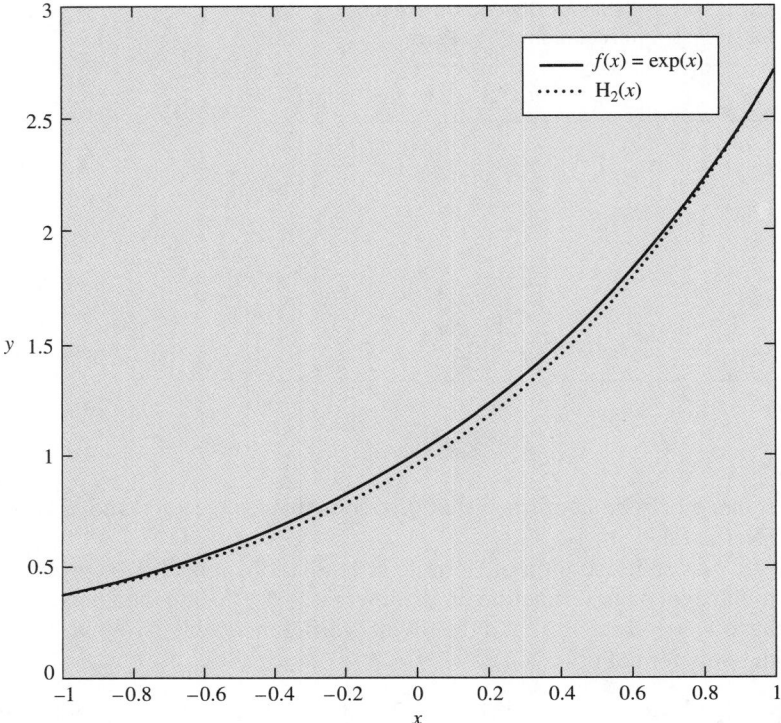

FIGURE 4.6 Hermite interpolation to $f(x) = e^x$.

FIGURE 4.7 Error in Hermite interpolation to $f(x) = e^x$.

The accuracy of Hermite interpolation is based on a result very similar to Theorem 1.3.

Theorem 4.5 (Hermite Interpolation Error Theorem)   Let $f \in C^{2n}([a,b])$ and let the nodes $x_k \in [a,b]$ for all k, $1 \leq k \leq n$. Then, for each $x \in [a,b]$, there is a $\xi_x \in [a,b]$ such that

$$f(x) - H_n(x) = \frac{\psi_n(x)}{(2n)!} f^{(2n)}(\xi_x), \tag{4.29}$$

where

$$\psi_n(x) = \prod_{k=1}^{n} (x - x_k)^2.$$

Proof   The proof follows essentially the same argument as Theorem 1.3; see Appendix A.1 for details.   ☐

EXERCISES

1. Show that H_2, as defined in (4.28), is the cubic Hermite interpolant to f at the nodes $x = a$ and $x = b$.

2. Construct the cubic Hermite interpolant to $f(x) = \sin x$ using the nodes $a = 0$ and $b = \pi$. Plot the error between the interpolant and the function.

3. Construct the cubic Hermite interpolant to $f(x) = \sqrt{1+x}$ using the nodes $a = 0$ and $b = 1$. Plot the error between the interpolant and the function.

4. Show that the error in cubic Hermite interpolation at the nodes $x = a$ and $x = b$ is given by

$$\|f - H_2\|_\infty \leq \frac{(b-a)^4}{384} \|f^{(4)}\|_\infty.$$

5. Construct the cubic Hermite interpolant to $f(x) = \sqrt{x}$ on the interval $[\frac{1}{4}, 1]$. What is the maximum error on this interval as predicted by theory? What is the maximum error that actually occurs (as determined by observation—no need to do a complete calculus max/min problem)?

6. Construct the cubic Hermite interpolant to

$f(x) = 1/x$ on the interval $[\frac{1}{2}, 1]$. What is the maximum error as predicted by theory? What is the actual (observed) maximum error?

7. Construct the cubic Hermite interpolant to $f(x) = x^{1/3}$ on the interval $[\frac{1}{8}, 1]$. What is the maximum error as predicted by theory? What is the actual (observed) maximum error?

8. Construct the cubic Hermite interpolant to $f(x) = \ln x$ on the interval $[\frac{1}{2}, 1]$. What is the maximum error as predicted by theory? What is the actual (observed) maximum error?

9. Extend the divided-difference table for cubic Hermite interpolation to *quintic* Hermite interpolation, using the three nodes $x = a$, $x = b$, and $x = c$.

10. Construct the quintic Hermite interpolant to $f(x) = \ln x$ on the interval $[\frac{1}{2}, 1]$; use $x = 3/4$ as the third node.

11. What is the error in quintic Hermite interpolation?

12. Extend the ideas of Section 4.5 to allow us to compute second derivative approximations using Hermite interpolation.

4.7   PIECEWISE POLYNOMIAL INTERPOLATION

The results in Figure 4.3 suggest that interpolation might be unsatisfactory as an approximation tool. This is true if we insist on letting the *order* of the polynomial get larger and larger (see Section 4.11). However, if we keep the order of the polynomial fixed and use different polynomials over different intervals, with the length of the intervals getting smaller and smaller, then interpolation can be a very accurate and powerful approximation tool. These ideas reach their full potential in Section 4.8, on splines; here we confine ourselves to the basics.

Consider the problem of constructing an approximation to a function $f(x)$, defined only by 101 equally spaced points on the interval $[-1,1]$. If we use ordinary interpolation, we will have to construct a 100-degree polynomial, and we suspect that this might produce unsatisfactory results. However, we could use linear interpolation over each subinterval $[x_{k-1}, x_k]$ or quadratic interpolation over each subinterval $[x_{2k-2}, x_{2k}]$. That is, we define the approximating function as

$$q_d(x) = p_{d,k}(x), \qquad x_{d(k-1)} \leq x \leq x_{dk},$$

where each $p_{d,k}$ is a polynomial of degree d. The accuracy of the approximation comes from the error estimate (4.7), (4.11), or (4.13), since the distance between nodes is small.

We illustrate this by approximating $f(x) = (1 + 25x^2)^{-1}$ over $[-1,1]$ using the seven equally spaced points $x_k = -1 + k/3$, $k = 0,1,\ldots,6$. The results are shown in Figure 4.8 for piecewise linear, quadratic, and cubic interpolation. Thus, for the piecewise quadratic case, we have (see Exercise 3)

$$q_2(x) = \begin{cases} 0.0385 + 0.1323(x-x_0) + 0.6211(x-x_0)(x-x_1) & x_0 \leq x \leq x_2 \\ 0.2647 + 2.2059(x-x_2) - 6.6176(x-x_2)(x-x_3) & x_2 \leq x \leq x_4 \qquad (4.30) \\ 0.2647 - 0.5464(x-x_4) + 0.6211(x-x_4)(x-x_5) & x_4 \leq x \leq x_6. \end{cases}$$

Note that all of these piecewise approximations are much more accurate than the 16-degree interpolating polynomial that was constructed in Figure 4.3(c). This contrast is heightened even more when we take lots of points. Figures 4.9 and 4.10 show the results using 33 equally spaced points and piecewise quadratic approximation; compare this with Figure 4.3(c).

The use of different polynomials over different intervals complicates the entire process somewhat. We now have one set of points (the nodes, x_k, $0 \leq k \leq n$) that define the interpolation conditions, and a second set of points, called *knots*, that define the subintervals on which the separate polynomial pieces are defined. (In our presentation here we assume that the knots are a subset of the nodes, but this doesn't have to be the case.) For a piecewise polynomial of degree d, the knots will be x_{dj}, $0 \leq j \leq m$, where m is the number of polynomial pieces in the approximation. Thus, for the quadratic example defined in (4.30), the knots are x_0, x_2, x_4, x_6.

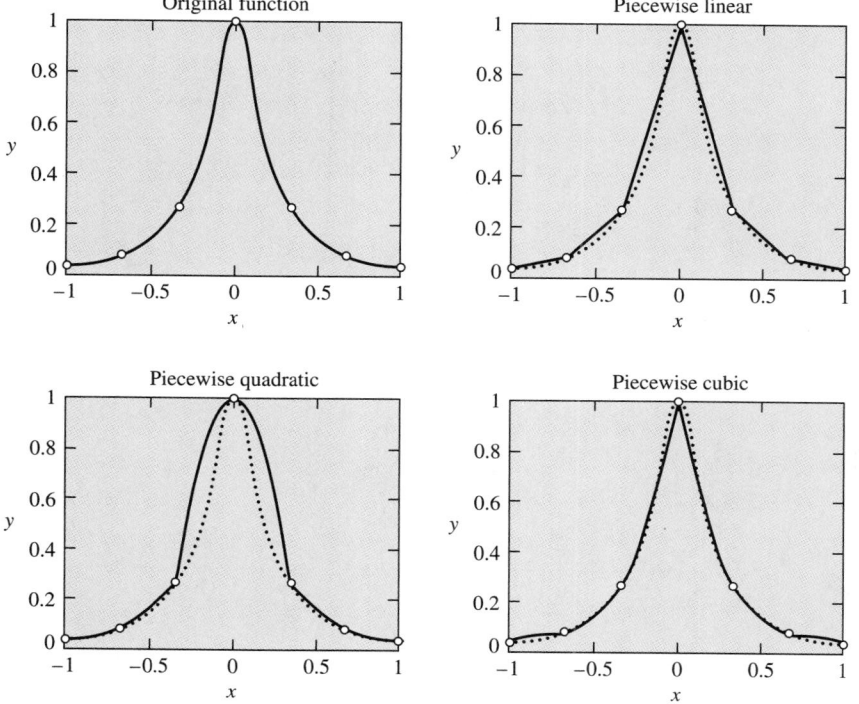

FIGURE 4.8 Piecewise interpolation to $f(x) = (1 + 25x^2)^{-1}$. The nodes are denoted by circles.

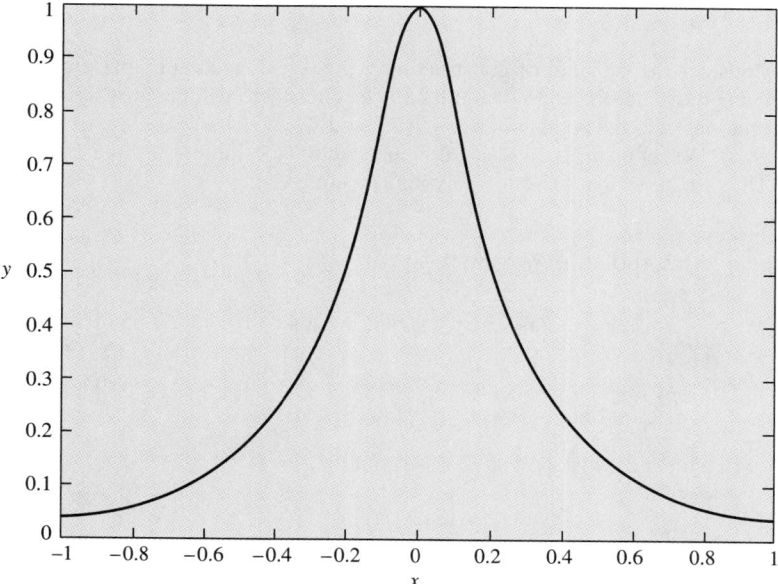

FIGURE 4.9 Piecewise quadratic interpolation to $f(x) = (1 + 25x^2)^{-1}$ using 33 nodes.

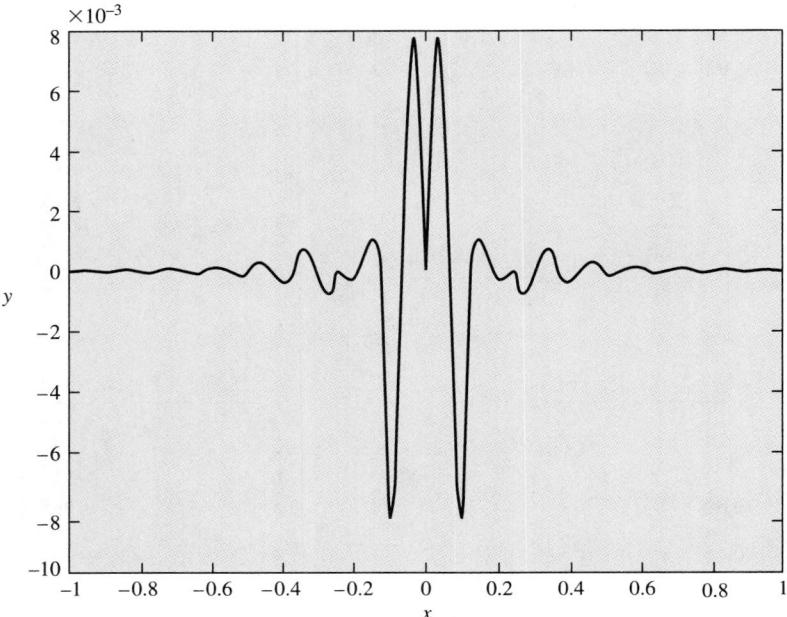

FIGURE 4.10 Error in 33-node piecewise quadratic interpolation to $f(x) = (1 + 25x^2)^{-1}$.

EXAMPLE 4.6

Consider the problem of constructing a piecewise quadratic approximation to $f(x) = \sin \pi x$, using the nodes $x_k = k/6$, $k = 0, 1, 2, \ldots, 6$. Thus there will be three separate polynomial approximations, and the knots will be x_0, x_2, x_4, and x_6.

We set up three divided-difference tables (see Table 4.18), one for each polynomial piece. Thus we know that the three polynomials are given by

$$p_{2,1}(x) = 3.0000000x - 2.4115427x \left(x - \frac{1}{6} \right)$$

$$p_{2,2}(x) = 0.8660254 + 0.8038476 \left(x - \frac{1}{3} \right) - 4.8230855 \left(x - \frac{1}{3} \right) \left(x - \frac{1}{2} \right)$$

$$p_{2,3}(x) = 0.8660254 - 2.1961524 \left(x - \frac{2}{3} \right) - 2.4115427 \left(x - \frac{2}{3} \right) \left(x - \frac{5}{6} \right),$$

so that the piecewise polynomial is given by

$$q_2(x) = \begin{cases} 3.0000000x - 2.4115427x(x - \frac{1}{6}) & 0 \leq x \leq \frac{1}{3} \\ 0.8660254 + 0.803847576(x - \frac{1}{3}) - 4.8230855(x - \frac{1}{3})(x - \frac{1}{2}) & \frac{1}{3} \leq x \leq \frac{2}{3} \\ 0.8660254 - 2.1961524(x - \frac{2}{3}) - 2.4115427(x - \frac{2}{3})(x - \frac{5}{6}) & \frac{2}{3} \leq x \leq 1. \end{cases} \quad (4.31)$$

TABLE 4.18   Divided Difference Tables for Piecewise Polynomial Approximation to $f(x) = \sin \pi x$

x_k	$f_0(x_k)$	$f_1(x_k)$	$f_2(x_k)$
0	0.0000000		
		3.0000000	
$\frac{1}{6}$	0.5000000		−2.4115427
		2.1961524	
$\frac{1}{3}$	0.8660254		

x_k	$f_0(x_k)$	$f_1(x_k)$	$f_2(x_k)$
$\frac{1}{3}$	0.8660254		
		0.80384758	
$\frac{1}{2}$	1.0000000		−4.8230855
		−0.080384758	
$\frac{2}{3}$	0.8660254		

x_k	$f_0(x_k)$	$f_1(x_k)$	$f_2(x_k)$
$\frac{2}{3}$	0.8660254		
		−2.1961524	
$\frac{5}{6}$	0.5000000		−2.4115427
		−3.0000000	
1	0.0000000		

Figure 4.11 shows a plot of this function, using a different line style for each polynomial piece. The nodes and knots are also marked on the graph.   ■

An algorithm for piecewise polynomial approximation will generally consist of two parts: one that constructs the approximation, usually in the form of an array of polynomial coefficients or divided-difference coefficients (which is what we did in the example above), and a second part that uses the output of the first part to evaluate the approximation. Pseudocodes for these are given in Algorithms 4.3 and 4.4.

The underlying error theorem is easy to state and prove.

Theorem 4.6 (Piecewise Polynomial Interpolation Error Estimate)   Let f be sufficiently differentiable on the closed interval $[a,b]$, and let q_d be the piecewise polynomial interpolant of degree d to f on $[a,b]$, using $n+1$ equally spaced nodes x_k, $0 \le k \le n$, $n = md$. Then

$$\|f - q_d\|_\infty \le C_d h^{d+1} \|f^{(d+1)}\|_\infty, \quad d = 1,2,3,$$

where

$$C_1 = 1/8 \quad C_2 = 1/9\sqrt{3}, \quad C_3 = 1/24,$$

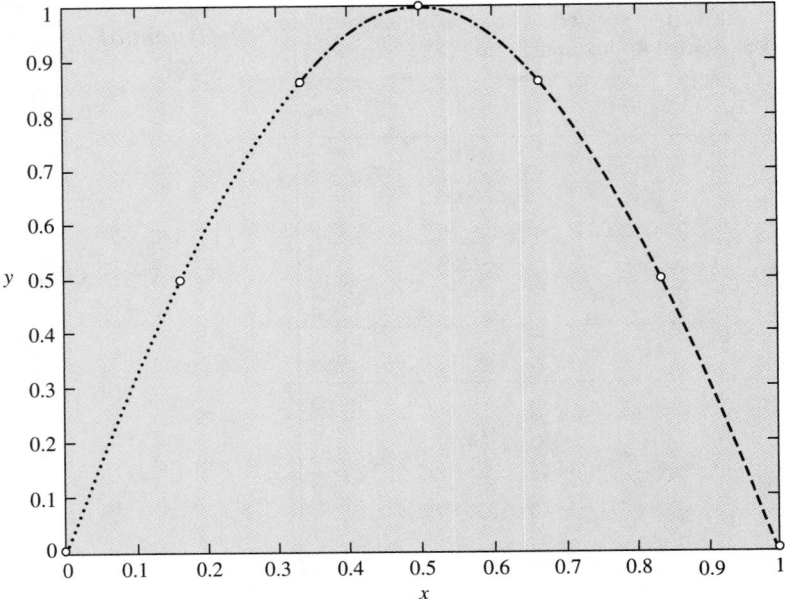

FIGURE 4.11   Piecewise polynomial approximation to $y = \sin \pi x$ over $[0,1]$; o denotes a node, and + denotes a knot.

Algorithm 4.3: Pseudocode for constructing piecewise polynomial approximation

```
input n, (x(i),y(i),i=0,n), d
!
kount = 0
for k=0 to n by d do
     kount = kount + 1
     for i=0 to d do
             xc(i) = x(k+i)
             yc(i) = y(k+i)
     endfor
!
!    compute divided-difference coefficients
!    for the xc and yc arrays
!
     ac(0) = yc(0)
     for i=1 to d do
             w = 1
             p = 0
             for j=0 to i-1 do
                     p = p + ac(j)*w
                     w = w*(xc(i) - xc(j))
             endfor
             ac(i) = (yc(i) - p)/w
     endfor
     for i=0 to d do
             a(kount,i) = ac(i)
     endfor
```

Algorithm 4.4: *Pseudocode for evaluating piecewise polynomial approximation at a given x value*

```
input n, d, kount, (x(i),i=0,n), (a(k,i),k=1,kount,i=0,d), xx
!
!    This code evaluates the pcw poly at the point xx
!
!    First, find the subinterval containing the evaluation
!    point xx
!
!    j = search(x,xx)
!
!    Next, extract the correct nodes and DD coefficients
!    (Not really necessary, but the code is easier to
!    follow this way.)
!
     for k=0 to d
            ac(k) = a(j,k)
            xc(k) = x(j+k)
     endfor
     yy = ac(j,d)
     for k = d-1 to 0 by -1
            xd = xx - xc(k)
            yy =  ac(k) + yy*xd
     endfor
```

Programming Hints: Note the line `j = search(x,xx)` in the evaluation algorithm. Evaluating a piecewise polynomial requires, first of all, that we find out which "piece" of the polynomial to use for each given x value for which we want to compute $q_d(x)$, and this pseudocode assumes that this is done in the separate routine called `search`. If the mesh spacing is uniform, then the computation

$$j = \left\lfloor \frac{x - x_0}{h} \right\rfloor$$

guarantees that $x_j \le x \le x_{j+1}$. Here $\lfloor \cdot \rfloor$ is the "floor" function, defined as the greatest integer less than or equal to the argument. From this it is an easy task to figure out which polynomial piece to use to compute q_d. If the mesh spacing is not uniform, then a more general search procedure must be used to find the indices such that $x_{(d-1)k} \le x \le x_{dk}$.

Proof   The proof follows directly from the separate error estimates (4.8), (4.12), and (4.13). We have

$$\|f - q_d\|_\infty = \max_{a \le x \le b} |f(x) - q_d(x)| = \max_{1 \le k \le n} \left(\max_{x_{k-1} \le x \le x_k} |f(x) - q_d(x)| \right)$$

$$= \max_{1 \le k \le n} \left(\max_{x_{k-1} \le x \le x_k} |f(x) - p_{d,k}(x)| \right)$$

$$= \max_{1 \le k \le n} \|f - p_{d,k}\|_{\infty,[x_{k-1},x_k]} \le \max_{1 \le k \le n} C_d h^{d+1} \|f^{(d+1)}\|_{\infty,[x_{k-1},x_k]}$$

$$= C_d h^{d+1} \|f^{(d+1)}\|_{\infty,[a,b]} = C_d h^{d+1} \|f^{(d+1)}\|_\infty,$$

and we are done. $\square$

While piecewise polynomial approximation is heavily used in a variety of applications (we already used it, essentially, to construct the composite trapezoid rule back in Chapter 2), it is most important, perhaps, as a precursor to the development of spline approximations, our next main topic.

One interesting and useful form of piecewise polynomial approximation is the use of piecewise cubic Hermite interpolation. Here we use cubic Hermite polynomials in the obvious way to define the polynomial approximation between each pair of nodes x_{k-1} and x_k. Because we are also interpolating to the derivative values, the resulting approximation is much smoother than ordinary piecewise interpolation and is less prone to having "kinks" in the graph. Some of the exercises ask the student to look at this kind of approximation.

EXERCISES

1. Use divided-difference tables to construct the separate parts of the piecewise quadratic polynomial $q_2(x)$ that interpolates to $f(x) = \sin \frac{1}{2}\pi x$ at $x = 0, \frac{1}{4}, \frac{1}{2}, \frac{3}{4}, 1$. Plot the approximation and the error $\sin \frac{1}{2}\pi x - q_2(x)$.

2. Repeat the preceding using $f(x) = \sqrt{x}$ with the nodes $x = \frac{1}{5}, \frac{2}{5}, \frac{3}{5}, \frac{4}{5}, 1$.

3. Confirm that (4.30) is the correct piecewise quadratic approximation to $f(x) = 1/(1+25x^2)$ using the nodes $x_0 = -1$, $x_1 = -2/3$, $x_2 = -1/3$, $x_3 = 0$, $x_4 = 1/3$, $x_5 = 2/3$, and $x_6 = 1$.

4. Using the data in Table 4.8, construct a piecewise cubic interpolating polynomial to the gamma function, using the nodes $x = 1.0, 1.2, 1.3, 1.5$ for one piece and the nodes $x = 1.5, 1.7, 1.8, 2.0$ for the other piece. Use this approximation to estimate $\Gamma(x)$ for $x = 1.1, 1.4, 1.6$, and 1.9. How accurate is the approximation?

5. Using the results of Exercise 6 from Section 4.6 together with the data of function values from Exercise 10 of Section 4.2, construct a piecewise cubic Hermite interpolating polynomial for the gamma function, using the nodes $x = 1.0, 1.3, 1.7, 2.0$. Test the accuracy of the interpolation by using it to approximate $\Gamma(x)$ for $x = 1.1, 1.2, 1.4, 1.5, 1.6, 1.8, 1.9$.

6. Construct a piecewise cubic interpolating polynomial to $f(x) = \ln x$ on the interval $[\frac{1}{2}, 1]$ using the nodes

$$x_k = \frac{1}{2} + \frac{k}{18}, \quad 0 \le k \le 9.$$

Compute the value of the error $\ln x - p(x)$ at 500 equally spaced points on the interval $[\frac{1}{2}, 1]$, and plot the error. What is the maximum sampled error?

7. Repeat the preceding, using piecewise cubic Hermite interpolation over the same grid.

8. Construct piecewise polynomial approximations of degree $d = 1, 2, 3$ to the data in Table 4.12, using only the nodes $\log_{10} \theta_k = -6, -4, -2, 0, 2, 4, 6$. Plot the resulting curve and compare it with the ordinary interpolating polynomial found in Exercise 14 of Section 4.2. Test how well this approximation matches the tabulated values at $\log_{10} \theta = -5, -3, -1, 1, 3, 5$.

9. Show that the error in piecewise cubic Hermite interpolation satisfies

$$\|f - H_2\|_\infty \le \frac{1}{384} h^4 \|f^{(4)}\|_\infty,$$

where we have assumed uniformly spaced nodes with $x_k - x_{k-1} = h$.

10. Given a grid of points

$$a = x_0 < x_1 < x_2 < \cdots < x_n = b,$$

define the piecewise linear functions ϕ_k^h, $1 \le k \le n-1$, according to

$$\phi_k^h(x) = \begin{cases} \dfrac{x - x_{k-1}}{x_k - x_{k-1}} & x_{k-1} \le x \le x_k \\ \dfrac{x_{k+1} - x}{x_{k+1} - x_k} & x_k \le x \le x_{k+1} \\ 0 & \text{otherwise.} \end{cases}$$

Define the function space

$$S_0^h = \{f \in C([a,b]), f(a) = f(b) = 0,$$
$$f \text{ is piecewise linear on the given grid}\}.$$

Show that the ϕ_k^h are a basis for S_0^h, i.e., that every element of the space S_0^h can be written as a linear combination of the ϕ_k^h functions.

11. Implement a routine for approximating the natural logarithm using piecewise polynomial interpolation, i.e., a table look-up scheme. Assume the table of (known) logarithm values is uniformly distributed on the interval $[\frac{1}{2},1]$. Choose enough points in the table to guarantee 10^{-10} accuracy for any x. Use

(a) Piecewise linear interpolation
(b) Piecewise cubic Hermite interpolation

Test your routine against the intrinsic logarithm function on your computer by evaluating the error in your approximation at 5000 equally spaced points on the interval $[\frac{1}{10},10]$. Use the way that the computer stores floating-point numbers to reduce the logarithm computation to the interval $[\frac{1}{2},1]$, so long as $\ln 2$ is known to high accuracy.

4.8 AN INTRODUCTION TO SPLINES

Piecewise polynomial approximation (Section 4.7) allows us to construct highly accurate approximations, but it does not always produce approximations that are pleasing to the eye, and this is actually a problem in some applications. This occurs because the approximating function is not smooth at the junctions between separate pieces of the piecewise polynomial approximation. While we have that q_d is continuous, it is *not* continuously differentiable on the entire interval of approximation. The graph of the interpolant can have "kinks" in it, and some of these are apparent in Figure 4.8.

Splines are an attempt to address this problem. The basic idea is to construct a piecewise polynomial approximation that not only interpolates given data or function values, but is also "smooth," meaning continuously differentiable to some degree.

A complete theory of splines is beyond the scope of this text. Interested readers are referred to the book by Carl de Boor, *A Practical Guide to Splines* [8]. Another good reference is Paddy Prenter's *Splines and Variational Methods* [13]. We will outline two different spline constructions here and state some of the more basic theoretical results.[11]

4.8.1 Definition of the Problem

We have to properly define the problem before we can solve it. Suppose we are given a set of *nodes* $\{x_i, 0 \leq i \leq n\}$ at which we wish to interpolate a given function $f(x)$ with a spline of degree d. The problem is then to find a piecewise polynomial function, q_d,

---

[11]The history of the development of splines—and even the origin of the word itself—might be of interest to readers.

Many years ago, one of the author's students suggested that the word *splines* was constructed from "spliced lines." While this is indeed an interesting observation, the truth of the matter is that the mathematical use of the word *spline* comes from the days when being an engineer often meant being an accomplished draftsman. One draftsman's tool was known as a "spline"—a thin, very flexible, piece of metal that was used to draw a smooth curved line based on a few discrete points marked on a drawing.

One of the historical uses of splines was in the automobile industry, where they were used to define the shape of sheet-metal pieces for car bodies. Much of the mathematical development work was done at General Motors in the early 1960s, and also at Renault and Citroen in France at about the same time. Curiously, the original paper on splines was written with applications to actuarial data. Some of the impetus for the development of the theory also came from the British aircraft industry during World War II, as well as the ship-building industry.

The work at GM is detailed nicely in the article [2] and the retrospective [16]. Paul Davis summarized some of this material in *SIAM News* in 1996 [5]. Carl de Boor maintains a spline bibliography on the Internet at http://www.cs.wisc.edu/deboor/bib/bib.html.

that satisfies the following conditions:

1. *Interpolation:*

$$q_d(x_k) = f(x_k), \qquad 0 \le k \le n.$$

Here d is called the *degree of approximation* of the spline.

2. *Smoothness:*

$$\lim_{x \to x_k^-} q_d^{(i)}(x) = \lim_{x \to x_k^+} q_d^{(i)}(x)$$

for $0 \le i \le N$. We call N the *degree of smoothness* of the spline.

3. *Interval of definition:* q_d is a polynomial of degree $\le d$ on each subinterval $[x_{k-1}, x_k]$.

We need to investigate the relationship (if any) between the degree of approximation, d, and the degree of smoothness, N. The degree of the polynomials is related to the number of unknown coefficients (i.e., the degrees of freedom) in the problem, whereas N is related to the number of constraints. We expect that the degrees of freedom and the number of constraints have to balance in order for the spline to be well defined.

Since there are n subintervals, each being the domain of definition for a separate polynomial of degree d, we have a total of $K_f = n(d+1)$ degrees of freedom. On the other hand, there are $n+1$ interpolation conditions and $n-1$ junction points (again called *knots*) between subintervals (note that in the spline construction, neither x_0 nor x_n is considered a knot), with $N+1$ continuity conditions being imposed at each of them. Thus there are $K_c = n+1+(n-1)(N+1)$ constraints. If we consider the difference $K_f - K_c$, we get

$$K_f - K_c = n(d+1) - [n+1+(n-1)(N+1)] = nd - n - nN + N = n(d-1-N) + N.$$

We can make the first term vanish by setting $d-1-N = 0$. This establishes a relationship between the polynomial degree of the spline and the smoothness degree. For example, if we consider the common case of cubic splines, then $d = 3$ and $N = 2$. However, we will not have the number of constraints equal the number of degrees of freedom: $K_f - K_d = N$. Thus we need to add N additional constraints. Partly for this reason, odd-polynomial-order splines are preferred, because if d is odd, then $N = d - 1$ is even and the additional constraints can be imposed equally at the two endpoints (which is the typical choice) of the interval.

It is worth noting, briefly, that we can consider a continuous piecewise linear function to be a spline with degree of smoothness $N = 0$. See Exercise 18.

4.8.2 Cubic B-Splines

The construction we use here is based on the *B-spline*. This idea uses a single exemplar function from which a basis of splines is formed, and the approximation is then defined in terms of this basis. The notion of B-splines is much more general than is presented here, where we restrict ourselves to the (widely used) case of cubic B-splines, with coincident nodes and knots. For simplicity's sake, we assume at first a uniform grid

$$a = x_0 < x_1 < x_2 < \cdots < x_{n-1} < x_n = b \tag{4.32}$$

with $x_k - x_{k-1} = h$ for all $k \geq 1$. We will also need to define the extra grid points

$$x_{-3} = a - 3h, \quad x_{-2} = a - 2h, \quad x_{-1} = a - h, \tag{4.33}$$

and

$$x_{n+1} = b + h, \quad x_{n+2} = b + 2h, \quad x_{n+3} = b + 3h. \tag{4.34}$$

Then define the function

$$B(x) = \begin{cases} 0 & x \leq -2 \\ (x+2)^3 & -2 \leq x \leq -1 \\ 1 + 3(x+1) + 3(x+1)^2 - 3(x+1)^3 & -1 \leq x \leq 0 \\ 1 + 3(1-x) + 3(1-x)^2 - 3(1-x)^3 & 0 \leq x \leq 1 \\ (2-x)^3 & 1 \leq x \leq 2 \\ 0 & 2 \leq x \end{cases} \tag{4.35}$$

and note that B is a cubic spline with nodes/knots at the points $x = -2, -1, 0, 1, 2$.

Wait a minute. How do we know that $B(x)$ is a cubic spline function? Well, a spline function is a piecewise polynomial with a certain amount of derivative continuity between the pieces, the amount being related to the polynomial degree by $N = d - 1$, as we showed above. Thus, to check that $B(x)$ is a cubic spline, we simply note that it clearly is a piecewise cubic polynomial, and then we compute the one-sided derivatives at the knots:

$$B'_-(x_k) = \lim_{x \to x_k^-} B'(x), \quad B'_+(x_k) = \lim_{x \to x_k^+} B'(x),$$

and similarly for the second derivative (and the function value). If the one-sided values are equal to each other, then the first and second derivatives are continuous, and hence B is a cubic spline. If only the first derivative is continuous, it would fail to be a spline because *cubic* splines require second-derivative continuity.

Figure 4.12 shows a graph of B; note the bell shape of the curve. Note also that

$$B(0) = 4, \quad B(\pm 1) = 1, \quad B(\pm 2) = 0 \tag{4.36}$$

$$B'(0) = 0, \quad B'(\pm 1) = \mp 3, \quad B'(\pm 2) = 0 \tag{4.37}$$

$$B''(0) = -12, \quad B''(\pm 1) = 6, \quad B''(\pm 2) = 0. \tag{4.38}$$

Finally, note that B is only "locally defined," meaning that it is nonzero on only a small interval.[12] This local definition property is important in the utility of B-splines as an approximation tool.

---

[12]The correct terminology is that B is a function of "compact support," but the phrase "locally defined" conveys the meaning more clearly at this level.

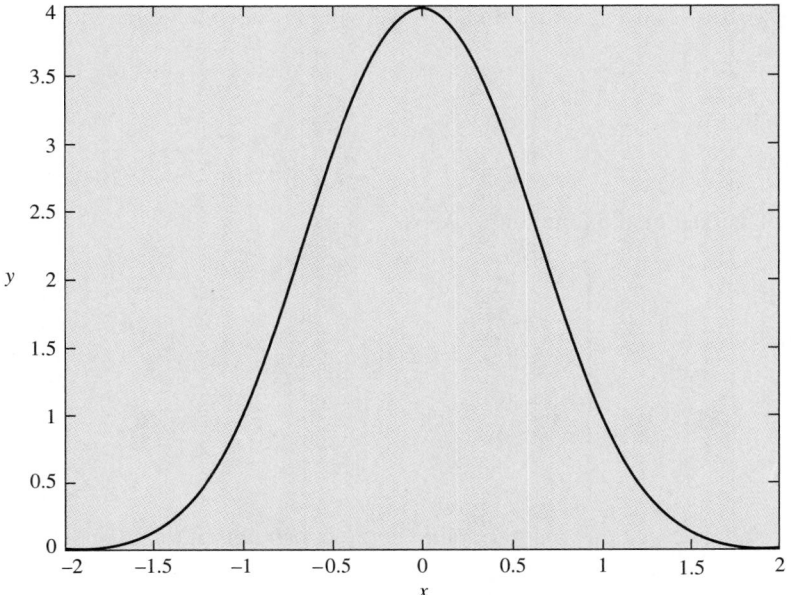

FIGURE 4.12 Original B-spline.

We can use B to construct a spline approximation to an arbitrary function f, using the grid defined in (4.32), (4.33), and (4.34). Define the sequence of functions

$$B_i(x) = B((x - x_i)/h), \quad -1 \le i \le n+1. \tag{4.39}$$

Each B_i is similar in shape to B but centered at x_i instead of at 0. Figure 4.13 shows plots of some of the B_i for a specific grid. Note that

$$B_i(x_i) = B(0), \qquad B_i(x_{i\pm1}) = B(\pm1), \qquad B_i(x_{i\pm2}) = B(\pm2), \tag{4.40}$$

and similarly for the derivatives:

$$B_i'(x_i) = 0, \qquad B_i'(x_{i\pm1}) = \mp3/h, \qquad B_i'(x_{i\pm2}) = 0, \tag{4.41}$$

$$B_i''(x_i) = -12/h^2, \qquad B_i''(x_{i\pm1}) = 6/h^2, \qquad B_i''(x_{i\pm2}) = 0. \tag{4.42}$$

To construct a spline interpolant to a given function f, we define the spline as a linear combination of the B_i functions, $-1 \le i \le n+1$,

$$q_3(x) = \sum_{i=-1}^{n+1} c_i B_i(x), \tag{4.43}$$

and seek the coefficients c_i, $-1 \le i \le n+1$, such that

$$f(x_k) = \sum_{i=-1}^{n+1} c_i B_i(x_k), \quad 0 \le k \le n. \tag{4.44}$$

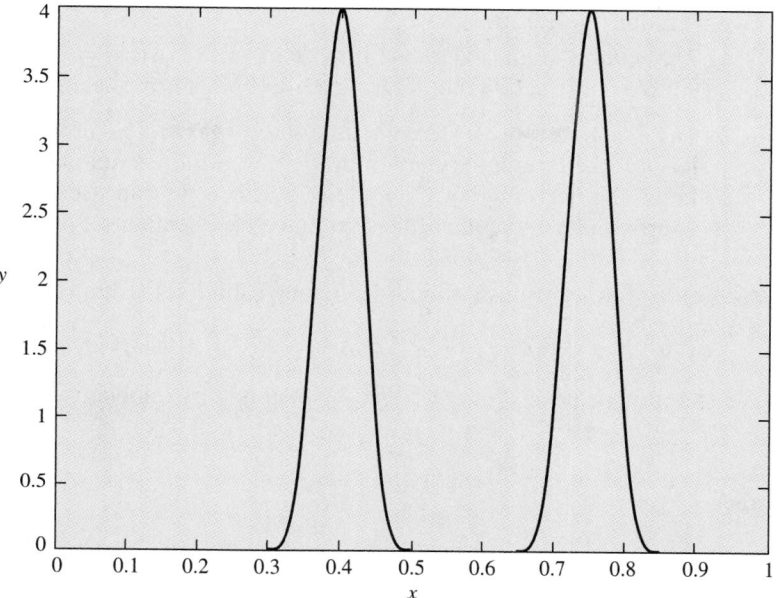

FIGURE 4.13   B-splines on a grid.

It is at this point that the local definition of B becomes so useful. If we look at (4.44) for $k = 0$, we see that we have

$$f(x_0) = c_{-1}B_{-1}(x_0) + c_0B_0(x_0) + c_1B_1(x_0)$$

since $B_i(x_0) = 0$ for all $i \geq 2$. In general, then, we have

$$f(x_k) = c_{k-1}B_{k-1}(x_k) + c_kB_k(x_k) + c_{k+1}B_{k+1}(x_k).$$

Thus, each equation in the system defined by (4.44) has only three nonzero terms—it is a *tridiagonal* system, just the kind we learned how to solve back in Section 2.6. Moreover, we can use (4.36) to show that (4.44) simplifies to

$$f(x_k) = c_{k-1} + 4c_k + c_{k+1}.$$

Written in matrix-vector form, the system is

$$\begin{bmatrix} 1 & 4 & 1 & & & \\ & 1 & 4 & 1 & & \\ & & \ddots & \ddots & \ddots & \\ & & & 1 & 4 & 1 \end{bmatrix} \begin{bmatrix} c_{-1} \\ c_0 \\ \vdots \\ c_{n+1} \end{bmatrix} = \begin{bmatrix} f(x_0) \\ f(x_1) \\ \vdots \\ f(x_n) \end{bmatrix}. \qquad (4.45)$$

of course, this is still a system of $n + 1$ equations in $n + 3$ unknowns; we need to come up with two additional constraints in order to eliminate two of the unknowns.

Two common choices are

1. *The natural spline.* Here we impose $q_3''(x_0) = q_3''(x_n) = 0$. This leads to a very simple construction, but it also leads to higher error near the endpoints.

2. *The complete spline.* Here we impose $q_3'(x_0) = f'(x_0)$ and $q_3'(x_n) = f'(x_n)$. This leads to better approximation properties and does not actually require that we know the derivative at the endpoints, since we can use the function values to approximate it via one of the formulas from Section 4.5.

Natural Spline   We can directly compute, using (4.43) and (4.37), that

$$q_3''(x_0) = h^{-2}(6(c_{-1} + c_1) - 12c_0), \qquad q_3''(x_n) = h^{-2}(6(c_{n+1} + c_{n-1}) - 12c_n).$$

We therefore achieve $q_3''(x_0) = q_3''(x_n) = 0$ simply by imposing

$$c_{-1} + c_1 = 2c_0 \quad \Rightarrow \quad c_{-1} = 2c_0 - c_1 \tag{4.46}$$

and

$$c_{n+1} + c_{n-1} = 2c_n \quad \Rightarrow \quad c_{n+1} = 2c_n - c_{n-1}. \tag{4.47}$$

If we substitute these into the system (4.45), we find that the first and last equations simplify to

$$6c_0 = f(x_0), \qquad 6c_n = f(x_n) \tag{4.48}$$

so that we have the new system

$$
\begin{array}{ccccccl}
4c_1 & + & c_2 & & & = & f(x_1) - \tfrac{1}{6} f(x_0) \\
c_1 & + & 4c_2 & + & c_3 & = & f(x_2) \\
& \ddots & & \ddots & & \vdots & \\
& & c_{n-2} & + & 4c_{n-1} & = & f(x_{n-1}) - \tfrac{1}{6} f(x_n).
\end{array}
\tag{4.49}
$$

Not only is this now square, in addition to being tridiagonal, but it is also diagonally dominant. Therefore, as discussed in Section 2.6, the matrix is nonsingular and the solution algorithm of Section 2.6 can be applied to compute the coefficients.

Complete Spline   Here we impose derivative data at the endpoints. Direct calculation shows that

$$q_3'(x_0) = -3h^{-1}(c_{-1} - c_1), \qquad q_3'(x_n) = -3h^{-1}(c_{n-1} - c_{n+1}),$$

so that the conditions $q_3'(x_0) = f'(x_0)$ and $q_3'(x_n) = f'(x_n)$ imply

$$c_{-1} = c_1 - \frac{1}{3} h f'(x_0), \qquad c_{n+1} = c_{n-1} + \frac{1}{3} h f'(x_n). \tag{4.50}$$

When we substitute into the system (4.45), we again lose the dependence on c_{-1} and c_{n+1} and get the square system

$$4c_0 \quad + \quad 2c_1 \qquad \cdots \qquad\qquad = \quad f(x_0) + \tfrac{1}{3}hf'(x_0)$$

$$\ddots \qquad \ddots \qquad \ddots \qquad\qquad \vdots$$

$$c_{k-1} \quad + \quad 4c_k \quad + \quad c_{k+1} \quad = \quad f(x_k) \tag{4.51}$$

$$\ddots \qquad \ddots \qquad\qquad \vdots$$

$$2c_{n-1} \quad + \quad 4c_n \quad = \quad f(x_n) - \tfrac{1}{3}hf'(x_n),$$

which is (again) diagonally dominant, so the tridiagonal solution algorithm can be applied. Note that in the natural spline case, the system had $n-1$ equations in the $n-1$ unknowns c_i, $1 \le i \le n-1$, whereas here we have a system of $n+1$ equations in the $n+1$ unknowns c_i, $0 \le i \le n$.

Evaluation   The natural or complete spline can be easily constructed, in the sense that the coefficients c_k are defined, by solving (4.49) or (4.51). How do we use these values to evaluate the spline q_3, defined by (4.43)?

For any $x \in [x_{k-1}, x_k]$, the fact that each B_i is only locally nonzero means that we have

$$q_3(x) = c_{k-2}B_{k-2}(x) + c_{k-1}B_{k-1}(x) + c_kB_k(x) + c_{k+1}B_{k+1}(x), \tag{4.52}$$

so evaluation of $q_3(x)$ requires only that we find the index k such that $x_{k-1} \le x \le x_k$. For a uniform grid this is easily accomplished:

$$k = \left\lfloor \frac{x - x_0}{h} \right\rfloor + 1.$$

For a nonuniform grid a more general search routine would have to be employed.

Programming Hint   Note that (4.52) implies that

$$q_3(x_k) = f(x_k) = c_{k-1}B_{k-1}(x_k) + c_kB_k(x_k) + c_{k+1}B_{k+1}(x_k) = c_{k-1} + 4c_k + c_{k+1}.$$

This can be a useful check on the correctness of the spline coefficients.

EXAMPLE 4.7

We start with a simple example that can be carried out by hand. Suppose we want to compute a cubic spline approximation to $f(x) = \sin \pi x$ on the interval $[0, \tfrac{1}{2}]$, using the nodes $x_0 = 0, x_1 = \tfrac{1}{4}$, and $x_2 = \tfrac{1}{2}$. Thus $n = 2$ and the spline will be defined by

$$q_3(x) = \sum_{k=-1}^{3} c_kB_k(x).$$

For the natural spline, (4.48) implies that

$$c_0 = \frac{1}{6}\sin 0 = 0$$

and

$$c_2 = \frac{1}{6}\sin\frac{\pi}{2} = \frac{1}{6},$$

and instead of a system of equations to solve, we have the single equation

$$c_0 + 4c_1 + c_2 = f(x_1) \quad \Rightarrow \quad c_1 = \frac{1}{4}\left(f(x_1) - \frac{1}{6}f(x_0) - \frac{1}{6}f(x_2)\right),$$

from which we get

$$c_1 = \frac{1}{4}\left(\frac{\sqrt{2}}{2} - \frac{1}{6}\right) = 0.1351100286.$$

Finally, then, (4.46) and (4.47) imply that

$$c_{-1} = \frac{1}{4}\left(\frac{1}{6} - \frac{\sqrt{2}}{2}\right) = -0.1351100286, \quad c_3 = \frac{1}{3} - \frac{\sqrt{2}}{8} + \frac{1}{24} = 0.1982233048.$$

So the spline approximation is now given by

$$q_3(x) = c_{-1}B_{-1}(x) + c_1B_1(x) + c_2B_2(x) + c_3B_3(x).$$

To compute values of the spline is fairly simple. Suppose we want to find $q_3(1/6) \approx \sin\pi/6 = 1/2$. We first determine that $x = 1/6$ is between $x_0 = 0$ and $x_1 = \frac{1}{4}$; this means that we want to compute

$$q_3(1/6) = c_{-1}B_{-1}(1/6) + c_1B_1(1/6) + c_2B_2(1/6).$$

The B_3 term vanishes because B_3 is identically zero on the subinterval $[x_0, x_1]$. We then have

$$\begin{aligned}
q_3(1/6) &= c_{-1}B_{-1}(1/6) + c_1B_1(1/6) + c_2B_2(1/6) \\
&= c_{-1}B(5/3) + c_1B(-1/3) + c_2B(-4/3) \\
&= (-0.1351100286)(0.0370370370) + (0.1351100286)(3.4444444444) \\
&\quad + (0.1666666667)(0.2962962967) \\
&= 0.5097576284,
\end{aligned}$$

which is a decent approximation to the exact value. The entire spline is graphed in Figure 4.14. The error is given in Figure 4.15. Recall, however, that using a natural spline imposes the condition $q_3''(x_0) = q_3''(x_n) = 0$, which might not be the correct value to use (and, in the case of our example, isn't even a good approximation to the correct value at $x_2 = b = \frac{1}{2}$). If we construct the complete spline approximation, then we are led, via (4.50) and (4.51), to the 3×3 linear system

$$\begin{bmatrix} 4 & 2 & 0 \\ 1 & 4 & 1 \\ 0 & 2 & 4 \end{bmatrix}\begin{bmatrix} c_0 \\ c_1 \\ c_2 \end{bmatrix} = \begin{bmatrix} \frac{1}{12}\pi \\ \frac{1}{2}\sqrt{2} \\ 1 \end{bmatrix},$$

which we can easily solve to get

$$c_0 = 0.00017369124366, \quad c_1 = 0.13055231141225, \quad c_2 = 0.18472384429387.$$

FIGURE 4.14 Natural spline approximation to part of the sine function.

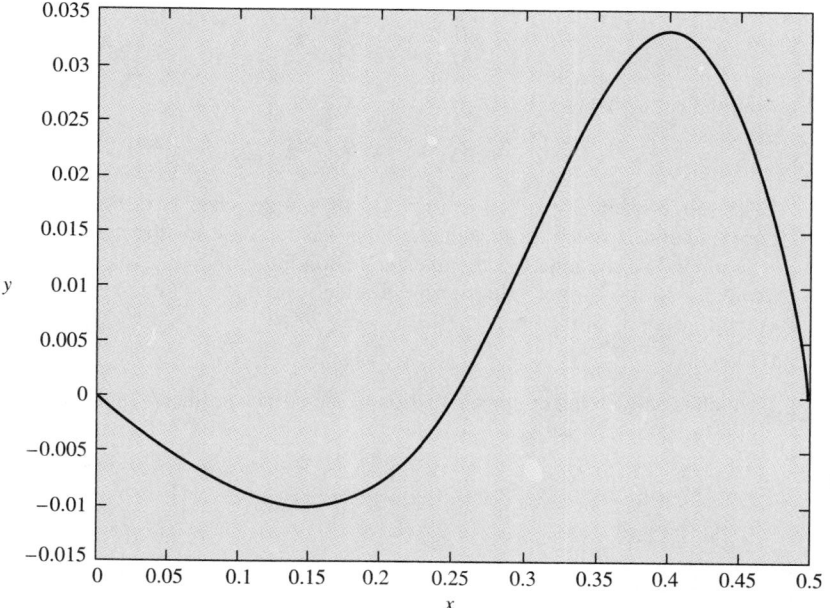

FIGURE 4.15 Error in natural spline approximation to part of the sine function.

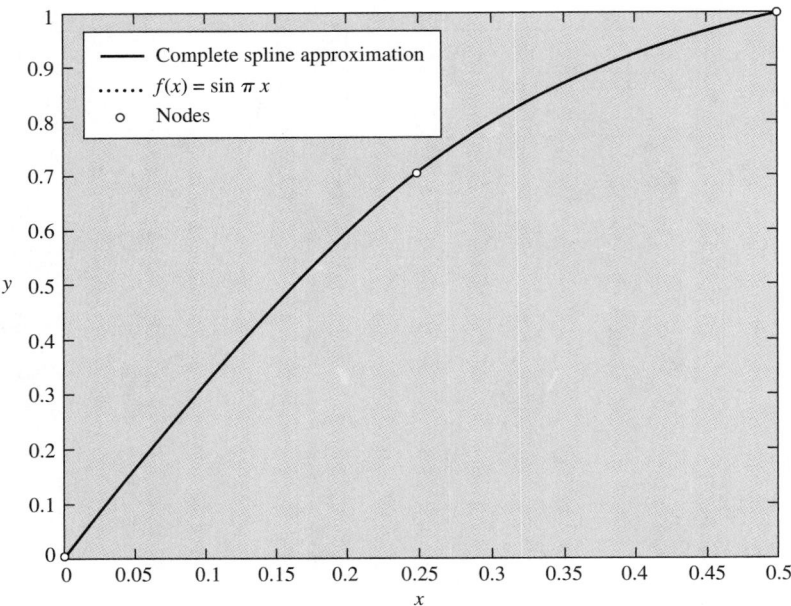

FIGURE 4.16 Complete spline approximation to part of the sine function.

Thus the complete set of coefficients is

$$c_{-1} = -0.13124707638690, \qquad c_0 = 0.00017369124366, \qquad c_1 = 0.13055231141225,$$
$$c_2 = 0.18472384429387, \qquad c_3 = 0.13055231141225,$$

and the spline is given by

$$q_3(x) = c_{-1}B_{-1}(x) + c_0B_0(x) + c_1B_1(x) + c_2B_2(x) + c_3B_3(x).$$

The spline itself is graphed in Figure 4.16 and the error is in Figure 4.17. Note that the approximation is much more accurate than was the case for the natural spline. This is because the natural spline imposed the inaccurate condition $q_3''(x) = 0$ at $x = \frac{1}{2}$, whereas the complete spline used the correct values of the first derivative at both endpoints. Just as a single point of comparison, this time we get

$$q_3(1/6) = 0.4999381524,$$

which is a much better approximation to the correct value of $\frac{1}{2}$ than we got for the natural spline.

■

EXAMPLE 4.8

Consider now the function $f(x) = (1 + 25x^2)^{-1}$, which we have had trouble approximating with ordinary interpolation. We can construct both natural and complete spline approximations to this function, using equally spaced points on the interval $[-1,1]$, with $h^{-1} = 2,4,8,16$. Figures 4.18 and 4.19 show the plots of the natural spline and associated error, and Figures 4.20 and 4.21 are the plots for the complete spline and its error. Compare these plots with the simple piecewise polynomial fits shown in Figures 4.8–4.10.

■

FIGURE 4.17 Error in complete spline approximation to part of the sine function (same vertical scale as Figure 4.15).

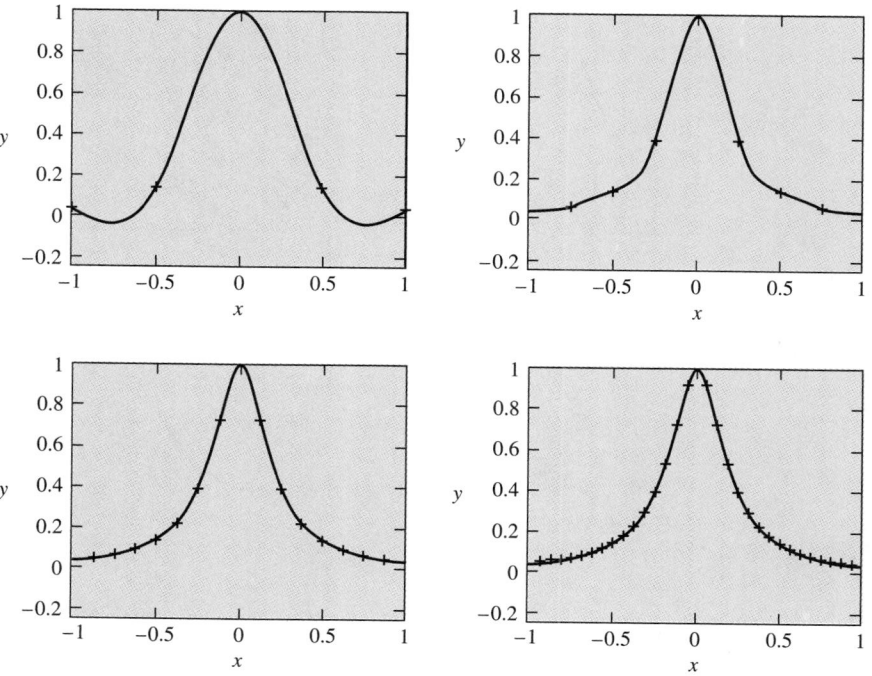

FIGURE 4.18 Natural spline interpolation to $f(x) = (1 + 25x^2)^{-1}$. The nodes are marked by + signs.

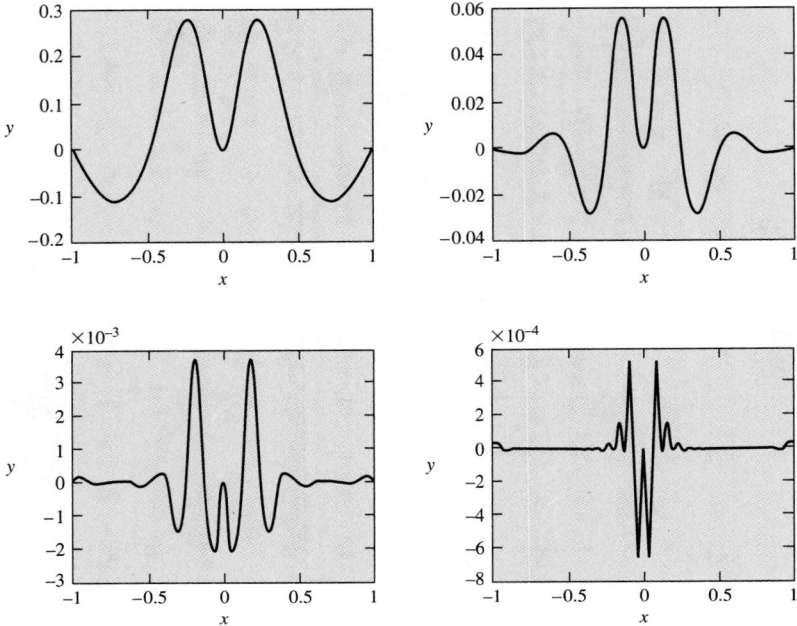

FIGURE 4.19 Error in natural spline interpolation to $f(x) = (1 + 25x^2)^{-1}$.

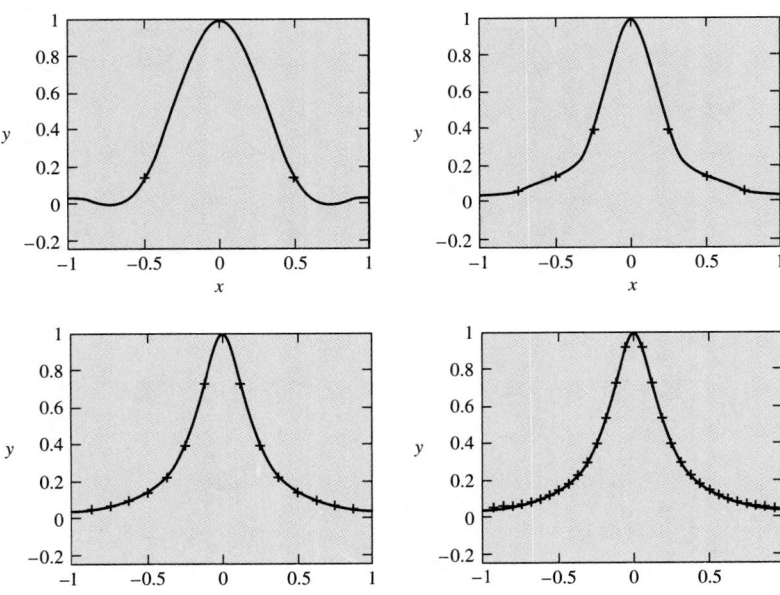

FIGURE 4.20 Complete spline interpolation to $f(x) = (1 + 25x^2)^{-1}$. The nodes are marked by + signs.

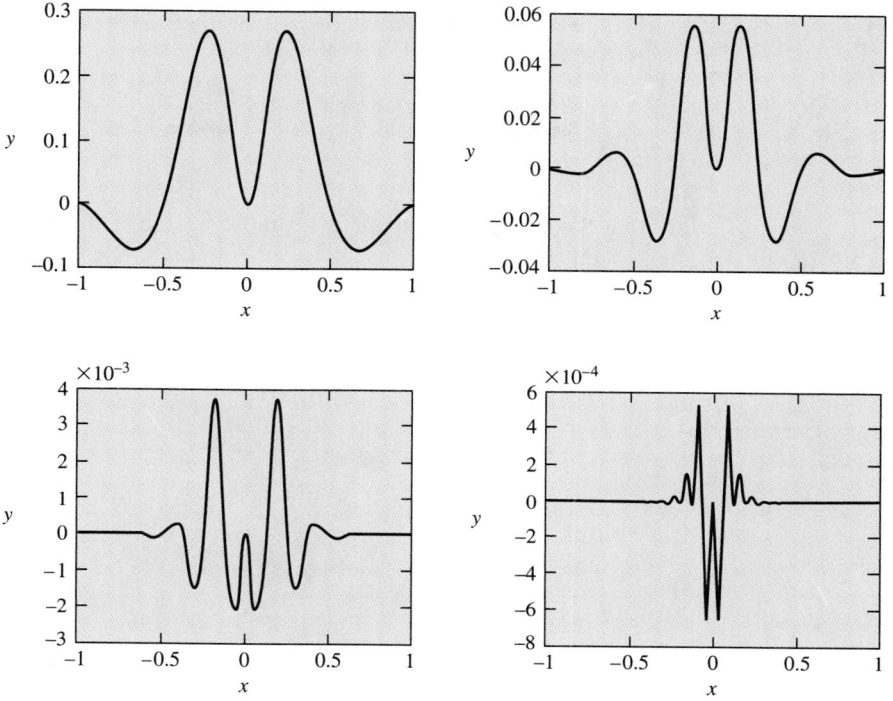

FIGURE 4.21 Error in complete spline interpolation to $f(x) = (1 + 25x^2)^{-1}$.

Our construction here assumes a uniform grid spacing, but splines can be constructed using non-uniform grids. The basic principles are the same, but the governing equations become substantnatially more involved. The interested reader should consult the references at the end of the chapter, especially deBoor's book [8], for the details.

Theoretical Results   The error theory for spline approximations is somewhat more involved than that for ordinary piecewise polynomial interpolation, so here we will reference the main result without proof and then comment on it. See [10] for many of the details.

Theorem 4.7 (Spline Approximation)   *If $f \in C^4([a,b])$ and q is a cubic spline interpolant to f on a grid*

$$a = x_0 < x_1 < x_2 < \cdots < x_{n-1} < x_n = b$$

with $\max(x_i - x_{i-1}) \le h$ for all i, then

$$\|f - q\|_\infty \le \frac{5}{384} h^4 \|f^{(4)}\|_\infty.$$

Moreover, there exist constants C_k, $1 \leq k \leq 3$, such that

$$\|f^{(k)} - q^{(k)}\|_\infty \leq C_k h^{4-k} \|f^{(4)}\|_\infty.$$

Note that the spline interpolant is no more accurate, in terms of the exponent on the mesh size, than ordinary piecewise polynomial approximation (note that the constant in the estimate is smaller). The advantage of spline interpolation lies in the smoothness of the approximation.

EXERCISES

1. Given the set of nodes $x_0 = 0$, $x_1 = 1/2$, $x_2 = 1$, $x_3 = 3/2$, and $x_4 = 2$, we construct the cubic spline function

$$q_3(x) = 0.15B_{-1}(x) + 0.17B_0(x) + 0.18B_1(x)$$
$$+ 0.22B_2(x) + 0.30B_3(x) + 0.31B_4(x) + 0.32B_5(x),$$

where each B_k is computed from the exemplar B spline according to (4.39). Compute q_3 and its first derivative at each node.

2. Is the function

$$p(x) = \begin{cases} 0 & x \leq 0 \\ x^2 & 0 \leq x \leq 1 \\ -2x^2 + 6x + 3 & 1 \leq x \leq 2 \\ (x-3)^2 & 2 \leq x \leq 3 \\ 0 & x \geq 3 \end{cases}$$

a spline function? Why/why not?

3. For what value of k is the following a spline function?

$$q(x) = \begin{cases} kx^2 + (3/2) & 0 \leq x \leq 1 \\ x^2 + x + (1/2) & 1 \leq x \leq 2 \end{cases}$$

4. For what value of k is the following a spline function?

$$q(x) = \begin{cases} x^3 + 3x^2 + 1 & -1 \leq x \leq 0 \\ -x^3 + kx^2 + 1 & 0 \leq x \leq 1 \end{cases}$$

5. For what value of k is the following a spline function?

$$q(x) = \begin{cases} x^3 - x^2 + kx + 1 & 0 \leq x \leq 1 \\ -x^3 + (k+2)x^2 - kx + 3 & 1 \leq x \leq 2 \end{cases}$$

6. Construct the natural cubic spline that interpolates to $f(x) = 1/x$ at the nodes $1/2, 5/8, 3/4, 7/8, 1$. Do this as a hand calculation. Plot your spline function and f on the same set of axes, and also plot the error.

7. Repeat the preceding using the complete spline.

8. Construct a natural spline interpolant to the mercury thermal conductivity data (Table 4.7), using the 300 K, 500 K, and 700 K values. How well does this predict the values at 400 K and 600 K?

9. Confirm that the function $B(x)$, defined in (4.35), is a cubic spline.

10. Construct a natural cubic spline to the gamma function, using the data in Table 4.8 and the nodes $x = 1.0, 1.2, 1.4, 1.6, 1.8$, and 2.0. Use this approximation to estimate $\Gamma(x)$ at $x = 1.1, 1.3, 1.5, 1.7$, and 1.9.

11. Repeat the preceding using the complete spline approximation, and use the derivative approximations from Section 4.5 for the required derivative endpoint values.

12. Repeat Exercise 6 of Section 4.7, except this time construct the complete cubic spline interpolant to $f(x) = \ln x$, using the same set of nodes. Plot the approximation and the error. What is the maximum sampled error in this case?

TABLE 4.20   Data for Exercise 21

Temperature (K), u	200	300	400	500	600	700	800	900	1000
Conductivity (W/cm K), k	1.06	0.94	0.905	0.801	0.721	0.655	0.653	0.674	0.696

TABLE 4.19   Data for Exercise 7

x	0.0	0.25	0.50	0.75	1.00	1.25
t	0.0	25.0	49.4	73.0	96.4	119.4

13. Recall Exercise 7 from Section 4.2, in which we constructed polynomial interpolants to timing data from the 1973 Kentucky Derby, won by the horse Secretariat. For simplicity, we repeat the data in Table 4.19.

Here t is the elapsed time (in seconds) since the race began and x is the distance (in miles) that Secretariat has traveled.
(a) Construct a natural cubic spline that interpolates this data.
(b) What is Secretariat's speed at each quarter-mile increment of the race? (Use miles per hour as your unit.)
(c) What is Secretariat's "initial speed" according to this model? Does this make sense?

(Note: It is possible to do this problem using a uniform grid. Construct the spline that interpolates t as a function of x, then use your knowledge of calculus to find $x'(t)$ from $t'(x)$.)

14. Show that the complete cubic spline problem can be solved (approximately) without having an explicit expression for f' at the endpoints. (Hint: Consider the material from Section 4.5.)

15. Construct the natural cubic spline that interpolates the data in Table 4.12 at the nodes defined by $\log_{10}\theta = -6, -4, \ldots, 4, 6$. Test the accuracy of the approximation by computing $q_3(x)$ for $x = \log_{10}\theta = -5, -3, \ldots, 3, 5$ and comparing against the actual data in the table.

16. Construct the exemplar quadratic B-spline; i.e., construct a piecewise quadratic function that is C^1 over the nodes/knots $x = -1, 0, 1, 2$, and that vanishes for x

outside the interval $[-1, 2]$.

17. Construct the exemplar *quintic* B-spline.

18. For a linear spline function we have $d = 1$, which forces $N = 0$. Thus a linear spline has no derivative continuity, only function continuity, and no additional conditions are required at the endpoints. Show that the exemplar B-spline of first degree is given by

$$B(x) = \begin{cases} 0 & x \le -1 \\ x+1 & -1 \le x \le 0 \\ 1-x & 0 \le x \le 1 \\ 0 & 1 \le x \end{cases}$$

Write out, in your own words, how to construct and evaluate a linear spline interpolant using this function as the basic B-spline.

19. Discuss, in your own words, the advantages or disadvantages of spline approximation compared with ordinary piecewise polynomial approximation.

20. Write an essay that compares and contrasts piecewise cubic Hermite interpolation with cubic spline interpolation.

21. Table 4.20 gives the actual thermal conductivity data for the element nickel. Construct a natural spline interpolant to this data, using only the data at 200 K, 400 K, ..., 1000 K. How well does this spline predict the values at 300 K, 500 K, etc.?

22. Construct a natural spline interpolant to the pennsyltuckium thermal conductivity data from opening ing of this chapter. Plot the spline and the nodal data. If you did Exercise 8 from Section 4.2, compare your results.

4.9 APPLICATION: SOLUTION OF BOUNDARY VALUE EXERCISES

The range of application of spline approximations is incredibly wide, including such diverse things as scalable computer printer fonts, which are typically stored as spline approximations and modern computer special effects in movies, which are often done by using spline approximations to complicated surfaces, which can then be easily manipulated and moved about on the screen by the digital special effects artist. But some of the oldest and most important applications of spline functions are to the accurate solution of differential equations. In Section 2.7 we briefly touched on the approximate solution of boundary value problems using finite difference approximations. In this section we will build on that experience by using splines to construct our approximation.

Consider the two-point boundary value problem

$$-u'' + a^2 u = f(x), \quad 0 \leq x \leq 1,$$
$$u(0) = g_0, \qquad u(1) = g_1,$$

which is slightly more general than the one we looked at in Section 2.7. We construct the grid of points

$$0 = x_0 < x_1 < x_2 < \cdots < x_{n-1} < x_n = 1,$$

where we assume (for simplicity only) that the grid spacing is uniform, i.e., $x_k - x_{k-1} = h$ for all k, $1 \leq k \leq n$. We now look for our approximation in the form of a cubic spline defined on this grid. That is, we consider the function

$$u_h(x) = \sum_{k=-1}^{n+1} c_k B_k(x)$$

where the coefficients c_k are to be determined. The advantage of this approach over what we did in Section 2.7 is that here we get a continuous *smooth* function as our approximation, whereas in Section 2.7 all we got was a set of discrete points. We find the coefficients by requiring that u_h satisfy the differential equation at each of the nodes x_k, $0 \leq k \leq n$. (Note that the smoothness of the spline function is essential for this to even make sense. We must have $u_h \in C^2$ to even talk about it satisfying the ODE.) Because we know the values of B_k and its derivatives at each of the nodes, we can easily reduce this to the system of equations

$$T'c' = b' \tag{4.53}$$

where

$$
T' = \begin{bmatrix}
-6h^{-2}+a^2 & 12h^{-2}+4a^2 & -6h^{-2}+a^2 & 0 & \cdots & & 0 \\
0 & -6h^{-2}+a^2 & 12h^{-2}+4a^2 & -6h^{-2}+a^2 & & & \vdots \\
\vdots & & \ddots & & \ddots & 0 & \ddots \\
0 & \cdots & 0 & -6h^{-2}+a^2 & 12h^{-2}+4a^2 & & -6h^{-2}+a^2
\end{bmatrix}
$$

$$
c' = \begin{bmatrix} c_{-1} \\ c_0 \\ \vdots \\ c_{n+1} \end{bmatrix}, \qquad
b' = \begin{bmatrix} f(x_0) \\ f(x_1) \\ \vdots \\ f(x_n) \end{bmatrix}.
$$

Since this is a system of only $n+1$ equations in $n+3$ unknowns, we can't construct a solution yet. We can eliminate the two extra unknowns by imposing the boundary conditions on the approximation:

$$
u_h(0) = g_0 \quad \Rightarrow \quad c_{-1} + 4c_0 + c_1 = g_0 \quad \Rightarrow \quad c_{-1} = g_0 - 4c_0 - c_1 \tag{4.54}
$$

and

$$
u_h(1) = g_1 \quad \Rightarrow \quad c_{n-1} + 4c_n + c_{n+1} = g_1 \quad \Rightarrow \quad c_{n+1} = g_1 - 4c_n - c_{n-1}. \tag{4.55}
$$

If we substitute these into the first and last equations of the rectangular system, we get

$$
(-6h^{-2}+a^2)(g_0 - 4c_0 - c_1) + (12h^{-2}+4a^2)c_0 + (-6h^{-2}+a^2)c_1 = f(x_0)
$$

and

$$
(-6h^{-2}+a^2)c_{n-1} + (12h^{-2}+4a^2)c_n + (-6h^{-2}+a^2)(g_1 - 4c_n - c_{n-1}) = f(x_n).
$$

The c_1 and c_{n-1} terms drop out here, so we can conclude that

$$
c_0 = \frac{h^2}{36}\left(f(x_0) + \left(\frac{6}{h^2} - a^2\right) g_0 \right), \qquad c_n = \frac{h^2}{36}\left(f(x_n) + \left(\frac{6}{h^2} - a^2\right) g_1 \right). \tag{4.56}
$$

With these two values known, we can look at the $(n-1) \times (n-1)$ system created by using (4.56) to define c_0 and c_n, and dropping the first and last two equations in the rectangular system. We are then left with the square system

$$
Tc = b,
$$

where

$$T = \begin{bmatrix} 12h^{-2}+4a^2 & -6h^{-2}+a^2 & & \\ -6h^{-2}+a^2 & 12h^{-2}+4a^2 & -6h^{-2}+a^2 & \\ & \ddots & & \ddots \\ & & -6h^{-2}+a^2 & 12h^{-2}+4a^2 \end{bmatrix}$$

(which is diagonally dominant),

$$c = \begin{bmatrix} c_1 \\ c_2 \\ \vdots \\ c_{n-1} \end{bmatrix}, \text{ and } b = \begin{bmatrix} f(x_1)+\frac{1}{6}(1-\frac{1}{6}a^2h^2)f(x_0)+h^{-2}(1-\frac{1}{6}h^2a^2)^2 g_0 \\ f(x_2) \\ \vdots \\ f(x_{n-2}) \\ f(x_{n-1})+\frac{1}{6}(1-\frac{1}{6}a^2h^2)f(x_n)+h^{-2}(1-\frac{1}{6}h^2a^2)^2 g_1 \end{bmatrix}.$$

Programming Note   We have used a grid of $n+1$ points, with $h=1/n$, but the system we solve is $n-1 \times n-1$.

The diagonal dominance of T means that we can use the solution algorithm of Section 2.6. Once the system is solved, we can get the rest of the coefficients from (4.54), (4.55), and (4.56).

EXAMPLE 4.9

Consider the example problem

$$-u'' + \pi^2 u = 2\pi^2 \cos(\pi x), \qquad 0 \le x \le 1$$
$$u(0) = 1, \qquad u(1) = -1$$

which has exact solution $u(x) = \cos \pi x$. The linear system for $h = \frac{1}{8}$ is

$$\begin{bmatrix} 768+4\pi^2 & -384+\pi^2 & & & & & \\ -384+\pi^2 & 768+4\pi^2 & -384+\pi^2 & & & & \\ & -384+\pi^2 & 768+4\pi^2 & -384+\pi^2 & & & \\ & & -384+\pi^2 & 768+4\pi^2 & -384+\pi^2 & & \\ & & & -384+\pi^2 & 768+4\pi^2 & -384+\pi^2 & \\ & & & & -384+\pi^2 & 768+4\pi^2 & -384+\pi^2 \\ & & & & & -384+\pi^2 & 768+4\pi^2 \end{bmatrix}$$

$$\begin{bmatrix} c_1 \\ c_2 \\ c_3 \\ c_4 \\ c_5 \\ c_6 \\ c_7 \end{bmatrix} = \begin{bmatrix} b_1 \\ b_2 \\ b_3 \\ b_4 \\ b_5 \\ b_6 \\ b_7 \end{bmatrix},$$

where

$$
\begin{bmatrix} b_1 \\ b_2 \\ b_3 \\ b_4 \\ b_5 \\ b_6 \\ b_7 \end{bmatrix} = \begin{bmatrix} 2\pi^2 \cos\left(\dfrac{\pi}{8}\right) + 64\left(1 - \dfrac{\pi^2}{384}\right)\left(1 + \dfrac{\pi^2}{192}\right) \\ 2\pi^2 \cos\left(\dfrac{2\pi}{8}\right) \\ 2\pi^2 \cos\left(\dfrac{3\pi}{8}\right) \\ 2\pi^2 \cos\left(\dfrac{4\pi}{8}\right) \\ 2\pi^2 \cos\left(\dfrac{5\pi}{8}\right) \\ 2\pi^2 \cos\left(\dfrac{6\pi}{8}\right) \\ 2\pi^2 \cos\left(\dfrac{7\pi}{8}\right) + 64\left(1 - \dfrac{\pi^2}{384}\right)\left(1 + \dfrac{\pi^2}{192}\right) \end{bmatrix}.
$$

This has solution

$$
c = \begin{bmatrix} 0.15763927907189 \\ 0.12053536253339 \\ 0.06520274709146 \\ 0 \\ -0.06520274709146 \\ -0.12053536253339 \\ -0.15763927907189 \end{bmatrix}.
$$

From this we get

$$c_0 = 0.17095034913242, \qquad c_{-1} = 0.15855932439844, \qquad c_8 = -0.17095034913242,$$
$$c_9 = -0.15855932439844.$$

The spline approximation is therefore given by

$$
\begin{aligned}
u_h(x) = {} & 0.15855932439844 B_{-1}(x) + 0.17095034913242 B_0(x) + 0.15763927907189 B_1(x) \\
& + 0.12053536253339 B_2(x) + 0.06520274709146 B_3(x) \\
& + 0 B_4(x) - 0.06520274709146 B_5(x) \\
& - 0.12053536253339 B_6(x) - 0.15763927907189 B_7(x) \\
& - 0.17095034913242 B_8(x) - 0.15855932439844 B_9(x),
\end{aligned}
$$

and the error $\cos(\pi x) - u_h(x)$ is plotted in Figure 4.22. If we take a sequence of grids with $h^{-1} = 8, 16, 32, \ldots, 128$ and compute the approximate solutions, we find that the norm of the error, as estimated by sampling at 200 discrete points on the interval, is as indicated in Table 4.21; note that the error goes down like a factor of 4 as h is cut in half, indicating (but not proving) that this scheme is $O(h^2)$ accurate. Note that this means the spline solution of the boundary value problem is *less* accurate than a direct spline approximation of the exact solution (which, according to the estimate given in the previous section, would be $O(h^4)$ accurate). ∎

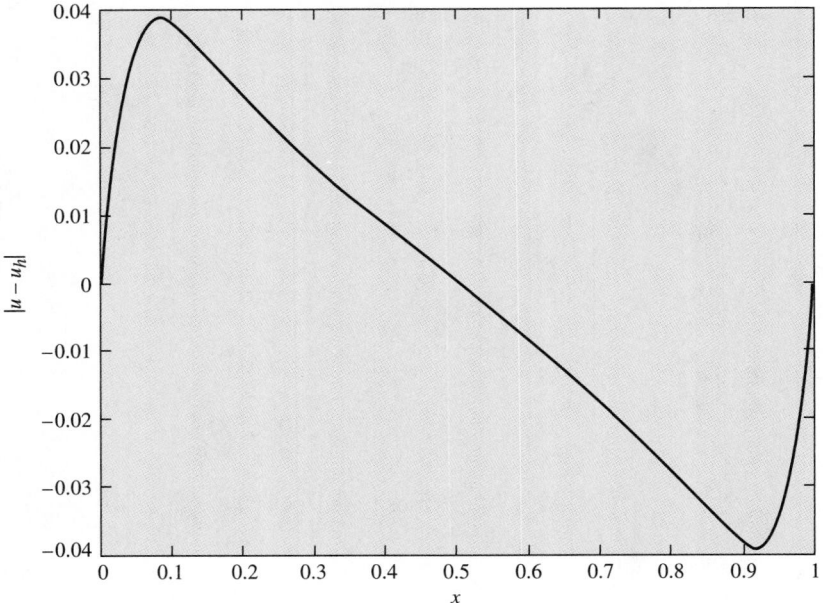

FIGURE 4.22 Error in the spline approximation to the boundary value problem for $h = 1/8$.

TABLE 4.21 **Estimated Error for Spline Approximation to BVP Solution**

h^{-1}	Estimated error
8	0.03881321103858
16	0.01100251675792
32	0.00295824984969
64	0.00076798902568
128	0.00019487261990
256	0.00004945300189

It can be shown that this approximation is indeed $O(h^2)$ accurate, but the analysis is beyond the intended scope of the text. We can get more accuracy by using higher-degree splines—using quintic splines gives us an error that is $O(h^4)$—but this makes the linear algebra problem slightly more difficult, as it results in a *pentadiagonal* matrix system to solve.

What we have done in this section is an example of what is called *collocation*, a broadly used technique for solving differential equations (approximately) using expansions in terms of basis functions, and then solving for the coefficients in the expansion. Another method that uses a basis function expansion (and that uses a more complicated means of constructing the system for the expansion coefficients) is the finite element method. The book by Prenter [13] has some discussion of collocation by splines.

EXERCISES

1. Set up the linear system for solving the boundary value problem

$$-u'' + u = 1, \qquad u(0) = 1, \qquad u(1) = 0,$$

using $h = \frac{1}{4}$. You should get

$$\begin{bmatrix} 196 & -95 & 0 \\ -95 & 196 & -95 \\ 0 & -95 & 196 \end{bmatrix} \begin{bmatrix} c_1 \\ c_2 \\ c_3 \end{bmatrix} = \begin{bmatrix} 101/6 \\ 1 \\ 671/576 \end{bmatrix}.$$

2. Solve the system in the previous exercise and find the coefficients for the spline expansion of the approximate solution. The exact solution is

$$u(x) = 1 - \frac{e}{e^2 - 1} e^x + \frac{e}{e^2 - 1} e^{-x}.$$

Plot your approximation and the error.

3. Use the algorithm from this section to approximate the solution to each of the following boundary value problems using $h^{-1} = 8,16,32$. Estimate the maximum error in each case by sampling the difference between the exact and approximate solutions at 200 equally spaced points on the interval.

(a) $-u'' + u = (\pi^2 + 1)\sin \pi x, u(0) = u(1) = 0$; $u(x) = \sin \pi x$

(b) $-u'' + u = 4e^{-x} - 4xe^{-x}, u(0) = u(1) = 0$; $u(x) = x(1-x)e^{-x}$

(c) $-u'' + u = \pi(\pi \sin \pi x + 2\cos \pi x)e^{-x}, u(0) = u(1) = 0$; $u(x) = e^{-x}\sin \pi x$

(d) $-u'' + u = 3 - \frac{1}{x} - (x^2 - x - 2)\log x, u(0) = u(1) = 0$; $u(x) = x(1-x)\log x$

(e) $-u'' + \pi^2 u = 2\pi^2 \sin(\pi x), u(0) = 0, u(1) = 0, u(x) = \sin(\pi x)$

(f) $-u'' + u = \frac{x^2 + 2x - 1}{(1+x)^3}, u(0) = 1, u(1) = 1/2, u(x) = (1+x)^{-1}$

4. Try to extend the method from this section to the more general two-point boundary value problem defined by

$$-a(x)u'' + u = f(x)$$
$$u(0) = u(1) = 0.$$

Is the resulting linear system diagonally dominant for all choices of $a(x)$?

5. Try to extend the method from this section to the more general two-point boundary value problem defined by

$$-u'' + bu' + u = f(x)$$
$$u(0) = u(1) = 0.$$

Is the resulting linear system diagonally dominant for all values of b?

4.10 LEAST SQUARES CONCEPTS IN APPROXIMATION

4.10.1 An Introduction to Data Fitting

An important area in approximation is the problem of fitting a curve to experimental data. Since the data is experimental, we must assume that it is polluted with some degree of error—most commonly measurement error—so we do not necessarily want to construct a curve that goes through every data point. (In fact, the material in Section 4.11.1 suggests that this would be a disastrous way to proceed.) Rather, we want to construct a function that represents the "sense of the data" and is, in some sense, a close approximation to the data.

The most common approach is known as *least squares* data fitting. Consider Figure 4.23; this shows an example set of data for which the general trend is clearly a straight line. But *which* straight line do we use to represent the data?

The least squares approach defines the "correct" straight line as the one that minimizes the sum of the squares of the distances between the data points and the line.

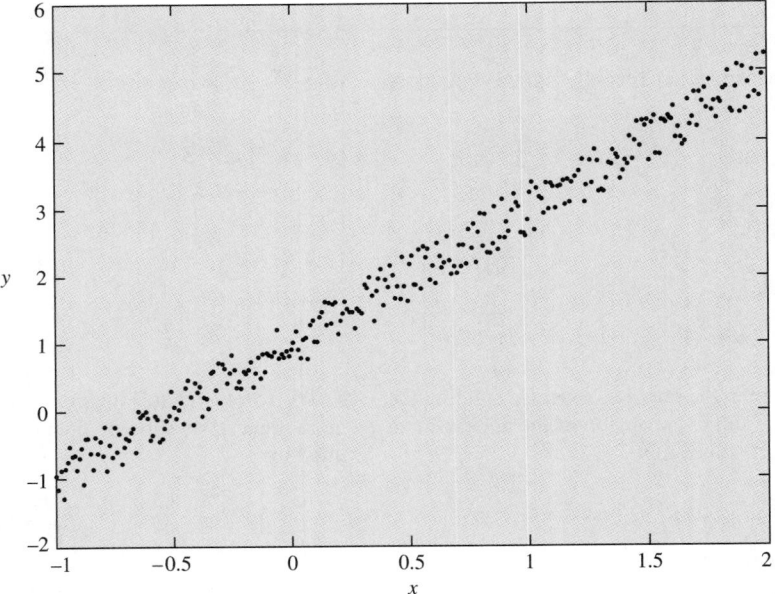

FIGURE 4.23 Example plot of data.

Let the experimental data be defined as pairs (x_k, y_k), $1 \le k \le n$ for some n. Thus, we want to find the coefficients m and b in the equation $y = mx + b$ such that

$$F(m,b) = \sum_{k=1}^{n} (y_k - (mx_k + b))^2$$

is minimized. This is a straightforward problem from multivariable calculus: We compute the partial derivatives F_m and F_b and find where they both vanish, and this will define a critical point. It can be shown that this critical point defines a global minimum for F. Thus m and b are defined by the two equations

$$\frac{\partial F}{\partial m} = -2 \sum_{k=1}^{n} (y_k - (mx_k + b)) x_k = 0$$

$$\frac{\partial F}{\partial b} = -2 \sum_{k=1}^{n} (y_k - (mx_k + b)) = 0,$$

which can be simplified to the system of two equations in two unknowns

$$\sum_{k=1}^{n} (y_k - (mx_k + b)) x_k = 0$$

$$\sum_{k=1}^{n} (y_k - (mx_k + b)) = 0,$$

FIGURE 4.24   Straight-line fit to the data.

or

$$m\left(\sum_{k=1}^{n} x_k^2\right) + b\left(\sum_{k=1}^{n} x_k\right) = \sum_{k=1}^{n} x_k y_k$$

$$m\left(\sum_{k=1}^{n} x_k\right) + b\left(\sum_{k=1}^{n} 1\right) = \sum_{k=1}^{n} y_k.$$

The solution is then

$$m = \frac{n\sum_{i=1}^{n} x_k y_k - \left(\sum_{i=1}^{n} x_k\right)\left(\sum_{i=1}^{n} y_k\right)}{n\sum_{k=1}^{n} x_k^2 - \left(\sum_{i=1}^{n} x_k\right)^2}$$

$$b = \frac{\left(\sum_{i=1}^{n} x_k^2\right)\left(\sum_{i=1}^{n} y_k\right) - \left(\sum_{i=1}^{n} x_k\right)\left(\sum_{i=1}^{n} x_k y_k\right)}{n\left(\sum_{k=1}^{n} x_k^2\right) - \left(\sum_{i=1}^{n} x_k\right)^2},$$

which, for the data in Figure 4.23, produces the straight-line graph shown in Figure 4.24.

The notion of a least squares data fit can be generalized beyond simply fitting a straight line to data. We can look at higher-degree polynomials, and we can also look at higher-dimensional data sets. Exercises 1–?? contain some examples of more involved least squares data fit problems.

TABLE 4.22 Data for Example 4.10

x	0.0	1.0	2.00	3.00	4.00	5.00
y	10.0	25.0	51.0	66.0	97.0	118

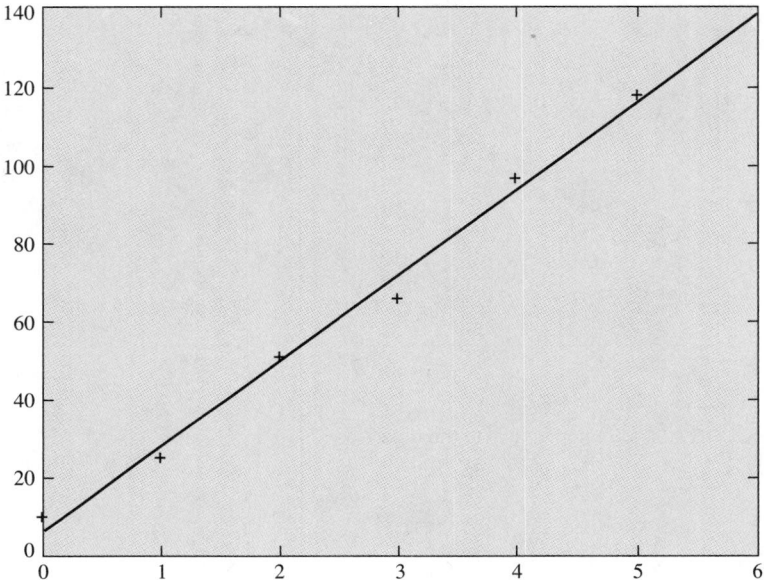

FIGURE 4.25 Least squares fit to the data in Table 4.22.

EXAMPLE 4.10

Consider the data in Table 4.22. If we plot this, we get what appears to be a straight line as the general trend of the data, so we look for the equation of the line $y = mx + b$ that best fits this data in the least squares sense. Forming the separate sums gives us

$$\sum_{i=1}^{6} x_i = 15, \qquad \sum_{i=1}^{6} y_i = 367, \qquad \sum_{i=1}^{6} x_i^2 = 55, \qquad \sum_{i=1}^{6} x_i y_i = 1303,$$

from which it follows that

$$m = 22.0286, \qquad b = 6.0952.$$

The line is plotted, along with the data, in Figure 4.25. ∎

EXAMPLE 4.11

We don't have to restrict ourselves to linear or quadratic models to make good use of the idea of least squares data fitting. Consider the data in Table 4.23. When we plot this data, we get the curve shown in Figure 4.26. Generally, this looks like an exponential growth curve. Ordinarily this would require us to do a fit to a curve of the form $y_k = A e^{r x_k}$ which will lead to a nonlinear

TABLE 4.23   Raw Data for Example 4.11

k	x_k	y_k
1	0	0.716
2	10	0.893
3	20	1.055
4	30	1.134
5	40	1.167
6	50	1.281
7	60	1.994
8	70	2.500
9	80	3.151
10	90	4.300
11	100	5.308
12	110	4.966
13	120	10.919

FIGURE 4.26   Plot of raw data for Example 4.11.

system for the parameters A and r. However, if the raw data is exponential, then the logarithm of the data is linear, since we have

$$z_k = \ln y_k = r x_k + (\ln A),$$

and we can fall back on our existing algorithm to do a fit to the log data.

TABLE 4.24   Logarithmic Data for Example 4.11

k	x_k	$\ln y_k$
1	00	−0.3341
2	10	−0.1132
3	20	0.0535
4	30	0.1258
5	40	0.1544
6	50	0.2476
7	60	0.6901
8	70	0.9163
9	80	1.1477
10	90	1.4586
11	100	1.6692
12	110	1.6026
13	120	2.3905

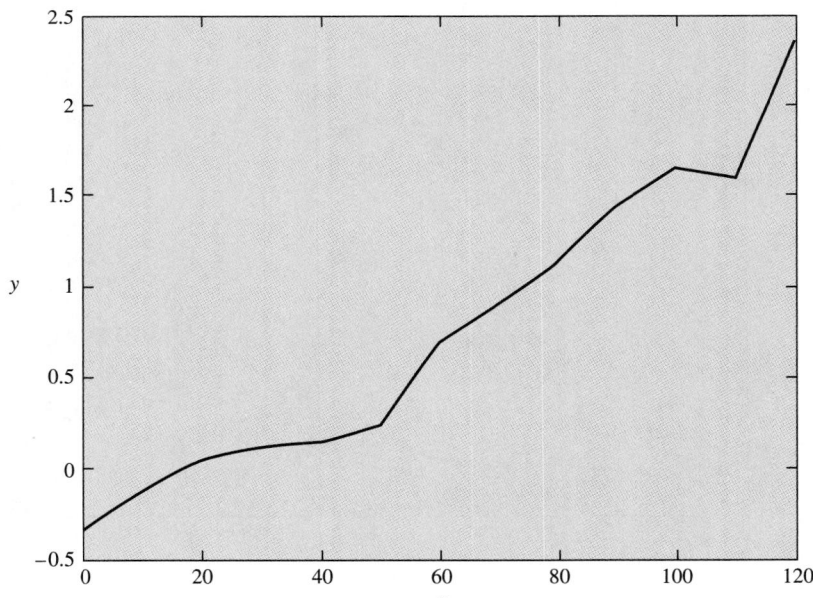

FIGURE 4.27   Plot of log data for Example 4.11.

To verify this, we look at the logarithm of our example data; this is provided in Table 4.24 and plotted in Figure 4.27, and the general trend is indeed linear. So we do a least squares fit to the log data, getting the straight line

$$z = 0.0209x - 0.4842.$$

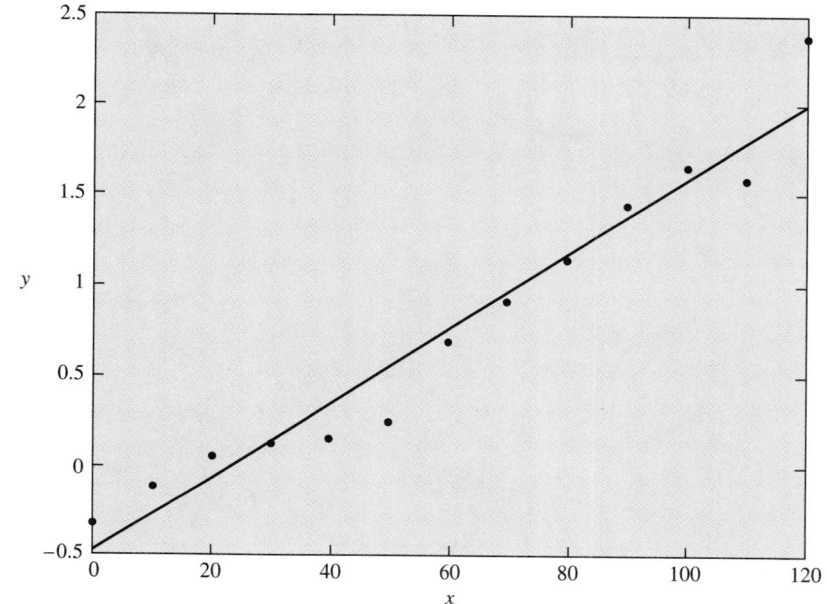

FIGURE 4.28 Log data plus fitted curve for Example 4.11.

This is plotted, along with the raw (logarithm) data, in Figure 4.28. It follows, then, that our curve fit to the original data is

$$y = e^{(0.0209x - 0.4842)},$$

which is plotted in Figure 4.29 along with the original data. Note that, except for the last two points, this is a pretty good fit to the data set. ∎

4.10.2   Least Squares Approximation and Orthogonal Polynomials

The notion of least squares approximation can be extended beyond the data-fitting problem. Consider the problem of finding an approximation to a given function f in terms of a set of basis functions $\{\phi_k, 1 \le k \le n\}$. How do we find the coefficients in the expansion

$$f(x) \approx q_n(x) = \sum_{k=1}^{n} c_k \phi_k(x)?$$

One way to do this is to require that the coefficients c_k produce an approximation that minimizes the error

$$r_n = f - q_n$$

in the least squares sense, i.e., in the sense of the integral 2-norm. Thus we seek c_k such that

$$R_n = \|f - q_n\|_2^2$$

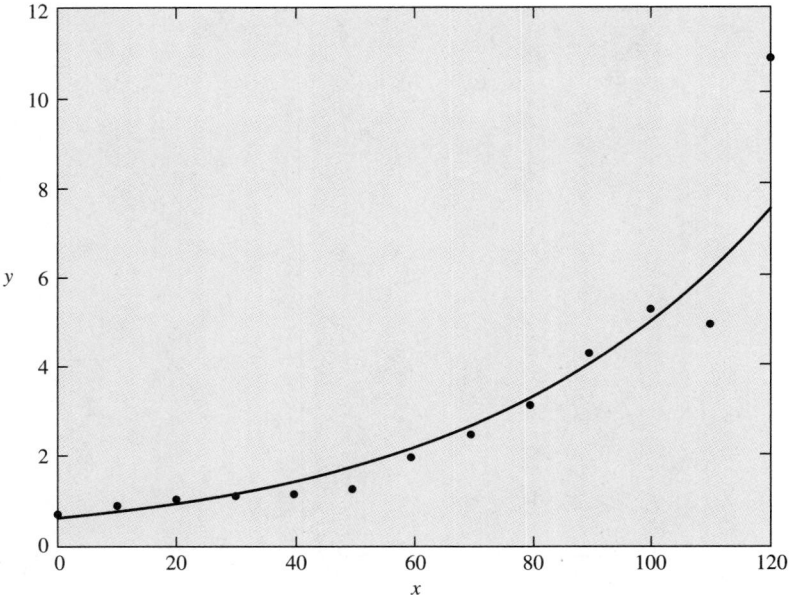

FIGURE 4.29   Raw data plus fitted curve for Example 4.11.

is minimized. Before we do this, it will be convenient to introduce the notion of an *inner product*. It is best to do this thoroughly, so we pause in our development of approximations, but only briefly.

Inner Products of Functions   The reader should be familiar with the notion of the dot product of two vectors in $\mathbf{R}^n$:

$$x \cdot y = \sum_{i=1}^{n} x_i y_i.$$

This is an example of a more general operation called an *inner product*, which can be defined on general vector spaces, including spaces of functions, rather than just Euclidean n-space. The formal definition is as follows.

Definition 4.2 (Inner Product on Real Vector Spaces)   *Let f and g be elements of a real vector space V. Let (f,g) denote any operation on f and g which satisfies the following xthree properties:*

    1. $(f,f) > 0$ *for all nonzero $f \in V$.*

    2. $(f,\alpha g_1 + \beta g_2) = \alpha(f,g_1) + \beta(f,g_2)$ *for all $f,g \in V$ and all scalars α and β.*

    3. $(f,g) = (g,f)$ *for all $f,g \in V$.*

Then (f,g) is called an inner product.

If we want to consider our overlying vector space to be $C([a,b])$, i.e., continuous functions on a closed interval, then we can easily establish that the positively weighted integral of a product of two functions will be an inner product.

Theorem 4.8 *Let w be integrable on $[a,b]$ and nonnegative, i.e., $\int_a^b w(x)\, dx$ is defined and $w(x) \geq 0$ for all $x \in [a,b]$. For given f and g in $C([a,b])$, define $(f,g)_w$ as*

$$(f,g)_w = \int_a^b w(x)f(x)g(x)\, dx. \tag{4.57}$$

Then $(\cdot,\cdot)_w$ defines on inner product on $C([a,b])$.

Proof   See Exercise 6.   □

Note that it therefore follows that if $(f,g)_w$ is an inner product, then $\|f\|_w = (f,f)_w^{1/2}$ defines a norm. In the common case when $w(x) \equiv 1$, we will simply write $(f,g)_w = (f,g)$, i.e., we will drop the subscript, and the norm is the ordinary 2-norm for functions that we defined earlier in this chapter. Exercise 7 offers some practice with a norm defined by a weighted inner product.

This definition of inner product will allow us to apply a number of ideas from linear algebra to the construction of approximations, as we shall soon see. Of more immediate interest is the fact that we can use the inner product notation to write the residual, R_n, in a very convenient form:

$$R_n = \|f\|^2 - 2\sum_{k=1}^{n} c_k(f,\phi_k) + \sum_{i=1}^{n}\sum_{j=1}^{n} c_i c_j(\phi_i,\phi_j).$$

Note that we can regard R_n as a function of the n variables c_k, and thus apply ordinary calculus to the problem of minimizing R_n. After some manipulations (see Exercise 8) we find that the c_k are defined by simultaneously solving the set of equations

$$
\begin{array}{ccccccc}
(\phi_1,\phi_1)c_1 & + & (\phi_1,\phi_2)c_2 & + & \cdots & (\phi_1,\phi_n)c_n & = & (f,\phi_1) \\
(\phi_2,\phi_1)c_1 & + & (\phi_2,\phi_2)c_2 & + & \cdots & (\phi_2,\phi_n)c_n & = & (f,\phi_2) \\
\vdots & & \vdots & & \cdots & \vdots & \vdots & \vdots \\
(\phi_n,\phi_1)c_1 & + & (\phi_n,\phi_2)c_2 & + & \cdots & (\phi_n,\phi_n)c_n & = & (f,\phi_n).
\end{array}
$$

This system can be organized along matrix-vector lines as

$$
\begin{bmatrix}
(\phi_1,\phi_1) & (\phi_1,\phi_2) & \cdots & (\phi_1,\phi_n) \\
(\phi_2,\phi_1) & (\phi_2,\phi_2) & \cdots & (\phi_2,\phi_n) \\
\vdots & \vdots & \cdots & \vdots \\
(\phi_n,\phi_1) & (\phi_n,\phi_2) & \cdots & (\phi_n,\phi_n)
\end{bmatrix}
\begin{bmatrix}
c_1 \\ c_2 \\ \vdots \\ c_n
\end{bmatrix}
=
\begin{bmatrix}
(f,\phi_1) \\ (f,\phi_2) \\ \vdots \\ (f,\phi_n)
\end{bmatrix}. \tag{4.58}
$$

Solving a system of linear equations is a problem that we do not encounter in the general case until Chapter 7. We can avoid it altogether at this point *if* our basis functions satisfy the *orthogonality* condition

$$(\phi_i, \phi_j) = 0 \quad \text{for all } i \neq j. \tag{4.59}$$

In this case, the matrix in (4.58) is a diagonal matrix, and we very easily have

$$c_k = \frac{(f, \phi_k)}{(\phi_k, \phi_k)}.$$

To summarize what we have done so far, we can construct an approximation to a given function f from a given basis set $\{\phi_1, \phi_2, \ldots, \phi_n\}$, and the construction is very easy, *if the basis satisfies condition* (4.59) So the question becomes, When can we find a basis that satisfies (4.59), and how good is the resulting approximation?

The answer is that we can always find such a basis if we consider polynomial functions for our basis elements, and the resulting approximations are usually quite good. The special basis functions that satisfy (4.59) are called *orthogonal polynomials*. To be more specific, we have to introduce some new concepts and notation, and recall a major theorem from linear algebra. But first, one more definition.

Definition 4.3 (Vector Space of Polynomials of Degree $\leq N$)   For any $N \geq 0$ define $\mathcal{P}_N$ as the vector space of polynomials of degree $\leq N$. Note that this space has a standard basis consisting of $\{1, x, x^2, x^3, \ldots, x^N\}$, and thus is an $N + 1$–dimensional space. And now the theorem:

Theorem 4.9   Let w be a given nonnegative weight function on an interval $[a, b]$ and $(\cdot, \cdot)_w$ the associated inner product, defined as in (4.57). Then there exists a family of orthogonal polynomials $\{\phi_k\}$, $\phi_k \in \mathcal{P}_k$, $0 \leq k \leq N$, such that

$$(\phi_i, \phi_j)_w = 0, \quad i \neq j,$$

and

$$(\phi_i, \phi_i)_w > 0, \quad i \geq 0.$$

In addition, the ϕ_k satisfy

1. The set $\{\phi_0, \phi_1, \ldots, \phi_N\}$ is a basis for $\mathcal{P}_N$.

2. If q_k is an arbitrary element of $\mathcal{P}_k$, for $k < N$, then $(q_k, \phi_N)_w = 0$ for all $N > k$ (thus, orthogonal polynomials are orthogonal to all polynomials of strictly lower degree).

3. For $j \geq 1$, the roots of each ϕ_j are all in $[a, b]$ and are all distinct.

Proof   The proof is somewhat lengthy, in part because of the length of the theorem, but it is not difficult.

To establish that the family $\{\phi_k\}$ exists, we will construct it directly, using the Gram-Schmidt process from linear algebra. Take $\phi_0(x) = 1$, and define $\xi_k(x) = x^k$, for

$0 \leq k \leq N$. Then the subsequent ϕ_k can be found according to

$$\phi_k(x) = \xi_k(x) - \sum_{j=0}^{k-1} \frac{(\phi_j, \xi_k)_w}{(\phi_j, \phi_j)_w} \phi_j(x), \tag{4.60}$$

and an inductive argument shows very quickly that the orthogonality holds. Exercise 9 asks the student to fill in the details of this part of the proof.

Having proved that the family of orthogonal polynomials exists, we turn our attention to proving each of the results (1)–(3).

1. The space $\mathcal{P}_N$ is finite-dimensional with dimension $N+1$. It therefore follows that any set of $N+1$ independent elements of $\mathcal{P}_N$ will be a basis. The orthogonality condition (4.59) forces the members of the family $\{\phi_k\}$ to be independent (see Exercise 10); therefore, the set $\{\phi_0, \phi_1, \ldots, \phi_N\}$ is a basis for $\mathcal{P}_N$.

2. Let q_k be an arbitrary polynomial of degree $k < N$. Then we can write

$$q_k = \sum_{i=0}^{k} a_i \phi_i$$

because $\{\phi_0, \phi_1, \ldots, \phi_k\}$ is a basis for $\mathcal{P}_k$. Therefore,

$$(q_k, \phi_N) = \sum_{i=0}^{k} a_i (\phi_i, \phi_N) = 0$$

since ϕ_N is orthogonal to each element of $\{\phi_0, \phi_1, \ldots, \phi_k\}$. In fact, we can write (see Exercise 11)

$$q_k = \sum_{i=0}^{k} \frac{(q_k, \phi_i)}{(\phi_i, \phi_i)} \phi_i. \tag{4.61}$$

3. First, suppose that ϕ_j, $j \geq 1$, has no roots in $[a, b]$. This means that ϕ_j does not change sign on the interval, thus

$$(\phi_0, \phi_j) = \int_a^b w(x) \phi_0(x) \phi_j(x) \, dx \neq 0,$$

since the integrand does not change sign on $[a, b]$. But the orthogonality requires that $(\phi_0, \phi_j) = 0$; hence we have a contradiction, and hence there must be at least one root in $[a, b]$.

Now, let x_1 be any root of ϕ_j that lies in $[a, b]$, and suppose it is a multiple root. Then it follows that

$$\phi_j(x) = (x - x_1)^2 q(x)$$

for some polynomial q; thus

$$q(x) = (x - x_1)^{-2} \phi_j(x)$$

is a polynomial of degree $j - 2$; therefore, $(q, \phi_j) = 0$, by a previous part of this theorem. But

$$(q, \phi_j) = \int_a^b w(x)(x - x_1)^{-2}(\phi_j(x))^2 \, dx > 0,$$

so we have another contradiction. Thus any roots that lie in $[a, b]$ must be simple roots.

Suppose now that only some of the roots lie in $[a, b]$. Call these roots $x_i, 1 \leq i < j$, and note that we can write ϕ_j as

$$\phi_j(x) = \Psi_j(x)(x - x_1) \cdots (x - x_i)$$

where $\Psi_j(x)$ does not change sign in $[a, b]$ and is a polynomial of degree $j - i$. Therefore,

$$\phi_j(x)(x - x_1) \cdots (x - x_i) = \Psi_j(x)(x - x_1)^2 \cdots (x - x_i)^2$$

is also a polynomial that does not change sign in $[a, b]$. Hence, the integral

$$I = \int_a^b \Psi_j(x)(x - x_1)^2 \cdots (x - x_i)^2 \, dx$$

cannot be zero. However,

$$I = \int_a^b \Psi_j(x)(x - x_1)^2 \cdots (x - x_i)^2 \, dx = (\phi_j, q),$$

where $q(x) = (x - x_1) \cdots (x - x_i)$, and q is a polynomial of degree $i < j$. Therefore, $(\phi_j, q) = 0$, and we have a contradiction. Thus $i \geq j$, and since a polynomial of degree j cannot have more than j roots, we must have $i = j$. $\qquad \square$

Families of Orthogonal Polynomials   At this point it might be useful to look at some examples of orthogonal polynomial families. Four of the most common ones are discussed next.

Note that the orthogonality condition (4.59) means that an orthogonal polynomial can be multiplied by an arbitrary nonzero constant and still satisfy (4.59). To avoid the problems of nonuniqueness that this can lead to, it is common to impose a specific scaling on the elements of each family.

   1. *Legendre polynomials:* The Legendre[13] polynomials are the orthogonal polynomials on $[-1, 1]$ with no weight function (more correctly, the unit weight

---

[13]Adrien-Marie Legendre (1752–1833) was born and educated in Paris. He contributed greatly to what we now call number theory as well as elliptic function theory. He was the first to publish, in 1805, a description of the method of least squares, although it appears that Gauss had previously worked out much of the same material.

Legendre introduced the polynomials that bear his name in a 1785 paper on the gravitational attraction of spherical bodies. They arise in this context because they are the solutions to an ordinary differential equation that occurs as part of the solution process for the equations of motion in a spherical coordinate system.

function); thus we have

$$\int_{-1}^{1} P_i(x) P_j(x) \, dx = 0, \quad i \neq j.$$

The usual scaling is to take $P_n(1) = 1$. The first five Legendre polynomials are

$$P_0(x) = 1$$
$$P_1(x) = x$$
$$P_2(x) = \frac{1}{2}(3x^2 - 1)$$
$$P_3(x) = \frac{1}{2}(5x^3 - 3x)$$
$$P_4(x) = \frac{1}{8}(35x^4 - 30x^2 + 3).$$

2. *Chebyshev polynomials:* The common notation for the Chebyshev[14] polynomials is $T_n(x)$, and the interval and weight function are defined in the orthogonality relation

$$\int_{-1}^{1} \frac{T_i(x) T_j(x)}{\sqrt{1 - x^2}} \, dx = 0, \quad i \neq j.$$

The common scaling is to set the leading coefficient equal to 2^{n-1}. The first five Chebyshev polynomials are

$$T_0(x) = 1$$
$$T_1(x) = x$$
$$T_2(x) = 2x^2 - 1$$
$$T_3(x) = 4x^3 - 3x$$
$$T_4(x) = 8x^4 - 8x^2 + 1.$$

It can be shown that the Chebyshev polynomials are related in a very simple way to cosines; see Theorem 4.10.

---

[14]Pafnuty Lvovich Chebyshev (1821–1894) was born in Borovsk, Russia, and educated at the University of Moscow, from which he was graduated in 1841. From 1847 until his death he lived in St. Petersburg.

Many areas of mathematics, from number theory to the theory of equations, were touched by Chebyshev, but he is perhaps best known, at least in applied mathematics, for his work in approximation theory.

Because of the many different ways that the Russian alphabet can be transliterated into the Roman alphabet, there are several different ways to spell Chebyshev's name. The most common alternative is "Tschebyscheff." A marvelously engaging discussion of the perils of transliterating between the Roman and the Russian (Cyrillic) alphabets, as well as a substantial treatment of Chebyshev's life, is contained in *The Thread* (Harcourt Brace Jovanovitch, New York, 1989), by Philip J. Davis, a leading mathematician in the area of interpolation and approximation.

3. *Hermite polynomials:* The Hermite polynomials are orthogonal on the entire real line, using the weight function $w(x) = e^{-x^2}$, i.e.,

$$\int_{-\infty}^{\infty} e^{-x^2} H_i(x) H_j(x) \, dx = 0, \quad i \neq j.$$

The common scaling is to set the leading coefficient equal to 2^n. The first five Hermite polynomials are

$$H_0(x) = 1$$
$$H_1(x) = 2x$$
$$H_2(x) = 4x^2 - 2$$
$$H_3(x) = 8x^3 - 12x$$
$$H_4(x) = 16x^4 - 48x^2 + 12.$$

4. *Laguerre polynomials*: The Laguerre[15] polynomials are orthogonal on the positive real line, using the weight function $w(x) = e^{-x}$, i.e.,

$$\int_0^{\infty} e^{-x} L_i(x) L_j(x) \, dx = 0, \quad i \neq j.$$

The common scaling is to set the leading coefficient equal to $(-1)^n/n!$. The first five Laguerre polynomials are then

$$L_0(x) = 1$$
$$L_1(x) = -x + 1$$
$$L_2(x) = \frac{1}{2}(x^2 - 4x + 2)$$
$$L_3(x) = \frac{1}{6}(-x^3 + 9x^2 - 18x + 6)$$
$$L_4(x) = \frac{1}{24}(x^4 - 16x^3 + 72x^2 - 96x + 24).$$

We can use any of these orthogonal polynomial families to construct approximations to functions defined on the appropriate interval. These approximations are the "best possible" in the sense that they minimize the error in the appropriate weighted 2-norm, i.e.,

$$\|f - q_n\|_w < \|f - p_n\|_w$$

for all $p_n \in \mathcal{P}_n$, $p_n \neq q_n$.

Consider, as illustrations, the following set of examples.

---

[15]Edmond Nicolas Laguerre (1834–1886) was born and died in Bar-le-Duc, France. He was educated at the École Polytechnique but did not graduate with a high ranking. He served in the military as an artillery officer for 10 years before returning to École as a faculty member, where he remained for the rest of his life.

Although best known for the orthogonal polynomials that bear his name, and their associated differential equation, Laguerre also published work in analysis, geometry, and abstract linear spaces.

EXAMPLE 4.12

Let's construct the fourth-degree least squares approximation to the exponential function, $f(x) = e^x$, over the interval $[-1,1]$, using Legendre polynomials. The approximation is defined by

$$p_4(x) = \sum_{k=0}^{4} \frac{(f, P_k)}{(P_k, P_k)} P_k(x)$$

where the P_k are the Legendre polynomials. We thus need to compute the integrals

$$I_0 = \int_{-1}^{1} e^x \, dx, \qquad I_1 = \int_{-1}^{1} x e^x \, dx, \qquad I_2 = \int_{-1}^{1} \frac{1}{2}(3x^2 - 1) e^x \, dx,$$

$$I_3 = \int_{-1}^{1} \frac{1}{2}(5x^3 - 3x) e^x \, dx, \qquad I_4 = \int_{-1}^{1} \frac{1}{8}(35x^4 - 30x^2 + 3) e^x \, dx,$$

and

$$J_0 = \int_{-1}^{1} dx, \qquad J_1 = \int_{-1}^{1} x^2 \, dx, \qquad J_2 = \int_{-1}^{1} \frac{1}{4}(3x^2 - 1)^2 \, dx,$$

$$J_3 = \int_{-1}^{1} \frac{1}{4}(5x^3 - 3x)^2 \, dx, \qquad J_4 = \int_{-1}^{1} \frac{1}{64}(35x^4 - 30x^2 + 3)^2 \, dx.$$

In practice, these integrals would be computed using some kind of numerical integration routine, such as the trapezoid rule or the more accurate methods we discuss in the next chapter. For this simple example, though, it is possible to use direct calculus methods or (much more attractive!) a computer algebra package such as Maple or Mathematica. However it is done, to eight digits the integrals are

$$I_0 = 2.3504024, \qquad I_1 = 0.73575888, \qquad I_2 = 0.14312574,$$
$$I_3 = 0.020130181, \qquad I_4 = 0.0022144731,$$

and

$$J_0 = 2.00000000, \qquad J_1 = 0.66666667, \qquad J_2 = 0.40000000,$$
$$J_3 = 0.28571429, \qquad J_4 = 0.22222222.$$

Thus the polynomial approximation is

$$p_4(x) = \frac{2.3504024}{2.00000000} P_0(x) + \frac{0.73575888}{0.66666667} P_1(x)$$

$$+ \frac{0.14312574}{0.40000000} P_2(x) + \frac{0.020130181}{0.28571429} P_3(x) + \frac{0.0022144731}{0.22222222} P_4(x),$$

which simplifies to

$$p_4(x) = 1.0000309 + 0.99795487x + 0.49935229x^2 + 0.17613908x^3 + 0.043597439x^4.$$

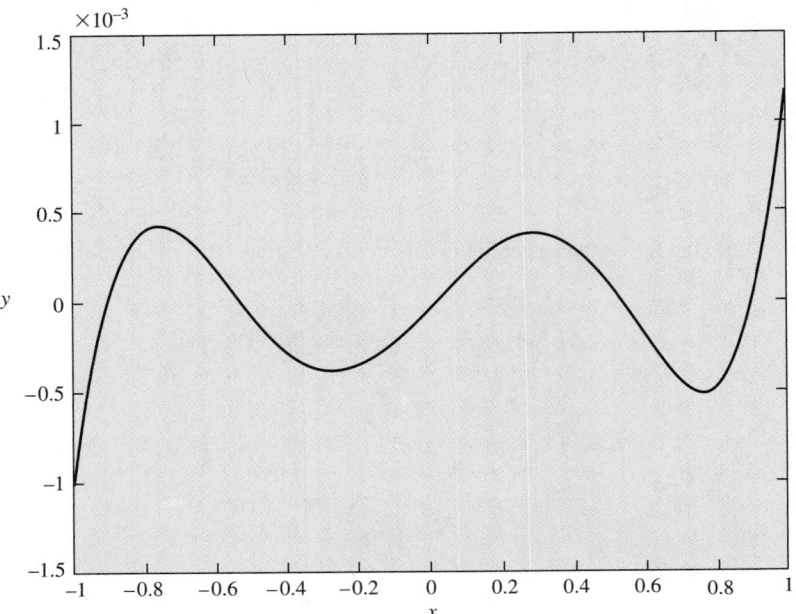

FIGURE 4.30   Error in fourth-degree least squares approximation to the exponential function.

Figure 4.30 shows a plot of the error $e^x - p_4(x)$. Compare this with the error plots for fourth-degree Taylor approximation and fourth-degree Lagrange/Newton interpolation from earlier in this chapter. Note, in particular, that the least squares error oscillates back and forth between its maximum and minimum values (or nearly so), several times. It can be shown that this is a necessary and sufficient condition for the approximating polynomial to be the "best" approximation to the function, and is one reason why least squares approximations are considered valuable: they are close to being the best possible approximations.   ∎

EXAMPLE 4.13

Here we construct a Legendre polynomial approximation to the function $f(x) = (1 + 25x^2)^{-1}$ using polynomials of degree $n = 8, 16$, and 32. Figure 4.31(a)–(c) shows the plots of both f and the least squares approximation; Figure 4.31(d) shows the error in the 32-degree approximation.   ∎

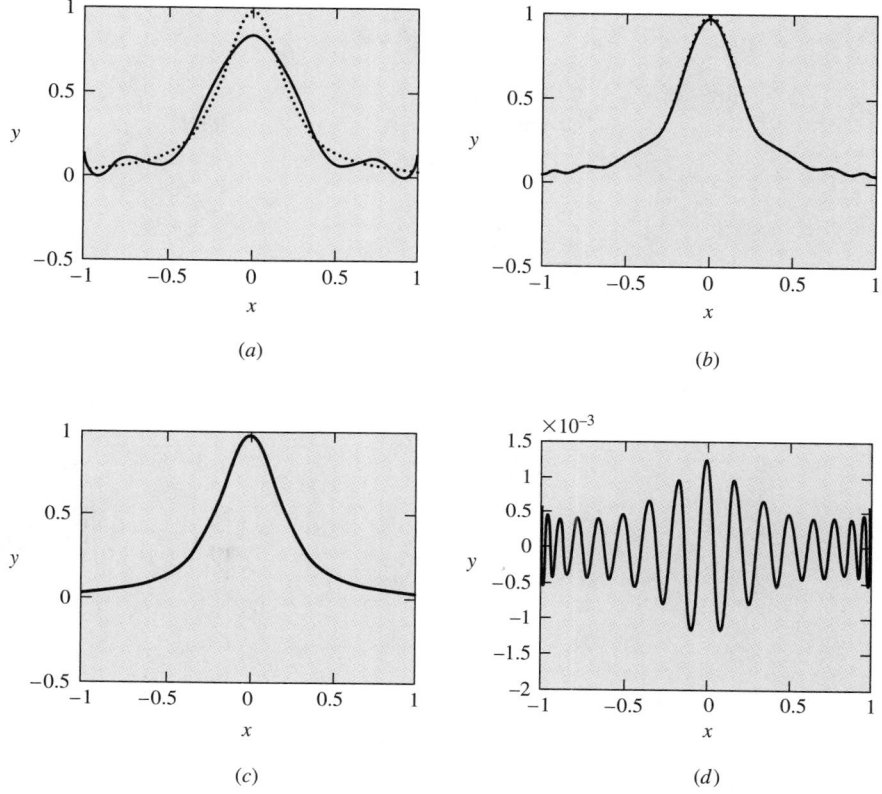

FIGURE 4.31 Legendre least squares approximation to $f(x) = 1/(1 + 25x^2)^{-1}$. A. $n = 8$. B. $n = 16$. C. $n = 32$. D. Error in the $n = 32$ case. In A and B, $f(x)$ is denoted by the dotted curve.

EXAMPLE 4.14

This time, we use a Chebyshev polynomial approximation to the same f as in the previous example, again using polynomials of degree $n = 8, 16$, and 32. Figure 4.32(a)–(c) shows the plots of both f and the least squares approximation; Figure 4.32(d) shows the error in the 32-degree approximation. The Chebyshev and Legendre approximations are very similar in performance, with the Chebyshev being slightly better. ∎

EXAMPLE 4.15

Legendre approximation to e^x. Here we construct a Legendre polynomial approximation to e^x on the interval $[-1, 1]$, in much the same way as was done for $f(x) = (1 + 25x^2)^{-1}$ in Example 4.13. However, the accuracy here is so much greater that we plot the errors for the $n = 2, 4, 8$ cases in Figure 4.33. ∎

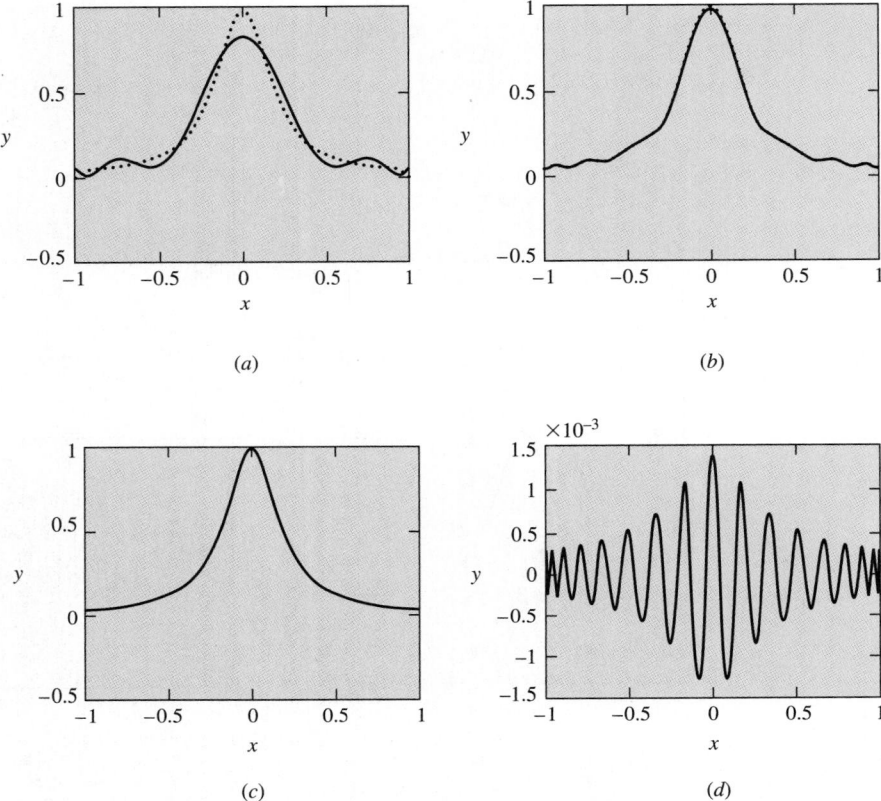

FIGURE 4.32   Chebyshev least squares approximation to $f(x) = 1/(1 + 25x^2)^{-1}$. (a). $n = 8$. (b). $n = 16$. (c). $n = 32$. (d). Error in the $n = 32$ case. In (a) and (b), $f(x)$ is denoted by the dotted curve.

EXAMPLE 4.16

Chebyshev approximation to e^x. This is the same as the previous example, except we use a Chebyshev expansion instead of a Legendre expansion. The error is plotted in Figure 4.34.   ■

A couple of comments are in order here. We first note how much easier it was to obtain a high degree of accuracy for the exponential than it was for $f(x) = (1 + 25x^2)^{-1}$. Second, although it might be difficult to discern in the plots, the results for the Chebyshev approximations were, in each case, very slightly better than for the Legendre approximations. A full discussion of this requires more mathematica machinery than we want to deal with right now, but it is generally true that Chebyshev least squares approximations are superior to those done with any other choice of basis.

Finally, note that in order to do a least-squares approximation, we have to be able to compute the inner products, which are integrals. Thus we need a tool like the trapezoid rule or, perhaps better, some of the more sophisticated methods to be developed in Chapter 5.

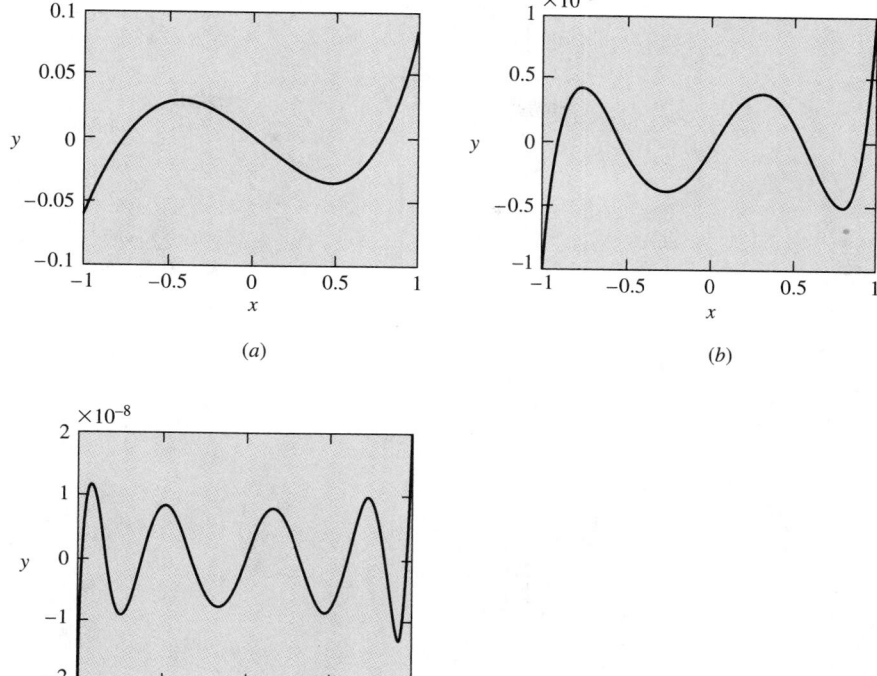

FIGURE 4.33   Error in the Legendre least squares approximation to $f(x) = e^x$. (a). Error for $n = 2$.
(b). Error for $n = 4$. (c). Error for $n = 8$.

EXERCISES

1. Modify the methods of Section 4.10.1 to compute the quadratic polynomial that gives the best least squares fit to the data in Table 4.25.

2. Table 4.26 gives the actual thermal conductivity data for the element iron. Construct a quadratic least squares fit to this data, and plot both the curve and the raw data. How well does your curve represent the data? Is the fit improved by using a cubic polynomial?

TABLE 4.25 Data for Exercises

Exercise 1		Exercise ??		
x_n	y_n	x_n	y_n	z_n
-1	0.9747	0	0	0.9573
0	0.0483	0	1	2.0132
1	1.0223	1	0	2.0385
2	4.0253	1	1	1.9773
3	9.0152	0.5	0.5	1.9936

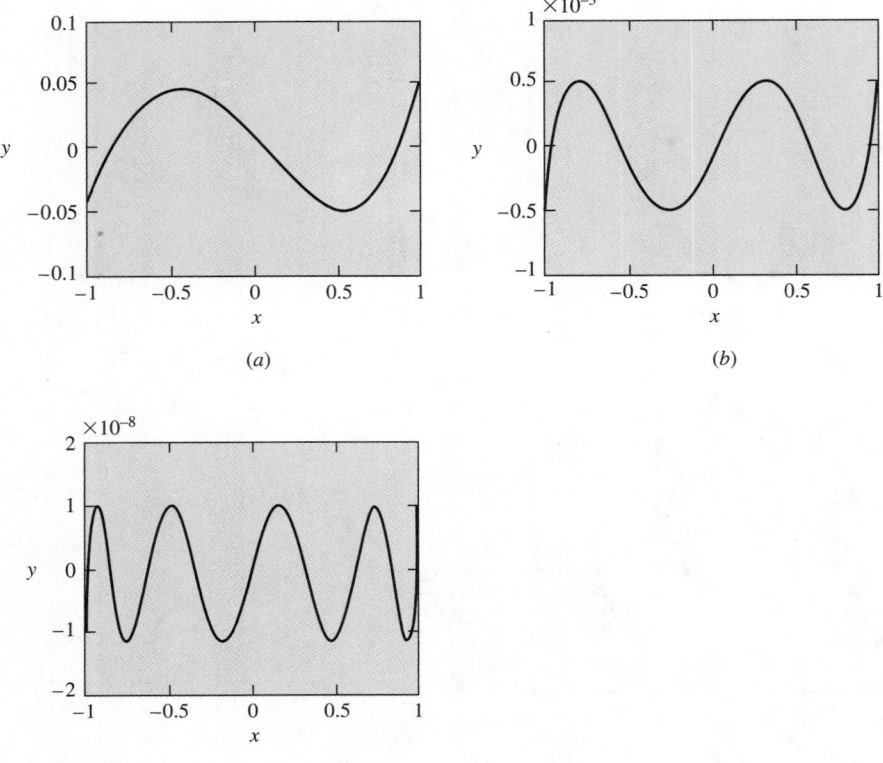

(c)

FIGURE 4.34 Error in Chebyshev least squares approximation to $f(x) = e^x$. (a). Error for $n = 2$. (b). Error for $n = 4$. (c). Error for $n = 8$.

TABLE 4.26   Data for Exercise 2

Temperature (K), u	100	200	300	400	500	600	700	800	900	1000
Conductivity (W/cm K), k	1.32	0.94	0.835	0.803	0.694	0.613	0.547	0.487	0.433	0.38

3. Repeat Exercise 2, this time using the nickel data from Exercise 21 in Section 4.8.

4. Modify the methods of Section 4.10.1 to compute the linear function of two variables that gives the best least squares fit to the data in Table 4.25.

5. An astronomical tracking station records data on the position of a newly discovered asteroid orbiting the Sun. The data is reduced to measure-ments of the radial distance from the Sun (in millions of kilometers) and angular position around the orbit (in radians), based on knowledge of the Earth's position relative to the Sun. In theory, these values should fit into the polar coordinate equation of an ellipse, given by

$$r = \frac{L}{2(1 + \epsilon \cos \theta)}$$

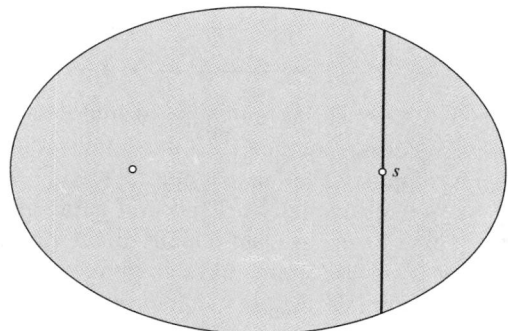

FIGURE 4.35 See Exercise 5. The closed curve is the elliptical orbit, and the vertical line has length L.

where ϵ is the eccentricity of the elliptical orbit and L is the width of the ellipse (sometimes known as the *latus rectum* of the ellipse) at the focus. (See Figure 4.35.) However, errors in the tracking process and approximations in the transformation to (r,θ) values perturb the data. For the data in Table 4.27, find the eccentricity of the orbit by doing a linear least squares fit to the data. (Hint: Write the polar equation of the ellipse as

$$2r(1+\epsilon\cos\theta) = L,$$

which can then be written as

$$2r = \epsilon(-2r\cos\theta) + L.$$

So $y_k = 2r_k$ and $x_k = -2r_k\cos\theta_k$.)

6. Prove Theorem 4.8.

7. Let $w(x) = x$ on the interval $[0,1]$; compute $\|f\|_w$ for each of the following functions:
(a) e^{-x}
(b) $1/\sqrt{x^2+1}$
(c) $1/\sqrt{x}$
Compare the results with the values obtained using the unweighted 2-norm.

8. Derive the linear system (4.58) as the solution to the least squares approximation problem.

9. Provide the missing details to show that the family of polynomials defined in (4.60) is indeed orthogonal.

10. Let $\{\phi_k\}$ be a family of orthogonal polynomials associated with a general weight function w and an interval $[a,b]$. Show that the $\{\phi_k\}$ are independent in the sense that

$$0 = c_1\phi_1(x) + c_2\phi_2(x) + \cdots + c_n\phi_n(x)$$

holds for all $x \in [a,b]$ if and only if $c_k = 0$ for all k.

11. Prove the expansion formula (4.61) for a polynomial.

TABLE 4.27 Data for Exercise 4.27

θ_n	r_n
−0.1289	42,895
−0.1352	42,911
−0.1088	42,851
−0.0632	42,779
−0.0587	42,774
−0.0484	42,764
−0.0280	42,750
−0.0085	42,744
0.0259	42,749
0.0264	42,749
0.1282	42,894

12. Construct the second-degree Laguerre least squares approximation to $f(x) = e^{\frac{1}{2}x}$ on the interval $[0,\infty)$.

13. Construct the second-degree Legendre least squares approximation to $f(x) = \cos\pi x$ over the interval $[-1,1]$.

14. Construct the third-degree Legendre least squares approximation to $f(x) = \sin\frac{1}{2}\pi x$ over the interval $[-1,1]$.

4.11 ADVANCED TOPICS IN INTERPOLATION ERROR

Based on the examples we have seen so far, the value of interpolation as an approximation tool seems unclear. When we applied it to the exponential function we got excellent results, but when we applied it to the example $f(x) = (1 + 25x^2)^{-1}$ (known as the *Runge*[16] *example*) we got substantially less accuracy for the same number of nodes. This is not simply a case of using insufficiently many points in a particular example. It can be shown that if we took more and more nodes and used higher- and higher-degree interpolating polynomials, the error in interpolating to the Runge example would continue to get larger and larger, when measured in the norm $\|f - p_n\|_\infty$. In fact, the trend that is suggested in Figure 4.3 (b) and (c) continues: For values of x in the middle of the interval $[-1, 1]$, $|f(x) - p_n(x)|$ actually goes to zero; but near the ends of the interval the polynomial interpolants oscillate wildly and do not converge. What is going on here?

A complete answer, while mathematically very interesting, is beyond the scope of this text, so we will skip the details and present only the broader ideas. We will also look at a couple of different ways in which the potential problems with interpolation at equally spaced nodes can impact a calculation, and we will derive a better set of interpolating nodes.

4.11.1 Stability of Polynomial Interpolation

One problem with interpolation using high-degree polynomials is that it really is a potentially unstable process, in the sense that small changes to the data can lead to large changes in the interpolating polynomial. To see this, consider the effects of rounding error on the interpolation problem. Let $f(x)$ be the exact function that we wish to interpolate, and let $\hat{f}(x)$ be the function polluted by rounding error, and assume that $f(x) - \hat{f}(x) = \epsilon(x)$, where

$$\|f - \hat{f}\|_\infty = \|\epsilon\|_\infty = \epsilon_0$$

for some small ϵ_0.

The polynomial interpolant that we compute is based on the function that is polluted by rounding error, so we have the Lagrange form

$$\hat{p}_n(x) = \sum_{i=0}^{n} L_i^{(n)}(x)\hat{f}(x_i),$$

---

[16]Carle Runge (1856–1927) was born in Bremen and educated at the University of Munich. He originally intended to study literature but after less than two months switched to mathematics and physics. In 1877 he began attending the University of Berlin, where he received his doctorate in 1880 in differential geometry. He held academic positions at Hanover and Göttingen, and he retired in 1925. Much of his professional work was more in physics than in mathematics, but he did make contributions in the numerical solution of differential equations (Runge-Kutta methods) and polynomial interpolation theory. The paper in which he outlined the theory behind the Runge example appeared in 1901, under the title *Über empirische Funktionen und die Interpolation zwischen äquidistanten Ordinaten*.

and we want to estimate the error $f(x) - \hat{p}(x)$. For convenience we define the "ideal" interpolant, based on the exact function, f, as

$$p_n(x) = \sum_{i=0}^{n} L_i^{(n)}(x) f(x_i).$$

Theorem 4.3 gives us the error between f and p_n:

$$f(x) - p_n(x) = \frac{1}{(n+1)!} \left(\prod_{i=0}^{n} (x - x_i) \right) f^{(n+1)}(\xi_x).$$

We can relate this to the error we want to bound as follows:

$$f(x) - \hat{p}_n(x) = \underbrace{(f(x) - p_n(x))}_{\text{Error due to interpolation}} + \underbrace{(p_n(x) - \hat{p}_n(x))}_{\text{Error due to rounding}}.$$

We now turn our attention to analyzing the error due to rounding. We have

$$|p_n(x) - \hat{p}_n(x)| = \left| \sum_{i=0}^{n} L_i^{(n)}(x)(f(x_i) - \hat{f}(x_i)) \right|$$

$$\leq \|\epsilon\|_\infty \sum_{i=0}^{n} \|L_i^{(n)}\|_\infty$$

$$= \epsilon_0 \sum_{i=0}^{n} \Lambda_i,$$

where the norms are both taken over the interval defined by the nodes, and $\Lambda_i = \|L_i^{(n)}\|_\infty$. Let's assume that the rounding error is bounded and small—a very reasonable assumption based on our work in Chapter 1. Thus we take $\|\epsilon\|_\infty = \epsilon_0 = O(\mathbf{u})$.

We appear to be in good shape, having bounded the error due to rounding by something small times a sum depending on the norm of the Lagrange functions. The problem is that this sum can grow quite large as n increases. A general theorem (see Theorem 4.6 of [14]) is not necessary, since we can get the essentials by being experimental. Assume that the nodes are equidistant on the interval $[a,b]$, with $x_0 = a$ and $x_n = b$, and with $x_{j+1} - x_j = h$ being the uniform mesh spacing. Then we can write

$$x_j = a + jh, \quad 0 \leq j \leq n,$$

for each node, and

$$x = a + \eta_x h, \quad 0 \leq \eta_x \leq n,$$

where $\eta_x = (x - a)/h$ takes on *real* values between 0 and n. Therefore,

$$L_i^{(n)}(x) = \prod_{\substack{k=0 \\ k \neq i}}^{n} \frac{x - x_k}{x_i - x_k}, \tag{4.62}$$

$$= \prod_{\substack{k=0 \\ k \neq i}}^{n} \frac{\eta_x - k}{i - k}. \tag{4.63}$$

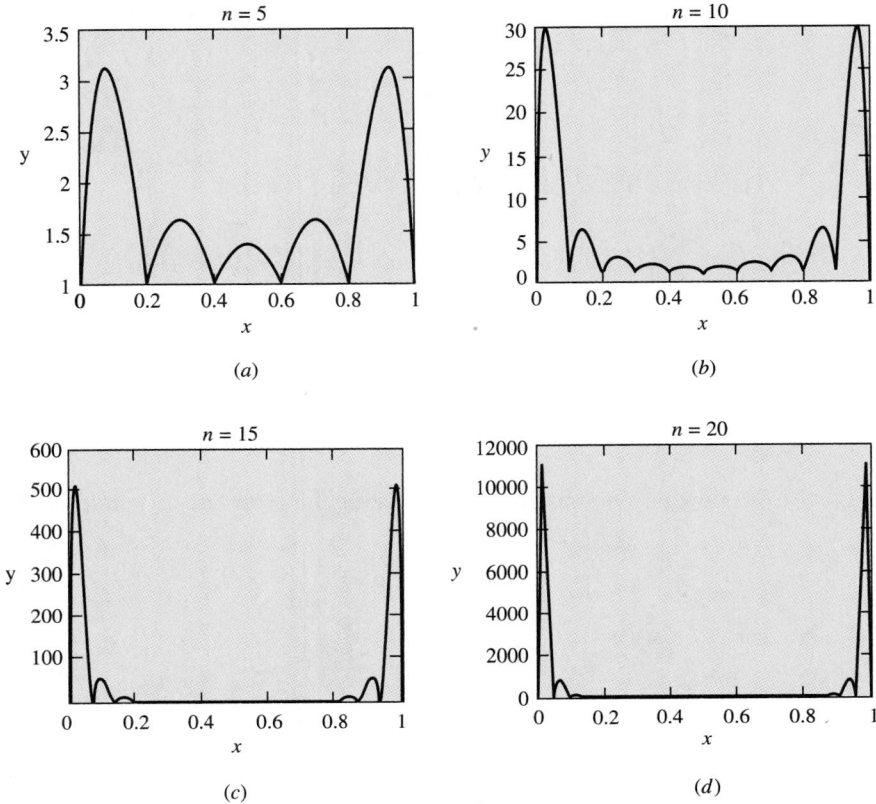

FIGURE 4.36   Plots of $L(x)$ versus x for $n = 5, 10, 15, 20$.

Hence the Lagrange functions (and therefore the Λ_i) are not dependent on the choice of a, b, or h. They depend entirely on n, η_x (which depends on x), and the distribution of the nodes. This means we ought to be able to plot them rather easily and get some idea of how big that sum term can be. Figure 4.36 shows plots of

$$L(x) = \sum_{i=0}^{n} |L_i^{(n)}(x)|$$

for various values of n (assuming equally spaced points on the interval $[0,1]$), and Figure 4.37 shows a plot of

$$M_n = \sum_{i=0}^{n} \Lambda_i$$

as a function of n. Note the *rapid* growth of M_n. If we take larger values of n, the trend continues and the growth becomes clearly exponential, as Figure 4.38 shows.

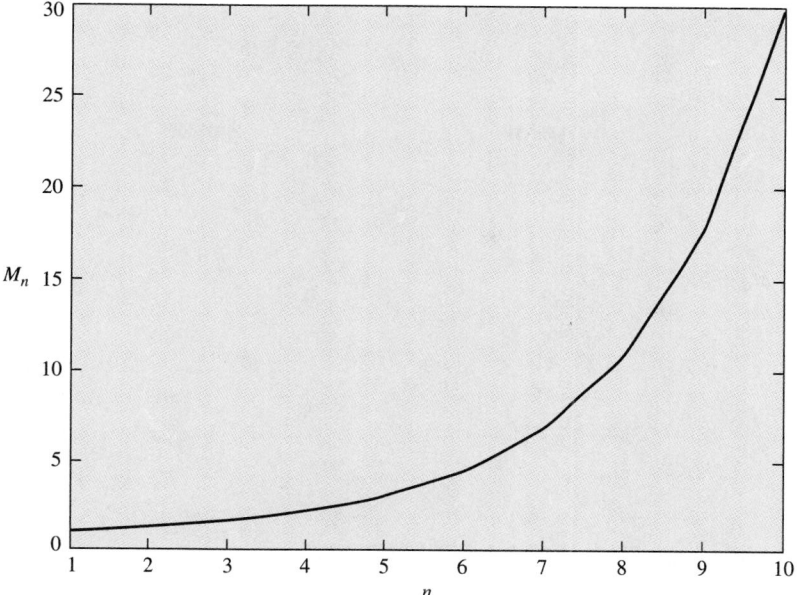

FIGURE 4.37   Plot of M_n versus n, for $n \leq 10$.

What does this mean for our original interpolation problem? Recall how the overall error was related to the error in interpolation and the error in rounding:

$$f(x) - \hat{p}_n(x) \ = \ (f(x) - p_n(x)) + (p_n(x) - \hat{p}_n(x));$$

from which it follows that

$$\|f - \hat{p}_n\|_\infty \leq \|f - p_n\|_\infty + \|p_n - \hat{p}_n\|_\infty \leq \|f - p_n\|_\infty + \epsilon_0 M_n. \qquad (4.64)$$

Since we now know that M_n can become quite large, we know that the presence of the small ϵ_0 multiplying the rounding error term is *not* enough to guarantee that the overall computation will not be badly corrupted by rounding error. On the other hand, if we keep the degree of the polynomial interpolation less than, say, 7, then we have that $M_n \leq 10$ (see Figure 4.37) and we know the overall error is not seriously corrupted by round-off.

Note that we are not claiming that polynomial interpolation of high degree *has* to be affected by large amounts of rounding error; the inequality in (4.64) goes the wrong way to support that conclusion. However, the estimate (4.64) does *allow* for large amounts of rounding error whenever n is large enough that M_n is very large, and we can avoid this potential problem by taking n so that M_n is small. Thus we avoid using polynomial interpolants of degree much higher than 7 or so.

FIGURE 4.38   Plot of M_n versus n, for $n \leq 20$.

4.11.2   The Runge Example

(Note: This section requires a bit more background than the others in this chapter, especially in complex arithmetic.)

One is tempted to look at the function $f(x) = (1 + 25x^2)^{-1}$ and say that it is a smooth, well-behaved function for all x, and this is indeed the case, so long as x is real. But if we allow x to be imaginary, then problems can occur; consider the value of $f(i/5)$, where $i = \sqrt{-1}$. Why does this matter, if we are considering the problem of interpolating to f as a function of the real variable x?

It turns out that it matters a great deal. It can be shown (see Section 3.4 of [11]) that for functions of the form $f(x) = (1 + a^2 x^2)^{-1}$, the error in interpolation can be expressed as

$$f(x) - p_n(x) = \left(\frac{r_n}{1 + a^2 x^2} \right) \left(\frac{w_n(x)}{w_n(ia^{-1})} \right),$$

where $r_n = x$ if n is even, $r_n = ia^{-1}$ if n is odd, and

$$w_n(x) = \prod_{k=0}^{n} (x - x_k).$$

Since

$$\left| \frac{r_n}{1 + a^2 x^2} \right| \leq C$$

for all x and n, the question of convergence then comes down to the size of the ratio

$$R_n(x) = \left| \frac{w_n(x)}{w_n(ia^{-1})} \right|.$$

Now we can write

$$R_n(x) = \left| \frac{\sigma_n(x)}{\sigma_n(ia^{-1})} \right|^{n+1}$$

for

$$\sigma_n(z) = \left(\prod_{k=0}^{n} |z - x_k| \right)^{1/(n+1)}.$$

But we can take logarithms of both sides to get

$$\log \sigma_n(z) = \frac{1}{n+1} \sum_{k=0}^{n} \log |z - x_i|.$$

Now, if the nodes are equally spaced on the interval $[-1,1]$, the sum can be interpreted as a Riemann sum for the integral

$$I(f) = \frac{1}{2} \int_{-1}^{1} \log |z - t| \, dt;$$

therefore, we can write

$$\log \sigma_n(z) \approx \frac{1}{2} \int_{-1}^{1} \log |z - t| \, dt,$$

and we define the right side of this as $\log \sigma(z)$, thus implicitly defining $\sigma(z)$:

$$\frac{1}{2} \int_{-1}^{1} \log |z - t| \, dt = \log \sigma(z).$$

The integral can be evaluated, but we have to be careful about doing it, since $z = x + iy$ is complex. Thus we have

$$\frac{1}{2} \int_{-1}^{1} \log |z - t| \, dt = \frac{1}{2} \int_{-1}^{1} \log \sqrt{(x-t)^2 + y^2} \, dt = \frac{1}{4} \int_{-1}^{1} \log[(x-t)^2 + y^2] \, dt.$$

Careful application of integration by parts yields

$$\frac{1}{2} \int_{-1}^{1} \log |z - t| \, dt = \frac{1}{4}(1-x)\log((x-1)^2 + y^2) + \frac{1}{4}(1+x)\log((x+1)^2 + y^2)$$
$$- \frac{1}{2} y \arctan \frac{x-1}{y} + \frac{1}{2} y \arctan \frac{x+1}{y}.$$

Thus

$$\sigma_n(z) \approx \sigma(z) = e^{-1} \left((x-1)^2 + y^2\right)^{(1-x)/4} \left((x+1)^2 + y^2\right)^{(1+x)/4}$$
$$\times \exp\left\{\frac{1}{2}y \left(\arctan\frac{x+1}{y} - \arctan\frac{x-1}{y}\right)\right\}.$$

Now the error satisfies

$$|f(x) - p_n(x)| = |R_n(x)| \approx C_n \left|\frac{\sigma(x)}{\sigma(ia^{-1})}\right|^{n+1},$$

so convergence depends, in the limit as $n \to \infty$, on the ratio $|\sigma(x)/\sigma(ia^{-1})|$. For real x we get

$$\sigma(x) = e^{-1}(x-1)^{(1-x)/2}(x+1)^{(1+x)/2},$$

whereas for imaginary arguments we get

$$\sigma(iy) = e^{-1}\sqrt{1+y^2}\exp\left\{y\arctan\frac{1}{y}\right\}.$$

To get convergence on the entire interval $[-1,1]$, we need to have

$$\max_{x\in[-1,1]} |\sigma(x)| = 2e^{-1} = 0.7357588824 < |\sigma(ia^{-1})|.$$

An elementary computation shows that $|\sigma(i/5)| = 0.4937581336...$; therefore, we will not get convergence on the entire interval $[-1,1]$ with $a = 5$, because

$$\frac{\max_{x\in[-1,1]} |\sigma(x)|}{\sigma(i/5)} = \frac{0.7357588824}{0.4937581336} > 1.$$

By decreasing a, we can finally get $0.7357588824 < |\sigma(ia^{-1})|$. For example, for $a = 3/4$ we have $\sigma(i/a) = 1.446....$

What this shows is that the problem really is that the *singularities* of the function (that is, the points $\pm i/5$) are "too close" to the interval of interpolation. If we looked at $g(x) = (1 + a^2x^2)^{-1}$ for a smaller value of a, we would get convergence on the entire interval from $[-1,1]$. An example of this is given in Figure 4.39, using g with $a = 3/4$, and $n = 4,8,16$ nodes; the last plot shows the error for the $n = 16$ case. (See also Exercise 3.) Note that the error still shows the development of "spikes" near the endpoints; the difference is that in this case, as $n \to \infty$, the amplitude of the spikes will eventually decay to zero.

It is also possible to work on a more arbitrary interval $[-b,b]$, in which case we learn that the length of this interval also plays a role in the convergence of the interpolation. (See [9] for a discussion of this.) Essentially, convergence will occur if $1 > Cab$, where $C = 0.5255....$

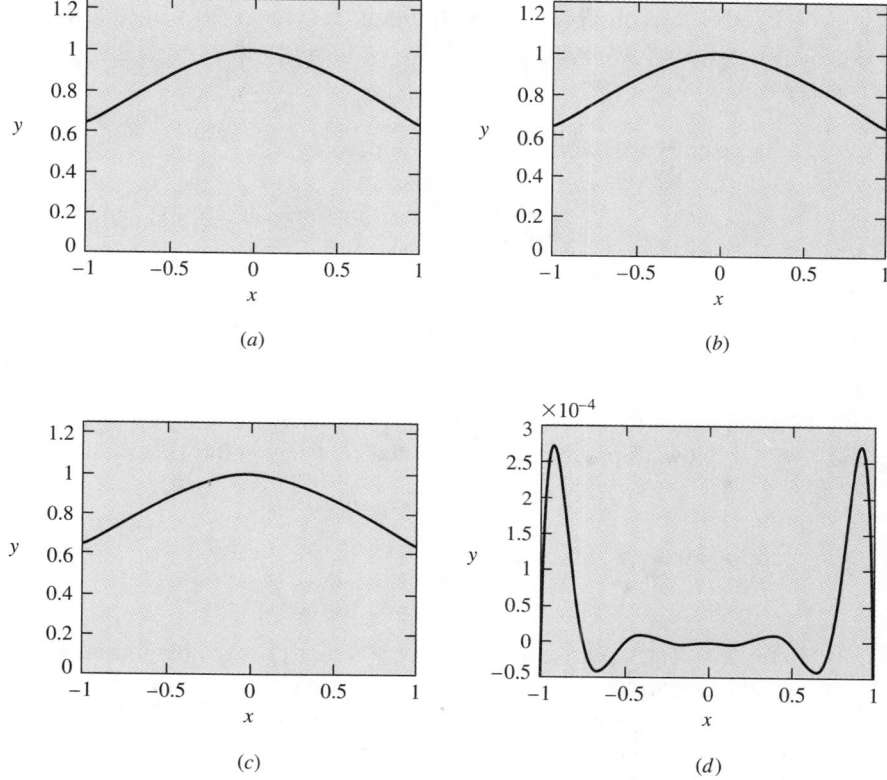

FIGURE 4.39 Successful interpolation to the Runge-like function, $g(x) = 1/(1 + (9/16)x^2)^{-1}$. (a). Four nodes. (b). Eight nodes. (c). Error in 16-node interpolation.

4.11.3 The Chebyshev Nodes

The previous two subsections should have made it clear that high-degree polynomial interpolation *at equidistant points* is a bad idea. Is there an alternative choice for the distribution of nodes that will produce better results? The answer is yes, and the nodes in question are called the *Chebyshev nodes*.

Define the family of functions $T_n(x)$, $n \geq 0$, by the formula

$$T_n(x) = \cos(n \arccos x), \quad x \in [-1, 1]. \tag{4.65}$$

Remarkably, these functions are polynomials (in fact, they are the Chebyshev polynomials introduced back in Section 4.10.2).

Theorem 4.10   The functions $T_n(x)$ satisfy the following properties:

1. *Each T_n is a polynomial of degree n.*

2. *For $n \geq 1$, $T_{n+1}(x) = 2xT_n(x) - T_{n-1}(x)$.*

3. *$T_n(x) = 2^{n-1}x^n +$ lower-order terms.*

Proof   We observe first of all that both (1) and (3) will follow quickly from (2), so we will start by proving (2). Write $x = \cos\theta$ so that (4.65) becomes

$$T_n(x) = \cos(n\theta).$$

Elementary trigonometry tells us that

$$\cos n\theta \cos\theta = \frac{1}{2}\left(\cos(n+1)\theta + \cos(n-1)\theta\right),$$

which can be solved to get

$$\cos(n+1)\theta = 2\cos\theta\cos n\theta - \cos(n-1)\theta,$$

from which (2) follows, using $T_n(x) = \cos n\theta$, etc.

Now, to prove (1), we note that we trivially have

$$\cos(n\arccos x) = 1$$

for $n = 0$ and

$$\cos(n\arccos x) = x$$

for $n = 1$. Now assume that $\cos(n\arccos x)$ is a polynomial for all $n \leq k$, and an inductive argument based on the recursion in (2) quickly finishes the proof.

The final part follows more or less directly from (2). We have that $T_0(x) = 1$ and $T_1(x) = x$, so that the recursion yields

$$T_2(x) = 2xT_1(x) - T_0(x) = 2x(x) - 1 = 2x^2 - 1,$$

and we can use induction to formally prove it for all n (see Exercise 5).   □

It should be noted that all orthogonal polynomial families satisfy a three-term recurrence relation similar to the one in part (2) of this theorem.

Assume now (for convenience) that we are interested in constructing approximations only on the interval $[-1, 1]$. As a polynomial of degree n, each T_n will have exactly n roots; we can show (it actually follows from Theorem 4.9) that these roots will be distinct and all lie in the interval $[-1, 1]$. Denote, then, the roots of the n-degree polynomial T_n as $z_k^{(n)}$, and note that (4.65) implies

$$z_k^{(n)} = \cos\left(\frac{(2k-1)\pi}{2n}\right), \quad 1 \leq k \leq n. \tag{4.66}$$

The Chebyshev nodes are simply the roots of the polynomials T_n. To construct an n-degree polynomial interpolant, use the $n+1$ roots of T_{n+1}; i.e., take the interpolation nodes to be

$$x_k = z_{k+1}^{(n+1)}, \quad 0 \leq k \leq n,$$

so that we have the following.

Definition 4.4 (Chebyshev Nodes)

$$x_k = \cos\left(\frac{(2k+1)\pi}{2(n+1)}\right), \quad 0 \le k \le n. \tag{4.67}$$

We will show why this works, but first let's look at some examples.

EXAMPLE 4.17

Let $f(x) = e^x$ on the interval $[-1,1]$. The Chebyshev nodes for the case $n = 1$ (linear interpolation) are given by

$$x_0 = \cos\left(\frac{(2 \times 0 + 1)\pi}{2(1+1)}\right) = \cos\frac{\pi}{4} = \frac{1}{2}\sqrt{2}$$

and

$$x_1 = \cos\left(\frac{(2 \times 1 + 1)\pi}{2(1+1)}\right) = \cos\frac{3\pi}{4} = -\frac{1}{2}\sqrt{2},$$

so the linear interpolant to the exponential is given by

$$p_1(x) = 1.260591837 + 1.085441641x.$$

EXAMPLE 4.18

Let's now look at higher-degree interpolation to the exponential; specifically, we will consider fourth-degree interpolation. The Chebyshev nodes now are

$$x_0 = \cos\left(\frac{(2 \times 0 + 1)\pi}{2(4+1)}\right) = \cos\frac{\pi}{10}$$

$$x_1 = \cos\left(\frac{(2 \times 1 + 1)\pi}{2(4+1)}\right) = \cos\frac{3\pi}{10}$$

$$x_2 = \cos\left(\frac{(2 \times 2 + 1)\pi}{2(4+1)}\right) = \cos\frac{5\pi}{10}$$

$$x_3 = \cos\left(\frac{(2 \times 3 + 1)\pi}{2(4+1)}\right) = \cos\frac{7\pi}{10}$$

$$x_4 = \cos\left(\frac{(2 \times 4 + 1)\pi}{2(4+1)}\right) = \cos\frac{9\pi}{10}.$$

Note that these values are somewhat more difficult to work with than the integer or simple-fraction nodes we have usually used. But the computer doesn't really care about that, and the Newton algorithm yields the polynomial

$$p_4(x) = 1 + 0.997317240x + 0.4995561859x^2 + 0.177334621x^3 + 0.043434107x^4.$$

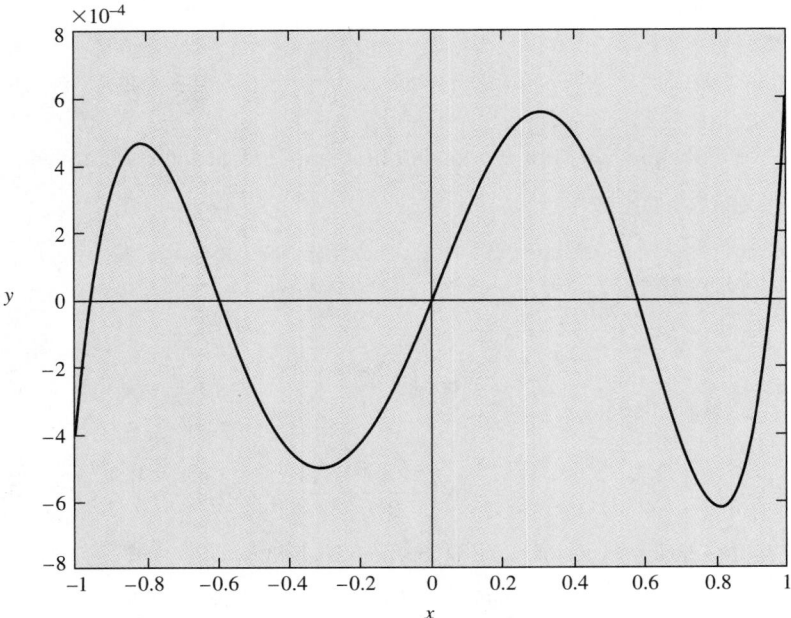

FIGURE 4.40 Error in fourth-degree Chebyshev interpolation to the exponential.

Figure 4.40 shows the plot of the error in this interpolation, over the interval $[-1, 1]$. Compare it with Figure 4.2(b), which shows the plot of the same error for fourth-degree interpolation using equally spaced points. It is difficult to tell because of the scales used, but the Chebyshev interpolation is slightly better in terms of the maximum error.

EXAMPLE 4.19

Consider again $f(x) = (1 + 25x^2)^{-1}$, the Runge example. If we use the Chebyshev nodes to compute interpolating approximations of degree 8, 16, and 32, we get the plots shown in Figure 4.41. Note that the accuracy of the approximation is much better than was the case using equally spaced nodes, as in Figure 4.3. Figure 4.41(c) shows that, with only 32 nodes, we cannot distinguish by eye between the interpolant and the original function. Figure 4.41(d) shows the error in the 32-degree approximation. ■

Why do these nodes work better? A complete answer would have to be phrased in the vague terms that we used in Section 4.11.2 to discuss the Runge example, or we would have to bring in a lot of additional mathematical machinery.

A fairly decent answer, however, can be obtained from the basic interpolation error theorem.

Theorem 4.11   *Let $f \in C^{n+1}([-1, 1])$ be given, and let p_n be the n-degree polynomial interpolant to f, using the Chebyshev nodes (4.67). Then*

$$\|f - p_n\|_\infty \leq \frac{1}{2^n(n+1)!}\|f^{(n+1)}\|_\infty.$$

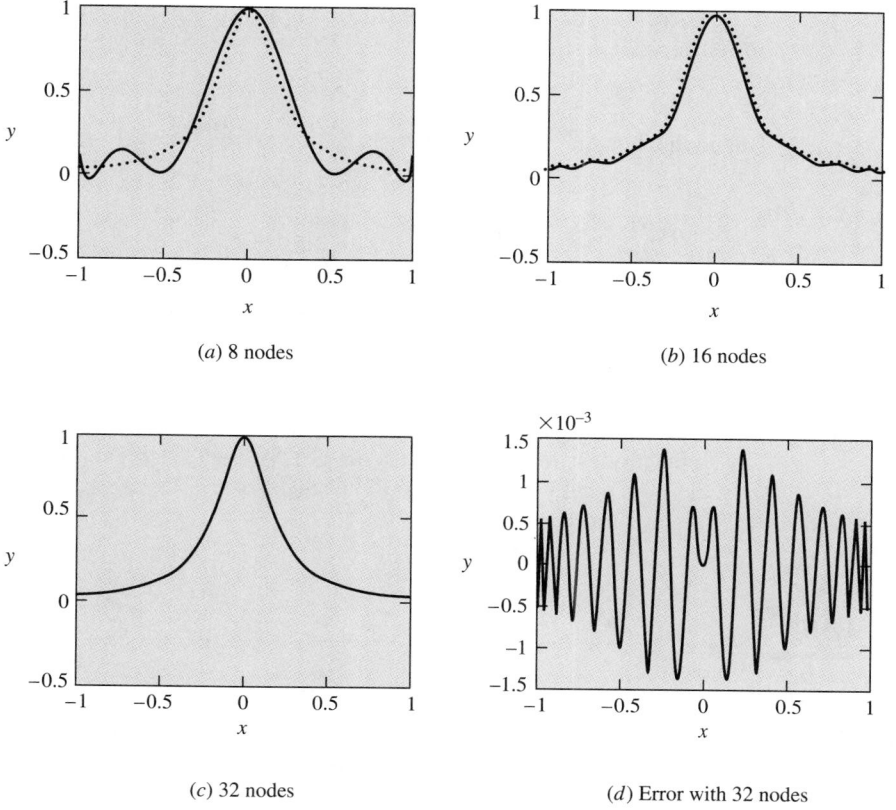

(a) 8 nodes

(b) 16 nodes

(c) 32 nodes

(d) Error with 32 nodes

FIGURE 4.41   Chebyshev interpolation to the Runge example.

Proof   Recall that we have

$$f(x) - p_n(x) = \frac{w_n(x)}{(n+1)!} f^{(n+1)}(\xi_x)$$

where

$$w_n(x) = \prod_{i=0}^{n} (x - x_i).$$

Since polynomials are uniquely defined by their roots (up to a constant multiple), and since w_n and T_{n+1} have the same roots, it follows that

$$w_n = c_n T_{n+1}$$

for some constant c_n. Now, w_n is what is known as a "monic" polynomial; i.e., the coefficient of the highest-order term is 1:

$$w_n(x) = x^{n+1} + \text{lower-order terms}.$$

Therefore, c_n is the *reciprocal* of the leading coefficient for T_{n+1}. Therefore, from Theorem 4.10, we have

$$c_n = 2^{-n}.$$

But each Chebyshev polynomial is a cosine on the interval $[-1,1]$; this means that

$$|T_{n+1}(x)| \leq 1, \quad -1 \leq x \leq 1.$$

Therefore,

$$|w_n(x)| = |c_n T_{n+1}(x)| \leq 2^{-n}$$

for all $x \in [-1,1]$, and the proof is complete. $\square$

What the theorem (and its proof) show is that the choice of the Chebyshev nodes allows us to bound the factor in the error theorem that depends on w_n. It can be shown, in fact, that any other choice of nodes would lead to a larger $\|w_n\|_\infty$, so in this sense the Chebyshev nodes are optimal. They are not "perfect," however. One can find functions for which the Chebyshev nodes fail, but these examples are much more involved than the Runge example.

EXERCISES

1. Use your computer's random number function to generate a set of random values r_k, $0 \leq k \leq 8$, with $|r_k| \leq 0.01$. Construct the interpolating polynomial to $f(x) = e^x$ on the interval $[-1,1]$, using nine equally spaced nodes; call this $p_8(x)$. Then construct the polynomial that interpolates the same function at the same nodes, *except* perturb the function values using the r_k values; call this $\hat{p}_8(x)$. How much difference is there between the divided-difference coefficients for the two polynomials? Plot both p_8 and $\hat{p}_8$, and comment on your results. (Note: It is important here that you look at x values *between* the nodes. Do not produce your plots based simply on the values at the nodes.)

2. Repeat the preceding, using $f(x) = \sin x$.

3. Use the Newton interpolating algorithm to construct interpolating polynomials of degree 4, 8, 16, and 32, using equally spaced nodes on the interval $[-1,1]$, to the function $g(x) = (1+4x^2)^{-1}$. Is the sequence of interpolants converging to g throughout the interval?

4. Use the Newton interpolating algorithm to construct interpolating polynomials of degree 4, 8, 16, and 32, using equally spaced nodes on the interval $[-1,1]$, to the function $g(x) = (1+100x^2)^{-1}$. Is the sequence of interpolants converging to g throughout the interval?

5. Write up a complete proof of Theorem 4.10, providing all the details left out in the text.

6. Construct interpolating polynomials of degree 4, 8, 16, and 32 on the interval $[-1,1]$, to $f(x) = e^x$ using equidistant and Chebyshev nodes. Sample the respective errors at 500 equally spaced points and compare your results.

7. The Chebyshev nodes are defined on the interval $[-1,1]$. Show that the change of variable

$$t_k = a + \frac{1}{2}(b-a)(x_k + 1)$$

will map the Chebyshev nodes to the interval $[a,b]$.

8. Use the preceding to find the Chebyshev nodes for linear interpolation on the interval $[\frac{1}{4},1]$. Use them to construct a linear interpolant to $f(x) = \sqrt{x}$. Plot the error in this interpolation. How does it compare with the error estimate used in Section 3.7?

9. Repeat the preceding for $f(x) = x^{-1}$ on the interval $[\frac{1}{2},1]$.

10. What is the error estimate for Chebyshev interpolation on a general interval $[a,b]$? In other words, how does the change of variable impact the error estimate?

LITERATURE AND SOFTWARE DISCUSSION

The subject of interpolation and approximation is a very broad one, with an extensive literature. In fact, "approximation theory" is a significant area of mathematical research, generally distinct from (although obviously related to) numerical analysis. The classic text on interpolation and approximation is Davis's book [6]; more recent and equally good works are those by Cheney [4], Powell [12], and Rivlin [14]. Generally, approximation theory is distinct from interpolation, which is one reason Davis's book is so useful, since it covers both. Modern work in what might be called computational approximation tends to center on splines, which is why de Boor's book [8] is a good reference. A rather large collection of FORTRAN codes is given there, which can be found at Netlib under the library `pppack`; MATLAB has incorporated them into its Splines Toolbox.

Several interesting ideas were not discussed here (the subject of interpolation and approximation is rich enough that an entire course could be based on this one subject). In Padé approximation, we look for *rational function* (i.e., ratios of polynomials) approximations to given functions. This is only marginally more involved than polynomial approximation, but it of course allows for more general functional behavior to be accurately approximated, since rational functions can exhibit asymptotic behavior, but polynomials cannot. A good introduction to Padé approximation is in Chapter 5 of [4].

Another missing topic is Fourier (i.e., trigonometric) approximation. Again, an entire course could be based on this subject, and this is done in many engineering departments.

A new approximation technique, called *wavelets*, is too young to be properly incorporated into a text at this level.

REFERENCES

[1] AHLBERG, J., NIELSON, E., & WALSH, J. *The Theory of Splines and Their Applications*. Academic Press, New York, 1967.

[2] BIRKHOFF, GARRETT, & DE BOOR, CARL. "Piecewise polynomial interpolation and approximation." *Proceedings of General Motors Symposium of 1964* (H.L. Garabedian, ed.). Elsevier, New York and Amsterdam, 1965, pp. 164–190.

[3] BIRKHOFF, GARRETT. "Fluid dynamics, reactor computations, and surface representation." In *A History of Scientific Computation* (Steve Nash, ed.). 1990.

[4] CHENEY, E.W. *Approximation Theory*. Chelsea, New York, 1966.

[5] DAVIS, PAUL. "B-splines and geometric design." *SIAM News*, vol. 29 (1996), no. 5.

[6] DAVIS, PHILIP. *Interpolation and Approximation*. Dover, New York, 1975 (originally published in 1963 by Blaisdell, Waltham, MA.)

[7] DE BOOR, CARL. "On uniform approximation by splines," *Journal of Approximation Theory*, vol. 1 (1968), pp. 219–235.

[8] DE BOOR, CARL. *A Practical Guide to Splines*. Springer-Verlag, New York, 1978.

[9] EPPERSON, JAMES. "On the Runge example," *American Mathematical Monthly*, 1987, pp. 329–341.

[10] HALL, C.A. "On error bounds for spline interpolation," *Journal of Approximation Theory*, vol. 1 (1968), pp. 209–218.

[11] ISAACSON, E., & KELLER, H.B. *Analysis of Numerical Methods*. Dover, New York, 1994 (originally published in 1966 by John Wiley & Sons).

[12] POWELL, M.J.D. *Approximation Theory and Methods*. Cambridge University Press, Cambridge, 1981.

[13] PRENTER, PADDY. *Splines and Variational Methods*. Wiley-Interscience, New York, 1975.

[14] RIVLIN, THEODORE J. *An Introduction to the Approximation of Functions*. Dover, New York, 1981 (originally published in 1969 by Blaisdell, Waltham, MA.)

[15] SCHOENBERG, I.J. "Contributions to the problem of approximation of equidistant data by analytic functions," *Quarterly of Applied Mathematics*, vol. 4 (1946), pp. 45–99, 112–141.

[16] YOUNG, DAVID. "Garrett Birkhoff and applied mathematics," *Notices of the AMS*, vol. 44 (1997), no. 11, pp. 1446–1449.

CHAPTER 5

NUMERICAL INTEGRATION

Pennsyltucky Porch and Patio makes prefabricated porch and deck additions for houses. One of their economy-line models uses a corrugated sheet-metal roof that PP&P manufactures at their home plant in Pittsfield. The cross-section of the metal sheets after they are stamped is sinusoidal, like the plot in the following figure. PP&P needs to know how much sheet metal to use in order to cover a roof of length L feet. Mathematically, this means they need to find the length of the sinusoidal curves that make up the corrugated sheets. If the "rib distance" in the corrugations is r and the depth of the corrugations is A (both measured in inches), then basic calculus says that they need to find the value of the integral

$$I = \frac{1}{12} \int_0^{12L} \sqrt{1 + (\pi A/2r)^2 \cos^2(\pi x/2r)}\ dx.$$

This integral cannot be evaluated analytically; thus we are compelled to use approximation techniques to find the value of the integral and hence to find how much sheet metal PP&P needs to order to make a corrugated roof of length L.

This chapter is concerned with techniques for approximating definite integrals like that given above. We look at a variety of methods, some of which are clear extensions of the basic ideas behind the trapezoid rule (Section 2.5), and some of which involve new concepts.

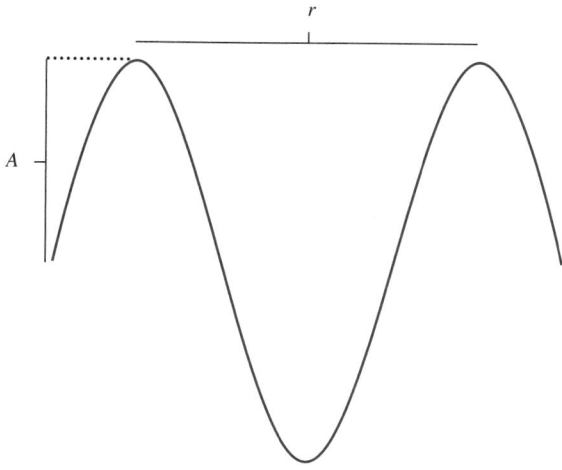

In this chapter we are concerned with a common problem from calculus: Find the value of the definite integral

$$I(f) = \int_a^b f(x)\, dx.$$

We are interested in finding numerical methods that yield an accurate approximation to the exact value of $I(f)$. Typically the approximation will be of the form

$$I_n(f) = \sum_{i=i_0}^n w_i f(x_i),$$

where $i_0 = 0$ or $i_0 = 1$ is almost always the case. The *weights* w_i and *nodes* or *abscissas* x_i define the method. Schemes for approximating integrals are often called *quadrature rules*, and different schemes use different rules for defining the weights and abscissas. In Chapter 2 we saw perhaps the simplest example of a reasonable quadrature scheme, the trapezoid rule. Here we will look at other methods, some of which are substantially more accurate than the trapezoid rule.

In addition to constructing various quadrature schemes and analyzing their accuracy, we will look at ways to estimate and improve the accuracy of existing quadrature methods, and we will also look at the idea of *adaptive quadrature*, in which we try to estimate the value of the integral to within a user-specified tolerance.

Note that we seem to be suggesting that the integral can be viewed (approximately, at least) as a sum of function values. This is not surprising, considering the theoretical foundations of the integral, which we now review briefly.

5.1 A REVIEW OF THE DEFINITE INTEGRAL

To construct the definite integral of a function, we first define the *mesh points* $x_i^{(n)}$ according to

$$a = x_0^{(n)} < x_1^{(n)} < x_2^{(n)} \cdots < x_{n-1}^{(n)} < x_n^{(n)} = b,$$

and the *evaluation points* η_i are then each taken from within the appropriate subinterval, i.e., $\eta_i \in [x_{i-1}, x_i], 1 \le i \le n$, arbitrarily. Now, if we assume that

$$\lim_{n \to \infty} \left[\max_{1 \le i \le n} \left(x_i^{(n)} - x_{i-1}^{(n)} \right) \right] = 0$$

(this simply means that the largest distance between adjacent points goes to zero as we take more and more points), then, under very mild conditions on f, it can be shown that the limit

$$L = \lim_{n \to \infty} \sum_{i=1}^n f(\eta_i)(x_i^{(n)} - x_{i-1}^{(n)})$$

exists and its value is independent of the choices made for the grid points $\{x_i^{(n)}\}$ and evaluation points $\{\eta_i\}$. When this happens, we call this limit value the *definite integral* of f, and we write

$$L = I(f) = \int_a^b f(x)\, dx.$$

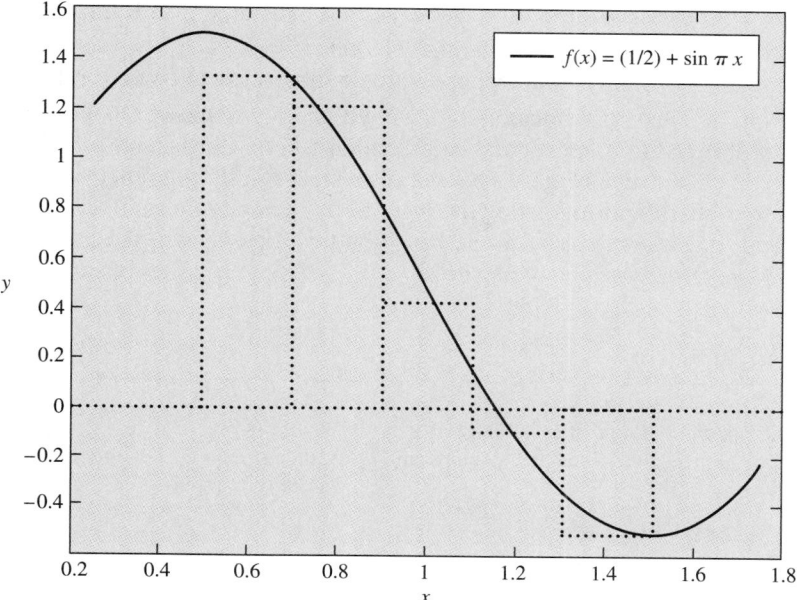

FIGURE 5.1   Illustration of a Riemann sum.

Note that this construction of the integral requires very little. We need only that f is continuous, that the sequence of meshes is such that the distance between adjacent points becomes arbitrarily small, and that the evaluation points are taken from anywhere in each subinterval. A summation of the form

$$R_n(f) = \sum_{i=1}^{n} f(\eta_i)\left(x_i^{(n)} - x_{i-1}^{(n)}\right)$$

is called a *Riemann sum*, after the German mathematician Georg Bernhard Riemann,[1] whose name is associated with the first development of much of the rigorous theory of the definite integral. Unfortunately, Riemann sums typically converge very slowly—meaning that it requires a large value of n for the summation value to be a good approximation of the limit value—and so we cannot use this as a practical means of approximating the value of the integral. Figure 5.1 shows a graph of the function

---

[1]Georg Friedrich Bernhard Riemann (1826–1866) was a child of poor health, the son of a Lutheran minister. He started university studies at Göttingen in 1846 but soon transferred to the University of Berlin. In 1849 he returned to Göttingen to finish his studies under the direction of Gauss, completing his doctoral dissertation in 1851. Albert Einstein would later use some of Riemann's ideas from this dissertation in the development of the theory of relativity. In 1862 Riemann's health began to decline, and he eventually came down with tuberculosis, which led to his death just four years later.

Riemann's brief life was nonetheless one of great mathematical significance. He made substantial contributions to the foundations of geometry and analysis, and virtually invented what we now call analytic number theory. His development of the Riemann sum came in his 1854 "Habilitationschrift" (sort of a postdoctoral dissertation, required to get an academic position in 19th-century Europe), *Über die Darstellbarkeit einer Function durch eine trigonometrische Reihe.*

$y = f(x) = \frac{1}{2} + \sin \pi x$ and the rectangles generated by a Riemann sum approximation to the integral of this f over the interval $[\frac{1}{2}, \frac{3}{2}]$.

We usually will dispense with the superscript notation for the abscissas from now on; i.e., instead of writing $x_i^{(n)}$, we will simply write x_i. On a couple of occasions we will revert to it, most notably in Section 5.6, on Gaussian quadrature.

Since the integral is essentially a limit of a sum of function values, it makes sense to think of approximating the integral by taking *finite* sums of function values. All of our quadrature rules will be based on this idea, as was the trapezoid rule, which we have already seen in Chapter 2.

EXERCISES

1. Basic properties of the definite integral show that it is a *linear operator*, i.e., it distributes across sums and multiplication by constants:

$$I(\alpha f + \beta g) = \alpha I(f) + \beta I(g).$$

Prove that if

$$I_n(f) = \sum_{i=0}^{n} w_i f(x_i),$$

then I_n is also linear:

$$I_n(\alpha f + \beta g) = \alpha I_n(f) + \beta I_n(g).$$

2. Assume that the quadrature rule I_n integrates all polynomials of degree less than or equal to N exactly:

$$I_n(p) = I(p)$$

for all $p \in \mathcal{P}_N$. Use this to prove that, for any integrand f, the error $I - I_n$ is equal to the error in integrating the Taylor remainder:

$$I(f) - I_n(f) = I(R_N) - I_n(R_N)$$

where $f(x) = p_N(x) + R_N(x)$. Does it really matter that we are using the Taylor polynomial and remainder? In other words, will this result hold for any polynomial approximation and its associated error?

3. Use Excercise 1 to prove the following: If a quadrature rule I_n is exact for all powers x^k for $k \leq d$, then it is exact for all polynomials of degree less than or equal to d.

5.2   IMPROVING THE TRAPEZOID RULE

In Section 2.5 we introduced the trapezoid rule for computing integrals. Recall that for equally spaced points, the approximation takes the form

$$T_n(f) = \frac{h}{2} \left(f(x_0) + 2f(x_1) + \cdots + 2f(x_{n-1}) + f(x_n) \right)$$

and the error is given by

$$I(f) - T_n(f) = -\frac{b-a}{12} h^2 f''(\xi_h), \tag{5.1}$$

where ξ_h is some value in the interval $[a,b]$.

We can improve the trapezoid rule by some deft analysis of the second-derivative term in the error. Let us return to the error estimate (5.1), which we can write in the form

$$I(f) - T_n(f) = -\frac{1}{12}h^3 \sum_{i=1}^{n} f''(\xi_{i,h}).$$

Now let's look at the summation term, carefully. We have

$$h^3 \sum_{i=1}^{n} f''(\xi_{i,h}) = h^2 \sum_{i=1}^{n} hf''(\xi_{i,h}),$$

and the sum can be interpreted as a Riemann sum for the integral

$$I(f'') = \int_a^b f''(x)\, dx = f'(b) - f'(a).$$

Thus, we know that

$$\lim_{n \to \infty} \sum_{i=1}^{n} hf''(\xi_{i,h}) = f'(b) - f'(a),$$

so that

$$\sum_{i=1}^{n} hf''(\xi_{i,h}) \approx f'(b) - f'(a).$$

Therefore,

$$I(f) - T_n(f) \approx -\frac{1}{12}h^2(f'(b) - f'(a)).$$

This relationship can be interpreted and used in two equally important and useful ways:

Error estimation: We can view the quantity $\hat{E}_n(f) = -\frac{1}{12}h^2(f'(b) - f'(a))$ as a "computable estimate of the error" and use it to predict how many points are required to obtain a given accuracy. Bear in mind that this approach does not *guarantee* that we will achieve the desired accuracy.

Improvement of the approximation: Define the new quadrature rule

$$T_n^C(f) = T_n(f) - \frac{1}{12}h^2(f'(b) - f'(a))$$

which we will call the *corrected trapezoid rule*. This value will tend to be much more accurate than the ordinary trapezoid rule and only marginally more expensive to compute.

EXAMPLE 5.1

To illustrate the first usage, consider the problem of trying to compute values of the integral

$$I(f) = \int_0^1 e^{-x^2} dx.$$

How do we know when we have computed this to within a specified accuracy? Suppose we want

$$|I(f) - I_n(f)| \leq 10^{-6};$$

then we could compute trapezoid rule values along with the values

$$\hat{E}_n(f) = -\frac{h^2}{12}(f'(b) - f'(a)) = \frac{h^2}{6e}.$$

When $|\hat{E}_n(f)| \leq 10^{-6}$, we would accept $T_n(f)$ as a sufficiently accurate approximation to $I(f)$; this occurs in this case for $n = 256$, since

$$|\hat{E}_n(f)| = \frac{h^2}{6e} \leq 10^{-6} \quad \Leftarrow \quad h \leq \sqrt{6e} \times 10^{-3} = 4.04 \times 10^{-3}$$

and $1/256 < 4.04 \times 10^{-3}$, but $1/128 > 4.04 \times 10^{-3}$. In practice, though, it would be better to use $\hat{E}_n(f)$ to estimate the error in $T_n(f)$, but then accept the more accurate value defined by $T_n^C(f)$. ∎

EXAMPLE 5.2

As an example of the second usage, consider Table 5.1, which shows the values generated by approximating the integral $I(f) = \int_0^1 e^x/dx$ using T_n^C. Note that the error is dropping by about a factor of 16 as the number of points doubles, suggesting that the error in the corrected rule goes like h^4. This is in fact the case, as we shall see in Section 5.8.1. Note that we lose the expected factor-of-16 decrease in the last two rows; this is not because there is a problem with the method or with the underlying theory, but because of the effects of finite-precision arithmetic. There is so much subtractive cancellation in the computation of $I(f) - T_n^C(f)$ that the value of the error ratio is being affected. We are starting to reach the limits of the accuracy which can be measured by the machine arithmetic. ∎

TABLE 5.1 Corrected Trapezoid Rule Applied to $f(x)$ $= e^x$, $[a,b] = [0,1]$

n	$T_n^C(f)$	$I(f) - T_n^C(f)$	**Error Ratio**
2	1.718133554	0.148274E-03	NA
4	1.718272520	0.930842E-05	15.9290
8	1.718281246	0.582426E-06	15.9822
16	1.718281792	0.364118E-07	15.9955
32	1.718281826	0.227589E-08	15.9989
64	1.718281828	0.142245E-09	15.9998
128	1.718281828	0.889111E-11	15.9986
256	1.718281828	0.556444E-12	15.9785
512	1.718281828	0.339728E-13	16.3791
1024	1.718281828	0.288658E-14	11.7692
2048	1.718281828	0.666134E-15	4.3333

TABLE 5.2 **"Approximate" Corrected Trapezoid Rule**
Applied to $f(x) = e^x$, $[a,b] = [0,1]$

n	$\tilde{T}_n^C(f)$	$I(f) - \tilde{T}_n^C(f)$	**Error Ratio**
2	1.718861152	−0.579323E−03	NA
4	1.718386631	−0.104802E−03	5.5278
8	1.718290593	−0.876407E−05	11.9582
16	1.718282447	−0.618963E−06	14.1593
32	1.718281869	−0.409496E−07	15.1152
64	1.718281831	−0.263078E−08	15.5656
128	1.718281829	−0.166666E−09	15.7847
256	1.718281828	−0.104863E−10	15.8938
512	1.718281828	−0.658362E−12	15.9278
1024	1.718281828	−0.406342E−13	16.2022
2048	1.718281828	−0.199840E−14	20.3333

Finally, we note that it is not even necessary to have a formula for the derivative of f in order to use this correction. Recall two of the derivative approximations from the previous chapter:

$$f'(x) = \frac{-3f(x) + 4f(x+h) - f(x+2h)}{2h} + \frac{h^2}{3}f'''(\xi)$$

and

$$f'(x) = \frac{3f(x) - 4f(x-h) + f(x-2h)}{2h} + \frac{h^2}{3}f'''(\xi).$$

Thus

$$f'(b) = f'(x_n) = \frac{3f(x_n) - 4f(x_{n-1}) + f(x_{n-2})}{2h} + \frac{h^2}{3}f'''(\xi_b) \tag{5.2}$$

and

$$f'(a) = f'(x_0) = \frac{-3f(x_0) + 4f(x_1) - f(x_2)}{2h} + \frac{h^2}{3}f'''(\xi_a), \tag{5.3}$$

so that we can write the corrected trapezoid rule as

$$T_n^C(f) = T_n(f) - \frac{h}{24}(3f(x_n) - 4f(x_{n-1}) + f(x_{n-2}) + 3f(x_0) - 4f(x_1) + f(x_2))$$
$$+ \frac{h^4}{36}(f'''(\xi_a) - f'''(\xi_b)).$$

Now define

$$\tilde{T}_n^C(f) = T_n(f) - \frac{h}{24}(3f(x_n) - 4f(x_{n-1}) + f(x_{n-2}) + 3f(x_0) - 4f(x_1) + f(x_2))$$

and note that, if the corrected trapezoid rule is $O(h^4)$ accurate, so is the new "approximate" trapezoid rule, since the additional error term due to the derivative approximations is $O(h^4)$. Table 5.2 shows the results of applying this approximation to the same example as before. Note that we do maintain the expected factor-of-16 decrease in the error for h sufficiently small.

EXERCISES

1. Apply the trapezoid rule and corrected trapezoid rule, with $h = \frac{1}{4}$, to approximate the integral

$$I = \int_0^1 x(1 - x^2) \, dx = \frac{1}{4}.$$

2. Apply the trapezoid rule and corrected trapezoid rule, with $h = \frac{1}{4}$, to approximate the integral

$$I = \int_0^1 \frac{1}{\sqrt{1 + x^4}} \, dx = 0.92703733865069.$$

3. Apply the trapezoid rule and corrected trapezoid rule, with $h = \frac{1}{4}$, to approximate the integral

$$I = \int_0^1 \ln(1 + x) \, dx = 2\ln 2 - 1.$$

4. Apply the trapezoid rule and corrected trapezoid rule, with $h = \frac{1}{4}$, to approximate the integral

$$I = \int_0^1 \frac{1}{1 + x^3} \, dx = \frac{1}{3}\ln 2 + \frac{1}{9}\sqrt{3}\pi.$$

5. Apply the trapezoid rule and corrected trapezoid rule, with $h = \frac{1}{4}$, to approximate the integral

$$I = \int_1^2 e^{-x^2} \, dx = 0.1352572580.$$

6. For each integral below, write a program to do the corrected trapezoid rule using the sequence of mesh sizes $h = \frac{1}{2}(b - a), \frac{1}{4}(b - a), \frac{1}{8}(b - a), \ldots, \frac{1}{2048}(b - a)$, where $b - a$ is the length of the given interval. Verify that the expected rate of decrease of the error is observed.

(a) $f(x) = 1/(1 + x^2)$, $[-5,5]$, $I(f) = 2\arctan(5)$
(b) $f(x) = x^2 e^{-x}$, $[0,2]$, $I(f) = 2 - 10e^{-2} = 0.646647168$
(c) $f(x) = \ln x$, $[1,3]$, $I(f) = 3\ln 3 - 2 = 1.295836867$
(d) $f(x) = \sqrt{1 - x^2}$, $[-1,1]$, $I(f) = \pi/2$
(e) $f(x) = e^{-x}\sin(4x)$, $[0,\pi]$, $I(f) = \frac{4}{17}(1 - e^{-\pi}) = 0.2251261368$

7. Apply the trapezoid rule and corrected trapezoid rule to the approximation of

$$I = \int_0^1 x^2 e^{-2x} dx = 0.0808308960\ldots$$

Compare your results in light of the expected error theory for both methods, and comment on what occurs. How does the error behave in each case, as a function of h? How should it have behaved?

8. Repeat the preceding for

$$I = \int_0^\pi \sin^2 x dx = \frac{1}{2}\pi.$$

9. In the example at the begining of this chapter, if the amplitude of the corrugations is half the distance between peaks, and if there are 12 corrugations per foot, how long a piece of (flat) sheet metal is required to make a corrugated piece that is 10 feet long?

10. Repeat the preceding when the amplitude of the corrugations is only one-fourth of the distance between peaks, but all else is the same.

11. The length of a curve $y = g(x)$, for x between a and b, is given by the integral

$$L(g) = \int_a^b \sqrt{1 + [g'(x)]^2} \, dx.$$

Use the corrected trapezoid rule to find the length of one "arch" of the sine curve.

12. Construct a version of the "quasi-corrected" trapezoid rule that uses the derivative approximations

$$f'(a) \approx \frac{f(a+h)-f(a)}{h},$$

$$f'(b) \approx \frac{f(b)-f(b-h)}{h}.$$

Explain why we should expect this to be less accurate than the rule using the approximations (5.2) and (5.3), and demonstrate that this is the case on the integral

$$I = \int_0^1 x(1-x^2)\, dx = \frac{1}{4}.$$

5.3   SIMPSON'S RULE AND DEGREE OF PRECISION

Common sense, plus the results of Chapter 4 on interpolation error, lead us to think that we might be able to do better than the trapezoid rule by using a higher-degree polynomial. Simpson's rule is nothing more than this next logical step, relying on quadratic interpolation to generate the quadrature rule. [2]

We follow the basic outline from our presentation for the trapezoid rule. Thus we start by approximating $I(f)$ by a single quadratic approximation. Let p_2 be the quadratic polynomial that interpolates f at the three points $x_0 = a$, $x_2 = b$, and $x_1 = c = (a+b)/2$, the midpoint. Define the basic Simpson's rule, then, as

$$S_2(f) = I(p_2) = \int_a^b \left(L_0(x)f(a)+L_1(x)f(c)+L_2(x)f(b)\right) dx,$$

where the Lagrange functions are here defined as

$$L_0(x) = \frac{(x-c)(x-b)}{(a-c)(a-b)}, \qquad L_1(x) = \frac{(x-a)(x-b)}{(c-a)(c-b)}, \qquad L_2(x) = \frac{(x-a)(x-c)}{(b-a)(b-c)}.$$

Then the quadrature rule is given by

$$S_2(f) = Af(a)+Cf(c)+Bf(b),$$

where

$$A = \int_a^b L_0(x)\, dx, \qquad C = \int_a^b L_1(x)\, dx, \qquad B = \int_a^b L_2(x)\,dx.$$

---

[2]Thomas Simpson (1710–1761) was born and died in Leicestershire, England. Little is known of his youth and education, and the legend persists to this day that he kept "low company" with whom he would "guzzle porter and gin." He published an early text on Newton's calculus (*The Doctrine and Application of Fluxions*) and also worked in probability theory. Most of his mathematics teaching was done privately.

The quadrature rule that bears his name was originally derived by the Italian Bonaventura Cavalieri in 1639, and was known to James Gregory in the late 1600s and to Roger Cotes in the early 1700s. It was rediscovered by Simpson and published in 1743 in his paper *Mathematical Dissertations on a Variety of physical and analytical subjects* ....

To compute these, we define $h = b - c = c - a = (b - a)/2$, so that we have

$$A = \int_a^{a+2h} L_0(x) \, dx, C = \int_a^{a+2h} L_1(x) \, dx, B = \int_a^{a+2h} L_2(x) \, dx.$$

Then

$$A = \int_a^{a+2h} L_0(x) \, dx$$

$$= \frac{1}{2h^2} \int_a^{a+2h} (x - (a+h))(x - (a+2h)) \, dx$$

$$= \frac{1}{2h^2} \int_{-h}^{h} u(u - h) \, du,$$

where we have used the change of variable $u = x - a - h$ in the last step. The computation is now fairly straighforward:

$$A = \frac{1}{2h^2} \int_{-h}^{h} u(u - h) \, du$$

$$= \frac{1}{2h^2} \left(\frac{1}{3} u^3 - \frac{1}{2} h u^2 \right) \Big|_{-h}^{h}$$

$$= \frac{1}{2h^2} \left(\left(\frac{1}{3} h^3 - \frac{1}{2} h^3 \right) - \left(-\frac{1}{3} h^3 - \frac{1}{2} h^3 \right) \right)$$

$$= \frac{1}{2h^2} \left(-\frac{1}{6} h^3 + \frac{5}{6} h^3 \right)$$

$$= \frac{h}{3}.$$

Similar computations (the student should verify these) yield $B = A$ and $C = 4h/3$. Thus, Simpson's rule becomes

$$S_2(f) = \frac{h}{3}(f(a) + 4f(c) + f(b)),$$

where $h = (b - a)/2$, i.e., h is the distance between points in the discretization of the interval $[a,b]$.

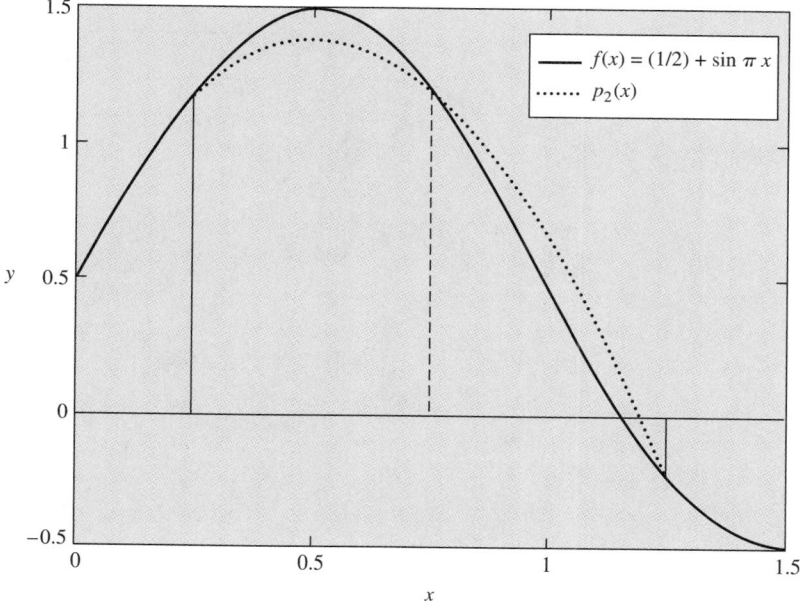

FIGURE 5.2 **Illustration of basic Simpson's rule.**

EXAMPLE 5.3

To apply this crude version of Simpson's rule to the function $f(x) = \frac{1}{2} + \sin \pi x$, over the interval $[\frac{1}{4}, \frac{5}{4}]$, we have

$$S_2(f) = \frac{1/2}{3}\left(\frac{1}{2} + \sin\frac{1}{4}\pi + 4\left(\frac{1}{2} + \sin\frac{3}{4}\pi\right) + \frac{1}{2} + \sin\frac{5}{4}\pi\right)$$

$$= \frac{1}{6}\left(3 + \frac{1}{2}\sqrt{2} + 2\sqrt{2} - \frac{1}{2}\sqrt{2}\right) = 0.9714045208.$$

The exact value is $I(f) = 0.9501581580$, showing that our approximation is decently accurate even though we used only a single approximating parabola. Figure 5.2 shows the application of Simpson's rule to this example. ∎

The composite rule is easily constructed by adding up individual instances of Simpson's rule applied to *pairs* of subintervals. (Note that this means that Simpson's rule requires an even number of subintervals.) We get

$$S_n(f) = \sum_{i=1}^{n/2} \frac{h_i}{3}\left[f(x_{2i-2}) + 4f(x_{2i-1}) + f(x_{2i})\right], \tag{5.4}$$

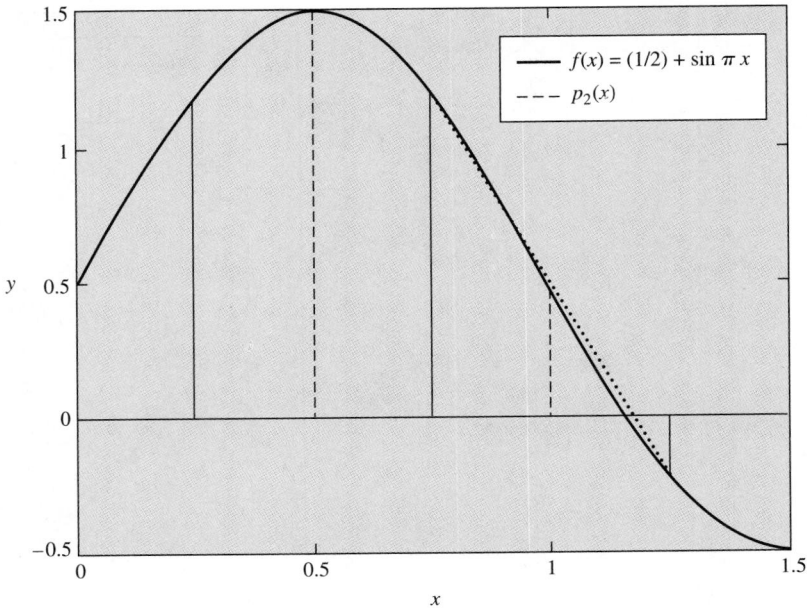

FIGURE 5.3 Illustration of Simpson's rule.

where $h_i = (x_{2i} - x_{2i-2})/2$. If we assume a uniform spacing of the mesh points, i.e., $h_i = h$ for all i, then all this simplifies a great deal and we have

$$S_n(f) = \frac{h}{3} \left(f(x_0) + 4f(x_1) + 2f(x_2) + 4f(x_3) + \cdots + 2f(x_{n-2}) + 4f(x_{n-1}) + f(x_n) \right).$$
(5.5)

EXAMPLE 5.4

If we consider the same example as previously, but now use two approximating parabolas, we get

$$S_4(f) = \frac{1/4}{3} \left(f(1/4) + 4f(1/2) + 2f(3/4) + 4f(1) + f(5/4) \right) = 0.9511844634,$$

which is much more accurate than the S_2 value. Figure 5.3 shows Simpson's rule applied in this case, and the decrease in the error is evident. ∎

Simpson's rule can be constructed for a nonuniform grid, but the formula corresponding to either (5.4) or (5.5) is somewhat more involved. See Exercise 12.

EXAMPLE 5.5

If we apply Simpson's rule to the example in which $f(x) = e^x$ over the interval $[0,1]$, using a sequence of decreasing grids, we get the results shown in Table 5.3. ∎

TABLE 5.3   Simpson's Rule Applied to $f(x) = e^x$, $[a,b] = [0,1]$

n	$S_n(f)$	$I(f) - S_n(f)$	Error ratio
2	1.718861151877	$-0.579323\text{E}{-03}$	NA
4	1.718318841922	$-0.370135\text{E}{-04}$	15.6517
8	1.718284154700	$-0.232624\text{E}{-05}$	15.9113
16	1.718281974052	$-0.145593\text{E}{-06}$	15.9777
32	1.718281837562	$-0.910273\text{E}{-08}$	15.9944
64	1.718281829028	$-0.568969\text{E}{-09}$	15.9986
128	1.718281828495	$-0.355611\text{E}{-10}$	15.9998
256	1.718281828461	$-0.222178\text{E}{-11}$	16.0057
512	1.718281828459	$-0.137890\text{E}{-12}$	16.1127
1024	1.718281828459	$-0.910383\text{E}{-14}$	15.1463
2048	1.718281828459	$0.444089\text{E}{-15}$	-20.5000

Note that the error here is decreasing by a factor of roughly 16 as h is halved.[3] Drawing on our experience with the trapezoid rule, we therefore suspect that

$$I(f) - S_n(f) = O(h^4).$$

However, our work on the interpolation error for quadratic interpolation (4.11), plus the definition of Simpson's rule as the *exact* integral of a piecewise quadratic interpolate, suggest that

$$f - p_2 = O(h^3) \quad \Rightarrow \quad I(f) - S_n(f) = I(f) - I(p_3) = \int_a^b (f(x) - p_2(x))\, dx = O(h^3).$$

Thus Simpson's rule appears to be *more* accurate than the underlying theory (from interpolation) would suggest. What's going on here that causes this?

It is possible to directly prove an error estimate for Simpson's rule, in a manner similar to that for the trapezoid rule. But a more illuminating explanation of *why* Simpson's rule is "more accurate than it ought to be" can result from looking at the extent to which it integrates polynomials exactly. This leads us to the notion of *degree of precision* for a quadrature rule.

Definition 5.1 (Degree of Precision)   *Let I_n be a quadrature rule, and assume it is exact for all polynomials of degree $\leq p$. Then we say that I_n has degree of precision p.*

We ought to expect to get better accuracy from integration methods that have a higher degree of precision.

---

[3]But again note that the last case is an exception. This is again because the method is, at this point, too accurate for the finite-precision arithmetic to properly resolve.

As an example, we note that Exercise 21 of Section 2.5 shows that the trapezoid rule has degree of precision $p = 1$, because it integrates all linear polynomials exactly. Simpson's rule, by construction, clearly ought to have degree of precision $p = 2$. We will show that it actually has degree of precision $p = 3$; i.e., Simpson's rule integrates cubic polynomials exactly, which is something of a bonus beyond what we ought to expect from the construction. We will then show that having degree of precision $p = 3$ leads to an error estimate that is $O(h^4)$.

Let q_3 be an arbitrary cubic polynomial, which we choose to write as

$$q_3(x) = Ax^3 + q_2(x),$$

where q_2 contains all the lower-order terms from q_3. Then the error in using Simpson's rule to integrate q_3 becomes

$$I(q_3) - S_2(q_3) = A(I(x^3) - S_2(x^3)) + (I(q_2) - S_2(q_2)).$$

But Simpson's rule will clearly integrate the quadratic part of q_3 exactly; thus Simpson's rule will be exact for all cubic polynomials if and only if it is exact for x^3. And direct computation gives us

$$
\begin{aligned}
S_2(x^3) &= \frac{(b-a)/2}{3}(a^3 + 4((a+b)/2)^3 + b^3) \\
&= \frac{b-a}{6}\left(a^3 + \frac{1}{2}(a+b)^3 + b^3\right) \\
&= \frac{b-a}{12}(2a^3 + (a+b)^3 + 2b^3) \\
&= \frac{b-a}{4}(a^3 + a^2b + ab^2 + b^3) \\
&= \frac{1}{4}(b^4 - a^4) \\
&= I(x^3).
\end{aligned}
$$

Thus Simpson's rule is exact for all cubic polynomials. More formally, we have proved the following lemma.

Lemma 5.1   *Let $q_3(x)$ be an arbitrary polynomial of degree less than or equal to 3. Then Simpson's rule integrates q_3 exactly, i.e.,*

$$S_2(q_3) = \int_a^b q_3(x)\, dx$$

for any interval $[a,b]$.

Now, how does this give Simpson's rule increased accuracy? Let $q_3 \approx f$, where q_3 now is some cubic polynomial approximation to f. Let the error in this approximation be given by R_3. Then we have

$$I(f) - S_2(f) = I(q_3 + R_3) - S_2(q_3 + R_3).$$

But both the integral and Simpson's rule are *linear*, in that their action distributes across sums (see the exercises at the end of Section 5.1). Thus we have $I(q_3 + R_3) = I(q_3) + I(R_3)$, and similarly for S_2. Therefore,

$$I(f) - S_2(f) = I(R_3) - S_2(R_3).$$

Thus, Simpson's rule is as accurate for f as it is for the *error* in *any* cubic polynomial approximation to f. Since we expect the best cubic approximation to be better than the best quadratic approximation, the fact that Simpson's rule integrates cubics exactly suggests a significant increase in accuracy.

All this is very vague and intuitive. Can we make it more precise? The answer is yes, as we see in the following theorem.

Theorem 5.1 (Simpson's Rule Error Estimate) If $f \in C^4([a,b])$, then there exists a point $\xi \in [a,b]$ such that

$$I(f) - S_2(f) = -\frac{1}{90}\left(\frac{b-a}{2}\right)^5 f^{(4)}(\xi). \tag{5.6}$$

Proof Let p_3 interpolate f at the nodes $x_0 = a, x_1 = c - \epsilon$, $x_2 = c + \epsilon, x_3 = b$, where $\epsilon > 0$ is a small parameter. Then for all $x \in [a,b]$, we have

$$f(x) - p_3(x) = R_3(x) = \frac{1}{4!}(x - x_0)(x - x_1)(x - x_2)(x - x_3)f^{(4)}(\xi_x)$$

for some $\xi_x \in [a,b]$. Since Simpson's rule integrates cubic polynomials exactly, we have

$$I(f) - S_2(f) = I(R_3) - S_2(R_3).$$

Note that the remainder will vanish at $x = a$ and $x = b$, so

$$S_2(R_3) = 4\left(\frac{b-a}{6}\right)\left(\frac{1}{4!}\right)(c - x_0)(c - x_1)(c - x_2)(c - x_3)f^{(4)}(\xi_c).$$

Recall that both x_1 and x_2 depend on the parameter ϵ, and that $x_1 \to c$ and $x_2 \to c$ as $\epsilon \to 0$. Thus, in the limit as $\epsilon \to 0$, $R_3(c) \to 0$, and hence $S_2(R_3)$ goes to zero.

On the other hand, as $\epsilon \to 0$ we have

$$I(R_3) \to \frac{1}{4!}\int_a^b (x - a)(x - c)^2(x - b)f^{(4)}(\xi_x)\,dx.$$

We therefore can say that

$$I(f) - S_2(f) = \frac{1}{4!}\int_a^b (x - a)(x - c)^2(x - b)f^{(4)}(\xi_x)\,dx,$$

and since $(x-a)(x-c)^2(x-b)$ does not change sign for $x \in [a,b]$, the Integral Mean Value Theorem can be used to simplify the integral, so that we have

$$I(f) - S_2(f) = \frac{1}{4!} f^{(4)}(\xi) \int_a^b (x-a)(x-c)^2(x-b)\, dx.$$

Direct evaluation of the integral completes the proof. □

The error for the composite rule now follows just as it did for the trapezoid rule.

Corollary 5.1   *If* $f \in C^4([a,b])$ *and the grid is uniform with mesh spacing h, then*

$$I(f) - S_n(f) = -\frac{(b-a)h^4}{180} f^{(4)}(\xi).$$

Proof   We have, for a uniform grid $(x_{i+1} - x_i = (b-a)/n = h)$,

$$I(f) - S_n(f) = \sum_{i=1}^{n/2} \left(\int_{x_{2i-2}}^{x_{2i}} f(x)\, dx - \left(\frac{x_{2i} - x_{2i-2}}{6} \right) (f(x_{2i}) + 4f(x_{2i-1}) + f(x_{2i})) \right),$$

$$= \sum_{i=1}^{n/2} \left(-\frac{1}{90} \left(\frac{2h}{2} \right)^5 f^{(4)}(\xi_i) \right) \quad \text{(from (5.6))}$$

$$= -\frac{h^4}{90} \sum_{i=1}^{n/2} h f^{(4)}(\xi_i)$$

$$= -\frac{h^4}{90} \sum_{i=1}^{n/2} \frac{b-a}{n} f^{(4)}(\xi_i)$$

$$= -\frac{(b-a)h^4}{180} \sum_{i=1}^{n/2} \frac{2}{n} f^{(4)}(\xi_i),$$

and the Discrete Average Value Theorem allows us to replace the sum with a single pointwise evaluation of the fourth derivative:

$$I(f) - S_n(f) = -\frac{(b-a)h^4}{180} f^{(4)}(\xi),$$

which completes the proof. □

EXAMPLE 5.6

While we cannot always use the error estimate to predict how small to take h, this is sometimes a useful exercise. Consider now our usual example in which $f(x) = e^x$ and the interval is $[0,1]$. Suppose we want the error to be less than 10^{-6}. How small should h be to guarantee that?

The absolute value of the error is

$$|E_n(f)| = |I(f) - S_n(f)| = \frac{1}{180} h^4 |f^{(4)}(\xi)|.$$

Therefore, to impose $|E_n(f)| \leq 10^{-6}$ we impose

$$\frac{1}{180} h^4 \max_{x \in [0,1]} |f^{(4)}(x)| \leq 10^{-6}$$

and solve for h. Thus we have

$$\frac{e}{180} h^4 \leq 10^{-6} \quad \Leftarrow \quad h \leq 0.09020788609.$$

So, taking $h \leq 0.0902...$ will be sufficient for the desired accuracy. Note that for the trapezoid rule we would require $h \leq 0.002101083838$, which is substantially smaller. ∎

Programming Hint One should always use the theory underlying a numerical method to help debug a program implementing that method. For example, since Simpson's rule is $O(h^4)$ accurate, we should see the error go down by a factor of 16 as the mesh size is halved. While this might not hold exactly for larger values of h, it will begin to hold as you take h smaller and smaller. If it doesn't, then it generally means that either the integrand is not smooth enough, or your program has an error in it. As was the case with the trapezoid rule, there exists a class of functions for which Simpson's rule is "super-accurate," and for these functions the error will not decrease as rapidly as we expect, because it starts out smaller than it should be. See Section 5.8.1 for details. Always use a few simple examples to predict the error and make sure the code is performing as expected.

The explanation so far for the extra accuracy of Simpson's rule might seem a bit less than satisfying. Yes, maybe we can show how all the mathematics fits together, but what is really happening here to *cause* the extra accuracy? The reason is hinted at in the proof of Theorem 5.1, and with a little work we can do a fairly complete exposition.

Consider the cubic polynomial

$$q_3(x) = f(a) + \left(\frac{f(c) - f(a)}{h} \right)(x - a) + \left(\frac{f'(c) - \frac{f(c) - f(a)}{h}}{h} \right)(x - a)(x - c)$$
$$+ A(x - a)(x - c)^2,$$

where

$$A = \frac{f(b) - f(a) - 2f'(c)(c - a)}{2h^3}$$

and

$$h = c - a = b - c.$$

Now this is an imposing formula, but it ought to be straightforward to show that q_3 is the cubic that interpolates to f at a, b, and c, and that also interpolates to f' at c:

$$q_3'(c) = f'(c).$$

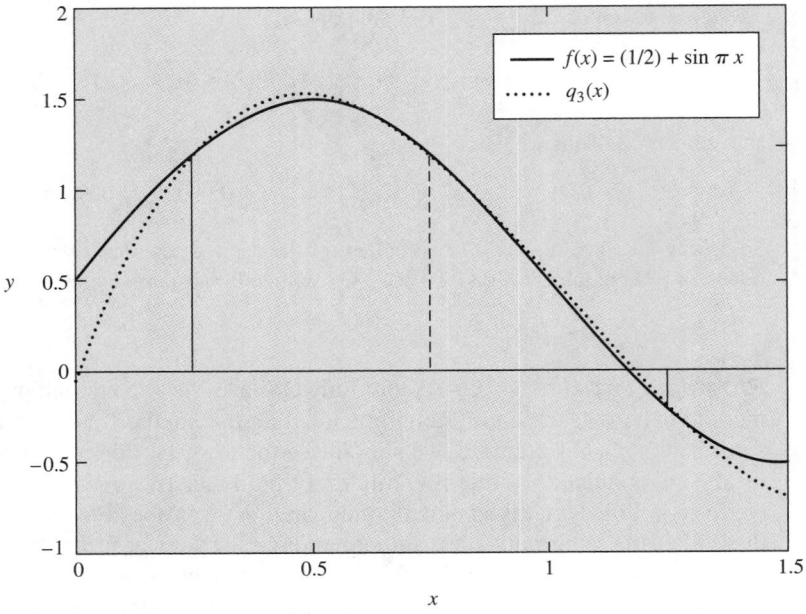

FIGURE 5.4 Another look at Simpson's rule.

Figure 5.4 shows the graph of this function for our example function $\frac{1}{2} + \sin \pi x$. Note that the requirement that the derivatives match at the midpoint results in a *much* closer approximation to the function than was the case for the quadratic p_2; thus we expect to get much better approximation to the integral.

Now for the payoff: If we use q_3 as the approximating polynomial to generate a quadrature rule, what do we get? In other words, what is

$$S_2^*(f) = I(q_3)$$

in terms of $f(a)$, $f(b)$, $f(c)$, and $f'(c)$? We will omit the details here (see Exercise 16, however), but the bottom line is the following: Using q_3 as the approximating polynomial *yields the same quadrature rule as when we use p_2: $I(q_3) = I(p_2)$. Or, to turn this around a bit, using the quadratic approximation p_2 produces the same quadrature rule as the slightly more accurate cubic q_3. This boost in accuracy is very much tied to using the midpoint of the interval as the third point. Although we can construct a version of Simpson's rule based on using any point c between a and b, it will only be $O(h^3)$ accurate unless c is the midpoint.

We close this section with a brief code segment that implements Simpson's rule over a uniform grid (see Algorithm 5.1.)

The extra accuracy that we get with Simpson's rule is an artifact of using an even-degree polynomial interpolant to define the integration rule. If we construct a method based on cubic interpolation (see Exercise 15), we still get degree of precision equal to 3, and an error estimate that is $O(h^4)$.

Algorithm 5.1: *Simpson's rule*

```
input a, b, n
external f
    simp = f(a) + f(b)
    h = (b - a)/n
    sum4 = 0.0
    for i=1 to n-1 by 2 do
        x = a + i*h
        sum4 = sum4 + f(x)
    endfor
    sum2 = 0.0
    for i=2 to n-2 by 2 do
        x = a + i*h
        sum2 = sum2 + f(x)
    endfor
    simp = (h/3.0)*(simp + 4*sum4 + 2*sum2)
end code
```

EXERCISES

1. Apply Simpson's rule with $h = \frac{1}{4}$ to approximate the integral

$$I = \int_0^1 x(1-x^2)\,dx = \frac{1}{4}$$

2. Apply Simpson's rule with $h = \frac{1}{4}$ to approximate the integral

$$I = \int_0^1 \frac{1}{\sqrt{1+x^4}}\,dx = 0.92703733865069.$$

3. Apply Simpson's rule with $h = \frac{1}{4}$ to approximate the integral

$$I = \int_0^1 \ln(1+x)\,dx = 2\ln 2 - 1.$$

How small does the error theory say that h must be to get the error less than 10^{-3}? 10^{-6}? How small does h have to be for the trapezoid rule to achieve this accuracy?

4. Apply Simpson's rule with $h = \frac{1}{4}$ to approximate the integral

$$I = \int_0^1 \frac{1}{1+x^3}\,dx = \frac{1}{3}\ln 2 + \frac{1}{9}\sqrt{3}\pi.$$

5. Apply Simpson's rule with $h = \frac{1}{4}$ to approximate the integral

$$I = \int_1^2 e^{-x^2}\,dx = 0.1352572580.$$

6. For each function below, write a program to carry out Simpson's rule using the sequence of mesh sizes $h = \frac{1}{2}(b-a), \frac{1}{4}(b-a), \frac{1}{8}(b-a), \ldots, \frac{1}{2048}(b-a)$, where $b - a$ is the length of the given interval. Verify that the expected rate of decrease of the error is observed. Comment on any anomalies observed.

(a) $f(x) = 1/(1+x^2)$, $[-5,5]$, $I(f) = 2\arctan(5)$
(b) $f(x) = x^2 e^{-x}$, $[0,2]$, $I(f) = 2 - 10e^{-2}$
$= 0.646647168$
(c) $f(x) = \ln x$, $[1,3]$, $I(f) = 3\ln 3 - 2 = 1.295836867$
(d) $f(x) = \sqrt{1-x^2}$, $[-1,1]$, $I(f) = \pi/2$
(e) $f(x) = e^{-x}\sin(4x)$, $[0,\pi]$, $I(f) = \frac{4}{17}(1 - e^{-\pi})$
$= 0.2251261368$

7. For each integral in the previous exercise, how small does h have to be to get accuracy, according to the error theory, of at least 10^{-3}? 10^{-6}? Compare with the value of h required by the trapezoid rule for this accuracy. (Feel free to use a computer algebra system to help with the computation of the derivatives.)

8. Since the area of the unit circle is $A = \pi$, it follows that

$$\frac{\pi}{2} = \int_{-1}^{1} \sqrt{1-x^2} \, dx.$$

Therefore, we can approximate π by approximating this integral. Use Simpson's rule to compute approximate values of π in this way, and comment on your results.

9. If we wanted to use Simpson's rule to approximate the natural logarithm function on the interval $[\frac{1}{2}, 1]$ by approximating

$$\ln x = \int_{1}^{x} \frac{1}{t} \, dt,$$

how many points would be needed to obtain an error of less than 10^{-6}? How many points for an error of less than 10^{-16}? What are the corresponding values for the trapezoid rule?

10. Use Simpson's rule to produce a graph of $E(x)$, defined to be the *length* of the exponential curve from 0 to x, for $0 \leq x \leq 3$. (See Exercise 14 of Section 2.5.)

11. Let $f(x) = |x|$; use the trapezoid rule (with $n = 1$), corrected trapezoid rule (with $n = 1$), and Simpson's rule (with $n = 2$) to compute

$$I(f) = \int_{-1}^{1} f(x) \, dx$$

and compare your results with the exact value. Explain what happens in the light of our error estimates for the trapezoid and Simpson's rules.

12. Write out the expression for Simpson's rule when c is *not* the midpoint of the interval $[a,b]$. To simplify matters, take $c - a = h$, $b - c = \theta h$.

13. What is the degree of precision of the corrected trapezoid rule, T_1^C? What about the n-subinterval version, T_n^C?

14. Prove that if we want to show that the quadrature rule $I_n(f)$ has degree of precision p, it suffices to show that it will exactly integrate x^k, $0 \leq k \leq p$, over the interval $(0,1)$.

15. Construct the analogue of Simpson's rule based on exactly integrating a cubic interpolant at equally spaced points on the interval $[a,b]$.

16. Show that $I(q_3) = S_2(f)$.

17. Consider the quadrature rule defined by exactly integrating a cubic Hermite interpolant:

$$I_1(f) = I(H_2).$$

Write down the quadrature formula for both the basic and composite settings, and state and prove an error estimate using the error results for Hermite interpolation from the previous chapter.

18. Consider a quadrature rule in the form

$$I_n(f) = \sum_{k=1}^{n} a_k f(x_k),$$

where the coefficients $a_k > 0$ and the grid points x_k are all known. Assume that I_n integrates the trivial function $w(x) = 1$ exactly:

$$I_n(w) = \sum_{k=1}^{n} a_k = I(w) = \int_{a}^{b} dx = b - a,$$

and that this holds for all intervals (a,b). Consider now the effects of rounding error on integrating an arbitrary function f. Let $\hat{f}(x) = f(x) + \epsilon(x)$ be f polluted by rounding error, with $|\epsilon(x)| \leq C\mathbf{u}$ for some constant $C > 0$, for all $x \in [a,b]$. Show that

$$|I_n(f) - I_n(\hat{f})| \leq C\mathbf{u}(b-a).$$

Comment on this in comparison with the corresponding result for numerical differentiation, as given in Section 2.2.

19. The *normal probability distribution* is defined as

$$p(x) = \frac{1}{\sigma\sqrt{2\pi}} e^{-(x-\mu)^2/2\sigma^2},$$

where μ is the mean, or average, and σ is the variance. This is the famous bell-shaped curve that one hears so much about; the mean gives the center of the bell, and the variance gives its width. If x is distributed in this fashion, then the probability that $a \leq x \leq b$ is given by the integral

$$P(a \leq x \leq b) = \int_{a}^{b} p(x) \, dx.$$

(a) Use the change of variable $z = (x - \mu)/\sigma$ to show that

$$P(-m\sigma \leq x \leq m\sigma) = \frac{1}{\sqrt{2\pi}} \int_{-m}^{m} e^{-z^2/2} \, dz.$$

(b) Compute values of $P(-m\sigma \leq x \leq m\sigma)$ for $m = 1, 2, 3$, using Simpson's rule.

20. Use Simpson's rule to solve Exercise 9 from Section 5.2.

21. Use Simpson's rule to solve Exercise 10 from Section 5.2.

5.4   THE MIDPOINT RULE

An interesting and very simple quadrature rule is the midpoint rule, based on exactly integrating a linear *Taylor* approximation to the integrand. This results in a rule that is actually *more* accurate than the trapezoid rule. We proceed in much the same manner as with the trapezoid and Simpson's rules, except that, since this is the third time, we skip some of the details.

First consider the integral

$$I(f) = \int_a^b f(x)\, dx$$

and the Taylor approximation

$$p_1(x) = f(c) + (x - c)f'(c),$$

where $c = (a + b)/2$, the midpoint of the interval. We define the numerical method by exactly integrating p_1:

$$M_1(f) = \int_a^b p_1(x)\, dx = (b - a)f(c).$$

The composite rule then becomes

$$M_n(f) = h \sum_{i=1}^n f(a + (i - 1/2)h).$$

Note that the term involving the derivative is not present in the quadrature rule. This is because we have

$$\int_a^b (x - c)f'(c)\, dx = 0$$

when c is the midpoint of the interval. This rule is called the *midpoint rule* because it evaluates the integrand at the midpoint of each subinterval in order to construct the quadrature rule. Figure 5.5 shows a single-interval example, and Figure 5.6 shows a composite example. Both illustrations use $f(x) = \frac{1}{2} + \sin \pi x$, integrated over the interval $[\frac{1}{4}, \frac{5}{4}]$.

The error estimate comes rather easily from the construction of the approximation. We have $M_1(f) = I(p_1)$, so

$$I(f) - M_1(f) = I(f) - I(p_1) = I(f - p_1) = I(R_1),$$

where $R_1(x) = \frac{1}{2}(x - c)^2 f''(\xi)$ is the Taylor remainder. We can apply the Integral Mean Value Theorem to get

$$I(f) - M_1(f) = \frac{1}{24}(b - a)^3 f''(\xi),$$

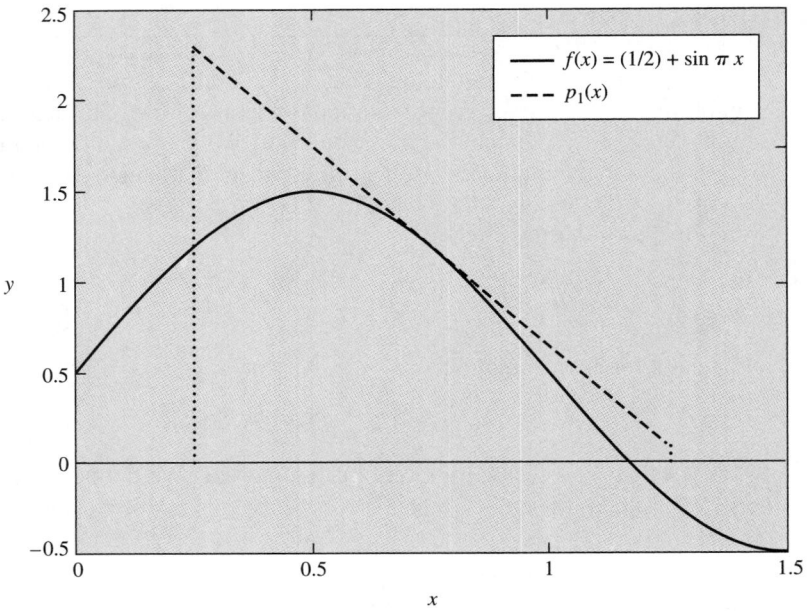

FIGURE 5.5 Illustration of single-interval midpoint rule.

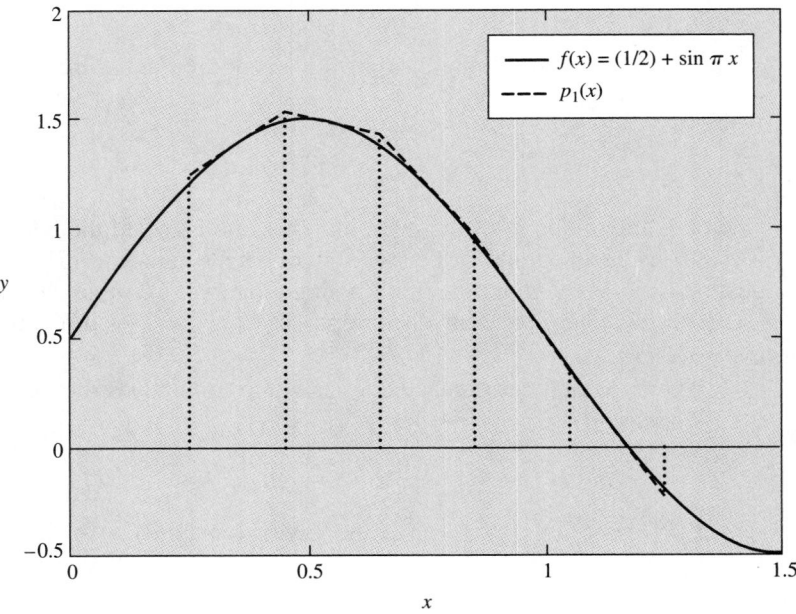

FIGURE 5.6 Illustration of midpoint rule.

TABLE 5.4   Midpoint Rule Applied to $f(x) = e^x$, $[a,b] = [0,1]$

n	$M_n(f)$	$I(f) - M_n(f)$	Error ratio
2	0.17005127166502E+01	0.1776911E–01	NA
4	0.17138152797711E+01	0.4466549E–02	3.9783
8	0.17171636649957E+01	0.1118163E–02	3.9945
16	0.17180021920527E+01	0.2796364E–03	3.9986
32	0.17182119133839E+01	0.6991508E–04	3.9997
64	0.17182643493169E+01	0.1747914E–04	3.9999
128	0.17182774586502E+01	0.4369809E–05	4.0000
256	0.17182807360054E+01	0.1092454E–05	4.0000
512	0.17182815553455E+01	0.2731135E–06	4.0000
1024	0.17182817601807E+01	0.6827838E–07	4.0000
2048	0.17182818113894E+01	0.1706960E–07	4.0000

which is essentially *half* the error of the single-interval trapezoid rule. For the composite midpoint rule, the error becomes

$$I(f) - M_n(f) = \frac{b-a}{24}h^2 f''(\xi).$$

EXAMPLE 5.7

Let's look at the use of the midpoint rule to approximate

$$I(f) = \int_0^1 e^x \, dx$$

using $h = \frac{1}{2}$. Then we will have

$$M_2(f) = \frac{1}{2} \times f(0 + (1/2)(1/2)) + \frac{1}{2} \times f(1/2 + (1/2)(1/2))$$

$$= \frac{1}{2} \times e^{1/4} + \frac{1}{2} \times e^{3/4} = 1.700512717.$$

If we continue with smaller and smaller values of h, we get the results presented in Table 5.4. ∎

EXERCISES

1. Apply the midpoint rule with $h = \frac{1}{4}$ to approximate the integral

$$I = \int_0^1 x(1 - x^2) \, dx = \frac{1}{4}.$$

How small does h have to be to get the error less than 10^{-3}? 10^{-6}?

2. Apply the midpoint rule with $h = \frac{1}{4}$ to approximate the integral

$$I = \int_0^1 \frac{1}{\sqrt{1 + x^4}} \, dx = 0.92703733865069.$$

How small does h have to be to get the error less than 10^{-3}? 10^{-6}?

3. Apply the midpoint rule with $h = \frac{1}{4}$ to approximate the integral

$$I = \int_0^1 \ln(1+x)\, dx = 2\ln 2 - 1.$$

How small does h have to be to get the error less than 10^{-3}? 10^{-6}?

4. Apply the midpoint rule with $h = \frac{1}{4}$ to approximate the integral

$$I = \int_0^1 \frac{1}{1+x^3}\, dx = \frac{1}{3}\ln 2 + \frac{1}{9}\sqrt{3}\pi.$$

How small does h have to be to get the error less than 10^{-3}? 10^{-6}?

5. Apply the midpoint rule with $h = \frac{1}{4}$ to approximate the integral

$$I = \int_1^2 e^{-x^2}\, dx = 0.1352572580.$$

How small does h have to be to get the error less than 10^{-3}? 10^{-6}?

6. Show that the midpoint rule can be derived by integrating, exactly, a polynomial interpolant of degree *zero*.

7. Apply the midpoint rule to each of the following functions, integrated over the indicated interval. Use a sequence of grids $h = (b-a), (b-a)/2, (b-a)/4, \ldots$ and confirm that the approximations are converging at the correct rate. Comment on any anomalies that you observe.

(a) $f(x) = 1/(1+x^2)$, $[-5,5]$, $I(f) = 2\arctan(5)$
(b) $f(x) = x^2 e^{-x}$, $[0,2]$, $I(f) = 2 - 10e^{-2} = 0.646647168$
(c) $f(x) = \ln x$, $[1,3]$, $I(f) = 3\ln 3 - 2 = 1.295836867$
(d) $f(x) = \sqrt{1-x^2}$, $[-1,1]$, $I(f) = \pi/2$
(e) $f(x) = e^{-x}\sin(4x)$, $[0,\pi]$, $I(f) = \frac{4}{17}(1 - e^{-\pi}) = 0.2251261368$

8. For each integral in the previous exercise, how small does h have to be to get accuracy, according to the error theory, of at least 10^{-3}? 10^{-6}?

9. State and prove a formal theorem concerning the error estimate for the midpoint rule over n subintervals. You may want to state and prove a formal theorem for the single-subinterval rule first, and then use this in the more general theorem.

10. Let T_1 be the trapezoid rule using a single subinterval, M_1 be the midpoint rule using a single subinterval, and S_2 be Simpson's rule using a single quadratic interpolant (hence, a single *pair* of subintervals). Show that, for any continuous function f and any interval $[a,b]$,

$$S_2(f) = \frac{1}{3}T_1(f) + \frac{2}{3}M_1(f).$$

5.5 APPLICATION: STIRLING'S FORMULA

Stirling's formula[4] is an interesting and useful way to approximate the factorial function, $n!$, for large values of n. Standard derivations require a substantial background in classical analysis, but in fact the rule can be derived in a fairly general form based on nothing more than the trapezoid rule and its error estimate.

Theorem 5.2 (Stirling's Formula)   *For all $n \geq 2$, there exists a value C_n, $2.37 < C_n < 2.501$, such that*

$$n! = C_n \sqrt{n}(n/e)^n.$$

---

[4]James Stirling (1692–1770), a Scotsman, began his studies at Oxford in 1710 but was expelled for political reasons before finishing his degree. Nonetheless, by 1717 he had already published his first paper, an extension of some of Newton's work on plane curves. For the next few years he was in Italy trying to obtain a position, and in 1722 he returned to Britain. In 1726, while living in London, he published a monograph on infinite series and other topics called *Methodus Differentialis*, which contains what we call Stirling's formula as one of its examples.

Proof   We start with the observation that

$$\ln n! = \sum_{k=1}^{n} \ln k,$$

$$= \sum_{k=1}^{n-1} \frac{1}{2} \left(\ln k + \ln(k+1) \right) + \frac{1}{2} \ln n,$$

and the summation is exactly the trapezoid rule for

$$I(\ln) = \int_{1}^{n} \ln x \, dx = n \ln n - n + 1$$

using $n-1$ subintervals of length 1. The error estimate for the trapezoid rule then gives

$$\ln n! = \left(n + \frac{1}{2} \right) \ln n - n + 1 - \frac{1}{12} \sum_{k=1}^{n-1} \xi_k^{-2}$$

for some set of $\xi_k \in [k, k+1]$. Now exponentiate both sides and let σ_n be the summation, to get

$$n! = \sqrt{n}(n/e)^n e^{1 - 1/12\sigma_n} = \sqrt{n}(n/e)^n C_n$$

and it remains only to establish the bounds on C_n. To this end, note that we can bound the sum as follows:

$$1 \le \sum_{k=1}^{n-1} \xi_k^{-2} \le \sum_{k=1}^{n-1} k^{-2} \le \sum_{k=1}^{\infty} k^{-2} = \frac{\pi^2}{6}.$$

Thus we have

$$2.37 < \exp\left(1 - \frac{1}{12}\frac{\pi^2}{6} \right) \le e^{1+\sigma_n} \le \exp\left(1 - \frac{1}{12} \right) < 2.501. \qquad \square$$

This estimate on C_n is not as sharp as others, but it is sufficient for most purposes. The sharpest result ([5], p. 467) states that there is a value $c_n \in [0,1]$ such that

$$n! = \sqrt{2\pi n} \left(\frac{n}{e} \right)^n e^{c_n/12n}.$$

EXERCISES

1. Use Stirling's formula to show that

$$\lim_{n\to\infty} \left(\frac{x^n}{n!} \right) = 0$$

for all x.

2. For $x = 10$ and $\epsilon = 10^{-3}$, how large does n have to be

for

$$\left| \frac{x^n}{n!} \right| \le \epsilon$$

to hold? Repeat for $\epsilon = 10^{-6}$. (Use Stirling's formula here; don't just plug numbers into a calculator.)

3. Use Stirling's formula to determine the value of

$$\lim_{n \to \infty} \frac{(n!)^p}{(pn)!},$$

where $p \geq 2$ is an integer.

4. Use Stirling's formula to get an upper bound on the ratio

$$R_n = \frac{1 \cdot 3 \cdot 5 \cdots (2n-1)}{2^{n+1} n!}$$

as a function of n.

5.6 GAUSSIAN QUADRATURE

Gaussian[5] quadrature is a very powerful tool for approximating integrals. It is derived in a very different way from the trapezoid, Simpson, and midpoint rules (although the midpoint rule can be considered a special case of Gaussian quadrature), and a complete discussion of Gaussian quadrature usually requires a substantial background in approximation theory. [6]

Before we get into the details of the derivation of the method, it might be a good idea to get some idea of how accurate Gaussian quadrature can be. The quadrature rules are all based on special values of *weights* and *abscissas* (evaluation points, commonly called "Gauss points") that are precomputed and stored. They are available in most standard mathematics tables and can also be computed using standard, available computer codes. Table 5.5 gives some of the values for a few cases, and the text website (http://wiley.com/college/epperson/) has several files which can be used to generate them.

The quadrature rule is written in the form

$$G_n(f) = \sum_{i=1}^{n} w_i^{(n)} f(x_i^{(n)}) \approx \int_{-1}^{1} f(x) \, dx, \tag{5.7}$$

where the weights $w_i^{(n)}$ and Gauss points $x_i^{(n)}$ will be determined later on. For now, we will simply assume that they are available, as is in fact the case.[7]

---

[5]Carl Friedrich Gauss (1777–1855) is widely regarded as one of the greatest mathematicians of all time, in a class with Archimedes, Newton, and Euler. A child prodigy who taught himself to read and do arithmetic before beginning elementary school, Gauss attended the Collegium Carolinum Brunswick and the University of Göttingen, graduating in 1800. His doctoral dissertation was accepted in 1801. Briefly supported by his patron, the Duke of Brunswick, Gauss accepted an appointment at Göttingen as director of the observatory in 1804 and, as professor of astronomy in 1807.

As with Euler, there are few areas of mathematics that are untouched by Gauss's mind and name. Unlike Euler, who published a lot, Gauss published sparingly, often not at all, leaving some of his most significant results in his personal diary. Nonetheless, Gauss has to his credit results in physics, astronomy (including the calculation of the orbit of the planetoid Ceres), number theory, and non-Euclidean geometry. His collected works come to 12 volumes. Gaussian quadrature bears his name because of a paper he presented to the Göttingen Society in 1814, titled *Methodus nova itegralium valores per approximationem inveniendi*.

[6]The treatment we use here—which avoids a lot of the need for approximation theory—is based on an idea due to Roy Mathias, published in the *International Journal of Mathematical Education in Science Technology*, vol. 28, pp. 134–137.

[7]Integrals on general intervals $[a,b]$ can be treated by a change of variable in the integrand, as we shall see later in this section.

TABLE 5.5   **Gaussian Quadrature Nodes and Weights**

n	$x_i^{(n)}$	$w_i^{(n)}$
1	0.0000000000000000E+00	0.2000000000000000E+01
2	±0.5773502691896257E+00	1.0000000000000000E+00
4	±0.3399810435848563E+00	0.6521451548625464E+00
	±0.8611363115940526E+00	0.3478548451374476E+00
8	±0.1834346424956498E+00	0.3626837833783620E+00
	±0.5255324099163290E+00	0.3137066458778874E+00
	±0.7966664774136268E+00	0.2223810344533745E+00
	±0.9602898564975362E+00	0.1012285362903697E+00
16	±0.9501250983763744E−01	0.1894506104550685E+00
	±0.2816035507792589E+00	0.1826034150449236E+00
	±0.4580167776572274E+00	0.1691565193950024E+00
	±0.6178762444026438E+00	0.1495959888165733E+00
	±0.7554044083550030E+00	0.1246289712555339E+00
	±0.8656312023878318E+00	0.9515851168249290E−01
	±0.9445750230732326E+00	0.6225352393864778E−01
	±0.9894009349916499E+00	0.2715245941175185E−01

EXAMPLE 5.8

Consider the example integral

$$I(f) = \int_{-1}^{1} e^x \, dx = e^1 - e^{-1} = 2.3504023872876029138....$$

Using $n = 2$, we get the approximation

$$G_2(f) = (1)e^{-0.5773502691896257} + (1)e^{0.5773502691896257} = 2.342696087910,$$

which is very easily computed—and very accurate as well, for the amount of effort put into the computation. Table 5.6 shows the approximate values given by using Gaussian quadrature at

TABLE 5.6   **Gaussian Quadrature Applied to Integrating $f(x) = e^x$, over the Interval $[a,b] = [-1,1]$**

n	$G_n(f)$	$I(f) - G_n(f)$	$S_n(f)$	$I(f) - S_n(f)$
2	2.342696087910	0.77062993778738E−02	2.362053756543	−0.11651369255893E−01
4	2.350402092156	0.29513124299996E−06	2.351194831880	−0.79244459265215E−03
8	2.350402387288	0.19539925233403E−13	2.350453017242	−0.50629954676307E−04
16	2.350402387288	0.15543122344752E−13	2.350405569305	−0.31820170360852E−05

4, 8, and 16 points, in addition to the $n = 2$ case we just did. For comparison's sake, the table also gives the approximate values (and associated errors) given by Simpson's rule. Note the extraordinary accuracy of Gaussian quadrature compared with Simpson's rule. How do we get such high accuracy? ∎

The trapezoid and Simpson rules—and the midpoint rule, as well—were derived by exactly integrating a polynomial approximation to the integrand. In Gaussian quadrature we will construct an approximate integral that is "best possible" in the sense that it has the highest possible degree of precision. Thus we pose the following question:

What are the coefficients
$\{w_i^{(n)}\}$ *and Gauss points* $\{x_i^{(n)}\}$ *such that the quadrature rule*

$$\int_{-1}^{1} f(x)\, dx \approx \sum_{i=1}^{n} w_i^{(n)} f(x_i^{(n)})$$

is exact *for as large a degree of polynomial as possible?*

To put it another way (see Exercise 3), we want to find $\{w_i^{(n)}\}$ and $\{x_i^{(n)}\}$ such that

$$\int_{-1}^{1} x^k\, dx = \sum_{i=1}^{n} w_i^{(n)} (x_i^{(n)})^k \tag{5.8}$$

for $n = 0, 1, 2, \ldots, N$, for N as large as possible.

The high accuracy of Gaussian quadrature then comes from the fact that it integrates very-high-degree polynomials exactly. In the trapezoid method, Simpson's rule, and the midpoint method, we use a fixed, predetermined set of grid points, and only the weights are defined by the method. A fixed degree of polynomial approximation is used over each subinterval, and so a fixed degree of polynomial is integrated exactly, no matter how many points are used. In Gaussian quadrature both the grid points (i.e., the abscissas or Gauss points) and the weights are chosen to maximize the degree of precision, i.e., the degree of polynomial that is integrated exactly. The degree of polynomial that is exactly integrated goes to infinity as the number of points used in the quadrature rule goes to infinity.

Since (5.8) is a system of (nonlinear) equations, we could just rely on that as a basis for computation, but this approach is unsatisfactory for a number of reasons, the most important one being that it is next to impossible to gain any broad understanding of the numerical method.

Note that we have to find not only the weights and abscissas, but also the value of N. Intuitively, we suspect that $N = 2n - 1$. This is because a polynomial of degree $2n - 1$ has $2n$ coefficients, and thus the number of unknowns (n weights plus n abscissas) equals the number of equations (one for each power in the polynomial). We will show that taking $N = 2n$ yields a contradiction, and then we will construct a solution for $N = 2n - 1$.

Lemma 5.2   *If $N = 2n$, then there are no weights and Gauss points such that (5.8) is satisfied for all $k = 0,1,2,\ldots,N$.*

Proof   Let $\{w_i^{(n)}\}$ and $\{x_i^{(n)}\}$ satisfy (5.8), and assume that (5.8) holds for all $k = 0,1,2,\ldots,N$; i.e., the quadrature rule is exact for all polynomials of degree less than or equal to $N = 2n$. Define

$$L(x) = \prod_{j=1}^{n}(x - x_j^{(n)})^2.$$

Thus $L(x) \geq 0$ and

$$\int_{-1}^{1} L(x)\, dx > 0.$$

However,

$$\sum_{i=1}^{n} w_i^{(n)} L(x_i^{(n)}) = 0.$$

Therefore, we have a contradiction, since the quadrature rule (5.7) is supposed to be exact for all polynomials of degree $2n$, and it is clearly not exact for the particular polynomial of degree $2n$ given by $L(x)$. Hence, our assumption that a solution exists must be false, and the lemma is proved.   □

Now we turn to the $N = 2n - 1$ case. We will derive necessary conditions for the weights and abscissas, and then show that these necessary conditions do imply that the rule is exact for all polynomials of degree $\leq 2n - 1$. Since the weights are the easier of the two, we will start there.

Lemma 5.3   *Let $\{w_i^{(n)}\}$ be a set of weights and $\{x_i^{(n)}\}$ a set of Gauss points, such that (5.8) is satisfied for $k = 0,1,2,\ldots,N = 2n - 1$. Then the weights must satisfy*

$$w_i^{(n)} = \int_{-1}^{1} L_i^{(n)}(x)\, dx, \tag{5.9}$$

where

$$L_i^{(n)}(x) = \prod_{\substack{k=1 \\ k \neq i}}^{n} \frac{x - x_k^{(n)}}{x_i^{(n)} - x_k^{(n)}}.$$

Proof   Note that the $L_i^{(n)}$ are of degree $n - 1 \leq 2n - 1$, and recall that $L_i^{(n)}(x_j) = \delta_{ij}$. Thus the fact that the quadrature rule is supposed to be exact for *all* polynomials of degree $\leq 2n - 1$ forces

$$\int_{-1}^{1} L_i^{(n)}(x)\, dx = \sum_{j=1}^{n} w_j^{(n)} L_i^{(n)}(x_j^{(n)}) = w_i^{(n)},$$

and this completes the proof. $\square$

Now that we understand how the weights are related to the Gauss points, we can concentrate on finding the Gauss points. To this end, define

$$P_n(x) = \prod_{i=1}^{n}(x - x_i^{(n)}), \qquad (5.10)$$

where the $x_i^{(n)}$ are the Gauss points, which we assume to be distinct, and note that P_n is a polynomial of degree n, since it has n roots. Let $Q(x)$ be *any* polynomial of degree $\leq n - 1$, so that the product $P_n Q$ has degree $\leq 2n - 1$. This means that the quadrature rule must integrate the product exactly, i.e.,

$$\int_{-1}^{1} P_n(x) Q(x)\, dx = \sum_{i=1}^{n} w_i^{(n)} P_n(x_i^{(n)}) Q(x_i^{(n)}).$$

But the construction of P_n implies that $P_n(x_i^{(n)}) = 0$; thus we have that

$$\int_{-1}^{1} P_n(x) Q(x)\, dx = 0$$

for *all* polynomials Q of degree $\leq n - 1$. Since the quadrature rule has to be exact when applied to the product $P_n Q$, this means that the polynomial P_n is one member of a family of *orthogonal polynomials* (see Section 4.10.2), and the Gauss points are the roots of P_n. At this point it might be appropriate to recall some results from Chapter 4 on orthogonal polynomials.

Theorem 5.3   *For any $k > 1$ there exists a family of polynomials $\phi_i \in \mathcal{P}_i$, $0 \leq i \leq k$, such that*

1. *For all $i \neq j$,*

$$\int_{-1}^{1} \phi_i(x) \phi_j(x)\, dx = 0.$$

2. *For all $q \in \mathcal{P}_j$, $j < i$,*

$$\int_{-1}^{1} \phi_i(x) q(x)\, dx = 0.$$

3. *All of the roots of each of the ϕ_i are distinct and lie in the interval $[-1, 1]$.*

With these results in hand, we can state and prove our main result.

Theorem 5.4 (Construction of Gaussian Quadrature) *For* $N = 2n - 1$, *there exists a set of Gauss points* $x_i^{(n)}$ *and weights* $w_i^{(n)}$, *with the weights given by* (5.9), *such that* (5.8) *holds for all* $k = 0, 1, 2, \ldots, N$.

Proof We begin by defining the Gauss points $x_i^{(n)}$ as the roots of the Legendre family of orthogonal polynomials.[8]

Let $P(x)$ be an arbitrary polynomial of degree $\leq 2n - 1$. Note that we can write

$$P(x) = P_n(x)Q(x) + R(x),$$

where both Q and R are polynomials of degree $\leq n - 1$ and P_n is defined in (5.10). (This follows from ordinary polynomial division considerations.) It follows, then (Exercise 9 of Section 4.1), that

$$R(x) = \sum_{i=1}^{n} R(x_i^{(n)}) L_i^{(n)}(x).$$

In addition, note that $P_n(x_i^{(n)}) = 0$ implies that $P(x_i^{(n)}) = R(x_i^{(n)})$. Then we have

$$\int_{-1}^{1} P(x)\, dx = \int_{-1}^{1} \left(P_n(x)Q(x) + R(x) \right)\, dx,$$

$$= \int_{-1}^{1} P_n(x)Q(x)\, dx + \int_{-1}^{1} R(x)\, dx.$$

The orthogonality property of P_n then implies that the first integral is zero. Hence,

$$\int_{-1}^{1} P(x)\, dx = \int_{-1}^{1} R(x)\, dx,$$

$$= \int_{-1}^{1} \sum_{i=1}^{n} R(x_i^{(n)}) L_i(x)\, dx$$

$$= \sum_{i=1}^{n} R(x_i^{(n)}) \int_{-1}^{1} L_i(x)\, dx$$

$$= \sum_{i=1}^{n} R(x_i^{(n)}) w_i^{(n)}$$

$$= \sum_{i=1}^{n} P(x_i^{(n)}) w_i^{(n)}$$

and we have proved that the quadrature rule (5.7) is exact for an arbitrary polynomial of degree $\leq 2n - 1$. This completes the proof. $\square$

---

[8]If Section 4.10.2 (Least Squares Approximation and Orthogonal Polynomials) was not covered prior to this, it would be wise at this point to review Theorem 4.9 for background material on orthogonal polynomials, and the subsequent discussion for material on the Legendre family.

We indicated before that the accuracy of Gaussian quadrature comes from its ability to integrate, exactly, polynomials of high degree. Let's give some indication of why this is so. What we know is that G_n will integrate any polynomial of degree $\leq 2n - 1$ exactly. So, given the problem of integrating f, let p_{2n-1} be *any* polynomial of degree $\leq 2n - 1$. Then

$$I(f) - G_n(f) = I(p_{2n-1}) + I(f - p_{2n-1}) - G_n(p_{2n-1}) - G_n(f - p_{2n-1}), \qquad (5.11)$$

where all we have done is added and subtracted p_{2n-1} and used the linearity of the integral and the quadrature rule. But since the quadrature rule is exact when applied to p_{2n-1}, we have

$$I(p_{2n-1}) - G_n(p_{2n-1}) = 0,$$

so that (5.11) becomes

$$I(f) - G_n(f) = I(f - p_{2n-1}) - G_n(f - p_{2n-1}).$$

Thus, the error in Gaussian quadrature is equal to the error in integrating the polynomial approximation *error*. Note that this will hold for *any* polynomial approximation of degree $\leq 2n - 1$.

The question then becomes, Can we find a specific polynomial approximation that yields a concise error formula? The answer is, yes, we can.

Let $p_{2n-1} = H_n$, the Hermite interpolant to f that uses the Gauss points as the nodes. Then from Theorem 4.5 we have that

$$f(x) - p_{2n-1}(x) = f(x) - H_n(x) = \frac{1}{(2n)!} \psi_n(x) f^{(2n)}(\xi_x),$$

where

$$\psi_n(x) = \prod_{i=1}^{n} (x - x_i^{(n)})^2. \qquad (5.12)$$

Therefore,

$$G_n(f - p_{2n-1}) = G_n(f - H_n) = \sum_{k=1}^{n} \frac{w_k^{(n)}}{(2n)!} \psi_n(x_k^{(n)}) f^{(2n)}(\xi_k).$$

But (5.12) implies that $\psi(x_k^{(n)}) = 0$; hence we have that $G_n(f - H_n) = 0$, and so

$$I(f) - G_n(f) = I(f - H_n) = \int_{-1}^{1} \frac{1}{(2n)!} \psi_n(x) f^{(2n)}(\xi_x) \, dx.$$

Since $\psi_n(x) \geq 0$ for all x, we can apply the Integral Mean Value Theorem to conclude that there exists a value $\eta_n \in [-1, 1]$ such that

$$I(f) - G_n(f) = \frac{1}{(2n)!} \left(\int_{-1}^{1} \psi_n(x) \, dx \right) f^{(2n)}(\eta_n).$$

Recall now that the Gauss points/nodes $x_i^{(n)}$ are the roots of the n-degree Legendre polynomial, P_n. Thus, $\psi_n(x) = (r_n P_n(x))^2$, where r_n is the reciprocal of the leading coefficient of P_n. We then have

$$I(f) - G_n(f) = \frac{r_n^2}{(2n)!} \left(\int_{-1}^{1} P_n^2(x)\, dx \right) f^{(2n)}(\eta_n) = \frac{r_n^2}{(2n)!} \|P_n\|_2^2 f^{(2n)}(\eta_n).$$

The values of r_n and $\|P_n\|_2$ can be determined with a little effort (see Exercise 14), and from this we get the final error statement

$$I(f) - G_n(f) = \frac{2^{2n+1}(n!)^4}{(2n+1)[(2n)!]^3} f^{(2n)}(\eta_n). \tag{5.13}$$

It is not immediately obvious how rapidly this decays, because of the competing effects of the factorials in the numerator and denominator. Thus, we appeal to Stirling's formula (Section 5.5) to gain some insight. We have

$$n! = C_n \sqrt{n}(n/e)^n$$

and

$$(2n)! = C_{2n}\sqrt{2n}(2n/e)^{2n},$$

so that

$$\frac{2^{2n+1}(n!)^4}{(2n+1)[(2n)!]^3} = \frac{2^{2n+1}C_n^4 n^2(n/e)^{4n}}{(2n+1)C_{2n}^3(2n)^{3/2}(2n/e)^{6n}} = K_n \frac{\sqrt{n}}{n+\frac{1}{2}} \left(\frac{e}{16n} \right)^{2n},$$

where $K_n = C_n^4/2^{3/2}C_{2n}^3$ lies between 0.7 and 1.04. Note that this estimate says that the error goes down exponentially with n, so long as the integrand is smooth enough. Just as an example, we note that for $n = 16$ we have

$$\frac{2^{2n+1}(n!)^4}{(2n+1)[(2n)!]^3} = K_n \frac{\sqrt{n}}{n+\frac{1}{2}} \left(\frac{e}{16n} \right)^{2n} = K_n \times 1.653181645 \times 10^{-64},$$

which goes a long way toward explaining the rapid convergence seen in Table 5.6.

Other Intervals, Other Rules   What if the integral is not posed over the interval $[-1,1]$? This is not an obstacle to the application of Gaussian quadrature, since we can apply a simple change of variable to rewrite any integral over $[a,b]$ as an integral over $[-1,1]$. We have

$$\int_a^b g(x)\, dx = \frac{1}{2}(b-a) \int_{-1}^{1} g\left(a + \frac{1}{2}(b-a)(z+1) \right) dz = \int_{-1}^{1} f(z)\, dz,$$

where

$$f(z) = \frac{1}{2}(b-a)g\left(a + \frac{1}{2}(b-a)(z+1) \right).$$

(See Exercise 15.) This affects the error estimate because of the factors of $\frac{1}{2}(b-a)$ in the change of variable. Instead of (5.13), we get

$$I(f) - G_n(f) = \frac{(b-a)^{2n+1}(n!)^4}{(2n+1)[(2n)!]^3} f^{(2n)}(\eta_n).$$

See Exercise 16 for the reduction of this via Stirling's formula to a more manageable expression.

EXAMPLE 5.9

Suppose we want to compute

$$I(f) = \int_1^2 x \ln x \, dx = 0.6362943610.$$

The change of variable gives us

$$x = 1 + \frac{1}{2}(z+1),$$

so the integral now is

$$I(f) = \frac{1}{2} \int_{-1}^1 \left(1 + \frac{1}{2}(z+1)\right) \ln\left(1 + \frac{1}{2}(z+1)\right) \, dz.$$

For simplicity's sake we define the new integrand as

$$F(z) = \frac{1}{2} \left(1 + \frac{1}{2}(z+1)\right) \ln\left(1 + \frac{1}{2}(z+1)\right).$$

The $n = 2$ Gaussian quadrature approximation is then computed quite simply:

$$G_2(F) = w_1 F(x_1) + w_2 F(x_2) = 0.6361494997.$$

■

There are also other Gaussian quadrature rules. In fact, any family of orthogonal polynomials will generate a Gaussian quadrature rule according to the following "template." The rules we have developed here are usually called Gauss-Legendre, because they are based on the Legendre family of orthogonal polynomials.

Theorem 5.5 (General Gaussian Quadrature Rule)   *Let $w(x) \geq 0$ be a weight function on the interval $[a,b]$, and let $\{\phi_k(x)\}$ be the family of orthogonal polynomials with respect to this weight function and this interval. Define the quadrature rule*

$$G_n(f) = \sum_{i=1}^n w_i^{(n)} f(x_i^{(n)})$$

for $x_i^{(n)}$ the roots of ϕ_n and $w_i^{(n)}$ given by

$$w_i^{(n)} = \int_a^b w(x) \left(\prod_{\substack{k=1 \\ k \neq i}}^n \frac{x - x_k^{(n)}}{x_i^{(n)} - x_k^{(n)}} \right) dx.$$

Then $G_n(p)$ is exact for all polynomials $p \in \mathcal{P}_{2n-1}$, and there exists $\xi_n \in [a,b]$ such that

$$\int_a^b w(x) f(x) \, dx - G_n(f) = \frac{1}{(2n)!} \left(\int_a^b \psi_n(x) \, dx \right) f^{(2n)}(\xi_n)$$

for all $f \in C^{2n}([a,b])$, where

$$\psi_n(x) = \prod_{k=1}^n (x - x_k^{(n)})^2.$$

Proof   See Exercise 17.   □

EXAMPLE 5.10

We can apply this theorem to construct a Gaussian quadrature rule for integrals of the form

$$I(f) = \int_0^\infty e^{-x} f(x) \, dx.$$

Here the weight function is $w(x) = e^{-x}$, so the orthogonal polynomial family that we need to use is the Laguerre family. To keep the task tractable, we will construct only the $n = 2$ rule. The second-order Laguerre polynomial is (see Section 4.10.2)

$$L_2(x) = \frac{1}{2}(x^2 - 4x + 2),$$

so the Gauss points are

$$x_1 = 2 - \sqrt{2} = 0.5857864376, \quad x_2 = 2 + \sqrt{2} = 3.414213562,$$

and the weights are

$$w_1 = \int_0^\infty e^{-x} \left(\frac{x - x_2}{x_1 - x_2} \right) dx = \frac{1}{4}(2 + \sqrt{2}) = 0.8535533903$$

and

$$w_2 = \int_0^\infty e^{-x} \left(\frac{x - x_1}{x_2 - x_1} \right) dx = \frac{1}{4}(2 - \sqrt{2}) = 0.1464466092.$$

We can check that these are the correct values by confirming that they can be used to integrate, exactly, any polynomial of degree ≤ 3 in $I(f)$. It suffices to use $f(x) = x^3$, so we look at

$$\int_0^\infty e^{-x} x^3 \, dx = 6.$$

Our Gauss-Laguerre approximation is

$$G_2(f) = w_1 f(x_1) + w_2 f(x_2)$$
$$= 0.8535533903 \times (0.5857864376)^3 + 0.1464466092 \times (3.414213562)^3$$
$$= 5.99999999$$

Considering the fixed precision of the computations for the Gauss points and weights, we ought to consider this adequate confirmation of our work. Note, by the way, that there is no explicit computation with the weight *function* in the evaluation of G_2; $w(x)$ does not appear in the formula, only $f(x)$.   ∎

EXERCISES

1. Apply Gaussian quadrature with $n = 4$ to approximate each of the following integrals.

(a) $I = \int_{-1}^1 (x^8 + 1) \, dx = 20/9$

(b) $I = \int_{-1}^1 \sin^2 \pi x \, dx = 1$

(c) $I = \int_{-1}^1 \ln(1 + x^2) \, dx = 2\ln 2 + \pi - 4$

(d) $I = \int_{-1}^1 e^{-x^2} \, dx = 1.493648266$

(e) $I = \int_{-1}^1 \frac{1}{1+x^4} \, dx = 1.733945974$

2. Apply Gaussian quadrature with $n = 4$ to approximate each of the following integrals. Remember that you have to do a change of variable to $[-1,1]$ first.

(a) $I = \int_0^1 x(1 - x^2) \, dx = \frac{1}{4}$

(b) $I = \int_0^1 \frac{1}{\sqrt{1+x^4}} \, dx = 0.92703733865069$

(c) $I = \int_0^1 \ln(1 + x) \, dx = 2\ln 2 - 1$

(d) $I = \int_0^1 \frac{1}{1+x^3} \, dx = \frac{1}{3}\ln 2 + \frac{1}{9}\sqrt{3}\pi$

(e) $I = \int_1^2 e^{-x^2} \, dx = 0.1352572580$

3. Show that (5.8) is both necessary and sufficient to make the quadrature exact for all polynomials of degree less than or equal to $2N - 1$.

4. Write a program that does Gaussian quadrature, using the weights and Gauss points given in the text. Apply this to each of the following integrals, and compare your results with those for the other quadrature methods in the exercises. Remember that you will have to do a change of variable if the interval is not $[-1,1]$. Be sure to run your program for more than one value of n.

(a) $f(x) = 1/(1+x^2)$, $[-5,5]$, $I(f) = 2\arctan(5)$

(b) $f(x) = x^2 e^{-x}$, $[0,2]$, $I(f) = 2 - 10e^{-2} = 0.646647168$

(c) $f(x) = \ln x$, $[1,3]$, $I(f) = 3\ln 3 - 2 = 1.295836867$

(d) $f(x) = \sqrt{1 - x^2}$, $[-1,1]$, $I(f) = \pi/2$

(e) $f(x) = e^{-x}\sin(4x)$, $[0,\pi]$, $I(f) = \frac{4}{17}(1 - e^{-\pi}) = 0.2251261368$

5. Let $P(x) = 6x^3 + 5x^2 + x$, and let $P_2(x) = 3x^2 - 1$ (this is the quadratic Legendre polynomial). Find linear polynomials $Q(x)$ and $R(x)$ such that $P(x) = P_2(x)Q(x) + R(x)$. Verify that $I(P) = I(R)$.

6. Let $P(x) = x^3 + x^2 + x - 1$, and repeat the preceding.

7. Let $P(x) = 3x^3 + x^2 - 6$, and repeat the preceding.

8. Verify that the weights for the $n = 2$ Gaussian quadrature rule satisfy formula (5.9).

9. Repeat the preceding for the $n = 4$ rule.

10. Show, by direct computation, that the $n = 2$ and $n = 4$ Gaussian quadrature rules are exact for the correct degree of polynomials.

11. The quadratic Legendre polynomial is $P_2(x) = (3x^2 - 1)/2$. Show that it is orthogonal (over $[-1,1]$) to all linear polynomials.

12. The cubic Legendre polynomial is $P_3(x) = (5x^3 - 3x)/2$. Show that it is orthogonal (over $[-1,1]$) to all polynomials of degree less than or equal to 2.

13. The quartic Legendre polynomial is $P_4(x) = (35x^4 - 30x^2 + 3)/8$. Show that it is orthogonal (over $[-1,1]$) to all polynomials of degree less than or equal to 3.

14. The first two Legendre polynomials are

$$P_0(x) = 1, \qquad P_1(x) = x,$$

and it can be shown that the others satisfy the recurrence relation

$$(n+1)P_{n+1}(x) = (2n+1)xP_n(x) - nP_{n-1}(x).$$

Use this to show (by induction) that the leading coefficient for the Legendre polynomials satisfies

$$k_n = \frac{(2n)!}{2^n (n!)^2}$$

and the 2-norm of the Legendre polynomials satisfies

$$\|P_n\|_2^2 = \frac{2}{2n+1}.$$

15. Let

$$I(f) = \int_a^b f(x)\, dx.$$

Show that the change of variable $x = a + \frac{1}{2}(b-a)(z+1)$ gives

$$I(f) = \int_{-1}^1 F(z)\, dz$$

for $F(z) = \frac{1}{2}(b-a)f(a + \frac{1}{2}(b-a)(z+1))$.

16. Show that the error for Gaussian quadrature applied to

$$I(f) = \int_a^b f(x)\, dx$$

is $O([(b-a)(e/8n)]^{2n})$.

17. Prove Theorem 5.5. (Hint: Simply generalize what was done in the text for the special case of

$$I(f) = \int_{-1}^1 f(x)\, dx.)$$

18. Once again, we want to consider the approximation of the natural logarithm function, this time using numerical quadrature. Recall that we have

$$\ln x = \int_1^x \frac{1}{t}\, dt.$$

Recall also that it suffices to consider $x \in [\frac{1}{2}, 1]$.
(a) How many grid points are required for 10^{-16} accuracy using the trapezoid rule? Simpson's rule?
(b) How many grid points are required if Gauss-Legendre quadrature is used?

19. Write a computer program that uses Gaussian quadrature for a specified number of points to compute the natural logarithm over the interval $[\frac{1}{2}, 1]$, to within 10^{-16} accuracy. Compare the accuracy of your routine to the intrinsic logarithm function on your system.

20. Write a brief essay, in your own words, of course, that explains the importance of the *linearity* of the integral and quadrature rule in the development of Gaussian quadrature.

5.7 EXTRAPOLATION METHODS

One of the most important ideas in computational mathematics is that we can take the information from a few approximations and use that to both estimate the error in the approximation and generate a significantly improved approximation. We have already seen examples of this with Aitken extrapolation for converging sequences (Section 3.10.1) and the corrected trapezoid rule (Section 5.2). In this section we will embark on a more detailed study of some of these ideas.

Given an integral $I(f)$, consider a generic quadrature rule, which we will denote by $I_n(f)$. Assume an error relationship of the form

$$I(f) - I_n(f) \approx Cn^{-p}. \tag{5.14}$$

(We note in passing that this is true for all three of the trapezoid ($p = 2$), midpoint ($p = 2$), and Simpson's ($p = 4$) rules, but *not* for Gaussian quadrature.) We want to use this approximate equality to do three different things:

1. Estimate the value of p.
2. Construct a computable estimate of the error $I(f) - I_n(f)$.
3. Construct an improved approximation of $I(f)$.

We can in fact do all three without a lot of work, and it leads us to a very efficient scheme for approximate integration.

Estimating p   We begin by applying (5.14) for three cases: n, $2n$, and $4n$ points.

$$I(f) - I_n(f) \approx Cn^{-p}$$
$$I(f) - I_{2n}(f) \approx C(2n)^{-p}$$
$$I(f) - I_{4n}(f) \approx C(4n)^{-p}.$$

Now, consider the ratio

$$r_{4n} = \frac{I_n - I_{2n}}{I_{2n} - I_{4n}},$$

where we have dropped the explicit argument f from the integration rule for notational simplicity. We have

$$r_{4n} = \frac{(I_n - I) + (I - I_{2n})}{(I_{2n} - I) + (I - I_{4n})} \approx \frac{-Cn^{-p} + C(2n)^{-p}}{-C(2n)^{-p} + C(4n)^{-p}}$$

$$= \frac{(2n)^{-p} - n^{-p}}{(4n)^{-p} - (2n)^{-p}} = \frac{2^{-p} - 1}{4^{-p} - 2^{-p}} = 2^p.$$

Thus, the computable ratio r_{4n} is approximately equal to 2^p. Therefore we can estimate the value of p by solving to get

$$p \approx \frac{\log r_{4n}}{\log 2}.$$

The importance of this ability to estimate p is that we can use it to verify that a computer program is working, or to estimate convergence rates when the integrand is not smooth enough to apply our error theory. Let's look at some examples that illustrate each of these ideas.

EXAMPLE 5.11

Consider the problem of evaluating the integral

$$I = \int_0^1 e^{-x^2} dx.$$

This integral cannot be evaluated in closed form—there is no simple antiderivative for e^{-x^2}—so we cannot compare our estimates to an exact solution. But we can observe that the integrand is infinitely smooth, so the error theory for Simpson's rule applies; thus approximations using Simpson's rule ought to yield that p is nearly 4. Table 5.7 shows the result of doing this

TABLE 5.7   Estimation of p

n	$S_n(f)$	p
4	0.746855379791E+00	NA
8	0.746826120527E+00	NA
16	0.746824257436E+00	3.973123
32	0.746824140607E+00	3.995232
64	0.746824133300E+00	3.998911
128	0.746824132843E+00	3.999734
256	0.746824132814E+00	3.999953
512	0.746824132813E+00	3.999232
1024	0.746824132812E+00	4.002964
2048	0.746824132812E+00	4.063215

computation, and we see that

$$r_{256} = \frac{0.746824133300 - 0.746824132843}{0.746824132843 - 0.746824132814} = 15.76...,$$

which implies that $p \approx 4$, as expected. If this had not been the case, we would have for the most part had one of two possible conclusions to draw: (1) The integrand is not as smooth as we thought it was, or (2) the program is not working properly. There is a third possibility—that the function is periodic—in which case Simpson's rule will be "super-accurate," and thus the error may not decrease very rapidly at all, because it starts out very small. (See Section 5.8.1 and Exercises 8 and 9 in this section.) Table 5.7 shows the complete set of computations for n ranging from 4 to 2048. (Note that we begin to lose accuracy at the end of the computation: The estimated value of p begins to wander away from 4. This occurs because $S_n(f)$ is so accurate as an approximation to $I(f)$ that the computations in estimating p are badly corrupted by rounding error for large n. The student ought to think about how this happens.) ∎

EXAMPLE 5.12

Next, consider the integral

$$I(f) = \int_0^1 \sqrt{x}\, dx = \frac{2}{3}.$$

Since the integrand does not have a bounded second derivative on the interval of integration, we cannot expect the trapezoid rule or Simpson's rule to be as accurate as theory predicts, something that is borne out by computational experiment (see Table 5.8). But this only begs the question, How accurate are these methods when applied to this problem? In other words, what sort of mesh exponent *do* we get?

Using the computation outlined above, we can estimate the exponent p for this example based on the data in Table 5.8. For both the trapezoid rule and Simpson's rule, we get that $p \approx 1.50$ (for this example, only in general, of course, we expect $p = 4$ for Simpson's rule and $p = 2$ for the trapezoid rule). See Table 5.9. ∎

TABLE 5.8 Trapezoid and Simpson's Rules Applied to $f(x) = \sqrt{x}$

n	$T_n(f)$	$S_n(f)$
2	0.6035533906E+00	0.6380711875E+00
4	0.6432830462E+00	0.6565262648E+00
8	0.6581302216E+00	0.6630792801E+00
16	0.6635811969E+00	0.6653981886E+00
32	0.6655589363E+00	0.6662181827E+00
64	0.6662708114E+00	0.6665081031E+00
128	0.6665256573E+00	0.6666106059E+00
256	0.6666165490E+00	0.6666468462E+00
512	0.6666488815E+00	0.6666596591E+00
1024	0.6666603622E+00	0.6666641891E+00
2048	0.6666644336E+00	0.6666657907E+00
4096	0.6666658761E+00	0.6666663570E+00

TABLE 5.9 Estimation of p from Data in Table 5.8

n	p (trapezoid rule)	p (Simpson's rule)
8	0.1420027799E+01	0.1420027799E+01
16	0.1445602214E+01	0.1445602214E+01
32	0.1462662030E+01	0.1462662030E+01
64	0.1474156297E+01	0.1474156297E+01
128	0.1481998892E+01	0.1481998892E+01
256	0.1487405107E+01	0.1487405107E+01
512	0.1491159895E+01	0.1491159895E+01
1024	0.1493781623E+01	0.1493781623E+01
2048	0.1495619062E+01	0.1495619062E+01
4096	0.1496910216E+01	0.1496910216E+01

Error Estimation and an Improved Approximation To directly estimate the error, we note that our assumptions imply that

$$I - I_{2n} \approx C(2n)^{-p} = 2^{-p}\left(Cn^{-p}\right) \approx 2^{-p}(I - I_n)$$

and we can then solve the approximate equality to get

$$I \approx \frac{2^p I_{2n} - I_n}{2^p - 1}.$$

We then define

$$R_{2n} = \frac{2^p I_{2n} - I_n}{2^p - 1}, \tag{5.15}$$

TABLE 5.10   Richardson Extrapolation Applied to $f(x) = e^x$, $[a,b] = [0,1]$

n	$R_n(f)$	$|I(f) - R_n(f)|$	$|I(f) - T_n(f)|$	$|E_n|$
8	1.718272532150	0.929631E–05	0.223676E–02	0.224606E–02
16	1.718281246223	0.582236E–06	0.559300E–03	0.559882E–03
32	1.718281792050	0.364088E–07	0.139832E–03	0.139868E–03
64	1.718281826183	0.227585E–08	0.349584E–04	0.349607E–04
128	1.718281828317	0.142247E–09	0.873962E–05	0.873977E–05
256	1.718281828450	0.889089E–11	0.218491E–05	0.218492E–05
512	1.718281828458	0.553779E–12	0.546227E–06	0.546228E–06
1024	1.718281828459	0.364153E–13	0.136557E–06	0.136557E–06
2048	1.718281828459	0.266454E–14	0.341392E–07	0.341392E–07

which we call *Richardson's*[9] *extrapolated value* and

$$E_{2n} = R_{2n} - I_{2n} = \frac{I_n - I_{2n}}{2^p - 1},$$ (5.16)

which is our *computable estimate of the error*.

EXAMPLE 5.13

Refer back to the data in Table 5.7. From this we have

$$S_8(f) = 0.746826120527, \qquad S_{16}(f) = 0.746824257436,$$

so that

$$R_{16}(f) = 0.7468241335$$

and

$$E_{16}(f) = 0.12421 \times 10^{-6}.$$

∎

EXAMPLE 5.14

We return once again to our example involving $\int_0^1 e^x \, dx$, using the trapezoid rule as the basic integration scheme. Table 5.10 shows the extrapolated approximation and the estimated error,

---

[9]Lewis Fry Richardson (1881–1953) was born in Newcastle upon Tyne in Great Britain and attended several different schools before finishing his education at King's College, Cambridge, in 1903. His professional career likewise spanned a number of different posts in industry, academia, and government science laboratories. He was the first person to suggest using mathematical techniques to predict the weather, by solving the fluid equations that would govern temperature, air pressure, etc. He did this in the 1920s, long before the development of modern high-speed computers, and it was because of this that he developed the notion of extrapolation methods for the accurate numerical approximation of solutions based on cruder approximations.

based on (5.15) and (5.16), along with the actual error. We note that the estimated error E_n tracks very well with the actual trapezoid error, $I(f) - T_n(f)$. Note that the error $I(f) - R_n(f)$ decreases by a factor of 16 as the mesh size is halved, suggesting that the $O(h^2)$ trapezoid rule has been improved to $O(h^4)$ by the extrapolation, a conclusion that can in fact be made rigorous; see Section 5.8.1.   ■

We can also use the extrapolation error E_{2n} as a computable estimate of the error in a computation, and consequently use this to decide when an approximation is sufficiently accurate.

EXAMPLE 5.15

Consider the integral

$$I(f) = \int_0^1 e^{-x^3} dx.$$

There is no closed-form antiderivative for the integrand, so there is no way to compare our approximation with an exact solution. However, if we compute Simpson's rule values for this integral, we get

$$I_{16}(f) = 0.807512351889, \qquad I_{32}(f) = 0.807511254956,$$

for which the estimated error is

$$E_{32} = 0.731 \times 10^{-7}.$$

We thus can say, with some confidence, that $|I(f) - I_{32}(f)| \leq 10^{-7}$, because the estimated error is that small. We expect, of course, that the extrapolated value $R_{32} = 0.807511182$ would be even more accurate, so we use E_{2n}, which is a computable estimate of the error $I(f) - I_{2n}(f)$, to tell us when to accept the extrapolated value $R_{2n}(f)$ as the approximate value of the integral.   ■

EXERCISES

1. Apply Simpson's rule with $h = \frac{1}{2}$ and $h = \frac{1}{4}$ to approximate the integral

$$I = \int_0^1 \frac{1}{\sqrt{1+x^4}} dx = 0.92703733865069,$$

and use Richardson extrapolation to obtain the improved value of the approximation. What is the *estimated* value of the error in S_4, compared with the actual error?

2. Repeat the preceding, for

$$I = \int_0^1 x(1 - x^2) \, dx = \frac{1}{4}.$$

3. Repeat the preceding, for

$$I = \int_0^1 \ln(1 + x) \, dx = 2\ln 2 - 1.$$

4. Repeat the preceding, for

$$I = \int_0^1 \frac{1}{1+x^3}\, dx = \frac{1}{3}\ln 2 + \frac{1}{9}\sqrt{3}\pi.$$

5. Repeat the preceding, for

$$I = \int_1^2 e^{-x^2}\, dx = 0.1352572580.$$

6. Write a trapezoid rule program to compute the value of the integral

$$I = \int_0^1 e^{-x^2}\, dx.$$

Take h small enough to justify a claim of accuracy to within 10^{-6}, and explain how the claim is justified. (There are several ways of doing this.)

7. Define

$$I(f) = \int_0^1 e^x\, dx$$

and consider the approximation of this integral using Simpson's rule together with extrapolation. By computing a sequence of approximate values S_2, S_4, S_8, etc., determine experimentally the accuracy of the extrapolated rule

$$R_{2n} = (16 S_{2n} - S_n)/15.$$

8. Consider the integral

$$I(f) = \int_0^\pi \sin^2 x\, dx = \frac{1}{2}\pi.$$

Write a trapezoid rule or Simpson's rule program to approximate this integral, using Richardson extrapolation to improve the approximations, and comment on your results. In particular, comment on the rate at which the error decreases as h decreases, and on the amount of improvement obtained by extrapolation.

9. Repeat the preceding, this time for the integral

$$I(f) = \int_0^{3\pi/4} \sin^2 x\, dx = \frac{1}{4} + \frac{3}{8}\pi.$$

10. The following table of values supposedly comes from applying the midpoint rule to a smooth, non-periodic function. Can we use this data to determine whether or not the program is working properly? Explain.

n	$M_n(f)$
4	-0.91595145
8	-0.95732875
16	-0.97850187
32	-0.98921026
64	-0.99459496
128	-0.99729494
256	-0.99864683
512	-0.99932326
1024	-0.99966159

11. The error function, defined as

$$\mathrm{erf}(x) = \frac{2}{\sqrt{\pi}} \int_0^x e^{-t^2}\, dt,$$

is an important function in probability theory and heat conduction. Use Simpson's rule with extrapolation to produce values of the error function that are accurate to within 10^{-8} for x ranging from 0 to 5 in increments of $1/4$. Check your values against the intrinsic error function on your computer or by looking at a set of mathematical tables such as those in [1].

12. Bessel functions appear in the solution of heat conduction problems in a circular or cylindrical geometry, and can be defined as a definite integral,

$$J_k(x) = \frac{1}{\pi} \int_0^\pi \cos(x\sin t - kt)\, dt.$$

Use Simpson's rule plus extrapolation to produce values of J_0 and J_1 that are accurate to within 10^{-8} over the interval $[0,6]$ in increments of $1/4$. Check your values by looking at a set of tables such as those in [1].

13. Apply the trapezoid rule to approximate the integral

$$I = \int_0^{\pi^2/4} \sin\sqrt{x}\, dx = 2.$$

Use Richardson extrapolation to estimate how rapidly your approximations are converging, and comment on your results in light of the theory for the trapezoid rule.

14. Show that, for any function f,

$$S_2(f) = (4T_2(f) - T_1(f))/3.$$

Comment on the significance of this result in light of this section.

5.8 SPECIAL TOPICS IN NUMERICAL INTEGRATION

5.8.1 Romberg Integration

If we restrict our attention now to the case of the trapezoid rule, we can make much of the previous section on extrapolation methods more precise and, at the same time, develop a very accurate recursive procedure for approximating integrals. A fundamental result is the Euler-Maclaurin formula,[10] which we now state without proof.

Theorem 5.6 (Euler-Maclaurin Formula)   If f is sufficiently differentiable, then for any $N > 0$ there exists a set of constants $c_k, 1 \leq k \leq N+1$, such that, for some $\xi \in [a,b]$, the error in the trapezoid rule satisfies

$$
\begin{aligned}
I(f) - T_n(f) \; = \;\; & \gamma_1 h^2 + \gamma_2 h^4 + \cdots + \gamma_N h^{2N} \\
& + c_{N+1}(b-a)h^{2N+2} f^{(2N+2)}(\xi),
\end{aligned}
\tag{5.17}
$$

where $\gamma_k = c_k(f^{(2k-1)}(b) - f^{(2k-1)}(a))$.

The significance of the Euler-Maclaurin formula is that it allows us to write the error in the trapezoid rule as a series of powers of the mesh spacing h, and then use this series to derive new quadrature rules that are more and more accurate. Note that the series expansion of the error can be carried out as far as f is differentiable, but has to be terminated; in general, it cannot be taken as an infinite series.

---

[10]Leonhard Euler (1707–1783) was one of the two greatest mathematicians of the post-Newton age, the other being Carl Friedrich Gauss. Euler was born in Basel, Switzerland, and educated at the University of Basel, at first with an eye towards following in his father's career as a minister. With the assistance of his tutor and mentor Johann Bernoulli, however, he was able to convince his father to let him pursue a career in mathematics. In 1727 Euler joined the St. Petersburg Academy of Sciences in Russia, where he remained until 1741, at which time he joined the Berlin Academy of Sciences at the invitation of the Prussian king, Frederick the Great. After some disputes with the monarch, Euler left Berlin in 1766 and returned to St. Petersburg.

Euler's contributions to mathematics are nearly unmatched in their breadth. He published an enormous amount of material in a wide variety of areas, including infinite series, special functions (a field of study that he practically invented), number theory, complex variables, and hydrodynamics. His name is attached to countless results in mathematics, from Euler's formula relating the trigonometric functions to complex exponentials, to the Euler-Cauchy differential equations, to Euler's formula relating the number of sides, edges, and vertices in a polyhedron. His influence on notation is still felt today, as it was Euler who introduced e, π, and $i = \sqrt{-1}$ into the literature as standard symbols, in addition to the use of Σ for denoting summations, and cos and sin for the cosine and sine of an angle. Euler's collected works, published between 1911 and 1975, encompass 72 volumes!

Euler's work on what we call the Euler-Maclaurin formula appears to have been done in about 1738, while he was in St. Petersburg. Although Maclaurin had first discovered the formula (see below), Euler's analysis of it went far beyond what the Scotsman had done.

Colin Maclaurin (1698–1746) was born and lived almost his entire life in Scotland. Educated at Glasgow University, he was professor of mathematics at Aberdeen from 1717 to 1725 and then went to Edinburgh. He worked in a number of areas of mathematics and is credited with writing one of the first textbooks based on Newton's calculus, *Treatise of fluxions* (1742). The Maclaurin series appears in this book as a special case of Taylor's series.

Maclaurin developed the Euler-Maclaurin formula independently of Euler, in about 1737, but did not publish it prior to including it in *Treatise of fluxions*.

The first few constants are

$$c_1 = -\frac{1}{12}, \qquad c_2 = \frac{1}{720}, \qquad c_3 = -\frac{1}{30,240}.$$

There is a general formula relating these to the so-called Bernoulli numbers.

One consequence of the Euler-Maclaurin formula is that the trapezoid rule is shown to be extraordinarily accurate when applied to periodic functions over full periods. In this case, the derivatives at the endpoints will be equal, and thus (5.17) becomes

$$I(f) - T_n(f) = c_{N+1}(b-a)h^{2N+2}f^{(2N+2)}(\xi), \tag{5.18}$$

where N is arbitrary, restrained only by the smoothness of f. See Exercise 10 for an illustration of this.

Note that if we look at the $N = 1$ case, we have

$$I(f) - T_n(f) = -\frac{1}{12}\left(f'(b) - f'(a)\right)h^2 + c_2(b-a)h^4 f^{(4)}(\xi),$$

which shows that the corrected trapezoid rule is $O(h^4)$ accurate, which we saw experimentally in Section 5.2.

To construct an even more accurate quadrature rule from the error expansion, simply write (5.17) for twice as many subintervals (replace h by $h/2$):

$$I(f) - T_{2n}(f) = \frac{1}{4}\gamma_1 h^2 + \frac{1}{16}\gamma_2 h^4 + \cdots + \frac{1}{4^N}\gamma_N h^{2N} + O(h^{2N+2}). \tag{5.19}$$

If we multiply (5.19) by 4, subtract it from (5.17), and then solve for $I(f)$, we get

$$I(f) - \frac{4T_{2n}(f) - T_n(f)}{3} = b_2 h^4 + b_3 h^6 + \cdots + b_N h^{2N} + O(h^{2N+2}), \tag{5.20}$$

where $b_k = -\frac{1}{3}\left(1 - 4^{1-k}\right)c_k(f^{(2k-1)}(b) - f^{(2k-1)}(a))$.

Note what we have done here. The value $\frac{1}{3}(4T_{2n}(f) - T_n(f))$ is nothing more than Richardson's extrapolated value (5.15) for $T_n(f)$, and the expression (5.20) shows that it is $O(h^4)$ accurate. But there is no reason to stop here. We rewrite (5.20) as

$$I(f) - R_{2n}(f) = b_2 h^4 + b_3 h^6 + \cdots + b_N h^{2N} + O(h^{2N+2}), \tag{5.21}$$

for which we also then have

$$I(f) - R_{4n}(f) = \frac{1}{16}b_2 h^4 + \frac{1}{128}b_3 h^6 + \cdots + \frac{1}{4^N}b_N h^{2N} + O(h^{2N+2}), \tag{5.22}$$

so that, multiplying (5.22) now by 16, subtracting from (5.21), and solving for $I(f)$ yields

$$I(f) - \frac{16R_{4n} - R_{2n}}{15} = a_3 h^6 + \cdots + a_k h^{2k} + \cdots + a_N h^{2N} + O(h^{2N+2}),$$

where $a_k = -\frac{1}{15}(1 - 4^{2-k})b_k$. Thus $\frac{1}{15}(16R_{4n} - R_{2n})$ can be viewed as yet another extrapolated approximation for the integral, and this one is $O(h^6)$ accurate.

There is no reason not to keep going. Each step of the *extrapolation* process yields a new quadrature method that is more accurate than the preceding one. The process can be systematized to yield an algorithm known as *Romberg integration*. Continuing much further requires that we define some notation.

Let $T_n^{(0)}(f)$ denote the trapezoid rule values; this is the first column of what will become a triangular array. We denote the second column by $T_n^{(1)}(f)$. These values are computed from the trapezoid values according to the Richardson extrapolation formula:

$$T_{2n}^{(1)}(f) = R_{2n}(f) = \frac{4T_{2n}^{(0)}(f) - T_n^{(0)}(f)}{3}.$$

Note that the indexing means that the second column will be one entry shorter than the first. Generally, then, each column is computed from the preceding one according to the formula

$$T_{2n}^{(j+1)}(f) = \frac{4^{j+1}T_{2n}^{(j)}(f) - T_n^{(j)}(f)}{4^{j+1} - 1}, \tag{5.23}$$

and the array looks like

$$
\begin{array}{ccccc}
T_n^{(0)}(f) \\
T_{2n}^{(0)}(f) & T_{2n}^{(1)}(f) \\
T_{4n}^{(0)}(f) & T_{4n}^{(1)}(f) & T_{4n}^{(2)}(f) \\
\vdots & \vdots & \vdots & \ddots \\
T_{2^k n}^{(0)}(f) & T_{2^k n}^{(1)}(f) & \cdots & \cdots & T_{2^k n}^{(k)}(f)
\end{array}
$$

If the integrand is smooth enough, each step across a row of the array eliminates another power of h^2 from the error expansion; i.e., the first column has error that is $O(h^2)$, the second column has error that is $O(h^4)$, the third column is $O(h^6)$, etc. Meanwhile, going *down* the array decreases h by a factor of 2 for each row: if we start with $h = 1/2$, then in the sixth row we have $h = 1/64$. The upshot of all this is that the diagonal elements of the Romberg array are very accurate approximations to the integral. A formal statement is the following, which we present without proof (see p. 328 of [3]).

Theorem 5.7 (Romberg Integration) *Assume that f is sufficiently differentiable on $[a,b]$, and let $\Theta_k(f)$ be the kth diagonal element of the Romberg array: $\Theta_k(f) = T_{2^k n}^{(k)}(f)$. If the mesh size for the initial trapezoid rule $T_n^{(0)}$ is h, then*

$$I(f) - \Theta_k(f) = O\left(4^{-k}h^{2k+2}\right).$$

In addition, at the same time that Romberg integration is producing accurate approximations, it can also be used to estimate the error, since for each entry in the

Romberg array we can compute an estimate of the error using Richardson extrapolation:

$$E_{2n}^{(k)} = \frac{T_n^{(k)} - T_{2n}^{(k)}}{4^k - 1}.$$

(This assumes, of course, that the integrand is sufficiently smooth.) This can be used to stop the Romberg process when the estimated error is sufficiently small.

Note that the bulk of the computational work in Romberg integration is involved in computing the first column—the trapezoid rule values—since the computation of the subsequent columns, done by (5.23), involves only a few simple operations. The ultimate efficiency therefore depends on our ability to rapidly compute the trapezoid rule values recursively, i.e., compute $T_{2n}^{(0)}(f)$ from $T_n^{(0)}(f)$ without wasted effort. Fortunately this is very easy, as the next result shows.

Theorem 5.8 *Let $T_n(f)$ denote the trapezoid rule applied to a given function f over a given interval $[a,b]$, using n subintervals, with uniform mesh spacing $h = (b-a)/n$. Then we can compute $T_{2n}(f)$, the trapezoid rule using twice as many subintervals, according to*

$$T_{2n}(f) = \frac{1}{2}T_n(f) + \left(\frac{b-a}{2n}\right)\sum_{j=1}^{n} f(a + (2j-1)(b-a)/(2n)). \qquad (5.24)$$

Proof The key step is to recognize that we can write the trapezoid rule for n subintervals as

$$T_n(f) = \left(\frac{b-a}{n}\right)\left(\frac{1}{2}f(a) + \frac{1}{2}f(b) + \sum_{i=1}^{n-1} f(a + i(b-a)/n)\right).$$

Thus, the rule for $2n$ subintervals is

$$T_{2n}(f) = \left(\frac{b-a}{2n}\right)\left(\frac{1}{2}f(a) + \frac{1}{2}f(b) + \sum_{i=1}^{2n-1} f(a + i(b-a)/(2n))\right).$$

Now manipulate:

$$T_{2n}(f) = \left(\frac{b-a}{2n}\right)\left(\frac{1}{2}f(a) + \frac{1}{2}f(b) + \sum_{i=1}^{2n-1} f(a + i(b-a)/(2n))\right)$$

$$= \frac{1}{2}\left(\frac{b-a}{n}\right)\left(\frac{1}{2}f(a) + \frac{1}{2}f(b) + \sum_{i=2,4,6,\ldots}^{2n-1} f(a + i(b-a)/(2n))\right.$$

$$\left. + \sum_{i=1,3,5,\ldots}^{2n-1} f(a + i(b-a)/(2n))\right)$$

$$= \frac{1}{2} \left(\frac{b-a}{n} \right) \left(\frac{1}{2} f(a) + \frac{1}{2} f(b) \sum_{j=1}^{n-1} f(a + 2j(b-a)/(2n)) \right.$$
$$\left. + \sum_{j=1}^{n} f(a + (2j-1)(b-a)/(2n)) \right)$$
$$= \frac{1}{2} \left(\frac{b-a}{n} \right) \left(\left[\frac{1}{2} f(a) + \frac{1}{2} f(b) + \sum_{j=1}^{n-1} f(a + j(b-a)/n) \right] \right.$$
$$\left. + \sum_{j=1}^{n} f(a + (2j-1)(b-a)/(2n)) \right)$$
$$= \frac{1}{2} T_n(f) + \left(\frac{b-a}{2n} \right) \left(\sum_{j=1}^{n} f(a + (2j-1)(b-a)/(2n)) \right),$$

and we are done. $\square$

The point of this result is that we can compute the entire first column of the Romberg array using the minimum number of function evaluations. A naive implementation, with $T_{2n}^{(0)}(f)$ computed from scratch, would require that we recompute n values of f that we had already computed in finding $T_n^{(0)}(f)$. As a result of this theorem, we can compute $T_{2^k n}^{(k)}(f)$ using no more function evaluations than for the ordinary trapezoid rule, yet we achieve much more accuracy.

Table 5.11 shows the application of Romberg integration to our standard example of integrating e^x over the interval $[0,1]$. Note the extremely rapid decay of the error.

We close this section with an algorithm (in outline form) for Romberg integration (see Algorithm 5.2).

EXAMPLE 5.16

To illustrate the computation with a minimum of extraneous effort, let's refer to Table 5.12, which gives trapezoid rule values (i.e., $T_n^{(0)}(f)$ values) for approximating

$$I(f) = \int_0^1 \frac{dx}{1+x^4} = 0.8669729871.$$

TABLE 5.11 Romberg Integration of $f(x) = e^x$, $[a,b] = [0,1]$

$n = 2^k$	$\Theta_k(f)$	$I(f) - \Theta_k(f)$
1	1.85914091422952	0.14085908577048E+00
2	1.71886115187659	0.57932341754729E–03
4	1.71828268792476	0.85946571215523E–06
8	1.71828182879453	0.33548497313518E–09
16	1.71828182845908	0.32640556923980E–13
32	1.71828182845905	0.44408920985006E–15
64	1.71828182845905	–0.44408920985006E–15

Algorithm 5.2 Romberg Integration

---

1. Compute $T_n^{(0)}(f) = T_n(f)$; this is the initial trapezoid rule computation, using n subintervals. Often $n = 1$, but this is not required.

2. For k from 1 to N

(a) Compute $T_{2^k n}(f) = T_{2^k n}^{(0)}(f)$; this is the first entry on a new row of the Romberg array.

(b) Extrapolate across the row: for j from 0 to $k-1$,

i. Compute $T_{2^k n}^{(j+1)}(f) = (4^{j+1} T_{2^k n}^{(j)}(f) - T_{2^{k-1} n}^{(j)}(f))/(4^{j+1} - 1)$.

Programming Hint: Note that it is easy to encode this using only enough storage for two rows in the Romberg array; we don't need to store the entire array. In fact, if we are careful, we can get away with storing only a single row of the array, and over-writing it with the new values as they are computed.

TABLE 5.12 Trapezoid Rule Integration of $f(x) = (1 + x^4)^{-1}$, $[a,b] = [0,1]$

n	$T_k(f)$
1	0.7500000000
2	0.8455882353
4	0.8576188079

The first extrapolation produces the first Romberg value:

$$\Theta_1(f) = T_2^{(1)}(f) = \frac{4T_2^{(0)}(f) - T_1^{(0)}(f)}{3} = 0.8774509800.$$

It takes two extrapolations to produce the second Romberg value:

$$T_4^{(1)}(f) = \frac{4T_4^{(0)}(f) - T_2^{(0)}(f)}{3} = 0.8616289989,$$

$$\Theta_2(f) = T_4^{(2)}(f) = \frac{16T_4^{(1)}(f) - T_2^{(1)}(f)}{15} = 0.8605742004.$$

The Richardson error estimate for $T_4^{(1)}(f)$ and $T_2^{(1)}(f)$ is given by

$$E_4^{(1)}(f) = \frac{T_2^{(1)}(f) - T_4^{(1)}(f)}{15} = 0.0010547987.4$$

Thus we are confident that the Romberg value $\Theta_2(f)$ is accurate to within about 10^{-3}, even though it was produced with only five function evaluations. ∎

5.8.2 Quadrature with Nonsmooth Integrands

So far we have assumed, in most of our developments, that the function being integrated (the *integrand*) was as smooth as we needed it to be. But this will not always be the case, and we need to understand the implications of this.

TABLE 5.13   **Integration of** $f(x) = x^{1/2}$, $[a,b] = [0,1]$

n	$T_n(f)$	$S_n(f)$	$G_n(f)$
4	0.643283046243E+00	0.656526264793E+00	0.667827645375E+00
8	0.658130221624E+00	0.663079280085E+00	0.666835580100E+00
16	0.663581196877E+00	0.665398188628E+00	0.666689631499E+00
32	0.665558936279E+00	0.666218182746E+00	0.666669667368E+00
64	0.666270811379E+00	0.666508103078E+00	0.666667050398E+00
128	0.666525657297E+00	0.666610605936E+00	0.666666715190E+00
256	0.666616548977E+00	0.666646846203E+00	0.666666672768E+00
512	0.666648881550E+00	0.666659659074E+00	0.666666667431E+00
1024	0.666660362219E+00	0.666664189109E+00	0.666666666762E+00
2048	0.666664433593E+00	0.666665790718E+00	0.666666666679E+00

We begin with an example. Let $f(x) = \sqrt{x}$ and consider

$$I(f) = \int_0^1 \sqrt{x}\, dx = \frac{2}{3}.$$

Let's apply the trapezoid, Simpson, and Gaussian quadrature rule to this integral; the results are summarized in Table 5.13. Note that none of the three methods is converging to the exact value as rapidly as we would expect, based on our theory and previous examples. Why?

Recall that all of our error estimates include a factor that is based on the evaluation of some derivative of the integrand at an unspecified point in the interval of integration. For the trapezoid rule, we require that f be twice continuously differentiable, for Simpson's rule we require that it be four times continuously differentiable, and for the n-point Gaussian rule we require that f be $2n$ times continuously differentiable. However, on the interval $[0,1]$, $f(x) = \sqrt{x}$ does not even have a continuous *first* derivative. The practical effect of this is that the contribution to the error estimate from the derivative of f can become arbitrarily large as we take more and more points. This growth will not be large enough to prevent convergence to the correct answer, but it will be large enough to destroy the expected convergence *rate*. Note that this is not a violation of our carefully developed error theory; the integrand in this case does not satisfy the hypotheses of the error theorems, so they do not apply.

An integrand that exhibits the kind of behavior as in our example is called *singular*, with $x = 0$ being the specific *point of singularity*. In this particular case we can actually remove the problem by making the change of variable $x = u^2$ in the integral. However, this kind of trick will not always work,[11] so we must be prepared to deal with singular integrands in a more sophisticated manner.

---

[11] However, the student is advised to always consider a clever change of variable as a way to avoid a singularity.

To be specific, we consider the integral

$$I(f) = \int_a^b f(x)\,dx,$$

where f is assumed to be smooth on the open interval (a,b) but to have some kind of singular (nonsmooth) behavior at one or both of the endpoints. Introduce the change of variable

$$x = \phi(t), \qquad \phi(t) = a + \frac{b-a}{\gamma}\Psi(t), \qquad \Psi(t) = \int_{-1}^t \psi(u)\,du, \qquad (5.25)$$

where

$$\psi(t) = \exp\left(-\frac{c}{1-t^2}\right), \qquad \gamma = \Psi(1),$$

and $c > 0$ is arbitrary.

Now this looks rather imposing, no doubt, so let's go over it a bit more carefully.

The first point to make is that ψ has been constructed to be extremely smooth near $t = \pm 1$; note that the argument to the exponential goes to $-\infty$ as $t \to \pm 1$, thus ψ and *all* its derivatives will vanish at the endpoints. This means that Ψ and hence ϕ will be very smooth at the endpoints—smooth enough to compensate for almost any nonsmooth behavior in f. Second, the change of variables from x to t will map the original interval $[a,b]$ to $[-1,1]$. In fact, the new integral becomes (see Exercise 5)

$$I(f) = \int_a^b f(x)\,dx = \int_{-1}^1 f(\phi(t))\phi'(t)\,dt.$$

Since all derivatives of ϕ vanish at the new endpoints, it follows from the Euler-Maclaurin formula that the trapezoid rule will be a very accurate quadrature rule to use here. The involved nature of the change of variable makes the application of the trapezoid rule appear difficult, but it really isn't. Let $\{t_k\}$ be a set of $n+1$ equally spaced grid points on $[-1,1]$, i.e.,

$$-1 = t_0 < t_1 < t_2 < \cdots < t_n = 1$$

with $t_k = -1 + 2/n$, $0 \le k \le n$. Then the trapezoid rule applied to the transformed integral becomes (remember that the integrand vanishes at the endpoints)

$$I(f) \approx I_n(f) = \frac{b-a}{2}\sum_{k=1}^{n-1} w_k f(x_k),$$

where

$$x_k = \phi(t_k), \qquad w_k = \frac{4}{n\gamma}\exp\left(-\frac{c}{1-t_k^2}\right).$$

The evaluation of the weights is a straightforward calculation with the exponential function. The evaluation of the x_k values requires the computation of $\phi(t_k)$, which is itself in terms of an integral. However, the values of that integral can be computed once and saved, much as the weights and Gauss points from Gaussian quadrature, and thus this is only a momentary difficulty. Alternatively, we could use the approximation techniques from Chapter 4 to construct a highly accurate polynomial approximation to Ψ, and simply evaluate that as needed. Table 5.14 shows the values of $\Psi(t)/\Psi(1)$ for the $n = 4, 8$, and 16 cases, computed using Romberg integration.

EXAMPLE 5.17

To illustrate the method, consider the following example:

$$I(f) = \int_0^1 (\log 1/x)^{1/2} \, dx = \frac{\sqrt{\pi}}{2} = 0.8862269255.$$

This has singularities at both endpoints, since $\log 0$ is undefined and the square root will have undefined derivatives at $x = 1$ since $\log 1 = 0$. The $n = 4$ version of this quadrature is written (using $c = 4$) quite simply as

$$I_4(f) = \frac{1}{2} \left(\frac{1}{\gamma} \right) \exp\left(\frac{-4}{1-t_1^2} \right) \left(\log \frac{1}{x_1} \right)^{1/2} + \frac{1}{\gamma} \exp\left(\frac{-4}{1-t_2^2} \right) \left(\log \frac{1}{x_2} \right)^{1/2}$$
$$+ \frac{1}{\gamma} \exp\left(\frac{-4}{1-t_3^2} \right) \left(\log \frac{1}{x_3} \right)^{1/2}.$$

Now we have (from (5.25) and Table 5.14)

$$x_1 = \Psi(t_1)/\Psi(1) = 0.3175495761647776 \times 10^{-1}$$
$$x_2 = \Psi(t_2)/\Psi(1) = 0.5000000000000000$$
$$x_3 = \Psi(t_3)/\Psi(1) = 0.9682450390677892$$

and

$$\gamma = \int_{-1}^1 \exp\left(\frac{-4}{1-t^2} \right) dt = 1.4059716861223 \times 10^{-2},$$

so the approximation becomes

$$I_4(f) = \left(\frac{0.5}{1.4059716861223 \times 10^{-2}} \right) \times (0.00482794999383144 \times 1.85733853145623$$
$$+ 0.0183156388887342 \times 0.8325546111576978$$
$$+ 0.00482794999383144 \times 0.179638760313930) = \frac{0.5 \times 0.0250831941163514}{1.4059716861223 \times 10^{-2}}$$
$$= 0.892023444139597,$$

for which the error is approximately 0.58×10^{-2}. Table 5.15 shows the result of applying the algorithm to this example (with $c = 4$) for larger values of n. Note that the high accuracy of the trapezoid rule for this kind of example is borne out. Note also that the accuracy stagnates a bit for the finer meshes. This is probably due to inaccuracies in the computation of the $\Psi(t_k)/\Psi(1)$ values; to avoid this kind of problem we would need to compute these in very high precision. ■

TABLE 5.14 $\Psi(t_k)/\Psi(1)$ for $-1 \leq t_k \leq 1$

n	k	t_k	$\Psi(t_k)/\Psi(1)$
4	0	$-0.1000000000000000E+01$	$0.0000000000000000E+00$
	1	$-0.5000000000000000E+00$	$0.3175495761647776E-01$
	2	$0.0000000000000000E+00$	$0.5000000000000001E+00$
	3	$0.5000000000000000E+00$	$0.9682450390677892E+00$
	4	$0.1000000000000000E+01$	$0.1000000000000000E+01$
8	0	$-0.1000000000000000E+01$	$0.0000000000000000E+00$
	1	$-0.7500000000000000E+00$	$0.1962922276951938E-03$
	2	$-0.5000000000000000E+00$	$0.3175495761647776E-01$
	3	$-0.2500000000000000E+00$	$0.2004317448343370E+00$
	4	$0.0000000000000000E+00$	$0.5000000000000001E+00$
	5	$0.2500000000000000E+00$	$0.7995682518334509E+00$
	6	$0.5000000000000000E+00$	$0.9682450390677892E+00$
	7	$0.7500000000000000E+00$	$0.9998037051191319E+00$
	8	$0.1000000000000000E+01$	$0.1000000000000000E+01$
16	0	$-0.1000000000000000E+01$	$0.0000000000000000E+00$
	1	$-0.8750000000000000E+00$	$0.1930534427638750E-07$
	2	$-0.7500000000000000E+00$	$0.1962922276951938E-03$
	3	$-0.6250000000000000E+00$	$0.5462420695964468E-02$
	4	$-0.5000000000000000E+00$	$0.3175495761647776E-01$
	5	$-0.3750000000000000E+00$	$0.9512852921910396E-01$
	6	$-0.2500000000000000E+00$	$0.2004317448343370E+00$
	7	$-0.1250000000000000E+00$	$0.3405226277340978E+00$
	8	$0.0000000000000000E+00$	$0.5000000000000001E+00$
	9	$0.1250000000000000E+00$	$0.6594773707084767E+00$
	10	$0.2500000000000000E+00$	$0.7995682518334509E+00$
	11	$0.3750000000000000E+00$	$0.9048714673344923E+00$
	12	$0.5000000000000000E+00$	$0.9682450390677892E+00$
	13	$0.6250000000000000E+00$	$0.9945375763409888E+00$
	14	$0.7500000000000000E+00$	$0.9998037051191319E+00$
	15	$0.8750000000000000E+00$	$0.9999999792694836E+00$
	16	$0.1000000000000000E+01$	$0.1000000000000000E+01$

Although this method was designed to work on finite intervals with integrands having endpoint singularities, it can be easily modified to compute integrals on infinite intervals as well. Suppose we want to compute

$$I(f) = \int_a^\infty f(x)\, dx.$$

TABLE 5.15   **Singular Quadrature Method
Applied to** $f(x) = (\log 1/x)^{1/2}$, $[a,b] = [0,1]$

n	$I_n(f)$	$I(f) - I_n(f)$
4	0.892023444140	0.579652E–02
8	0.885813277564	0.413648E–03
16	0.886218502855	0.842260E–05
32	0.886226915463	0.999072E–08
64	0.886226924278	0.117666E–08
128	0.886226924425	0.102882E–08
256	0.886226924497	0.957054E–09
512	0.886226924526	0.928343E–09
1024	0.886226924527	0.925976E–09

The change of variable

$$x = a + \frac{1+z}{1-z}$$

gives us the new integral

$$I(f) = \int_a^\infty f(x)\,dx = \int_{-1}^1 F(z)\,dz,$$

where $F(z) = 2f(a + (1+z)(1-z)^{-1})(1-z)^{-2}$. This new integral now has the kind of endpoint singularity (at $z = 1$) that the previous development was designed to attack, so we can apply the same ideas to get the quadrature rule

$$I(f) \approx I_n(f) = \sum_{k=1}^{n-1} W_k F(z_k) \tag{5.26}$$

with

$$z_k = \phi(t_k), \quad W_k = \frac{4}{n\gamma} \exp\left(-\frac{c}{1-t_k^2}\right), \tag{5.27}$$

where $\phi(t)$ is now given by

$$\phi(t) = -1 + \frac{2}{\gamma}\Psi(t).$$

TABLE 5.16 **Singular Quadrature Method Applied to** $f(x) = e^{-x^2}$, $[a,b] = [0,\infty)$

n	$I_n(f)$	$I(f) - I_n(f)$
4	1.141419627988	0.255193E+00
8	0.939027865252	0.528009E−01
16	0.884525323277	0.170160E−02
32	0.886229670193	0.274474E−05
64	0.886226922961	0.249179E−08
128	0.886226922944	0.250830E−08
256	0.886226922943	0.250951E−08
512	0.886226922942	0.251117E−08
1024	0.886226922941	0.251128E−08

EXAMPLE 5.18

Consider the following example:

$$I(f) = \int_0^\infty e^{-x^2} \, dx = \frac{1}{2}\sqrt{\pi}.$$

Table 5.16 shows the approximations as computed by (5.26)–(5.27). Note that the error once again stagnates. ∎

5.8.3 Adaptive Integration

It is possible—rather easy, in fact—to write a general-purpose quadrature routine based on any of the methods discussed so far in this chapter. However, we often want to have some firm knowledge that the error is "small enough" for the needs of the application. We could write a code that simply takes more and more points until some computable estimate of the error (such as we get from Richardson extrapolation) is less than τ, where τ is a user-supplied tolerance. Romberg integration is perhaps the best choice for this, incidentally.

However, such a process is often needlessly inefficient, for the simple reason that we do not want to take *too many* points in those regions of the interval where the integrand is smooth and therefore high accuracy is easily achieved. Consider the graph in Figure 5.7; to accurately integrate this function over the interval (say) $[-5,5]$, we will need to use many points near the origin, but fewer away from it. Using something like Romberg integration will put a lot of points uniformly throughout the interval, thus wasting effort.

For this reason, most automatic quadrature routines are *adaptive*; that is, they do not work with a uniform mesh, but adjust the mesh points to achieve a given overall

FIGURE 5.7 Function with rapid change near the origin.

accuracy. There are many ways to accomplish this task; what we outline here is one of the simpler techniques.

At the core of adaptive methods (whether applied to numerical integration or to other areas of computation) is the use of a computable estimate of the error to gauge the accuracy of the computation. Roughly speaking, if this computable estimate of the error is small enough, then we cease the calculation and move on. In the development presented here, we will use the Richardson extrapolation results from Section 5.7 to estimate the error.

An illustration might be useful at this point. Suppose we wish to compute

$$I = \int_a^b f(x)\,dx$$

to an accuracy of τ; that is, we want to find an approximate value $Q \approx I$ such that $|I - Q| \le \tau$. We will use Simpson's rule as the quadrature rule at the core of our method.

We start by computing two different approximations to the integral, one based on applying Simpson's rule over the entire interval and one based on applying it over two subintervals. Thus we have

$$S_1 = \frac{(b-a)/2}{3}\left(f(a) + 4f((a+b)/2) + f(b)\right)$$

and

$$S_2 = \frac{(b-a)/4}{3} \left(f(a) + 4f((a+b)/4) + f((a+b)/2) \right)$$
$$+ \frac{(b-a)/4}{3} \left(f((a+b)/2) + 4f(3(a+b)/4) + f(b) \right)$$

We can apply Richardson extrapolation to estimate the absolute error in S_2:

$$E_2 = |S_2 - S_1|/16.$$

If $E_2 \leq \tau$, then we use the Richardson extrapolated value $R_2 = (16S_2 - S_1)/15$ as our approximation to I, and we are confident that this approximation is to the specified accuracy.[12]

However, we do not really expect to compute our integral to any realistic accuracy with only two approximations; we are going to have to continue the process. Basically, we next try to compute, accurately, half of the integral,

$$I' = \int_{(a+b)/2}^{b} f(x)\, dx,$$

using the same ideas, then half of this integral, and so on. Describing this in any sort of precise, formal manner that is amenable to summarization as an algorithm is going to require some notation. Define

$$J(\alpha,\beta) = \int_{\alpha}^{\beta} f(x)\, dx;$$

thus the value we want to compute is $I = J(a,b)$. We apply a similar notation to the Simpson's rule approximations:

$$S(\alpha,\beta) = \frac{(\beta-\alpha)/2}{3} [f(\alpha) + 4f((\alpha+\beta)/2) + f(\beta)].$$

It will also be useful to define $\theta_k = 2^{-k+1}(b-a)$, so $\theta_1 = (b-a)$, $\theta_2 = (b-a)/2$, etc. Thus, our first step involves the computation of

$$S_1 = S(a,b) = S(b-\theta_1,b)$$

and

$$S_2 = S(a,(a+b)/2) + S((a+b)/2,b)$$
$$= S(b-\theta_1,b-\theta_2) + S(b-\theta_2,b).$$

---

[12] Note that we are only "confident," not "sure." Since we are using a computable *estimate* of the error, we do not *know* that the error is less than τ.

The complete algorithm continues this process, recursively, until we have computed *part* of the integral to the specified accuracy. It is simplest to organize the algorithm so that the values of the integral accumulate from right to left; i.e., we next try to approximate

$$J(b - \theta_2, b) = J((a+b)/2, b) = \int_{(a+b)/2}^{b} f(x) \, dx$$

using the two approximate values

$$S_1 = S(b - \theta_2, b) = S((a+b)/2, b)$$

and

$$S_2 = S(b - \theta_2, b - \theta_3) + S(b - \theta_3, b).$$

Once again we compute the Richardson extrapolation values $E = |S_1 - S_2|/16$ and $R = (16S_2 - S_1)/15$. The step is evaluated in the following way:

If $E \leq \frac{1}{2}\tau$, then we accept R as the approximate value of $J((a+b)/2, b)$.

We use $\frac{1}{2}\tau$ because the integral in question involves only *half* the full interval of integration. We continue this process, at the kth step trying to approximate $J(b - \theta_k, b)$ to an estimated accuracy of $2^{-k+1}\tau$.

Eventually we will find a subinterval small enough that the desired fractional error tolerance is satisfied. When this happens, we accumulate the approximate value of the integral and move on to try and compute the next portion of the integral.

A key element in making this an efficient process is the reuse of function values that were previously computed. As the computation proceeds, we maintain a list of grid points and the associated function values that have already been computed. Since new points are introduced as the algorithm proceeds, the grid is getting redefined and reindexed at each step. What makes this a difficult algorithm to encode (and explain!) is the recursion and the need for careful management of the data to ensure that we do not use any unnecessary function evaluations. This algorithm has been implemented in a number of places, most notably in QUADPACK and in the MATLAB routine `quad`. In languages that support recursion it is much easier to encode, but much harder to follow.

EXAMPLE 5.19

Let's continue the illustration with a specific example, rather than a generic one. We will try to approximate the value of the integral

$$I = J(-5, 1) = \int_{-5}^{1} \frac{1}{1 + e^{-4x}} \, dx = 1.0045374814641640301,$$

which is the function graphed in Figure 5.7, using a tolerance of $\tau = 10^{-2}$. Thus $f(x) = 1/(1 + e^{-4x})$, $a = -5$, $b = 1$, and $\tau = 0.01$. (Since we are using such a crude error for this illustration, we will display our answers using only 5 digits, although the computations are being done with 10 digits.)

Step 1: We compute the two Simpson's rule approximations

$$S_1 = \left(\frac{(1-(-5))/2}{3} \right) (f(-5) + 4f(-2) + f(1)) = 0.98336$$

$$S_2 = \left(\frac{(-2-(-5))/2}{3} \right) (f(-5) + 4f(-3.5) + f(-2))$$

$$+ \left(\frac{(1-(-2))/2}{3} \right) (f(-2) + 4f(-0.5) + f(1)) = 0.72975$$

and the associated Richardson extrapolated value

$$R = (16S_2 - S_1)/15 = 0.71284$$

and Richardson error estimate

$$E = |S_2 - S_1|/15 = 0.016907.$$

Since E is *not* less than the specified tolerance $\tau = 0.01$, we do not accept R as an accurate approximation to the integral I. We thus must look at a portion of the integral. Note that we have so far used function values at the grid points

$$x_k = -5, -3.5, -2, -0.5, 1.$$

The total count of function calls is thus $K = 5$.

Step 2: We are now concerned with approximating the "subintegral" defined by

$$J(-2,1) = \int_{-2}^{1} \frac{1}{1+e^{-4x}} \, dx$$

Note that we can compute the S_1 approximation using function values that have already been computed:

$$S_1 = \left(\frac{(1-(-2))/2}{3} \right) (f(-2) + 4f(-0.5) + f(1)) = 0.72958.$$

Computing S_2, the more accurate approximation to $J(-2,1)$, requires two new function calls, at $x = 0.25$ and $x = -1.25$:

$$S_2 = \left(\frac{(-0.5-(-2))/2}{3} \right) (f(-2) + 4f(-1.25) + f(-0.5))$$

$$+ \left(\frac{(1-(-0.5))/2}{3} \right) (f(-0.5) + 4f(0.25) + f(1)) = 1.0429.$$

The Richardson values are

$$R = 1.0638$$

and

$$E = 0.020890.$$

Since we are looking at *half* the original integral, the error comparison is to *half* the original tolerance, i.e., $\tau/2 = 0.005$; thus we still do not have sufficient accuracy. The grid now consists of the points

$$x_k = -5, -3.5, -2, -1.25, -0.5, 0.25, 1.$$

The total number of function calls is now $K = 7$.

Step 3: The subintegral of interest is now

$$J(-0.5, 1) = \int_{-0.5}^{1} \frac{1}{1 + e^{-4x}} \, dx$$

and the S_1 approximation can be found using previously computed function values:

$$S_1 = \left(\frac{(1 - (-0.5))/2}{3} \right) (f(-0.5) + 4f(0.25) + f(1)) = 1.0064.$$

We have

$$S_2 = \left(\frac{(0.25 - (-0.5))/2}{3} \right) (f(-0.5) + 4f(-0.125) + f(0.25))$$

$$+ \left(\frac{(1 - (0.25))/2}{3} \right) (f(0.25) + 4f(0.625) + f(1)) = 0.97126,$$

and the Richardson values are

$$R = 0.96892$$

and

$$E = 0.0023403.$$

Since $E \leq \frac{1}{4}\tau = 0.0025$, we accept R as the approximate value of $J(-0.5, 1)$. The grid is now

$$x_k = -5, -3.5, -2, -1.25, \underline{-0.5, -0.125, 0.25, 0.625, 1},$$

where we have underlined those grid points used in the computation of $J(-0.5, 1)$, i.e., that part of the integral that has been computed to the desired accuracy. The function call count is now $K = 9$.

Step 4: We now look at trying to approximate

$$J(-2, -0.5) = \int_{-2}^{-0.5} \frac{1}{1 + e^{-4x}} \, dx.$$

This choice is made because this is the subintegral defined by the two grid points immediately to the left of the previous interval, together with the left-hand endpoint of the previous interval. We proceed as we have before, getting

$$S_1 = 0.036577, \qquad S_2 = 0.032022,$$

with

$$R = 0.031719, \qquad E = 0.00030368.$$

We are again looking at trying to approximate over one-fourth of the interval of integration, so once again the comparison is to $\tau/4 = 0.0025$. The estimated error is easily less than this, so we accept R as the approximate value of $J(-2, -0.5)$. Thus we have

$$J(-2,1) = J(-2, -0.5) + J(-0.5,1) = 1.0006$$

The grid is now

$$x_k = -5, -3.5, \underline{-2}, -1.625, -1.25, -0.875, -0.5, -0.125, 0.25, 0.625, 1.$$

The function call count is now $K = 11$.

Step 5: In the final step we look at the subintegral

$$J(-5, -2) = \int_{-5}^{-2} \frac{1}{1 + e^{-4x}} \, dx.$$

We get the values

$$S_1 = 0.00016934, \qquad S_2 = 0.00010100,$$

with

$$R = 0.000096440, \qquad E = 0.45562 \times 10^{-5}.$$

This subintegral now represents *half* of the interval of integration, so we are now comparing with $\tau/2 = 0.005$; thus the estimated error is much smaller than the specified tolerance, so we accept R as the approximate value of $J(-5, -2)$, which completes the integration process. The final value is

$$I = 1.0007$$

and the final grid is

$$x_k = \underline{-5}, -4.25, -3.5, -2.75, -2, -1.625, -1.25, -0.875, -0.5, -0.125, 0.25, 0.625, 1.$$

The final function call count is $K = 13$. ∎

Of course, we used a rather crude error tolerance in this example (and thus did not get much variation in the grid spacing), but this was only to make it more practical as a demonstration problem. If we repeat the same computation, but this time specify that $\tau = 10^{-4}$, we get the results shown in Table 5.17. Rather than display all the detail that we did with the example, this table simply shows the accumulated value of the integral at each step, along with the lower limit of integration for which we have accurately approximated the integral. That is, we define

$$I_c = \int_c^1 \frac{1}{1 + e^{-4x}} \, dx,$$

where the adaptive quadrature scheme has determined that Figure 5.8

$$|I - I_c| \le \tau \frac{1 - c}{6}.$$

plots the integrand again, along with a mark on the curve at each point used in the computation. This shows the distribution of function evaluations needed to adequately

TABLE 5.17 **Adaptive Quadrature Applied to** $f(x) = 1/(1 + e^{-4x})$

Step	I_c	c
1	0.000000000000	1.000000000000
2	0.000000000000	1.000000000000
3	0.000000000000	1.000000000000
4	0.000000000000	1.000000000000
5	0.676243647403	0.250000000000
6	0.676243647403	0.250000000000
7	0.886039349337	−0.125000000000
8	0.972826902351	−0.500000000000
9	0.972826902351	−0.500000000000
10	0.972826902351	−0.500000000000
11	0.997121163206	−0.875000000000
12	1.002879936750	−1.250000000000
13	1.004475327568	−2.000000000000
14	1.004571768033	−5.000000000000

FIGURE 5.8 Adaptive quadrature.

approximate the integral. Note that a total of 29 function evaluations were used to obtain this degree of accuracy; if we had applied Romberg integration to this function, we would have needed 64 function evaluations. Gaussian quadrature achieves this accuracy with only 16 points, but we have no way of estimating the accuracy of this approximation without doing the next one and observing that the change in the approximation is negligible, which makes the Gaussian estimate more costly. Recall that the adaptive approach gives us a solid basis for being confident that we have achieved a certain degree of accuracy.

5.8.4 Peano Estimates for the Trapezoid Rule

Recall that the trapezoid rule is constructed by exactly integrating a linear polynomial approximation to the integrand. We can use this to derive an interesting error formula, known as a *Peano*[13] *estimate*.

Consider

$$I(f) = \int_a^b f(x)\,dx$$

and

$$T_1(f) = \frac{1}{2}(b-a)(f(b)+f(a)),$$

the basic trapezoid rule. Since T_1 is derived from exactly integrating a linear approximation, it is *exact* when applied to any and all linear functions:

$$T_1(p) = I(p)$$

for all functions p of the form $p(x) = Ax + B$. (See Exercise 12.) In addition, we can view both I and T_1 as *linear operators*, meaning that

$$I(\alpha f + \beta g) = \alpha I(f) + \beta I(g)$$

for constants α and β, and similarly for $T_1(\alpha f + \beta g)$. (See Exercise 13.) Now, for any given integrand f, use Taylor's Theorem to write

$$f(x) = p_1(x) + R_1(x),$$

where p_1 is the linear Taylor polynomial and R_1 is the integral form of the remainder. Since T_1 is exact for linear polynomials, we know that

$$I(p_1) = T_1(p_1).$$

---

[13] Giuseppe Peano (1858–1932) entered the University of Turin, in Italy, in 1876 and graduated with a doctorate in mathematics in 1880. Known primarily for his work in mathematical logic and the foundations of mathematics, he published the development of the Peano kernel in two papers that appeared in Italian journals in 1913–1914.

Therefore,

$$
\begin{aligned}
I(f) - T_1(f) &= I(p_1 + R_1) - T_1(p_1 + R_1) \\
&= I(R_1) - T_1(R_1) \\
&= \int_a^b \int_a^x (x - t) f''(t) \, dt \, dx - \frac{1}{2}(b - a) \int_a^b (b - t) f''(t) \, dt.
\end{aligned}
$$

We can interchange the order of integration in the double integral to get

$$
\begin{aligned}
\int_a^b \int_a^x (x - t) f''(t) \, dt \, dx &= \int_a^b \int_t^b (x - t) f''(t) \, dx \, dt \\
&= \int_a^b \left(\int_t^b (x - t) \, dx \right) f''(t) \, dt \\
&= \int_a^b \frac{1}{2}(b - t)^2 f''(t) \, dt,
\end{aligned}
$$

so that we can finally get

$$
\begin{aligned}
I(f) - T_1(f) &= \int_a^b \frac{1}{2}(b - t)^2 f''(t) \, dt - \frac{1}{2}(b - a) \int_a^b (b - t) f''(t) \, dt \\
&= \frac{1}{2} \int_a^b (b - t)[(b - t) - (b - a)] f''(t) \, dt \\
&= \frac{1}{2} \int_a^b (a - t)(b - t) f''(t) \, dt.
\end{aligned}
$$

This is known as the *Peano estimate* for the error, and the function $K(t) = \frac{1}{2}(a - t) \cdot (b - t)$ is known as the *Peano kernel* for the (basic) trapezoid rule. When the result is extended to the composite case (Exercise 14), we have

$$
I(f) - T_n(f) = \int_a^b K(t) f''(t) \, dt,
$$

where

$$
K(t) = \frac{1}{2}(x_{i-1} - t)(x_i - t), \quad x_{i-1} \le t \le x_i. \tag{5.28}
$$

There are a couple of reasons for looking at the Peano form of the error estimate. One is that it allows us to bound the error when the integrand is not smooth in the pointwise sense. Note that we can write

$$
\begin{aligned}
|I(f) - T_n(f)| &= \left| \int_a^b K(t) f''(t) \, dt \right| \\
&\le \max_{a \le t \le b} |K(t)| \int_a^b |f''(t)| \, dt \\
&\le \frac{1}{8} h^2 \int_a^b |f''(t)| \, dt
\end{aligned}
$$

TABLE 5.18 Trapezoid Rule Applied to $f(x) = x^{3/2}$, $[a,b] = [0,1]$

$n = h^{-1}$	$T_n(f)$	$I(f) - T_n(f)$
2	0.4267766953E+00	−0.2677669530E−01
4	0.4070181109E+00	−0.7018110858E−02
8	0.4018124648E+00	−0.1812464800E−02
16	0.4004634013E+00	−0.4634013020E−03
32	0.4001176712E+00	−0.1176712098E−03
64	0.4000297399E+00	−0.2973986253E−04
128	0.4000074919E+00	−0.7491908991E−05
256	0.4000018830E+00	−0.1883044172E−05
512	0.4000004725E+00	−0.4725406820E−06
1024	0.4000001184E+00	−0.1184497712E−06
2048	0.4000000297E+00	−0.2966805679E−07

where we have assumed a uniform mesh spacing of h. (See Exercise 17.) Thus, even if f is not pointwise twice continuously differentiable, we can still get an error estimate if f'' is integrable in the absolute sense. For example, consider the problem of integrating $f(x) = x^{3/2}$ over the interval $[0,1]$. We know that $f''(x)$ will be proportional to $x^{-1/2}$, and therefore the pointwise estimates of Section 2.5 will not hold. However, $x^{-1/2}$ is absolutely integrable, so the Peano form of the error tells us we can expect second-order accuracy when using the trapezoid rule. Table 5.18 confirms this conclusion.

A second reason to use the Peano form is that it allows us to obtain lower-order estimates. If f has only a single derivative (whether in the pointwise or absolutely integrable sense), then we can use integration by parts to obtain the estimate

$$I(f) - T_1(f) = \frac{1}{2} \int_a^b (a-t)(b-t) f''(t)\, dt = \frac{1}{2} \int_a^b (t-c) f'(t)\, dt,$$

where $c = (a+b)/2$. This suggests the composite kernel in this case is

$$K(t) = \left(t - \frac{1}{2}(x_i + x_{i-1}) \right), \quad x_{i-1} \leq t \leq x_i,$$

from which an estimate that is $O(h)$ accurate follows quickly. (See Exercise 16.)

In fact, this explains the order of convergence that we observe when using the trapezoid rule to integrate $f(x) = \sqrt{x}$; although the second derivative is not integrable, the first derivative $f'(x) = \frac{1}{2} x^{-1/2}$ is, and so we see that the error is $O(h)$, as confirmed by the data in Table 5.13. (See Exercise 15.)

Peano estimates can be constructed for any of the other quadrature rules, but they are sometimes a bit more complex. See Exercise 18.

EXERCISES

1. Use the Euler-Maclaurin formula to state and prove a formal theorem that the corrected trapezoid rule, T_n^C, is $O(h^4)$ accurate.

2. Using a hand calculator, compute $T_1^{(0)}(f)$, $T_2^{(0)}(f)$, $T_4^{(0)}(f)$, and $T_8^{(0)}(f)$ for each of the following functions, and then use Romberg integration to compute $\Theta_3(f)$. (Note: Be sure to use Theorem 5.8 to minimize the work in computing the first column of the Romberg array.)
(a) $I = \int_0^1 x(1-x^2)\,dx = \frac{1}{4}$
(b) $I = \int_0^1 \frac{1}{\sqrt{1+x^4}}\,dx = 0.92703733865069$
(c) $I = \int_0^1 \ln(1+x)\,dx = 2\ln 2 - 1$
(d) $I = \int_0^1 \frac{1}{1+x^3}\,dx = \frac{1}{3}\ln 2 + \frac{1}{9}\sqrt{3}\pi$
(e) $I = \int_1^2 e^{-x^2}\,dx = 0.1352572580$

3. Write a program to do Romberg integration. Be sure to use Theorem 5.8 to minimize the number of function evaluations. Test your program by applying it to the following example problems.
(a) $f(x) = 1/(1+x^2)$, $[-5,5]$, $I(f) = 2\arctan(5)$
(b) $f(x) = x^2 e^{-x}$, $[0,2]$, $I(f) = 2 - 10e^{-2} = 0.646647168$
(c) $f(x) = \ln x$, $[1,3]$, $I(f) = 3\ln 3 - 2 = 1.295836867$
(d) $f(x) = \sqrt{1-x^2}$, $[-1,1]$, $I(f) = \pi/2$
(e) $f(x) = e^{-x}\sin(4x)$, $[0,\pi]$, $I(f) = \frac{4}{17}(1 - e^{-\pi}) = 0.2251261368$
For each example, compute

$$N_f = \frac{\text{Number of function evaluations}}{-\log_{10}|\text{error}|}.$$

This measures the number of function evaluations needed to produce each correct decimal digit in the approximation.

4. Write a computer program that uses Romberg integration for a specified number of points to compute the natural logarithm over the interval $[\frac{1}{2},1]$, to within 10^{-16} accuracy. Compare the accuracy of your routine to the intrinsic logarithm function on your system.

5. Show that the change of variable $x = \phi(t)$, where ϕ is as given in Section 5.8.2, transforms the integral

$$I(f) = \int_a^b f(x)\,dx$$

into the integral

$$I(f) = \int_{-1}^1 f(\phi(t))\phi'(t)\,dt.$$

6. Apply the singular integral technique of Section 5.8.2, with $n = 4$, to estimate the value of each of the following integrals. Do this with a hand calculator, using the values in Table 5.14.
(a) $I(f) = \int_0^1 \left(\ln\frac{1}{x}\right)^{-1/2} dx = \sqrt{\pi}$
(b) $I(f) = \int_0^1 x\ln(1-x)\,dx = -\frac{3}{4}$
(c) $I(f) = \int_0^1 \frac{\ln x}{1-x}\,dx = -\frac{\pi^2}{6}$
(d) $I(f) = \int_0^1 \frac{\ln x}{1-x^2}\,dx = -\frac{\pi^2}{8}$

7. Repeat the previous problem, except this time use a computer program together with the values in Table 5.14 to compute the $n = 16$ approximations.

8. We looked at the gamma function in a number of exercises in previous chapters. The formal definition of $\Gamma(x)$ is

$$\Gamma(x) = \int_0^\infty e^{-t} t^{x-1}\,dt.$$

Use the infinite interval algorithm from Section 5.8.2 to construct a table of values for the gamma function over the interval $[1,2]$. Compare your results with the values you get for $\Gamma(x)$ on your computer or from a standard book of tables (such as [1]).

9. Modify your Romberg integration program to compute values of $\Psi(t_k)/\Psi(1)$ for $t_k \in [-1,1]$, and use this to extend the values in Table 5.14 to the $n = 64$ case.

10. Apply the trapezoid rule to each of the following functions, integrated over the indicated intervals, and interpret the results in terms of the Euler-Maclaurin formula.
(a) $f(x) = 1 + \sin \pi x$, $[a,b] = [0,2]$
(b) $f(x) = \sin^2 x$, $[a,b] = [0,\pi]$

11. Using a hand calculator and τ as indicated, perform the adaptive quadrature scheme outlined in Section 5.8.3 on each of the following integrals. Be sure to present your results in an orderly fashion so that the progress of the calculation can be followed.
(a) $\tau = 10^{-4}$,

$$I = \int_0^1 \frac{1}{\sqrt{1+x^4}}\,dx = 0.92703733865069$$

(b) $\tau = 5 \times 10^{-6}$,

$$I = \int_0^1 \ln(1+x)\,dx = 2\ln 2 - 1$$

(c) $\tau = 10^{-5}$,

$$I = \int_0^1 \frac{1}{1+x^3}\,dx = \frac{1}{3}\ln 2 + \frac{1}{9}\sqrt{3}\pi$$

(d) $\tau = 10^{-5}$

$$I = \int_1^2 e^{-x^2}\,dx = 0.1352572580$$

12. Show that the trapezoid rule $T_1(f)$ is exact for all functions of the form $f(x) = Ax+B$.

13. Show that

$$I(\alpha f + \beta g) = \alpha I(f) + \beta I(g)$$

for constants α and β, and similarly for $T_n(\alpha f + \beta g)$.

14. Confirm that (5.28) is the correct Peano kernel for the composite trapezoid rule.

15. Show that the data in Table 5.13 confirms that the trapezoid and Simpson's rules applied to $f(x) = \sqrt{x}$ are both $O(h)$ accurate.

16. Show that if f' is integrable over $[a,b]$, but f'' is not, then the trapezoid rule is $O(h)$ accurate.

17. Show that the Peano theorem implies an error estimate for the trapezoid rule of the form

$$|I(f) - T_n(f)| \le \frac{1}{8}h^2 \int_a^b |f''(t)|\,dt$$

Be sure to provide all details missing from the development in the text.

18. Derive the Peano kernel for Simpson's rule.

LITERATURE AND SOFTWARE DISCUSSION

The quadrature problem has a long history and a fairly mature literature. The standard monograph is perhaps [3], and decent treatments can be found in any good numerical analysis text.

The trapezoid, Simpson, and midpoint rules are special cases of a broad class of methods known as Newton-Cotes formulas, for which a unified error theory is available. See [2] for a standard treatment of this.

The main deficiency of Gaussian quadrature has been the need to recompute f at all grid points whenever the grid is refined. A variation of Gaussian quadrature, known as Gauss-Kronrod quadrature, addresses this issue and has become popular in recent years. Extensive work on a similar idea was done by Patterson, and some of these ideas are incorporated in the QUADPACK package. See [6] or [3].

Multiple integrals are usually treated by obvious extensions of the ideas presented here, although the fact that higher-dimensional geometry is potentially much more involved than the simple intervals in **R** does complicate matters.

REFERENCES

[1] ABRAMOWITZ, M., & STEGUN, I.A. *Handbook of Mathematical Functions*. U.S. Govt. Printing Office, Washington, DC, 1964.

[2] ATKINSON, KENDALL. *An Introduction to Numerical Analysis*. John Wiley & Sons, New York, 1987 (2nd ed.).

[3] DAVIS, P., & RABINOWITZ, P. *Methods of Numerical Integration*. Academic Press, New York, 1984 (2nd ed.).

[4] DE DONCKER, E., & PIESSENS, R. "Algorithm 32: Automatic computation of integrals with singular integrand, over a finite or infinite range," *Computing*, vol. 17 (1976), pp. 265–279.

[5] JEFFREYS, HAROLD, & JEFFREYS, BERTHA, *Methods of Mathematical Physics*. Cambridge University Press, Cambridge, 1972.

[6] PIESSENS, R., ET AL. *QUADPACK: A Subroutine Package for Automatic Integration*. Springer-Verlag, New York, 1983.

CHAPTER 6

NUMERICAL METHODS FOR ORDINARY DIFFERENTIAL EQUATIONS

Pennsyltucky Pharmaceutical is involved in an effort to model the growth of certain kinds of tumors. Based on some simplifying assumptions, they are able to arrive at the differential equation[1]

$$R'(t) = -\frac{1}{3}S_i R + \frac{2\lambda\sigma}{\mu R + \sqrt{\mu^2 R^2 + 4\sigma}}, \quad R(0) = a.$$

Here $R(t)$ is the radius of the tumor (assumed spherical); λ and μ are scale parameters, both $O(1)$; S_i measures the rate at which cells at the core of the tumor die; and σ is a nutrient level. To go forward with this model, the scientists at Pennsyltucky Pharmaceutical must be able to solve this differential equation accurately for a variety of parameter sets.

In this chapter we will study a number of techniques for solving ordinary differential equations, approximately but accurately. We started doing this in Chapter 2, with Euler's method, but here we will look at some more sophisticated ideas. Some of the exercises will refer back to this model.

We are concerned here with the problem of numerically solving differential equations. First we concentrate on the so-called initial value problem (IVP): Find a function $y(t)$ such that

$$\frac{dy}{dt} = f(t, y(t)), \quad y(t_0) = y_0,$$

where f is a known function of two variables, and t_0 and y_0 are known values. This is called the initial value problem because (as the notation suggests) we can view the independent variable t as time, and the equation as modeling a process that moves forward from some initial time t_0 with initial state y_0. (Very often $t_0 = 0$.) The dependent variable y, the unknown function, may be a scalar function or possibly a vector function defined as

$$y(t) = (y_1(t), y_2(t), \ldots, y_N(t))^T.$$

---

[1] See H.P. Greenspan, "On the growth and stability of cell cultures and solid tumors," *Journal of Theoretical Biology*, vol. 56 (1976), pp. 229–242. This equation is derived on p. 234.

In Section 2.3 we developed Euler's method for approximating solutions to initial value problems; in this chapter we will not only review Euler's method, but also look at more sophisticated (and therefore, we hope, more accurate) methods for solving this kind of problem. Later we will tackle the boundary value problem (BVP), which can be written as

$$-\frac{d^2u}{dx^2} = F\left(x, u, \frac{du}{dx}\right), \quad a < x < b \tag{6.1}$$

$$u(a) = g_0, \qquad u(b) = g_1. \tag{6.2}$$

Here the unknown function is u with independent variable x, F is a known function of three variables, and g_0 and g_1 are known data values. Very often the interval $(a, b) = (0, 1)$.

In both cases we want to find an unknown function. We will do this by approximating individual points on the graph of the function, as we did in Section 2.7 with Euler's method for initial value problems. Thus, we will seek (in the case of the IVP) a set of values y_k such that $y_k \approx y(t_k)$ for some set of (known) grid points t_k, or (in the case of the BVP) a set of values u_k such that $u_k \approx u(x_k)$ for some (known) grid points x_k. Note that this means that our approximation is defined only on the grid points, although we could use the approximation methods of Chapter 4 to construct continuous approximate solutions to differential equations; this, in fact, is something that is often done, and we showed one example of it in Section 4.9, where we used splines to solve two-point boundary value problems.

It should be noted that the numerical solution of ordinary differential equations—whether IVPs or BVPs—is a very active area of ongoing research. What we present here is a selection of the basic algorithms and underlying theory. The reader is referred to the list of references at the end of the chapter for more in-depth treatments of this material.

6.1 THE INITIAL VALUE PROBLEM: BACKGROUND

Consider the ordinary differential equation

$$\frac{dy}{dt} = f(t, y(t)), \quad y(t_0) = y_0, \tag{6.3}$$

where f is a function from $\mathbf{R} \times \mathbf{R}^N$ into $\mathbf{R}^N$ for some $N > 0$ (if $N = 1$, we have a scalar equation; otherwise we have a vector equation); t_0 is a given scalar value, often taken to be $t_0 = 0$, known as the *initial point*; and y_0 is a known vector in $\mathbf{R}^N$, known as the *initial value*. We want to find the unknown function $y(t)$ that solves (6.3) in the sense that

$$y'(t) - f(t, y(t)) = 0$$

for all $t > t_0$, and $y(t_0) = y_0$.

Some examples might be useful at this point.

EXAMPLE 6.1

Consider the simple problem

$$y' = -2ty, \quad y(0) = 1.$$

Here $f(t,y) = -2ty, t_0 = 0, y_0 = 1$, and we can apply methods from a standard ordinary differential equations (ODE) course to show that $y(t) = e^{-t^2}$ is the solution. ∎

EXAMPLE 6.2

Consider now the simple *system of equations*

$$y'_1 = -4y_1 + y_2, \qquad\qquad y_1(0) = 1$$
$$y'_2 = y_1 - 4y_2, \qquad\qquad y_2(0) = 0.$$

In this case, $y(t) = (y_1(t), y_2(t))^T$ is a *vector* of two components; $f(t,y)$ is defined by

$$f(t,y) = (f_1(t,y_1,y_2), f_2(t,y_1,y_2))^T,$$

where

$$f_1(t,y_1,y_2) = -4y_1 + y_2$$
$$f_2(t,y_1,y_2) = y_1 - 4y_2;$$

and the initial point is still $t_0 = 0$, but the initial value is now the vector $y_0 = (1,0)^T$. The solution, again obtained from a standard ODE course, is

$$y_1 = \frac{1}{2}e^{-3t} + \frac{1}{2}e^{-5t}$$
$$y_2 = \frac{1}{2}e^{-3t} - \frac{1}{2}e^{-5t}.$$

∎

EXAMPLE 6.3

What about the *second-order* equation

$$y'' + 3y' + y = 0, \quad y(0) = 1, \quad y'(0) = 0? \tag{6.4}$$

It does not appear to fit the standard form for our initial value problems, yet we can put it into this form without a lot of work. Define the vector $w(t) = (w_1(t), w_2(t))^T$, where $w_1(t) = y(t), w_2(t) = y'(t)$. Then $w'_1 = w_2$ and (6.4) becomes $w'_2 = -3w_2 - w_1$. Thus we have the *first-order system*

$$w'_1 = w_2, \qquad\qquad w_1(0) = 1$$
$$w'_2 = -w_1 - 3w_2, \qquad w_2(0) = 0.$$

∎

EXAMPLE 6.4

More generally, we can take any scalar equation of order N,

$$y^{(n)} = g(t, y, y', y'', \dots, y^{(n-1)}),$$

and write it as a first-order system of N equations in N unknowns, according to

$$w_1 = y, \quad w_2 = y', \quad \ldots, \quad w_n = y^{(n-1)},$$

where

$$
\begin{aligned}
w_1' &= w_2, \quad w_1(0) = y_0 \\
w_2' &= w_3, \quad w_2(0) = y_0' \\
w_3' &= w_4, \quad w_3(0) = y_0'' \\
&\vdots \qquad \vdots \\
w_n' &= g(t, w_1, w_2, \ldots, w_{n-1}, w_n), \quad w_n(0) = y_0^{(n-1)}.
\end{aligned}
$$

Thus we have the single vector equation

$$w' = f(t, w), \quad w(0) = w_0,$$

where

$$
f(t,w) = \begin{pmatrix} w_2 \\ w_3 \\ w_4 \\ \vdots \\ g(t, w_1, w_2, w_3, \ldots, w_n) \end{pmatrix}, \quad w_0 = \begin{pmatrix} y_0 \\ y_0' \\ y_0'' \\ \vdots \\ y_0^{(n-1)} \end{pmatrix}.
$$

■

To keep the presentation simple, we will state our results in language and notation appropriate for scalar equations; but the reader should be aware that corresponding results hold for vector equations, and we will spend some time discussing vector equations later in the chapter, in Section 6.8.

The nature of the solution of the differential equation—indeed, the very question of whether or not a solution exists—as well as our ability to approximate the solution accurately, is very much connected with the nature of the function f. Basically, if f is "smooth enough," then a solution will exist and be unique and we will therefore be able to approximate it accurately with a wide variety of methods. However, there are different ways of expressing what we mean by "smooth enough." We will use two of them: (1) Lipschitz continuity and (2) smooth and uniformly monotone decreasing. The former is the condition usually used to construct solutions of initial value problems; it is generally the weakest condition that we can get away with. The monotone decreasing condition describes a wide class of physically significant problems, where "monotone decreasing" corresponds to dissipation or decay in the physical system being modeled.

Definition 6.1 (Lipschitz Continuity)   *Let g be a given function from* **R** *to* **R**. *We say that g is* Lipschitz[2] *continuous on an interval I if there exists a constant K such*

---

[2]Rudolf Otto Sigismund Lipschitz (1832–1903) was born in Königsberg, Germany. He entered the University of Königsberg at the age of 15 and completed his studies at the University of Berlin, from which he was awarded a doctorate in 1853. By 1864 he was a full professor at the University of Bonn, where he remained the rest of his professional life.

The Lipschitz condition first appeared in a paper on the existence of solutions to differential equations, which was published in book form as part of an 1877 treatise on analysis.

that

$$|g(x_1) - g(x_2)| \le K|x_1 - x_2|$$

for all $x_1, x_2 \in I$.

Definition 6.2 (Smooth and Uniformly Monotone Decreasing)   Let g be a given function from $\mathbf{R}$ to $\mathbf{R}$. We say that g is smooth and uniformly monotone decreasing *if g is differentiable and the derivative satisfies*

$$-M \le g'(x) \le -m < 0$$

for all x, where M and m are given positive constants.

We have actually already seen the idea of Lipschitz continuity; this is the same condition that was required in Section 3.9 for uniqueness and convergence of fixed-point iterations.

EXAMPLE 6.5

Just to illustrate, consider the two initial value problems

$$y' = 4y - e^{-t}, \quad y(0) = 1 \tag{6.5}$$

and

$$y' = -(1 + t^2)y + \sin t, \quad y(0) = 1. \tag{6.6}$$

For (6.5) we have $f(t,y) = 4y - e^{-t}$ so that

$$f(t,y_1) - f(t,y_2) = (4y_1 - e^{-t}) - (4y_2 - e^{-t}) = 4(y_1 - y_2);$$

hence

$$|f(t,y_1) - f(t,y_2)| \le 4|y_1 - y_2|.$$

Hence, in this case f is Lipschitz continuous in y with constant $K = 4$. However, f is not smooth and uniformly monotone decreasing, because $f_y(t,y) = 4 > 0$ for all t and y. On the other hand, for (6.6) we have $f(t,y) = -(1 + t^2)y + \sin t$, so

$$f(t,y_1) - f(t,y_2) = (-(t^2 + 1)y_1 + \sin t) - (-(t^2 + 1)y_2 + \sin t) = -(t^2 + 1)(y_1 - y_2);$$

thus, for $0 \le t \le 1$,

$$|f(t,y_1) - f(t,y_2)| \le (t^2 + 1)|y_1 - y_2| \le 2|y_1 - y_2|.$$

Therefore, this f is Lipschitz continuous with constant $K = 2$. In addition, we have (again, for $0 \le t \le 1$)

$$f_y(t,y) = -(t^2 + 1) \quad \Rightarrow \quad -2 \le f_y(t,y) \le -1 < 0,$$

showing that this f is also smooth and uniformly monotone decreasing. ∎

The significance of these definitions lies in how they impact what we know about the solution of the differential equation. The standard result for ordinary differential equations is the following.

Theorem 6.1 (Existence-Uniqueness of Solutions for IVPs, Version I)   Let $f(t,y)$ be continuous for all (t,y) in an open rectangle $R = \{(t,y) : a < t < b, c < y < d\}$ and Lipschitz continuous in y, with constant K. Then for all $(t_0, y_0) \in R$ there exists a unique solution to the initial value problem

$$y' = f(t,y), \quad y(t_0) = y_0.$$

Moreover, if $z(t)$ is the solution to the same problem with initial data $z(t_0) = z_0$, then

$$|y(t) - z(t)| \le e^{K(t-t_0)}|y_0 - z_0|. \tag{6.7}$$

Proof   This result is proved in some texts on ordinary differential equations; see, for instance, [5]. Interestingly, the original proof was done by constructing a discrete solution and showing that it converged, in the limit as $h \to 0$, to a differentiable function that solved the IVP. See Goldstine's book [9] for some more details on the history behind this result.   □

If we impose only slightly more on the function f, then we get a very different bound on the change in solutions due to a change in the initial data.

Theorem 6.2 (Existence-Uniqueness of Solutions for IVPs, Version II)   Let $f(t,y)$ be continuous for all (t,y) in a rectangle R, and smooth and uniformly monotone decreasing in y. Then for all $(t_0, y_0) \in R$ there exists a unique solution to the initial value problem

$$y' = f(t,y), \quad y(t_0) = y_0.$$

Moreover, if $z(t)$ is the solution to the same problem with initial data $z(t_0) = z_0$, then

$$|y(t) - z(t)| \le e^{-m(t-t_0)}|y_0 - z_0|. \tag{6.8}$$

Here m is the upper monotonicity constant, as given in Definition 6.2.

Proof   Since f is smooth and uniformly monotone decreasing in y, we can apply the Mean Value Theorem to easily prove that it is Lipschitz in y, as well:

$$|f(t,y_1) - f(t,y_2)| = |f_y(t,\eta)(y_1 - y_2)| \le M|y_1 - y_2|.$$

Thus we can apply Theorem 6.1 to get the existence and uniqueness of the solution. The inequality (6.8) takes a bit more work, and its proof is deferred to the Appendix.   □

The inequalities (6.7) and (6.8) are known as *stability results* because they measure how stable the solutions to the initial value problem are, i.e., the extent to which small changes in the problem will lead to small changes in the solution. In the first case, if all we know is that f is Lipschitz, then small changes in the initial data can lead to very large changes in the solutions, if we consider large enough t. Note, however, that it still is the case that the differences in the two solutions can be bounded for all $t \leq T$, where T is a fixed value, and that $z(t) \to y(t)$ on this interval, uniformly, as $z_0 \to y_0$. In the second instance, because we know that f is smooth and monotone decreasing, we know that the perturbation in the solution due to the initial error in the solution will *actually decay to zero as* $t \to \infty$. This will have similar implications for the error associated with numerical methods applied to the solution of the differential equation.

EXERCISES

1. For each initial value problem below, verify (by direct substitution) that the given function y solves the problem.

(a) $y' + 4y = 0$, $y(0) = 3$; $y(t) = 3e^{-4t}$
(b) $ty' - y = t^2$, $y(1) = 4$; $y(t) = 3t + t^2$
(c) $y' = t^2/y$, $y(0) = 1$; $y(t) = \sqrt{1 + \frac{2}{3}t^3}$

2. For each initial value problem in Exercise 1, write down the definition of f.

3. For each scalar equation below, write out explicitly the corresponding first-order system. What is the definition of f?

(a) $y'' + 9y = e^{-t}$

(b) $y'''' + y = 1$
(c) $y'' + \sin y = 0$

4. For each initial value problem below, determine the Lipschitz constant, K, for the given rectangle.
(a) $y' = 1 - 3y$, $y(0) = 1$, $R = (-1,1) \times (0,2)$
(b) $y' = y(1-y)$, $y(0) = \frac{1}{2}$, $R = (-1,1) \times (0,2)$
(c) $y' = y^2$, $y(0) = 1$, $R = (-1,1) \times (0,2)$

5. Are any of the initial value problems in the previous exercise smooth and uniformly monotone decreasing over the given rectangle? If so, determine the values of M and m.

6.2   EULER'S METHOD

Euler's method is the natural starting point for any discussion of numerical methods for IVPs. Although it is not the most accurate of the methods we will study, it is by far the simplest, and much of what we learn from analyzing Euler's method in detail carries over to other methods without a lot of difficulty. Even though we treated Euler's method in Chapter 2, we are going to cover it again here, but this time more fully.

There are two main derivations; one is geometric, the other is analytic, and a third derivation (the one we used in Chapter 2) is given in Section 6.4. We start with the geometric derivation.

Consider Figure 6.1. This shows the graph of the solution $y(t)$ to the initial value problem

$$y' = f(t,y), \quad y(t_0) = y_0$$

near the initial point t_0. All we know about y is that it passes through the point $(t_0, y_0) = (0,0)$ and that it has slope $f(t_0, y_0)$ (why?) at that point. We therefore can approximate $y(t+h)$ for some small h by using the tangent line approximation:

$$y(t_0 + h) \approx y(t_0) + hf(t_0, y(t_0)). \tag{6.9}$$

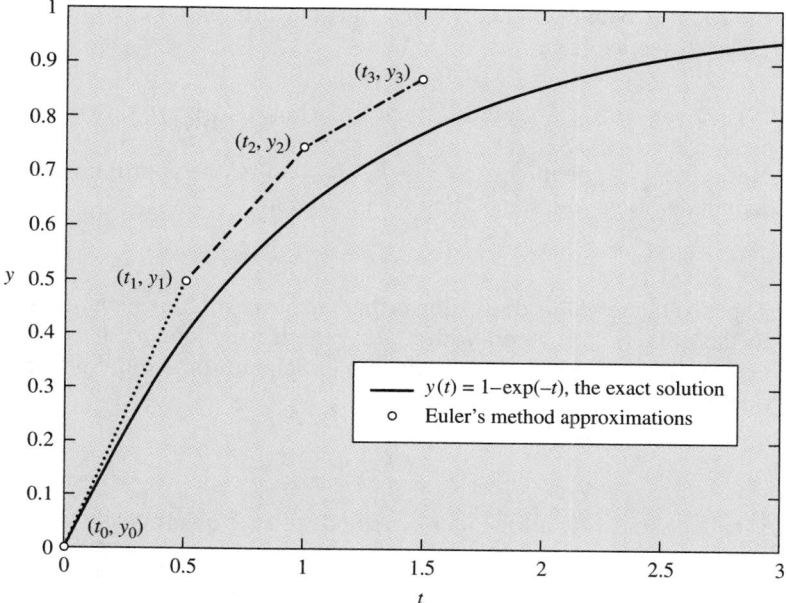

FIGURE 6.1 Geometric derivation of Euler's method.

For small values of h this will be a decent approximation, and if we define the next point in our grid by $t_1 = t_0 + h$, then it is very natural to extend (6.9) to get

$$y(t_1 + h) \approx y(t_1) + hf(t_1, y(t_1)) \qquad (6.10)$$

or, more generally,

$$y(t_n + h) \approx y(t_n) + hf(t_n, y(t_n)). \qquad (6.11)$$

The numerical method is then defined from (6.11) according to

$$y_{n+1} = y_n + hf(t_n, y_n),$$

with y_0 given and $y_n \approx y(t_n)$.

This is a reasonable and easy-to-follow derivation, but it does not admit of much error analysis, and it also opens up a lot of questions. Perhaps the most important question concerns the validity of extending the tangent line approximation from the first computed point (t_1, y_1) to the next one. From the graph in Figure 6.1 it is clear that there is substantial error between y_1 and $y(t_1)$. How valid, then, is (6.10) in defining the next approximate point?

The analytic derivation is based, not surprisingly, on Taylor's Theorem. We have

$$y(t + h) = y(t) + hy'(t) + \frac{1}{2}h^2 y''(\theta)$$

for some θ between t and $t + h$. But the differential equation implies that $y'(t) = f(t,y(t))$, so we have

$$y(t+h) = y(t) + hf(t,y(t)) + \frac{1}{2}h^2 y''(\theta). \tag{6.12}$$

Now set $t = t_0$, define $t_{n+1} = t_n + h$, set y_n to be the approximate value, and drop the remainder to get the same thing we had before:

$$y_{n+1} = y_n + hf(t_n,y_n). \tag{6.13}$$

The advantage of this derivation is that we have (6.12), which relates the exact solution to the numerical method with a very precise remainder. It is the key to getting an error estimate for Euler's method. To anticipate some more general terminology, the quantity

$$R(t,h) = \frac{1}{2}h^2 y''(\theta)$$

is called the *residual* for Euler's method, and the quantity

$$\tau(t,h) = \frac{1}{h}R(t,h)$$

is called the *truncation error*.

Before we get into the analysis (see the next section), some examples are in order.

EXAMPLE 6.6

Consider the two differential equations

$$y' + y = 1, \quad y(0) = 0 \tag{6.14}$$

and

$$y' = y, \quad y(0) = 1. \tag{6.15}$$

The exact solutions can be found by typical ODE methods to be

$$y(t) = 1 - e^{-t}$$

for (6.14) and

$$y(t) = e^t$$

for (6.15). Table 6.1 shows the error made by Euler's method in solving each of these, using a sequence of decreasing meshes. The entries in the table give the maximum error over the interval $[0,1]$, i.e.,

$$E_n = \max_{k \le n} |y(t_k) - y_k|.$$

The table shows that the errors in both cases are going down by about a factor of 2, from which we infer that Euler's method is $O(h)$ accurate, only. Figures 6.2 and 6.3 show plots of

TABLE 6.1   Euler's Method Applied to (6.14) and (6.15)

$n = h^{-1}$	Error in approximating (6.14)	Error in approximating (6.15)
16	0.118053E–01	0.803533E–01
32	0.582415E–02	0.412917E–01
64	0.289292E–02	0.209369E–01
128	0.144173E–02	0.105428E–01
256	0.719686E–03	0.529020E–02
512	0.359550E–03	0.264983E–02
1024	0.179702E–03	0.132610E–02

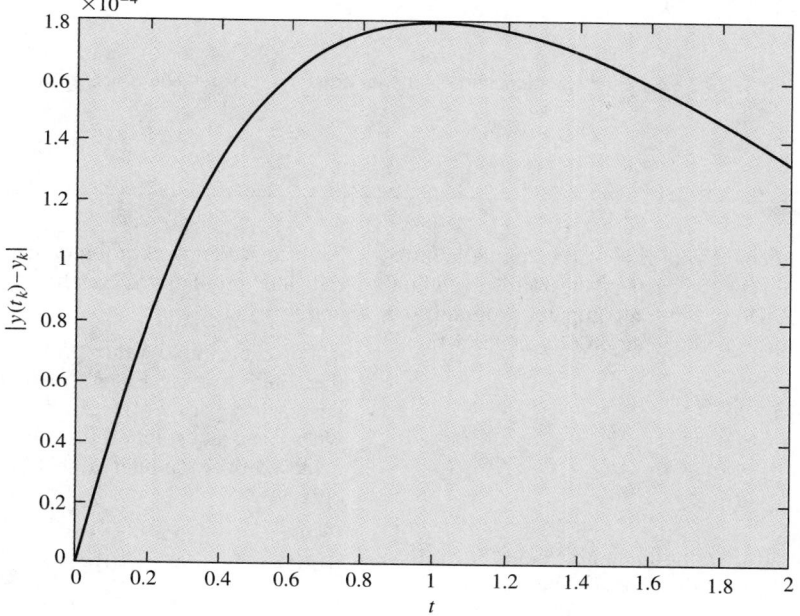

FIGURE 6.2   Error as a function of time for equation (6.14), using $h = 1/1024$.

the absolute error as a function of t for $n = 1024$; note that for (6.14) the worst error occurs near the beginning of the computation and afterward the error actually decreases, whereas for (6.15) the worst error is always the most recent error. In fact, the error for (6.15) appears to be growing exponentially. We will touch on this issue in the next section, and try to explain it. ∎

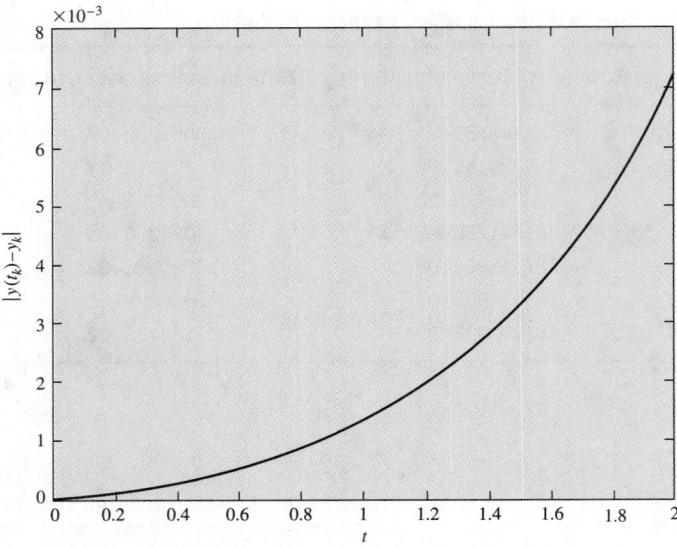

FIGURE 6.3   Error as a function of time for equation (6.15), using $h = 1/1024$.

EXERCISES

1. Use Euler's method with $h = \frac{1}{4}$ to compute approximate values of $y(1)$ for each of the following initial value problems. Don't write a computer program; use a hand calculator to produce an orderly table of (t_k, y_k) pairs.

(a) $y' + 2y = 1$, $y(0) = 2$

(b) $y' = y(1 - y)$, $y(0) = \frac{1}{2}$

(c) $y' = y(1 + e^{2t})$, $y(0) = 1$

(d) $ty' = y(\sin t)$, $y(0) = 2$

2. For each initial value problem above, use the differential equation to produce approximate values of y' at each of the grid points t_k, $k = 0,1,2,3,4$.

3. Write a computer program that solves each of the initial value problems in Exercise 1, using Euler's method and $h = 1/16$.

4. For each initial value problem below, approximate the solution using Euler's method with a sequence of decreasing grids $h^{-1} = 2,4,8,\ldots$. For those problems where an exact solution is given, compare the accuracy achieved over the interval $[0,1]$ with the theoretical accuracy.

(a) $y' + 4y = 1$, $y(0) = 1$; $y(t) = \frac{1}{4}(3e^{-4t} + 1)$

(b) $y' + y = \sin 4\pi t$, $y(0) = \frac{1}{2}$

(c) $y' = -y \ln y$, $y(0) = 3$; $y(t) = e^{(\ln 3)e^{-t}}$

(d) $y' + \sin y = 0$, $y(0) = 1$

6.3   ANALYSIS OF EULER'S METHOD

In this section we will prove two results that establish the convergence and error estimate for Euler's method. We will provide a fair amount of detail in this section, in order to avoid going into so much detail with more sophisticated methods that we will derive later. Throughout the section we will be concerned with the approximate solution, via Euler's method, of the initial value problem

$$y' = f(t,y), \quad y(t_0) = y_0.$$

The first theorem shows that Euler's method is, indeed, first-order accurate.

Theorem 6.3 (Error Estimate for Euler's Method, Version I)   *Let f be Lipschitz continuous, with constant K, and assume that the solution $y \in C^2([t_0,T])$ for some $T > t_0$. Then*

$$\max_{t_k \leq T} |y(t_k) - y_k| \leq C_0 |y(t_0) - y_0| + Ch\|y''\|_{\infty,[t_0,T]},$$

where

$$C_0 = e^{K(T-t_0)}$$

and

$$C = \frac{e^{K(T-t_0)} - 1}{2K}. \tag{6.16}$$

Proof   The key element in the proof of this result is the fact that the exact solution satisfies the same relationship as does the approximate solution, except for the addition of a remainder term. Thus we have (from (6.12) and (6.13)),

$$y(t_{n+1}) = y(t_n) + hf(t_n, y(t_n)) + \frac{1}{2}h^2 y''(\theta_n),$$

$$y_{n+1} = y_n + hf(t_n, y_n),$$

for which we use subtraction to get

$$y(t_{n+1}) - y_{n+1} = y(t_n) - y_n + hf(t_n, y(t_n)) - hf(t_n, y_n) + \frac{1}{2}h^2 y''(\theta_n).$$

Take absolute values and apply the Lipschitz continuity of f to get

$$|y(t_{n+1}) - y_{n+1}| \leq |y(t_n) - y_n| + Kh|y(t_n) - y_n| + \frac{1}{2}h^2|y''(\theta_n)|,$$

which we write as

$$e_{n+1} \leq \gamma e_n + R_n$$

where $e_n = |y(t_n) - y_n|$, $\gamma = 1 + Kh$, and $R_n = \frac{1}{2}h^2|y''(\theta_n)|$, for notational simplicity. This is a simple recursive inequality, which we can "solve" as follows. We have

$$e_1 \leq \gamma e_0 + R_0$$
$$e_2 \leq \gamma e_1 + R_1 \leq \gamma^2 e_0 + \gamma R_0 + R_1$$
$$e_3 \leq \gamma e_2 + R_2 \leq \gamma^3 e_0 + \gamma^2 R_0 + \gamma R_1 + R_2,$$

and so on. An inductive argument can be applied to get the general result

$$e_n \leq \gamma^n e_0 + \frac{1}{2}h^2 \sum_{k=0}^{n-1} \gamma^k |y''(\theta_{n-1-k})|.$$

It remains only to simplify the summation term. We have

$$|y''(t)| \leq \|y''\|_{\infty,[t_0,T]}$$

for all $t \in [t_0,T]$, so we then have

$$e_n \leq \gamma^n e_0 + \frac{1}{2}h^2\|y''\|_{\infty,[t_0,T]}\sum_{k=0}^{n-1}\gamma^k.$$

We can explicitly produce the sum now, to get

$$e_n \leq \gamma^n e_0 + \frac{1}{2}h^2\|y''\|_{\infty,[t_0,T]}\frac{\gamma^n-1}{\gamma-1},$$

but $\gamma = 1+Kh$ so that $\gamma - 1 = Kh$; thus we have

$$e_n \leq \gamma^n e_0 + \frac{\gamma^n-1}{2K}h\|y''\|_{\infty,[t_0,T]}.$$

Finally, we note that it follows from Taylor's Theorem (see Exercise 1) that

$$(1+x)^n \leq e^{nx}, \quad x > -1,$$

so that

$$\gamma^n = (1+Kh)^n \leq e^{nKh} = e^{K(nh)} = e^{K(t_n-t_0)} \leq e^{K(T-t_0)},$$

and we are done. $\square$

This estimate shows that Euler's method is first-order (i.e., the error is $O(h)$), but it also shows that the constants multiplying the terms in the error estimate can become quite large. If we now assume that f is smooth and uniformly monotone decreasing, then we get an estimate that, while still first-order, involves much smaller constants.

Theorem 6.4 (Error Estimate for Euler's Method, Version II)   Let f be smooth and uniformly monotone decreasing in y, and assume that the solution $y \in C^2([t_0,T])$ for some $T > t_0$. Then, for h sufficiently small,

$$\max_{t_k \leq T}|y(t_k) - y_k| \leq C_0|y(t_0) - y_0| + Ch\|y''\|_{\infty,[t_0,T]},$$

where $C_0 \leq 1, C_0 \to 0$ as $k \to \infty$, and

$$C = \frac{1}{2m}.$$

Proof   The proof here is much akin to the previous one, at least in the beginning. We have

$$y(t_{n+1}) - y_{n+1} = y(t_n) - y_n + hf(t_n,y(t_n)) - hf(t_n,y_n) + \frac{1}{2}h^2y''(\theta_n). \tag{6.17}$$

Now, instead of taking absolute values and inequalities, we use the Mean Value Theorem to write

$$f(t_n, y(t_n)) - f(t_n, y_n) = \frac{\partial f}{\partial y}(t_n, \eta_n)(y(t_n) - y_n),$$

where η_n is a value between $y(t_n)$ and y_n. For simplicity we write this as

$$\frac{\partial f}{\partial y}(t_n, \eta_n)(y(t_n) - y_n) = k_n(y(t_n) - y_n),$$

where we note that the smooth and uniformly monotone decreasing hypothesis implies that $-M \le k_n \le -m < 0$. Then (6.17) becomes

$$y(t_{n+1}) - y_{n+1} = \gamma_n(y(t_n) - y_n) + \frac{1}{2}h^2 y''(\theta_n),$$

where $\gamma_n = 1 - k_n h$. Now take absolute values and use the triangle inequality to get

$$e_{n+1} \le |\gamma_n| e_n + R_n,$$

where e_n and R_n are as in the previous proof. Assume now that h is small enough that $1 - Mh > 0$; i.e., assume that $h < M^{-1}$. Then $\gamma > 0$ for all n, so that

$$|\gamma_n| = \gamma_n = 1 - k_n h \le \gamma = 1 - mh < 1.$$

This is the crucial difference between the two theorems and their proofs. By making a stronger assumption about the way f behaves, we are able to show that the rate constant in the error recursion is less than 1. We then can write

$$e_{n+1} \le \gamma e_n + R_n,$$

from which

$$e_n \le \gamma^n e_0 + \frac{1}{2}h^2 \|y''\|_{\infty,[t_0,T]} \frac{\gamma^n - 1}{\gamma - 1} \tag{6.18}$$

now follows as before. We now observe that

$$0 \le \frac{\gamma^n - 1}{\gamma - 1} = \frac{1 - \gamma^n}{1 - \gamma} \le \frac{1 - \gamma^n}{1 - (1 - mh)} = \frac{1 - \gamma^n}{mh} \le \frac{1}{mh},$$

so that (6.18) becomes

$$e_n \le \gamma^n e_0 + \frac{1}{2m}h\|y''\|_{\infty,[t_0,T]}$$

and we are done. □

Both error theorems show that Euler's method is only first-order accurate; i.e., the error is $O(h)$. The difference between the theorems lies in the way that the estimates depend on the function f: If f is only Lipschitz continuous, then the constants multiplying the initial error and the mesh parameter can be quite large, rapidly growing; however, if f is smooth and uniformly monotone decreasing in y—which means that the differential equation is modeling a decay process, essentially—then the constants in the error estimate are bounded for all n. Note, in particular, how the initial error is affected by this. If f is monotone decreasing in y, then the initial error is multiplied by γ^n, where $0 \leq \gamma < 1$; thus the effect of the initial error decreases rapidly as the computation progresses. However, if f is only Lipschitz continuous, then (6.16) suggests that any initial error that is made could be amplified to something exponentially large. While the initial error is typically quite small—usually it is $O(\mathbf{u})$—if the term multiplying it grows exponentially, then the effects of this small initial error could dominate the calculation at a later point in time.

Now let's look back at the two examples we did in the previous section. For one example ($y' = y$), the error grew almost exponentially with time, and if we look at the definition of f in this example, we see that it is only Lipschitz continuous. For the second example ($y' = 1 - y$), however, we have that f is smooth and uniformly monotone decreasing in y; thus the error growth with time is much more benign, as predicted by the theory and confirmed by the computation.

EXERCISES

1. Use Taylor's Theorem to prove that

$$(1+x)^n \leq e^{nx}$$

for all $x > -1$. (Hint: Expand e^x in a Taylor series, throw away the unnecessary terms, and then take powers of both sides.)

2. For each initial value problem below, use the error theorems of this section to estimate the value of

$$E(h) = \frac{\max_{t_k \leq 1} |y(t_k) - y_k|}{\|y''\|_{\infty, [0,1]}}$$

using $h = 1/16$, assuming Euler's method was used to approximate the solution.

(a) $y' + 4y = 1$, $y(0) = 1$
(b) $y' = y(1 - y)$, $y(0) = 1/2$
(c) $y' = \sin y$, $y(0) = \frac{1}{2}\pi$

3. Consider the initial value problem

$$y' = e^{-t} - 16y, \quad y(0) = 1.$$

(a) Confirm that this is smooth and uniformly monotone decreasing in y. What is M? What is m?

(b) Approximate the solution using Euler's method and $h = \frac{1}{8}$. Do we get the expected behavior from the approximation? Explain.

6.4 VARIANTS OF EULER'S METHOD

Euler's method, of course, is not the only or even the best scheme for approximating solutions to initial value problems, and what we need to do now is look at other methods that we might employ. Several ideas can be considered based on some simple extensions of one derivation of Euler's method.

Our third derivation of Euler's method, which also gives us a remainder term, is based on our difference methods for derivative approximation, from Section 2.2. We start with the differential equation

$$y'(t) = f(t, y(t))$$

and replace the derivative with the simple difference quotient derived in (2.1). This yields

$$\frac{y(t+h)-y(t)}{h} = f(t,y(t)) + \frac{1}{2}hy''(\theta_{t,h}),$$

where the subscripts on θ remind us that the value depends on both t and h. Euler's method is then obtained by simply dropping the remainder and replacing t with t_n, $y(t)$ with y_n, etc. This is the derivation we used in Chapter 2.

This prompts an obvious question: What happens if we use other approximations to the derivative? For example, if we use

$$y'(t) = \frac{y(t)-y(t-h)}{h} - \frac{1}{2}hy''(\theta),$$

then we get the *backward* Euler method,

$$y_{n+1} = y_n + hf(t_{n+1}, y_{n+1}) : \tag{6.19}$$

and if we use

$$y'(t) = \frac{y(t+h)-y(t-h)}{2h} - \frac{1}{6}h^2y'''(\theta_{t,h}),$$

we get what is commonly referred to as the *midpoint* method:

$$y_{n+1} = y_{n-1} + 2hf(t_n, y_n). \tag{6.20}$$

Or we could use the methods based on interpolation (Section 4.5),

$$y'(t) \approx \frac{1}{2h}\left(-y(t+2h) + 4y(t+h) - 3y(t)\right)$$

$$y'(t+2h) \approx \frac{1}{2h}\left(3y(t+2h) - 4y(t+h) + y(t)\right),$$

to get the two numerical methods

$$y_{n+1} = 4y_n - 3y_{n-1} - 2hf(t_{n-1}, y_{n-1}) \tag{6.21}$$

and

$$y_{n+1} = \frac{4}{3}y_n - \frac{1}{3}y_{n-1} + \frac{2}{3}hf(t_{n+1}, y_{n+1}). \tag{6.22}$$

Finally, we note that yet another set of methods can be derived by integrating the differential equation. We have that the exact solution satisfies

$$y(t+h) = y(t) + \int_t^{t+h} f(s, y(s))\, ds. \tag{6.23}$$

We can therefore apply the trapezoid rule to (6.23) to get

$$y(t+h) = y(t) + \frac{1}{2}h[f(t+h, y(t+h)) + f(t, y(t))] - \frac{1}{12}h^3 y'''(\theta_{t,h}), \tag{6.24}$$

where $\theta_{t,h} \in [t, t+h]$, and we remind the reader that

$$f(t,y(t)) = y'(t) \quad \Rightarrow \quad \frac{d^2}{dt^2} f(t,y(t)) = y'''(t).$$

Dropping the remainder from (6.24) leads to the numerical method (commonly called the trapezoid method, for obvious reasons)

$$y_{n+1} = y_n + \frac{1}{2} h \left(f(t_{n+1}, y_{n+1}) + f(t_n, y_n) \right). \tag{6.25}$$

Alternatively, we can use a midpoint rule approximation to integrating (6.23). This leads to

$$y(t+h) = y(t) + h f\left(t + \frac{1}{2} h, y\left(t + \frac{1}{2} h \right) \right) - \frac{1}{24} h^3 y'''(\theta_{t,h}),$$

which suggests the numerical method

$$y_{n+1} = y_n + h f(t_{n+1/2}, y_{n+1/2}), \tag{6.26}$$

where $t_{n+1/2} = t_n + \frac{1}{2} h$ and $y_{n+1/2} \approx y(t_n + \frac{1}{2} h)$.

What about these methods? Are any of them any good?

Several observations can be made immediately. The methods (6.20), (6.21), and (6.22) are all based on derivative approximations that are $O(h^2)$, whereas Euler's method is based on a derivative approximation that is only $O(h)$ (as is the backward Euler method (6.19)). This suggests (but does not, of course, prove) that (6.20), (6.21), and (6.22) should be more accurate than Euler (and backward Euler). Similarly, the methods (6.25) and (6.26) are based on integral approximations that are more accurate.

A second observation involves the midpoint method (6.20) and the two methods (6.21) and (6.22). Note that here we have formulas for y_{n+1} in terms of y_n *and* y_{n-1}. These are not *single-step* methods, but *multistep* methods; that is, they depend on information from more than one previous approximate value of the unknown function. How do we actually implement these methods? The differential equation gives us only a single initial value, y_0; we need more in order to even start the recursion here.

A third observation concerns backward Euler and the methods (6.25) and (6.22). Note that all of these formulas involve $f(t_{n+1}, y_{n+1})$; we cannot explicitly solve for the new approximate value y_{n+1}. This is why these methods (and others like them) are called *implicit* whereas methods such as Euler, midpoint, and (6.21) are called *explicit*, because they define y_{n+1} *explicitly* in terms of information from previous steps.

We would like to address the issue of accuracy, at least experimentally, but we can't even implement several of the methods until we address the other problems. However, it will be useful at this point to introduce some terminology associated with the accuracy of the various methods.

6.4.1 The Residual and Truncation Error

We are, of course, interested in the accuracy of the methods we develop here. All of the numerical methods for IVPs that we will study can be written in the general format

$$y_{n+1} = \sum_{k=0}^{p} a_k y_{n-k} + hF(y_{n+1}, y_n, \ldots, y_{n-p}; f_{n+1}, f_n, \ldots, f_{n-p}) \qquad (6.27)$$

where we have used $f_k = f(t_k, y_k)$ for simplicity. Thus, for example, in Euler's method we have

$$y_{n+1} = y_n + hf(t_n, y_n)$$

so that

$$a_0 = 1$$
$$a_k = 0, \quad \text{all } k \geq 1,$$

and

$$F(y_{n+1}, y_n, \ldots, y_{n-p}; f_{n+1}, f_n, \ldots, f_{n-p}) = f_n = f(t_n, y_n).$$

For the trapezoid rule method (6.25), we have

$$a_0 = 1$$
$$a_k = 0, \quad \text{all } k \geq 1,$$

but

$$F(y_{n+1}, y_n, \ldots, y_{n-p}; f_{n+1}, f_n, \ldots, f_{n-p}) = \frac{1}{2}(f_n + f_{n+1}) = \frac{1}{2}(f(t_n, y_n) + f(t_{n+1}, y_{n+1})).$$

Crucial to the accuracy of these methods is the extent to which the exact solution $y(t)$ satisfies the numerical method (6.27). To this end we define the residual, the truncation error (sometimes called the local truncation error), and the concept of consistency.

Definition 6.3 (Residual, Truncation Error, and Consistency)   For a numerical method written as in (6.27), define the residual *as R_n, given by*

$$R_n = y(t_{n+1}) - \sum_{k=0}^{p} a_k y(t_{n-k}) - hF(y(t_{n+1}), y(t_n), \ldots, y(t_{n-p}); f(t_{n+1}, y(t_{n+1})), f(t_n, y(t_n)),$$
$$\ldots, f(t_{n-p}, y(t_{n-p}))).$$

The truncation error *is defined as*

$$\tau_n = \frac{1}{h} R_n,$$

and the method is said to be consistent *if*

$$\lim_{h \to 0} \max_{t_n \leq T} |\tau_n| = 0$$

for sufficiently smooth solutions y.

For any "reasonable" method, the error $\max_{t_k \leq T} |y(t_k) - y_k|$ is proportional to the truncation error, if the solution is smooth enough. Note that the residual and truncation error are both defined in terms of substituting the exact solution into the numerical method.[3]

For most methods that we develop, the residual and truncation error naturally come out of the construction process. This is the case, for example, with Euler's method and the several variants discussed above.

For example, consider some of the methods we have looked at so far.

EXAMPLE 6.7

Using either of the two analytic derivations for Euler's method, we can write

$$y(t_{n+1}) = y(t_n) + hf(t_n, y(t_n)) + \frac{1}{2}h^2 y''(\theta_n);$$

therefore, the residual is $\frac{1}{2}h^2 y''(\theta_n)$, and the truncation error is $\frac{1}{2}hy''(\theta_n)$. The method is consistent so long as the solution is C^2, because in this case we have

$$\lim_{h \to 0} \max_{t_n \leq T} \frac{1}{2}h^2 |y''(\theta_n)| = \lim_{h \to 0} \frac{1}{2}h^2 \max_{t \leq T} |y''(t)| = \lim_{h \to 0} \frac{1}{2}h^2 Y_2 = 0,$$

where Y_2 is the maximum absolute value of y'' for $t \leq T$.   ∎

EXAMPLE 6.8

Using what we know of the trapezoid rule for approximating integrals, we can write (see Section 2.5)

$$y(t_{n+1}) = y(t_n) + \frac{1}{2}h\left(f(t_n, y(t_n)) + f(t_{n+1}, y(t_{n+1}))\right) - \frac{1}{12}h^3 y'''(\theta_n);$$

therefore, the residual is $-\frac{1}{12}h^3 y'''(\theta_n)$, and the truncation error is $-\frac{1}{12}h^2 y'''(\theta_n)$. This time, the method is consistent if y is C^3.   ∎

Generally, the residual and truncation error are derived by expanding $y(t_{n+1})$ in a Taylor series about t_n and matching terms so that they vanish, or by applying some approximate integration or differentiation method.

EXAMPLE 6.9

Consider the method defined by

$$y_{n+1} = y_n + \frac{1}{2}h(3f(t_n, y_n) - f(t_{n-1}, y_{n-1})),$$

---

[3]There is a disturbing lack of consistency among the literature on the terminology used here. For example, Burden and Faires [4], Isaacson and Keller [11], and Allen and Isaacson [1] all define "truncation error" as we have here. Other books simply replace the word "truncation" with "discretization." However, Atkinson [2] and Lambert [14], among others, use "truncation error" or "local truncation error" for what we have called the *residual*.

So, when consulting references, be sure to check and see which definition is being used.

which we will encounter in Section 6.6 as one of the Adams-Bashforth methods. What are its residual and truncation error, and is it a consistent method?

We have

$$R_n = y(t_{n+1}) - y(t_n) - \frac{1}{2}h(3f(t_n, y(t_n)) - f(t_{n-1}, y(t_{n-1}))).$$

We write $y(t_{n+1})$ in a Taylor series as

$$y(t_{n+1}) = y(t_n) + hy'(t_n) + \frac{1}{2}h^2 y''(t_n) + \frac{1}{6}h^3 y'''(\theta),$$

and we use the fact that $y'(t) = f(t, y(t))$ to write the residual as

$$R_n = hy'(t_n) + \frac{1}{2}h^2 y''(t_n) + \frac{1}{6}h^3 y'''(\theta) - \frac{1}{2}h(3y'(t_n) - y'(t_{n-1}))$$

$$= \frac{1}{2}h\left[hy''(t_n) - (y'(t_n) - y'(t_{n-1}))\right] + \frac{1}{6}h^3 y'''(\theta).$$

Taylor's Theorem can again be applied, this time to y', to get

$$y'(t_{n-1}) = y'(t_n) - hy''(t_n) + \frac{1}{2}h^2 y'''(\eta),$$

so that we now have

$$R_n = \frac{1}{2}h\left[hy''(t_n) - hy''(t_n) + \frac{1}{2}h^2 y'''(\eta)\right] + \frac{1}{6}h^3 y'''(\theta) = h^3\left(\frac{1}{2}y'''(\eta) + \frac{1}{6}y'''(\theta)\right).$$

Therefore, the residual is $O(h^3)$ and the truncation error is $O(h^2)$. Moreover, the method is consistent whenever the exact solution is C^3. ∎

We close this section with a definition.

Definition 6.4 (Order of Accuracy) *If the truncation error for a numerical scheme for the solution of initial value problems is $O(h^k)$, then we say that the method has order of accuracy k.*

Thus, the trapezoid rule method has order of accuracy $p = 2$, and Euler's method has order of accuracy $p = 1$. It is commonplace to say that the trapezoid rule is second-order accurate and Euler's method is first-order accurate.

There is a potential for confusion here. Note that the definition of order of accuracy is based on what we defined as the *truncation error*, not the residual.

6.4.2 Implicit Methods and Predictor-Corrector Schemes

The methods (6.25), (6.26), and (6.22) are all examples of *implicit* methods, because in each instance it is not possible to solve for y_{n+1} in terms of y_n and other known values. It nonetheless is possible to make use of these methods; the trapezoid method, (6.25), is in fact very useful.

One way to implement methods of this type is to view the equations (6.25), (6.26), and (6.22) as instances of a single nonlinear equation in the single unknown y_{n+1}. That is, we *solve* for the unknown value of y_{n+1} using Newton's method or the secant method or a fixed-point iteration.

EXAMPLE 6.10

Consider the IVP

$$y' = -y \ln y, \quad y(0) = y_0. \tag{6.28}$$

Applying the trapezoid method (6.25) to this yields the computation

$$y_{n+1} = y_n - \frac{1}{2} h[y_{n+1} \ln y_{n+1} + y_n \ln y_n], \tag{6.29}$$

where we now view y_{n+1} as the (unknown) root of the function

$$F(y) = y - y_n + \frac{1}{2} h[y \ln y + y_n \ln y_n].$$

Alternatively, we can view (6.29) as defining y_{n+1} as the fixed point of the function

$$g(y) = y_n - \frac{1}{2} h[y \ln y + y_n \ln y_n].$$

Regardless of which view we take, we can apply the methods of Chapter 3 to solve for y_{n+1}. Denote the individual iterates as $y_{n+1,k}$, where k is the iteration counter.[4] Then Newton's method applied to $F(y)$ yields the iteration

$$y_{n+1,k+1} = y_{n+1,k} - \left(\frac{y_{n+1,k} - y_n - \frac{1}{2} h[f(t_{n+1}, y_{n+1,k}) + f(t_n, y_n)]}{1 - \frac{1}{2} h f_y(t_{n+1}, y_{n+1,k})} \right),$$

where

$$f_y(t,y) = \frac{\partial f}{\partial y}(t,y);$$

the secant method yields the iteration

$$y_{n+1,k+1} = y_{n+1,k} - \left(y_{n+1,k} - y_n - \frac{1}{2} h[f(t_{n+1}, y_{n+1,k}) + f(t_n, y_n)] \right)$$

$$\times \left(\frac{y_{n+1,k} - y_{n+1,k-1}}{y_{n+1,k} - \frac{1}{2} h f(t_{n+1}, y_{n+1,k}) - y_{n+1,k-1} + \frac{1}{2} h f(t_{n+1}, y_{n+1,k-1})} \right);$$

and a simple fixed-point iteration yields the iteration

$$y_{n+1,k+1} = y_n - \frac{1}{2} h[y_{n+1,k} \ln y_{n+1,k} + y_n \ln y_n]. \tag{6.30}$$

(Note that the fixed-point iteration is by far the simplest of these three, although it will also be the slowest to converge.) We know that all three of these will converge if h is sufficiently small and if we take the initial guess close enough to the actual value (for the fixed-point iteration this is a straightforward application of Theorem 3.6; see Exercise 17). It is very reasonable to

---

[4]At this point it is worthwhile to point out that some confusion can occur. We have two indices on y. One of them $(n+1)$ refers to the time step in the numerical solution of the differential equation. The other one (k) refers to the iteration count for the solution—within each time step—of the nonlinear equation for the next approximate value of y.

take $y_{n+1,0} = y_n$; i.e., take the value from the previous step of the iteration as the initial guess for the next step. An alternative (and usually better) means of initializing the iteration is to use some explicit method—Euler's method, for example—to generate $y_{n+1,0}$. ∎

While there are circumstances in which it is necessary to carry out this kind of iteration, for most problems it is in fact possible to use a much cruder means of estimating y_{n+1}. This leads us to the very important *predictor-corrector* idea.

Consider the trapezoid method, defined by (6.25); instead of actually solving for the exact value of y_{n+1} that satisfies this equation, we instead apply a simpler method to estimate y_{n+1} and then use this estimated value on the right side of (6.25). For the trapezoid method, it is common to use Euler's method as the estimator, or *predictor*:

$$\bar{y}_{n+1} = y_n + hf(t_n, y_n). \tag{6.31}$$

Then the *corrector* step is accomplished by

$$y_{n+1} = y_n + \frac{1}{2}h[f(t_{n+1}, \bar{y}_{n+1}) + f(t_n, y_n)]. \tag{6.32}$$

This combination is generally referred to as the trapezoid rule predictor-corrector method. Note that it is the same as using Euler's method to initialize a fixed-point iteration that we arbitrarily stop after one step.

EXAMPLE 6.11

Consider the IVP

$$y' = -y \ln y, \quad y(0) = \frac{1}{2}.$$

We will use the trapezoid rule predictor-corrector method to approximate the solution at $t = 1$, using the rather crude value $h = 0.25$. We have the following sequence of computations:

$$\bar{y} = y_0 + hf(t_0, y_0) = 0.5 - 0.25 \times 0.5 \ln 0.5 = 0.5866433976$$

$$y_1 = y_0 + \frac{1}{2}h[f(t_0, y_0) + f(t_1, \bar{y})] = 0.58243161136465$$

$$\bar{y} = y_1 + hf(t_1, y_1) = 0.66113901764113$$

$$y_2 = y_1 + \frac{1}{2}h[f(t_1, y_1) + f(t_2, \bar{y})] = 0.65598199856663$$

$$\bar{y} = y_2 + hf(t_2, y_2) = 0.72512609790319$$

$$y_3 = y_2 + \frac{1}{2}h[f(t_2, y_2) + f(t_3, \bar{y})] = 0.71968686944048$$

$$\bar{y} = y_3 + hf(t_3, y_3) = 0.77887015094459$$

$$y_4 = y_3 + \frac{1}{2}h[f(t_3, y_3) + f(t_4, \bar{y})] = 0.77360953103925.$$

∎

TABLE 6.2 Implicit Method Applied to $y' = -y \ln y$

	Predictor-corrector		Fixed-point solution	
n	y_n	Error	y_n	Error
1	0.521438685E+00	0.629350068E–05	0.521441082E+00	0.389659700E–05
2	0.542415401E+00	0.124265310E–04	0.542420677E+00	0.715031332E–05
3	0.562889543E+00	0.183888337E–04	0.562898148E+00	0.978351030E–05
4	0.582826911E+00	0.241665678E–04	0.582839249E+00	0.118283433E–04
5	0.602199375E+00	0.297435050E–04	0.602215794E+00	0.133243757E–04
6	0.620984504E+00	0.351021428E–04	0.621005290E+00	0.143164168E–04
7	0.639165163E+00	0.402246989E–04	0.639190535E+00	0.148526117E–04
8	0.656729097E+00	0.450939634E–04	0.656759208E+00	0.149827925E–04
9	0.673668505E+00	0.496939995E–04	0.673703441E+00	0.147570906E–04
10	0.689979615E+00	0.540106915E–04	0.690019401E+00	0.142247986E–04
11	0.705662271E+00	0.580321478E–04	0.705706870E+00	0.134334649E–04
12	0.720719531E+00	0.617489703E–04	0.720768874E+00	0.124059468E–04
13	0.735157278E+00	0.651544045E–04	0.735211228E+00	0.112040544E–04
14	0.748983859E+00	0.682443879E–04	0.749042237E+00	0.986672690E–05
15	0.762209745E+00	0.710175127E–04	0.762272333E+00	0.842946303E–05
16	0.774847210E+00	0.734749197E–04	0.774913760E+00	0.692422919E–05

EXAMPLE 6.12

To illustrate the difference between actually solving the nonlinear equation and using a predictor-corrector method, we continue with the same IVP (6.28) as used in the previous example. The exact solution here is $y = e^{(-\ln 2)e^{-t}}$. Table 6.2 shows the results of approximating the solution to this equation using the predictor-corrector method (6.31)–(6.32) (second column) and actually solving equation (6.29) (using a fixed-point iteration) for the exact value of y_{n+1} (fourth column). Clearly, the two methods are not producing wildly different approximations. Note also that the calculation based on exactly solving the recursion is more accurate; in fact, the error in the iterated approximation (fifth column) starts out at about 61% of the error for the predictor-corrector approximation, and ends up at around 9%. ∎

Generally speaking, unless the differential equation is *very* sensitive to changes in the data, a simple predictor-corrector method (see Algorithm 6.1) will be just as good as the more time-consuming process of solving for the exact value of y_{n+1} that satisfies the implicit recursion.

If the differential equation is linear, however, we can entirely avoid the problem of implicitness, as we will now illustrate. Write the general linear ODE as

$$y' = a(t)y + b(t), \quad y(t_0) = y_0,$$

Algorithm 6.1 *Pseudocode for the Trapezoid Predictor-Corrector*

```
input t0, y0, h, n
external f
for k = 1 to n do
!
!     First, predict:
!
      f0 = f(t0,y0)
      ybar = y0 + h*f0
!
!     Next, correct:
!
      y = y0 + 0.5*h*(f0 + f(t0+h,ybar))
!
!     Update for next pass through the loop
!
      y0 = y
      t0 = t
endfor
```

Programming Hint: Note how this code saves the function evaluation f(t0,y0) in order to reuse it in a later statementps.

where a and b are known functions of t only. If we apply the trapezoid method to this equation, we initially have the implicit recursion

$$y_{n+1} = y_n + \frac{1}{2}h\left[a(t_n)y_n + b(t_n) + a(t_{n+1})y_{n+1} + b(t_{n+1})\right].$$

However, we can make this explicit by using the linearity of the equation to solve for y_{n+1}:

$$y_{n+1} = \left(\frac{1 + \frac{1}{2}ha(t_n)}{1 - \frac{1}{2}ha(t_{n+1})}\right)y_n + \frac{1}{2}h\frac{b(t_n) + b(t_{n+1})}{1 - \frac{1}{2}ha(t_{n+1})}. \tag{6.33}$$

The significance of this will become apparent later, when we discuss stability and stiffness issues in more detail. For now, we content ourselves with an example.

EXAMPLE 6.13

Consider the IVP

$$y' = y, \quad y(0) = 1,$$

which we considered previously, using only Euler's method. This is a linear equation, so we can use the ideas of (6.33) to approximate this using the trapezoid rule. The recursion becomes

$$y_{n+1} = \frac{1+h}{1-h}y_n, \quad y_0 = 1;$$

TABLE 6.3 Trapezoid Rule Applied to $y' = y$

$n = h^{-1}$	$E(h)$
4	0.1432958E–01
8	0.3550064E–02
16	0.8855204E–03
32	0.2212558E–03
64	0.5530617E–04
128	0.1382606E–04
256	0.3456484E–05
512	0.8641191E–06
1024	0.2160295E–06

thus, for $h = \frac{1}{4}$ we very quickly get that

$$y_1 = \frac{5/4}{3/4} y_0 = 1.6667$$

$$y_2 = \frac{5}{3} y_1 = 2.7778$$

$$y_3 = \frac{5}{3} y_2 = 4.630$$

$$y_4 = \frac{5}{3} y_3 = 7.7160.$$

If we continue with smaller and smaller values of h, then the error $E(h) = \max_{t_k \le 1} |y(t_k) - y_k|$ is as given in Table 6.3. Note that the error is going down by a factor of 4 , as we expect from a second-order method.

The predictor-corrector idea can also be used to make (6.26) a usable method. If we use Euler's method as the predictor, then we have

$$\bar{y} = y_n + \frac{1}{2} hf(t_n, y_n) \tag{6.34}$$

$$y_{n+1} = y_n + hf\left(t_n + \frac{1}{2}h, \bar{y}\right). \tag{6.35}$$

Note that we use an Euler predictor to take a "half step," and use this value in the corrector to approximate $y(t_n + h/2)$. ∎

EXAMPLE 6.14

Let's compute approximate solutions to the IVP

$$y' = -e^{-(t+y)}, \quad y(0) = 1,$$

which has exact solution $y = \ln(e - 1 + e^{-t})$. Using $h = \frac{1}{4}$, we compute as follows:

$$\bar{y} = y_0 + \frac{1}{2} h\left(-e^{-(t_0 + y_0)}\right) = 1 - \frac{1}{8} e^{-1} = 0.9540150699,$$

so that

$$y_1 = y_0 + h\left(-e^{-(t_0+h/2+\bar{y})}\right) = 1 - \frac{1}{4}e^{-1/8-0.9540150699} = 0.957508729.$$

The second value is computed similarly:

$$\bar{y} = y_1 + \frac{1}{2}h\left(-e^{-(t_1+y_1)}\right) = 0.920141092$$

$$y_2 = y_1 + h\left(-e^{-(t_1+h/2+\bar{y})}\right) = 0.8890439235.$$

■

It can be shown (see the exercises) that this is also a second-order method.

6.4.3 Starting Values and Multistep Methods

Because they depend on more than just a single previous value, multistep methods such as (6.20), (6.21), and (6.22) require *starting values*. In essence, we must use some other method to compute approximations $y_1, y_2, \dots, y_p$ before we can use the multistep method. For the methods presented so far, we need only a single additional value, y_1. For some of the methods to be presented in Section 6.6, we will need several such starting values. One reason for the development of the predictor-corrector methods in the previous subsection is precisely to give us single-step methods that we can use to produce accurate starting values for the multistep methods such as (6.20), (6.21), and (6.22).

Let's look at an example.

EXAMPLE 6.15

We take as our initial value problem the same one we looked at in the beginning of the previous section,

$$y' = -y\ln y, \quad y(0) = \frac{1}{2}. \tag{6.36}$$

We want to apply, say, the midpoint method,

$$y_{n+1} = y_{n-1} + 2hf(t_n, y_n),$$

with $h = \frac{1}{4}$ to this problem. We will first use Euler's method to provide the estimate of $y(t_1)$ that is necessary to proceed with the computation. We get

$$y_1 = y_0 + hf(t_0, y_0) = \frac{1}{2} + \frac{1}{4}\left(-\frac{1}{2}\ln\frac{1}{2}\right) = 0.5866433976.$$

We can now compute using the midpoint method:

$$y_2 = y_0 + 2hf(t_1, y_1) = \frac{1}{2} + 2 \times \frac{1}{4}\left(-0.5866433976\,\ln 0.5866433976\right) = 0.6564396503$$

$$y_3 = y_1 + 2hf(t_2, y_2) = 0.7247991686,$$

and so forth.

TABLE 6.4   Midpoint Method Applied to $y' = -y \ln y$, $h = 1/16$

	y_1 computed by predictor-corrector		y_1 computed by Euler's method	
n	y_n	Error	y_n	Error
1	0.521438685E+00	0.629350068E−05	0.521660849E+00	-0.215870648E−03
2	0.542442735E+00	-0.149082970E−04	0.542433042E+00	-0.521501377E−05
3	0.562913366E+00	-0.543392852E−05	0.563136000E+00	-0.228068587E−03
4	0.582876067E+00	-0.249892345E−04	0.582854530E+00	-0.345267290E−05
5	0.602241522E+00	-0.124030017E−04	0.602465395E+00	-0.236276553E−03
6	0.621050404E+00	-0.307978504E−04	0.621015069E+00	0.453741300E−05
7	0.639220651E+00	-0.152631109E−04	0.639446837E+00	-0.241449481E−03
8	0.656807255E+00	-0.330637366E−04	0.656756294E+00	0.178973669E−04
9	0.673732972E+00	-0.147732929E−04	0.673962850E+00	-0.244651773E−03
10	0.690066203E+00	-0.325779206E−04	0.689997851E+00	0.357749024E−04
11	0.705732009E+00	-0.117057153E−04	0.705967262E+00	-0.246958289E−03
12	0.720811392E+00	-0.301122659E−04	0.720723877E+00	0.574032598E−04
13	0.735229212E+00	-0.678027590E−05	0.735471822E+00	-0.249390296E−03
14	0.749078472E+00	-0.263680981E−04	0.748969952E+00	0.821511474E−04
15	0.762281388E+00	-0.625682149E−06	0.762533643E+00	-0.252880568E−03
16	0.774942633E+00	-0.219485394E−04	0.774811136E+00	0.109548790E−03

Suppose now that we want to use the trapezoid rule predictor-corrector method (6.31)–(6.32) to provide the estimate of $y(t_1)$. Again using $h = \frac{1}{4}$, we compute

$$\bar{y} = y_0 + hf(t_0, y_0) = 0.5 - 0.25 \times 0.5 \ln 0.5 = 0.5866433976$$

$$y_1 = y_0 + \frac{1}{2}h[f(t_0, y_0) + f(t_1, \bar{y})] = 0.5824316114,$$

so that

$$y_2 = y_0 + 2hf(t_1, y_1) = \frac{1}{2} + 2 \times \frac{1}{4}(-0.5824316114 \ \ln 0.5824316114) = 0.6574148126$$

$$y_3 = y_1 + 2hf(t_2, y_2) = 0.7203046741,$$

etc. Regardless of which method is used to produce y_1, we can now compute y_k values using only the midpoint method.

Now, by taking $h = 1/16$ and computing out with $t = 1$, we get the results shown in Table 6.4. Compare this with Table 6.2. Note that, as expected, we are getting better accuracy when we use the more accurate starting value, although it should be noted that the discrepancy is not large.

This example prompts an interesting question. Note that we used Euler's method, which is only $O(h)$, to generate the starting values for the midpoint method, which is $O(h^2)$. Is the use of a lower-order method for the starting values going to affect the accuracy of the overall computation?

A complete explanation is beyond the scope of this text, but it can be shown that a $(p-1)$-order method can be used to generate the starting values for a p-order method without affecting the overall order of convergence. It would of course be more accurate to use a p-order method than to use a $(p-1)$-order method for the starting values, but the net order of accuracy of the method is not affected. ∎

6.4.4 The Midpoint Method and Weak Stability

Let's consider the midpoint method (6.20) as applied to the very simple differential equation

$$y' = -y, \quad y(0) = 1, \tag{6.37}$$

which has exact solution $y(t) = e^{-t}$. To minimize the effects of any error in the starting values, let's use

$$y_1 = y(t_1) = e^{-h}, \tag{6.38}$$

where h is the mesh spacing. Note that this is the *exact* value of $y(t_1)$; i.e., there is no error in using this starting value. Figure 6.4 shows the results of applying (6.20) to (6.37) using (6.38) as the starting value, for a sequence of mesh values $h^{-1} = 4, 8, 16, \ldots, 128$. We have plotted the exact solution curve and each approximate curve, as generated by the numerical method.

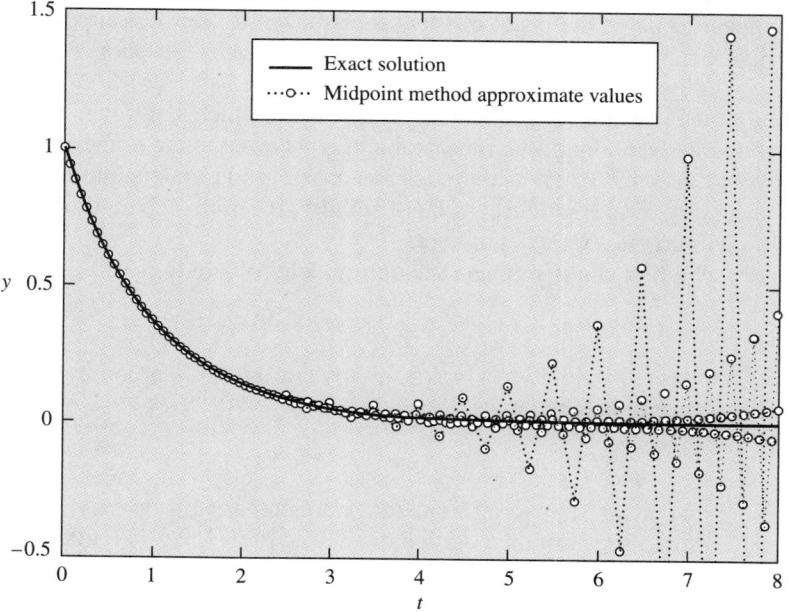

FIGURE 6.4   Illustration of weak stability, $h = 4^{-1}, 8^{-1}, 16^{-1}$:

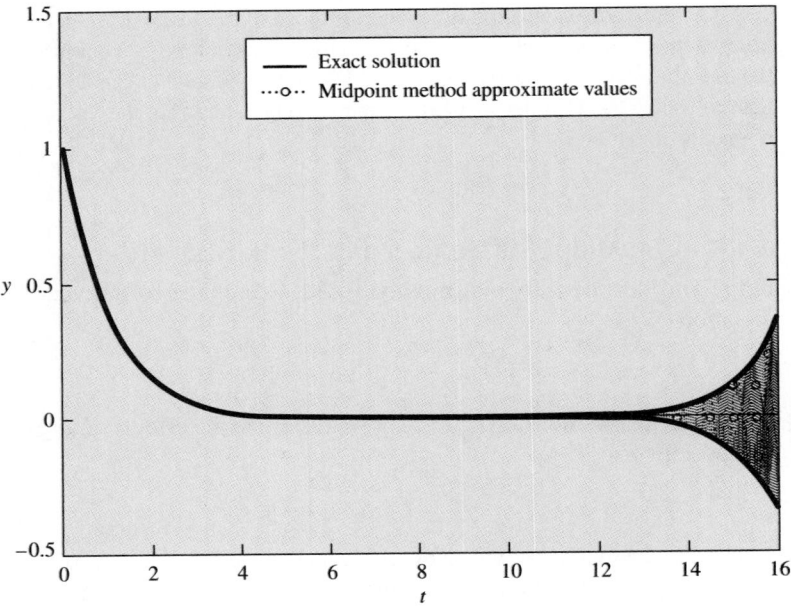

FIGURE 6.5   Illustration of weak stability, $h = 128^{-1}$.

We expect to get decent results, since the midpoint method is second-order, but instead we get terrible results for larger values of t. Taking a smaller value of h does appear to defer the onset of this problem, but it does not appear to eliminate it. As Figure 6.5 shows, even for $h^{-1} = 128$ we eventually get poor results.

What is going on here?

It is tempting to blame this problem—which is known as *weak stability*—on rounding error, but in fact the problem is inherent in the numerical method (6.20) and would occur in exact arithmetic. Because a good understanding of weak stability is necessary for the full discussion of stability for multistep methods (see Section 6.7.1), we will go into some detail here.

We first slightly generalize our model problem to

$$y' = \lambda y, \quad y(0) = 1,$$

which has exact solution $y = e^{\lambda t}$ for any constant λ. The midpoint method applied to this problem then becomes

$$y_{n+1} = y_n + 2h\lambda y_n, \quad y_0 = 1, y_1 = e^{h\lambda}.$$

Note that the mesh spacing h and the parameter λ show up only when multiplying each other; so, to simplify the notation, we will write $h\lambda = \xi$ hereafter. Thus we have

$$y_{n+1} = y_n + 2\xi y_n, \quad y_0 = 1, y_1 = e^{\xi}. \tag{6.39}$$

This is an example of what is called a *three-term recurrence relation*, and we can actually produce a formula to solve it, much as we would solve a second-order constant-coefficient differential equation. We first look for solutions in the form

$$y_n = r^n \tag{6.40}$$

for some value r. We will actually find *two* such values, r_1 and r_2, and the linearity of the recurrence then implies that

$$y_n = C_1 r_1^n + C_2 r_2^n$$

is a solution for any constants C_1 and C_2. The values of the constants are determined by the need to satisfy $y_0 = 1$ and $y_1 = e^\xi$, and that will complete the construction of the solution.

So much for the outline and preview. If we substitute (6.40) into (6.39), we get

$$r^{n+1} = r^{n-1} + 2\xi r^n.$$

Thus dividing by r^{n-1} yields the quadratic equation

$$r^2 - 2\xi r - 1 = 0,$$

which defines r_1 and r_2, according to the quadratic formula:

$$r_1 = (\xi + \sqrt{\xi^2 + 1}), \qquad r_2 = (\xi - \sqrt{\xi^2 + 1}). \tag{6.41}$$

The constants C_1 and C_2 are then found by solving the system

$$1 = C_1 + C_2$$
$$e^\xi = C_1 r_1 + C_2 r_2$$

to get

$$C_1 = \frac{r_2 - e^\xi}{r_2 - r_1} = \frac{(-\xi + \sqrt{\xi^2 + 1}) + e^\xi}{2\sqrt{\xi^2 + 1}}$$

and

$$C_2 = \frac{r_1 - e^\xi}{r_1 - r_2} = \frac{(\xi + \sqrt{\xi^2 + 1}) - e^\xi}{2\sqrt{\xi^2 + 1}}.$$

Hence, we have the solution

$$y_n = C_1 r_1^n + C_2 r_2^n, \tag{6.42}$$

which the student can verify by direct substitution (see Exercise 12) is the exact solution to the recurrence (6.39). We should repeat, for emphasis, that this solution formula is *exact*: this is what the computation will produce in the complete absence of rounding error.

Some modest work allows us to estimate the size of C_1 and C_2 (see Exercise 14):

$$C_1 = 1 + O(\xi^3), \qquad C_2 = O(\xi^3). \tag{6.43}$$

Consider, now, the behavior of the approximate solution as n grows, under the assumption that $\lambda < 0$. This assumption means that $\xi < 0$, and it also means that the exact solution will be a decaying exponential. However, the approximate solution satisfies

$$y_n = C_1 r_1^n + C_2 r_2^n,$$

where we have $0 < r_1 < 1$ and $r_2 < -1$ (see Exercise 13). Hence, as $n \to \infty$, $r_1^n \to 0$; but $|r_2|^n \to \infty$, thus destroying the accuracy of the computation. Since $C_2 \neq 0$, this term will be present in the calculation, although the fact that $C_2 = O(h^3)$ does mean that it will take more and more steps for the growing term to be noticed, for smaller and smaller values of h.

The fundamental issue here is that the exact solution to the recursion that defines the approximate solution contains solution components that do not correspond in any way to the exact solution of the ODE. The $C_1 r_1^n$ term, it can be shown, will converge to the exact solution and will be an $O(h^2)$ approximation to it. But the other term, usually called a *parasite* or *parasitic solution*, will eventually dominate and corrupt the approximation.

Parasitic solutions are an inherent feature of multistep methods, and the stability theory for multistep methods is designed to minimize their ability to corrupt the solution. See Section 6.7 for more details.

EXERCISES

1. Using the approximation

$$y'(t_n) \approx \frac{y(t_{n+1}) - y(t_{n-1})}{2h},$$

derive the numerical method (6.20) for solving initial value problems. What is the residual? What is the truncation error? Is it a consistent method?

2. Using the approximation

$$y'(t_{n-1}) \approx \frac{-y(t_{n+1}) + 4y(t_n) - 3y(t_{n-1})}{2h},$$

derive the numerical method (6.21) for solving initial value problems. What is the residual? What is the truncation error? Is it a consistent method?

3. Using the approximation

$$y'(t_{n+1}) \approx \frac{3y(t_{n+1}) - 4y(t_n) + y(t_{n-1})}{2h},$$

derive the numerical method (6.22) for solving initial value problems. What is the residual? What is the truncation error? Is it a consistent method?

4. Use the trapezoid rule predictor-corrector with $h = \frac{1}{4}$ to compute approximate values of $y(1)$ for each of the following initial value problems. Don't write a com-

puter program; use a hand calculator to produce an orderly table of (t_k, y_k) pairs.
(a) $y' + 2y = 1$, $y(0) = 2$
(b) $y' = y(1 - y)$, $y(0) = \frac{1}{2}$
(c) $y' = y(1 + e^{2t})$, $y(0) = 1$
(d) $ty' = y(\sin t)$, $y(0) = 2$

5. Repeat the preceding, using the method (6.22) as a predictor-corrector, with Euler's method as the predictor. Also use Euler's method to produce the starting value, y_1.

6. Write a computer program that solves each of the initial value problems in Exercise 4, using the trapezoid rule predictor-corrector and $h = 1/16$.

7. For each initial value problem below, approximate the solution using the trapezoid rule predictor-corrector with a sequence of decreasing grids $h^{-1} = 2, 4, 8, \ldots$. For those problems where an exact solution is given, compare the accuracy achieved over the interval $[0, 1]$ with the theoretical accuracy.
(a) $y' + 4y = 1$, $y(0) = 1$; $y(t) = \frac{1}{4}(3e^{-4t} + 1)$
(b) $y' + y = \sin 4\pi t$, $y(0) = \frac{1}{2}$
(c) $y' = -y \ln y$, $y(0) = 3$; $y(t) = e^{(\ln 3)e^{-t}}$
(d) $y' + \sin y = 0$, $y(0) = 1$

8. In this exercise we will consider the tumor growth model from the beginning of the chapter. Take $\lambda = \mu = 1$, $a = 0.25$, $S_i = 0.8$, and $\sigma = 0.25$. Use the trapezoid rule predictor-corrector to solve the differential equation, using $h = 1/16$, and show that the tumor radius approaches a limiting value as $t \to \infty$.

9. Repeat the previous exercise, but this time with $S_i = 0.90$, $\sigma = 0.05$, and $a = 0.50$. What happens now?

10. Now solve the differential equation using a variety of S_i, σ, and a values (your choices). What happens?

11. Now let's model some treatment for our mathematical tumor. In the previous three exercises, we assumed that the nutrient level was constant. Suppose we are able to decrease the nutrient level according to the model

$$\sigma(t) = \sigma_\infty + (\sigma_0 - \sigma_\infty)e^{-qt}.$$

Here σ_0 is the initial nutrient level, σ_∞ is the asymptotic nutrient level, and q measures the rate at which the nutrient level drops. Investigate the effect of various choices of these parameters on the growth of the tumor, based on your observations from the previous exercises. Again, use the trapezoid rule predictor-corrector with $h = 1/16$ to solve the differential equation.

12. Verify by direct substitution that (6.42) satisfies the recursion (6.39) for all $n \geq 2$.

13. For the midpoint method (6.20), show that if $\lambda < 0$ and h is sufficiently small, then $0 < r_1 < 1$ and $r_2 < -1$.

14. Show that (6.43) is valid. (Hint: First, show that

$C_1 = 1 - C_2$. Next, use Taylor expansions to write

$$\frac{\xi - e^\xi}{\sqrt{\xi^2 + 1}} = \frac{-1 - \frac{1}{2}\xi^2 + O(\xi^3)}{1 + \frac{1}{2}\xi^2 + O(\xi^4)}.\Bigg)$$

15. Use the midpoint rule predictor-corrector method (6.34)–(6.35) to solve each of the IVPs given in Exercise 4.

16. Show that the residual for the midpoint rule predictor-corrector is given by

$$R = y(t + h) - y(t) - hf(t + h/2, y(t) + (h/2)f(t, y(t))).$$

Then use Taylor's Theorem to show that $R = O(h^3)$ and hence that the midpoint rule predictor-corrector is a second-order method.

17. Assume that f is differentiable in y and that this derivative is bounded in absolute value for all t and y:

$$|f_y(t, y)| \leq F.$$

Show that using fixed-point iteration to solve for y_{n+1} in the trapezoid rule method will converge so long as h is sufficiently small. (Hint: Recall Theorem 3.6.)

18. Derive the numerical method based on using Simpson's rule to approximate the integral in

$$y(t + h) = y(t - h) + \int_{t-h}^{t+h} f(s, y(s)) \, ds.$$

What is the order of accuracy of this method? What is the truncation error? Is it implicit or explicit? Is it a single-step or multistep method?

6.5 SINGLE-STEP METHODS: RUNGE-KUTTA

The Runge-Kutta[5] family of methods is one of the most popular families of accurate solvers for initial value problems. The general derivation can become very involved; to avoid drowning in a sea of detail and notation, we will outline the basic ideas using the second-order case.

Recall the usual predictor-corrector formulation of the trapezoid method:

$$\bar{y}_{n+1} = y_n + hf(t_n, y_n)$$
$$y_{n+1} = y_n + \frac{1}{2}h[f(t_{n+1}, \bar{y}_{n+1}) + f(t_n, y_n)].$$

---

[5]Martin Wilhelm Kutta (1867–1944) studied at Breslau and Munich, in addition to a year spent in Britain at Cambridge. Most of his professional career was spent in Stuttgart. Building on Runge's original idea (first presented in an 1894 article), Kutta published his version of the Runge-Kutta methods in 1901.

We can directly substitute the predictor into the corrector to write this as a single recursion:

$$y_{n+1} = y_n + \frac{1}{2}h\left[f(t_{n+1}, y_n + hf(t_n, y_n)) + f(t_n, y_n)\right]. \tag{6.44}$$

Since the differential equation implies that f gives values of y', it follows that we can view (6.44) as defining y_{n+1} from y_n by advancing along a straight line defined by the simple average of the two slopes $f(t_n, y_n)$ and $f(t_{n+1}, \bar{y})$. This raises the question, of course, of whether or not a different average of two slopes might yield a more accurate method. To this end, we consider the more general method

$$y_{n+1} = y_n + c_1 hf(t_n, y_n) + c_2 hf(t_n + \alpha h, y_n + \beta hf(t_n, y_n)), \tag{6.45}$$

where c_1, c_2, α, and β are parameters as yet undetermined. We want to choose these so that the approximate solution defined by (6.45) is as accurate as possible; i.e., we want to make the truncation error as small as possible, in terms of powers of h. Thus we look at the (admittedly rather imposing) expression

$$R = y(t+h) - y(t) - c_1 hf(t, y(t)) - c_2 hf(t + \alpha h, y(t) + \beta hf(t, y(t))).$$

To reduce this so we can infer the correct values of c_1, c_2, α, and β that will yield the smallest residual R, we will need to use Taylor's Theorem in two variables, which we state here, without proof:

$$F(x+h, y+\eta) = F(x,y) + hF_x(x,y) + \eta F_y(x,y) +$$

$$\frac{1}{2}(h^2 F_{xx}(x,y) + h\eta F_{xy}(x,y) + \eta^2 F_{yy}(x,y)) + O(h^3 + \eta^3).$$

We can thus expand the last term in R as

$$f(t+\alpha h, y(t) + \beta hf(t, y(t))) = f(t, y(t)) + \alpha hf_t(t, y(t)) + \beta hf(t, y(t))f_y(t, y(t))$$

$$+ \frac{1}{2}(\alpha^2 h^2 f_{tt}(t, y(t)) + \alpha\beta h^2 f(t, y(t))f_{ty}(t, y(t))$$

$$+ \beta^2 h^2 [f(t, y(t))]^2 f_{yy}(t, y(t))) \quad + O(h^3).$$

We can also expand $y(t+h)$ in terms of $y(t)$ as follows:

$$y(t+h) = y(t) + hy'(t) + \frac{1}{2}h^2 y''(t) + O(h^3).$$

But the differential equation implies that $y' = f(t, y(t))$ and therefore, in addition,

$$y''(t) = \frac{d}{dt}f(t, y(t)) = f_t(t, y(t)) + f_y(t, y(t))y'(t)$$

$$= f_t(t, y(t)) + f_y(t, y(t))f(t, y(t)).$$

Now let us substitute both of these expansions into our expression for R, where we get

$$R = \left(y + hf + \frac{1}{2}h^2(f_t + f_y f) + O(h^3)\right) - y - c_1 hf$$
$$- c_2 h\left(f + \alpha h f_t + \beta h f f_y + \frac{1}{2}(\alpha^2 h^2 f_{tt} + \alpha\beta h^2 f f_{ty} + \beta^2 h^2 f^2 f_{yy}) + O(h^3)\right).$$

The reader should note that we have written this without explicit arguments to any of the f terms, including the partial derivatives, since all of the arguments are $(t, y(t))$. Remember that our goal here is to find the values of c_1, c_2, α, and β that make R as small (in terms of h) as possible. So, we rewrite our ponderous expression for R in terms of powers of h. We get

$$R = h(f - c_1 f - c_2 f)$$
$$+ h^2\left(\frac{1}{2}(f_t + f_y f) - c_2 \alpha f_t - c_2 \beta f f_y\right) + O(h^3)$$
$$= h(f - c_1 f - c_2 f)$$
$$+ h^2\left[\left(\frac{1}{2} - c_2 \alpha\right)f_t + \left(\frac{1}{2} - c_1 \alpha\right)f f_y\right] + O(h^3),$$

from which we can extract the equations

$$1 - c_1 - c_2 = 0, \quad \frac{1}{2} - c_2 \alpha = 0, \quad \frac{1}{2} - c_2 \beta = 0.$$

If these three equations are satisfied, then the residual satisfies $R = O(h^3)$; thus the truncation error is $O(h^2)$. Since this is a set of three equations in four unknowns, there is more than one solution. Typically, c_1 is regarded as arbitrary, and we then have

$$c_2 = 1 - c_1 \tag{6.46}$$
$$\alpha = \beta = \frac{1}{2c_2}. \tag{6.47}$$

Thus some possible solutions include

$$c_1 = c_2 = \frac{1}{2}, \quad \alpha = \beta = 1;$$

this yields the trapezoid rule predictor-corrector method:

$$y_{n+1} = y_n + \frac{1}{2}h[f(t_{n+1}, y_n + hf(t_n, y_n)) + f(t_n, y_n)].$$

Or

$$c_1 = 0, \quad c_2 = 1, \quad \alpha = \beta = \frac{1}{2};$$

this yields the midpoint rule predictor-corrector (6.34)–(6.35):

$$y_{n+1} = y_n + hf(t_n + \frac{1}{2}h, y_n + \frac{1}{2}hf(t_n, y_n)). \tag{6.48}$$

We could also try

$$c_1 = \frac{1}{4}, \quad c_2 = \frac{3}{4}, \quad \alpha = \beta = \frac{2}{3},$$

which yields a method sometimes called the method of Heun:[6]

$$y_{n+1} = y_n + \frac{1}{4}h \left[f(t_n, y_n) + 3f\left(t_n + \frac{2}{3}h, y_n + \frac{2}{3}hf(t_n, y_n)\right) \right]. \tag{6.49}$$

All three of these have second-order truncation errors, and it can be shown that all of them are second-order accurate in the sense that, for any $T > 0$,

$$\max_{t_k \leq T} |y(t_k) - y_k| \leq C_T h^2,$$

where C_T will depend on T but not on h.

The most commonly used Runge-Kutta method is the fourth-order method, derived in a manner very similar to what we did here, but using four slope values instead of only two. The required manipulations are necessarily much more involved, and so we skip them. The method is usually written as follows:

$$k_1 = hf(t_n, y_n) \tag{6.50}$$

$$k_2 = hf\left(t_n + \frac{1}{2}h, y_n + \frac{1}{2}k_1\right) \tag{6.51}$$

$$k_3 = hf\left(t_n + \frac{1}{2}h, y_n + \frac{1}{2}k_2\right) \tag{6.52}$$

$$k_4 = hf(t_n + h, y_n + k_3) \tag{6.53}$$

$$y_{n+1} = y_n + \frac{1}{6}(k_1 + 2k_2 + 2k_3 + k_4). \tag{6.54}$$

A pseudocode for implementing this is given at the end of the section.

EXAMPLE 6.16

For illustration, let's consider our usual example IVP

$$y' = -y \ln y, \quad y(0) = \frac{1}{2},$$

which we will first solve with a very coarse grid, using both the second-order method (6.49) and the fourth-order method (6.50)–(6.54). We then will show the results of solving the same

---

[6]This is perhaps a good time to point out that there is a lot of inconsistency in naming many of these methods. What we call the "trapezoid method" is called "improved Euler" or "modified Euler" by other authors, just as an example. We call the method in (6.49) Heun's method because Goldstine [9] says that Kutta used that attribution in his 1901 paper introducing what we call Runge-Kutta methods.

equation using the same methods, but with finer grids. Let $h = 1/2$; to compute out to $t = 1$, then, requires two steps with each method. For the method of Heun, we compute as follows:

1:

$$\bar{f} = f(t_0, y_0) = f(0, 1/2) = -(1/2)\ln(1/2) = 0.3465735903$$
$$\bar{y} = y_0 + (2/3)h\bar{f} = 0.5 + (1/3)\bar{f} = 0.6155245301$$
$$y_1 = y_0 + \frac{1/2}{4}\left[\bar{f} + 3f(t_0 + (2/3)h, \bar{y})\right]$$
$$= 0.5 + 0.125\left[0.3465735903 + 3(0.2987020395)\right] = 0.6553349636$$

2:

$$\bar{f} = f(t_1, y_1) = f(0.5, 0.6553349636) = 0.276950309$$
$$\bar{y} = y_1 + (2/3)h\bar{f} = 0.6553349636 + (1/3)(0.276950309) = 0.7476517333$$
$$y_2 = y_1 + \frac{1/2}{4}\left[\bar{f} + f(t_1 + (2/3)h, \bar{y})\right]$$
$$= 0.6553349636 + 0.125\left[0.276950309 + 3(0.2174305868)\right] = 0.7714902223$$

For fourth-order Runge-Kutta (RK4), the computation is longer, of course:

1:

$$k_1 = hf(t_0, y_0) = 0.17328679513999$$
$$z_1 = y_0 + (1/2)k_1 = 0.58664339756999$$
$$k_2 = hf(t_0 + (1/2)h, z_1) = 0.15643965031862$$
$$z_2 = y_0 + (1/2)k_2 = 0.57821982515931$$
$$k_3 = hf(t_0 + (1/2)h, z_2) = 0.15837474613445$$
$$z_3 = y_0 + k_3 = 0.65837474613445$$
$$k_4 = hf(t_0 + h, z_3) = 0.13759406302779$$
$$y_1 = y_0 + (1/6)(k_1 + 2k_2 + 2k_3 + k_4) = 0.65675160851232$$

2:

$$k_1 = hf(t_1, y_1) = 0.13806541027436$$
$$z_1 = y_1 + (1/2)k_1 = 0.72578431364950$$
$$k_2 = hf(t_1 + (1/2)h, z_1) = 0.11630780609087$$
$$z_2 = y_1 + (1/2)k_2 = 0.71490551155775$$
$$k_3 = hf(t_1 + (1/2)h, z_2) = 0.11996289517416$$
$$z_3 = y_1 + k_3 = 0.77671450368648$$
$$k_4 = hf(t_1 + h, z_3) = 9.8131054204024E - 02$$
$$y_2 = y_1 + (1/6)(k_1 + 2k_2 + 2k_3 + k_4) = 0.77487458634706$$

TABLE 6.5   Runge-Kutta Examples, $y' = -y \ln y$

$n = h^{-1}$	RK2 error	RK4 error
4	0.767315E–03	0.269490E–05
8	0.180964E–03	0.162549E–06
16	0.439226E–04	0.997759E–08
32	0.108187E–04	0.617969E–09
64	0.268460E–05	0.384481E–10
128	0.668655E–06	0.239708E–11
256	0.166852E–06	0.149658E–12
512	0.416742E–07	0.976996E–14

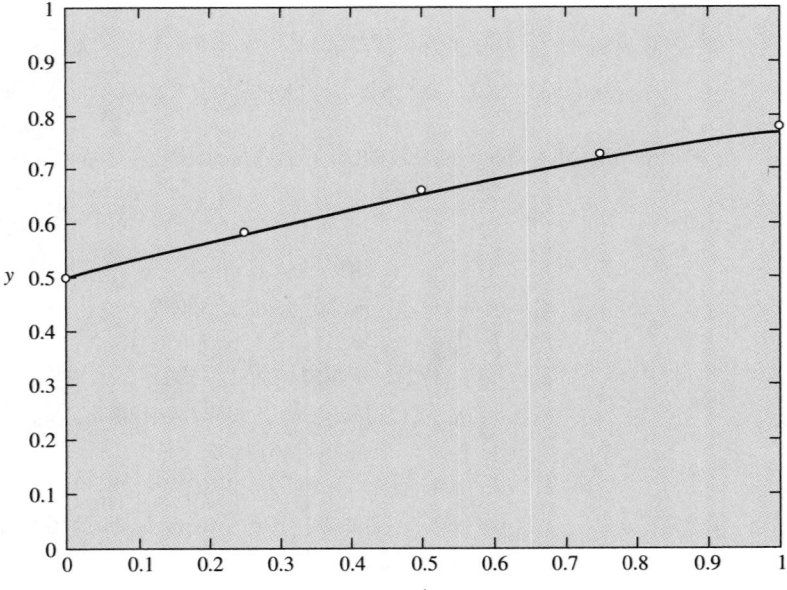

FIGURE 6.6   Solution of $y' = -y \ln y$, $y(0) = \frac{1}{2}$, along with second-order Runge-Kutta approximate values for $h = \frac{1}{4}$.

The exact value of the solution at $t = 1$ is $y(1) = 0.7749206846$, so we can see that the fourth-order method did very well even with this coarse grid, whereas the second-order method did less well.

If we apply the same two methods to the same problem, but with much finer grids, we of course will get better accuracy. Table 6.5 shows the maximum errors over the interval $[0,1]$ for a sequence of values of the mesh parameter. Note that the errors do decrease at the expected rate, in both cases.

Figure 6.6 shows the exact solution along with the second-order approximation for $h = \frac{1}{4}$ (denoted by the small circles). Note that the approximate values are right on

Algorithm 6.2 *Pseudocode for Fourth-Order Runge-Kutta*

```
input t0, y0, h, n
external f
for k = 1 to n do
      t = t0+k*h
      t2 = t0 + 0.5*h
      v1 = f(t0,y0)
      v2 = f(t2,y0+0.5*h*v1)
      v3 = f(t2,y0+0.5*h*v2)
      v4 = f(t,y0+h*v3)
      y = (h/6.0)*(v1 + 2.0*(v2 + v3) + v4)
      y0 = y
      t0 = t
endfor
```

top of the exact solution curve, even for this coarse value of h. For smaller h, or with the higher-order method, the accuracy would only improve, of course.

One major drawback of the Runge-Kutta methods is that they require more evaluations of the function f than other methods. For simple scalar equations like our example, this is not a big deal, but for larger systems it is an issue. To get methods that are as accurate as the Runge-Kutta methods, but less expensive, we have to turn to the multistep methods.

Finally, we include in Algorithm 6.2 a pseudocode implementation of the fourth-order Runge-Kutta method, just as an illustration. Note that four evaluations of f are required at each step.

EXERCISES

1. Use the method of Heun with $h = \frac{1}{4}$ to compute approximate values of $y(1)$ for each of the following initial value problems. Don't write a computer program; use a hand calculator to produce an orderly table of (t_k, y_k) pairs.

(a) $y' + 2y = 1$, $y(0) = 2$
(b) $y' = y(1 - y)$, $y(0) = \frac{1}{2}$
(c) $y' = y(1 + e^{2t})$, $y(0) = 1$
(d) $ty' = y(\sin t)$, $y(0) = 2$

2. Repeat the preceding, using the fourth-order Runge-Kutta.

3. Write a computer program that solves each of the initial value problems in Exercise 1, using the method of Heun and $h = 1/16$.

4. Write a computer program that solves each of the initial value problems in Exercise 1, using fourth-order Runge-Kutta and $h = 1/16$.

5. For each initial value problem below, approximate the solution using the method of Heun with a sequence of decreasing grids $h^{-1} = 2, 4, 8, \ldots$. For those problems where an exact solution is given, compare the accuracy achieved over the interval $[0, 1]$ with the theoretical accuracy.

(a) $y' + 4y = 1$, $y(0) = 1$; $y(t) = \frac{1}{4}(3e^{-4t} + 1)$
(b) $y' + y = \sin 4\pi t$, $y(0) = \frac{1}{2}$
(c) $y' = -y \ln y$, $y(0) = 3$; $y(t) = e^{(\ln 3)e^{-t}}$
(d) $y' + \sin y = 0$, $y(0) = 1$

6. Repeat the preceding, using fourth-order Runge-Kutta as the numerical method.

7. Repeat Exercise 8 from Section 6.4, except this time use fourth-order Runge-Kutta to solve the differential equation, with $h = 1/8$.

8. Repeat Exercise 9 from Section 6.4, except this time use fourth-order Runge-Kutta to solve the differential equation, with $h = 1/8$.

9. Repeat Exercise 10 from Section 6.4, except this time use fourth-order Runge-Kutta to solve the differential equation, with $h = 1/8$.

10. Repeat Exercise 11 from Section 6.4, except this time use fourth-order Runge-Kutta to solve the differential equation, with $h = 1/8$.

6.6   MULTISTEP METHODS

6.6.1   The Adams Families

Two of the most popular families of multistep methods are the so-called Adams families, which are based on the exact integration of interpolating polynomials. One family (Adams-Bashforth) leads to explicit methods; the other (Adams-Moulton) leads to implicit methods.[7]

Recall the differential equation

$$y'(t) = f(t, y(t)), \quad y(t_0) = y_0.$$

Starting at any value $t_n \geq t_0$, we can solve this, formally, by simple integration to get the value at the next step:

$$y(t_{n+1}) = y(t_n) + \int_{t_n}^{t_{n+1}} f(s, y(s)) \, ds.$$

---

[7]John Couch Adams (1819–1892) was born in Cornwall, England, and educated at St. John's College, Cambridge. Already something of a child prodigy in mathematics, at Cambridge he compiled an extraordinary record and was awarded several prizes. While still an undergraduate he decided to study the irregularities in the orbit of the planet Uranus, to see if they could be explained by the gravitational attraction of an as-yet-unknown eighth planet. Adams predicted the new planet's position but, probably because of his youth, the Cambridge Observatory took no action, and so the credit for the discovery of Neptune went to someone else.

Adams briefly held a position as professor of mathematics at St. Andrews College before being named professor of astronomy and director of the Cambridge Observatory.

Francis Bashforth (1819–1912) was born in Thurnscoe, England, the son of a farmer, and attended St. John's College of Cambridge at the same time as Adams. Although he had been ordained as an Anglican priest in 1851, upon graduation Bashforth worked first as a civil engineer and surveyor for a railroad company, and then (in 1864) he obtained a position as professor of applied mathematics at what evolved into the Royal Artillery College. Although Bashforth made numerous important contributions to the study of ballistics, in 1872 an army reorganization left him with such a reduced position that he resigned and became a parish rector.

The Adams-Bashforth method comes from a joint study of capillary action that the two men wrote in 1883.

Forest Ray Moulton (1872–1952), the youngest of eight children, was educated at Albion College in Michigan and received his Ph.D. in astronomy from the University of Chicago in 1899. He is credited, along with his colleague Thomas C. Chamberlain, with formulating the planetesimal hypothesis for the formation of the solar system.

During World War I, Moulton did ballistics research for the U.S. Army at Fort Sill, Oklahoma, and it was during this period that he refined the original work of Adams and Bashforth into what we now know as the Adams-Moulton method for solving initial value problems.

Suppose now that we have values $y(t_{n-k})$, $k = 0, 1, 2, \ldots, p$, of the exact solution. We can construct a polynomial of degree p that interpolates to $F(s) = f(s, y(s))$ at these points:

$$q_p(s) = \sum_{k=0}^{p} L_k(s) f(t_{n-k}, y(t_{n-k})),$$

where the L_k are essentially the Lagrange functions from Section 4.1 (although the notation and indexing are slightly different):

$$L_k(s) = \prod_{\substack{i=0 \\ i \neq k}}^{p} \frac{s - t_{n-i}}{t_{n-k} - t_{n-i}}. \tag{6.55}$$

We then have that

$$y(t_{n+1}) = y(t_n) + \int_{t_n}^{t_{n+1}} \sum_{k=0}^{p} L_k(s) f(t_{n-k}, y(t_{n-k})) ds + R_p(t_{n+1}), \tag{6.56}$$

where (recall that $f(t, y(t)) = y'(t)$)

$$R_p(t_{n+1}) = \int_{t_n}^{t_{n+1}} \frac{1}{(p+1)!} (t - t_n)(t - t_{n-1}) \ldots (t - t_{n-p}) y^{(p+2)}(\theta_{h,t}) \, dt.$$

Since the polynomial part of the integrand does not change sign on $[t_n, t_{n+1}]$, there exists a value ξ_n such that

$$R_p(t_{n+1}) = y^{(p+2)}(\xi_n) \int_{t_n}^{t_{n+1}} \frac{1}{(p+1)!} (t - t_n)(t - t_{n-1}) \ldots (t - t_{n-p}) \, dt$$

$$= \rho_p y^{(p+2)}(\xi_n),$$

for

$$\rho_p = \int_{t_n}^{t_{n+1}} \frac{1}{(p+1)!} (t - t_n)(t - t_{n-1}) \ldots (t - t_{n-p}) \, dt.$$

The expression (6.56) then simplifies to

$$y(t_{n+1}) = y(t_n) + \sum_{k=0}^{p} \lambda_k f(t_{n-k}, y(t_{n-k})) + \rho_p y^{(p+2)}(\xi_n)$$

where

$$\lambda_k = \int_{t_n}^{t_{n+1}} L_k(s) \, ds.$$

By dropping the residual and setting $y(t_k)$ to the approximate value y_k, we get the numerical method, known as the *Adams-Bashforth method* of order $p + 1$:

$$y_{n+1} = y_n + \sum_{k=0}^{p} \lambda_k f(t_{n-k}, y_{n-k}).$$

TABLE 6.6   **Adams-Bashforth Coefficients and Residual Terms**

Steps	p	λ_0	λ_1	λ_2	λ_3	R_p
1	0	h				$\frac{1}{2}h^2y''(\xi_n)$
2	1	$\frac{3}{2}h$	$-\frac{1}{2}h$			$\frac{5}{12}h^3y'''(\xi_n)$
3	2	$\frac{23}{12}h$	$-\frac{16}{12}h$	$\frac{5}{12}h$		$\frac{3}{8}h^4y''''(\xi_n)$
4	3	$\frac{55}{24}h$	$-\frac{59}{24}h$	$\frac{37}{24}h$	$-\frac{9}{24}h$	$\frac{251}{720}h^5y'''''(\xi_n)$

For simplicity's sake we will hereafter write $f_{n-k} = f(t_{n-k}, y_{n-k})$.

If we assume a uniform grid with mesh spacing h, then the formulas for the λ_k and ρ_p simplify substantially, and they are routinely tabulated; Table 6.6 lists them for the common range of $p \leq 3$. Note that they are explicit methods, with order of accuracy $p + 1$, using $p + 1$ steps.

EXAMPLE 6.17

We return to our example problem,

$$y' = -y\ln y, \quad y(0) = \frac{1}{2},$$

the solution to which we will approximate using the second-order Adams-Bashforth scheme, commonly denoted AB2, with $h = \frac{1}{8}$. We choose to use Euler's method to generate the starting value; [8] thus our first computation is

$$y_1 = y_0 + hf(t_0, y_0) = \frac{1}{2} + \frac{1}{8}\left(-\frac{1}{2} \times \ln\frac{1}{2}\right) = 0.5433216988.$$

Now we can do the first AB2 step, for y_2:

$$y_2 = y_1 + \frac{1}{2}h(3f(t_1, y_1) - f(t_0, y_0))$$

$$= 0.5433216988 + \frac{1}{2} \times \frac{1}{8}\left((-3(0.5433216988)\ln(0.5433216988) + \frac{1}{2}\ln\frac{1}{2}\right)$$

$$= 0.583808738.$$

This is followed by the second AB2 step, for y_3:

$$y_3 = y_2 + \frac{1}{2}\frac{1}{8}(3f(t_2, y_2) - f(t_1, y_1))$$

$$= 0.583808738 + \frac{1}{2} \times \frac{1}{8}\left(-3(0.583808738)\ln(0.583808738)\right.$$

$$\left. + 0.5433216988 \ln(0.5433216988)\right)$$

$$= 0.622004388.$$

---

[8]Recall the discussion in Section 6.4.3, that we can use an order $p - 1$ method to generate the starting values for a method of order p without a significant loss of accuracy.

Thus, $y(t_3) = y(0.375) \approx 0.622004388$. The exact solution is $y = e^{-(\ln 2)e^{-t}}$; thus $y(0.375) = 0.6210196063$, so our accuracy is not bad. We would have done better using a more accurate method for the starting value, of course. ∎

TABLE 6.7   Adams-Moulton Coefficients and Residual Terms

Steps	p	γ_{-1}	γ_0	γ_1	γ_2	R_p
1	0	h				$-\frac{1}{2}h^2 y''(\xi_n)$
2	1	$\frac{1}{2}h$	$\frac{1}{2}h$			$-\frac{1}{12}h^3 y'''(\xi_n)$
3	2	$\frac{5}{12}h$	$\frac{8}{12}h$	$-\frac{1}{12}h$		$-\frac{1}{24}h^4 y''''(\xi_n)$
4	3	$\frac{9}{24}h$	$\frac{19}{24}h$	$-\frac{5}{24}h$	$\frac{1}{24}h$	$-\frac{19}{720}h^5 y'''''(\xi_n)$

The Adams-Bashforth methods are based on an interpolating polynomial that is defined using the nodes $t_n, t_{n-1}, \ldots, t_{n-p}$. If we use the same number of nodes but include $t = t_{n+1}$, then (6.56) becomes

$$y(t_{n+1}) = y(t_n) + \int_{t_n}^{t_{n+1}} \sum_{k=-1}^{p-1} L_k(s) f(t_{n-k}, y(t_{n-k})) \, ds + R_p(t_{n+1}), \qquad (6.57)$$

where the Lagrange functions are now

$$L_k(s) = \prod_{\substack{i=-1 \\ i \neq k}}^{p-1} \frac{s - t_{n-i}}{t_{n-k} - t_{n-i}}, \qquad (6.58)$$

and so the numerical method—known as the *Adams-Moulton method* of order $p+1$— is

$$y_{n+1} = y_n + \sum_{k=-1}^{p-1} \gamma_k f(t_{n-k}, y_{n-k}),$$

where the constants are given by

$$\gamma_k = \int_{t_n}^{t_{n+1}} L_k(s) \, ds.$$

Table 6.7 summarizes these for the common range $p \leq 3$. Note that the Adams-Moulton methods are implicit, and that the $p+1$–step method has order of accuracy $p+1$.

Note that three methods that we have already studied (Euler, backward Euler, and trapezoid rule) are included in these two tables. While the Adams-Bashforth methods can be used by themselves (once starting values are generated), the Adams-Moulton methods require either the solution of a nonlinear equation or a predictor-corrector scheme. A very popular predictor-corrector scheme is to use the fourth-order Adams-Bashforth as the predictor and the fourth-order Adams-Moulton as the corrector.

TABLE 6.8 **Adams Family Example for** $y' = -y \ln y$

$n = h^{-1}$	Second-order AB $\max_{k \leq n} \lvert y(t_k) - y_k \rvert$	Fourth-order ABM PC $\max_{k \leq n} \lvert y(t_k) - y_k \rvert$
8	0.28474335730512E–02	0.19742002576040E–05
16	0.67170745938105E–03	0.12099205870530E–06
32	0.16212367000479E–03	0.72652883709168E–08
64	0.39755173993794E–04	0.44197434601045E–09
128	0.98386454340238E–05	0.27204238861600E–10
256	0.24469431793017E–05	0.16866508190105E–11
512	0.61013351659867E–06	0.10536016503693E–12
1024	0.15233231198675E–06	0.76605388699136E–14

EXAMPLE 6.18

Consider the example

$$y' = -y \ln y, \quad y(0) = y_0, \tag{6.59}$$

which has exact solution $y = e^{(-\ln 2)e^{-t}}$ when $y_0 = \frac{1}{2}$. We solve this in two ways:

1. Via the second-order Adams-Bashforth method, with the trapezoid rule predictor-corrector used to generate the starting value

2. Via the fourth-order Adams-Bashforth-Moulton predictor-corrector, using the fourth-order Runge-Kutta method to generate the starting values

The error results are summarized in Table 6.8.

Since they are such high-order methods, we need accurate techniques for getting the starting values for the Adams methods. Generally, the Runge-Kutta methods (Section 6.5) are used for this purpose. ∎

6.6.2 The BDF Family

The Adams families are relatively old numerical methods, with the original work going back to 1883. More recently, because of an interest in so-called stiff differential equations, another family of methods has become popular. Known formally as the "backward difference formula" family (BDF) family, they are derived from a different use of interpolating polynomials.

Again, we let the differential equation be

$$y'(t) = f(t, y(t)),$$

and this time we construct an interpolating polynomial to $y(t)$ (instead of $f(t, y(t)) = y'(t)$) at the nodes $t_{n-p+1}, t_{n-p+2}, \ldots, t_{n+1}$. Thus we have

$$q(t) = \sum_{k=-1}^{p-1} L_k(t) y(t_{n-k}) \approx y(t),$$

TABLE 6.9   Backward Difference Family Coefficients and Residual Terms

Steps	p	ν	μ_0	μ_1	μ_2	μ_3	R_p
1	1	h	1				$-\frac{1}{2}h^2 y''(\xi_n)$
2	2	$\frac{2}{3}h$	$\frac{4}{3}$	$-\frac{1}{3}$			$-\frac{2}{9}h^3 y'''(\xi_n)$
3	3	$\frac{6}{11}h$	$\frac{18}{11}$	$-\frac{9}{11}$	$\frac{2}{11}$		$-\frac{3}{22}h^4 y''''(\xi_n)$
4	4	$\frac{12}{25}h$	$\frac{48}{25}$	$-\frac{36}{25}$	$\frac{16}{25}$	$-\frac{3}{25}$	$-\frac{12}{125}h^5 y'''''(\xi_n)$

where the L_k are defined as in the case of Adams-Moulton. Now we use this polynomial to approximate $y'(t_{n+1})$, using the scheme outlined in Section 4.5. (This choice of argument to y' means that the BDF methods are all implicit.) Thus we have

$$y'(t_{n+1}) = q'(t_{n+1}) + \frac{1}{(p+1)!}w'_p(t_{n+1})y^{(p+1)}(\xi_p),$$

where

$$w_p(t) = \prod_{k=-1}^{p-1}(t - t_k),$$

We substitute this into the differential equation to get

$$\sum_{k=-1}^{p-1} L'_k(t_{n+1})y(t_{n-k}) = f(t_{n+1},y(t_{n+1})) - \frac{1}{(p+1)!}w'_p(t_{n+1})y^{(p+1)}(\xi_p),$$

which we can solve for $y(t_{n+1})$ to get

$$y(t_{n+1}) = \sum_{k=0}^{p-1} \mu_k y(t_{n-k}) + \nu f(t_{n+1},y(t_{n+1})) - \frac{\nu}{(p+1)!}w'_p(t_{n+1})y^{(p+1)}(\xi_p)$$

where the constants are defined as

$$\mu_k = -\frac{L'_k(t_{n+1})}{L'_{n+1}(t_{n+1})}, \qquad \nu = \frac{1}{L'_{n+1}(t_{n+1})}.$$

Table 6.9 summarizes these values for $p \le 4$, under the assumption that the grid spacing is uniform. Note that these methods have order of accuracy p.

A complete explanation of why the BDF family is particularly useful requires a little more discussion of the stability problem and the notion of *stiffness* for a differential equation, which we defer to Section 6.8.2. In general, they have better stability properties than the Adams methods. Right now it is worth noting that these methods can be used either as the corrector in a predictor-corrector scheme, or in conjunction with a nonlinear equation solver at each step, or to explicitly solve linear equations as was done in Section 6.4.2.

EXERCISES

1. Use second-order Adams-Bashforth with $h = \frac{1}{4}$ to compute approximate values of $y(1)$ for each of the following initial value problems. Don't write a computer program; use a hand calculator to produce an orderly table of (t_k, y_k) pairs. Use Euler's method to generate the needed starting value.

(a) $y' + 2y = 1$, $y(0) = 2$
(b) $y' = y(1 - y)$, $y(0) = \frac{1}{2}$
(c) $y' = y(1 + e^{2t})$, $y(0) = 1$
(d) $ty' = y(\sin t)$, $y(0) = 2$

2. Verify that the λ_k values and ρ_p are correct for second-order Adams-Bashforth.

3. Write a computer program that solves each of the initial value problems in Exercise 1, using second-order Adams-Bashforth and $h = 1/16$. Use Euler's method to generate the starting value.

4. Write a computer program that solves each of the initial value problems in Exercise 1, using fourth-order Adams-Bashforth, with fourth-order Runge-Kutta to generate the starting values.

5. For each initial value problem below, approximate the solution using the fourth-order Adams-Bashforth-Moulton predictor-corrector, with fourth-order Runge-Kutta to generate the starting values. Use a sequence of decreasing grids $h^{-1} = 2,4,8,\ldots$. For those problems where an exact solution is given, compare the accuracy achieved over the interval $[0,1]$ with the theoretical accuracy.

(a) $y' + 4y = 1$, $y(0) = 1$; $y(t) = \frac{1}{4}(3e^{-4t} + 1)$
(b) $y' + y = \sin 4\pi t$, $y(0) = \frac{1}{2}$
(c) $y' = -y \ln y$, $y(0) = 3$; $y(t) = e^{(\ln 3)e^{-t}}$
(d) $y' + \sin y = 0$, $y(0) = 1$

6. Derive the second-order Adams-Bashforth method under the assumption that the grid is *not* uniform. Assume that $t_{n+1} - t_n = h$, and $t_n - t_{n-1} = \eta$, with $\eta = \theta h$. What is the truncation error in this instance?

7. Repeat Exercise 8 from Section 6.4, this time using second-order Adams-Bashforth with $h = 1/16$. Use simple Euler to generate the starting value y_1. If you did the earlier problems of this type, compare your results now with what you got before.

8. Repeat Exercise 9 from Section 6.4, this time using second-order Adams-Bashforth with $h = 1/16$. Use simple Euler to generate the starting value y_1. If you did the earlier problems of this type, compare your results now with what you got before.

9. Repeat Exercise 10 from Section 6.4, this time using second-order Adams-Bashforth with $h = 1/16$. Use simple Euler to generate the starting value y_1. If you did the earlier problems of this type, compare your results now with what you got before.

10. Repeat Exercise 11 from Section 6.4, this time using second-order Adams-Bashforth with $h = 1/16$. Use simple Euler to generate the starting value y_1. If you did the earlier problems of this type, compare your results now with what you got before.

6.7   STABILITY ISSUES

6.7.1   Stability Theory for Multistep Methods

We have already seen an example of stability problems in the solution of IVPs, in Section 6.4.4, on the weak stability of the midpoint method. In this section we will outline the issues involved in a more general study of stability.

It is important to understand, at the outset, the distinction between the stability of the solution to the differential equation and the stability of the numerical method used to approximate that solution. To illustrate, consider the example initial value problem

$$y' = 100y + e^{-t}, \quad y(0) = -\frac{1}{101}.$$

The exact solution here is $y = -\frac{1}{101}e^{-t}$, which is a bounded, continuous, very well-behaved function. Yet, as a practical matter, almost any numerical method applied to this problem will result in a solution that grows rapidly and eventually blows up. The reason is that the *general* solution to the differential equation is

$$y = Ce^{100t} - \frac{1}{101}e^{-t}.$$

Note the presence of the e^{100t} term; this is missing from the exact solution to the IVP because the initial value is chosen to make the constant C in the general solution 0. However, the computed value y_1 will not be the exact value $y(t_1)$; thus, considering the new initial value problem

$$Y' = 100Y + e^{-t}, \quad Y(h) = y_1,$$

the exact solution here is

$$Y(t) = -\frac{1}{101}e^{-t} + e^{100(t-h)}\left(y_1 + \frac{1}{101}e^{-h}\right);$$

thus the e^{100t} term is multiplied by the error in the initial value y_1. Since this error will *not* be zero, any approximate solution will almost surely depend on the e^{100t} term; thus, the approximate values for $t > h$ will be affected by this term and will generate a huge growth of the approximate solution. While this is indeed a kind of instability— large growth of the solution due to a small change in the problem—it is an instability that is inherent in the differential equation itself, *not* in the numerical methods that we apply to compute solutions. We are interested in stability problems that come out of the numerical algorithms.

To motivate the larger discussion of stability, consider applying the $p = 1$ case of the BDF family to our standard initial value problem,

$$y' = f(t,y), \quad y(0) = y_0.$$

Thus we have that the exact solution satisfies

$$y(t_{n+1}) = \frac{4}{3}y(t_n) - \frac{1}{3}y(t_{n-1}) + \frac{2}{3}hf(t_{n+1}, y(t_{n+1})) - \frac{2}{9}h^3 y'''(\xi_n)$$

and the approximate solution satisfies

$$y_{n+1} = \frac{4}{3}y_n - \frac{1}{3}y_{n-1} + \frac{2}{3}hf(t_{n+1}, y_{n+1}).$$

By subtracting, we get that the error satisfies

$$e_{n+1} = \frac{4}{3}e_n - \frac{1}{3}e_{n-1} + \frac{2}{3}hf(t_{n+1}, y(t_{n+1})) - \frac{2}{3}hf(t_{n+1}, y_{n+1}) - \frac{2}{9}h^3 y'''(\xi_n).$$

The error, of course, depends on the mesh spacing h as well as the index n, so we can write the error as $e_n(h)$ to make this clear. Note, however, that if we consider the limiting case $h = 0$, we get

$$e_{n+1}(0) = \frac{4}{3}e_n(0) - \frac{1}{3}e_{n-1}(0). \tag{6.60}$$

Now, the fundamental issue—which we will not prove in any sort of formal sense—is the following:

If the error blows up as $n \to \infty$ when $h = 0$, then elementary considerations of continuity imply that it will still blow up for h small but nonzero. On the other hand, if the error for $h = 0$ does not blow up as $n \to \infty$, then the same continuity considerations imply that it will be bounded for at least some range of nonzero values of h, as $n \to \infty$.

We thus look at the recursion (6.60) and solve it using the same ideas as in Section 6.4.4: We look for solutions in the form

$$e_n(0) = r^n.$$

This leads to a second-degree polynomial in r,

$$\sigma(r) = r^2 - \frac{4}{3}r + \frac{1}{3},$$

which has roots $r_1 = 1$ and $r_2 = \frac{1}{3}$. Hence the $h = 0$ part of the error satisfies

$$e_n(0) = C_1 r_1^n + C_2 r_2^n,$$

where the coefficients depend on the values $e_0(0)$ and $e_1(0)$. From this we conclude that this particular BDF method is stable, because as $n \to \infty$ the $h = 0$ error does not blow up; quite to the contrary, one term goes to zero and the other is bounded. For this method, the parasitic solutions do not dominate the approximation for n large.

On the other hand, consider the method

$$y_{n+1} = 4y_n - 3y_{n-1} - 2hf(t_{n-1}, y_{n-1}), \tag{6.61}$$

which we derived in Section 6.4. This has a truncation error that is second-order in the mesh, comparable to the trapezoid method and the second-order Runge-Kutta methods, so we can call it an accurate method (it is consistent). However, the polynomial is, in this case,

$$\sigma(r) = r^2 - 4r + 3,$$

which has roots $r_1 = 1$ and $r_2 = 3$. Now, as $n \to \infty$ we see that one component of the exact solution to the approximate problem will blow up. Thus, we say that this method is unstable, and we expect that it will not perform well, as will be seen in some of the exercises.

The stability of a multistep method,[9] then, is determined by the roots of the polynomial corresponding to σ. A formal statement is based on the following definitions.

---

[9] Stability in this sense is not as much of an issue for single-step methods since they do not have parasitic solutions.

Definition 6.5 (Stability Polynomial)   *Consider a multistep method in the form*

$$y_{n+1} = \sum_{k=0}^{p} a_k y_{n-k} + \sum_{k=-1}^{p} b_k f(t_{n-k}, y_{n-k}).$$

Then the stability polynomial for this method is given by

$$\sigma(r) = r^{p+1} - \sum_{k=0}^{p} a_k r^{p-k}. \tag{6.62}$$

Definition 6.6 (Root conditions)   *Consider a multistep method with stability polynomial σ. Denote the roots of σ by $r_0, r_1, \ldots, r_p$. Define the following conditions.*

Root Condition *If*

$$|r_k| \leq 1, \quad 0 \leq k \leq p$$

and all roots that satisfy $|r_j| = 1$ are simple roots, then we say that the stability polynomial satisfies the root condition.

Strong Root Condition *If*

$$r_0 = 1, |r_k| < 1, \quad 1 \leq k \leq p,$$

then we say that the stability polynomial satisfies the strong root condition.

Then we have the following:

1. If the stability polynomial satisfies the root condition, then the method is stable, in the sense that for h sufficiently small it will deliver accurate results over a closed interval $[0, T]$.

2. If the stability polynomial satisfies the strong root condition, then the method is *relatively stable*, meaning that, for h sufficiently small, the parasitic solution components will go to zero as $n \to \infty$.

3. A method that is stable but not relatively stable is called weakly stable and will exhibit the kind of behavior seen in Section 6.4.4.

EXAMPLE 6.19

Consider the $(p+1)$-step Adams-Bashforth method:

$$y_{n+1} = y_n + \sum_{k=0}^{p} \lambda_k f(t_{n-k}, y_{n-k}).$$

The stability polynomial, based on the formal definition (6.62), will be

$$\sigma(r) = r^{p+1} - r^p;$$

thus the roots are $r = 1$ and $r = 0$ (multiplicity p). This satisfies the strong root condition, so we know the Adams-Bashforth methods are relatively stable. The same analysis holds for the Adams-Moulton methods, so we know they are relatively stable as well. ∎

6.7.2   Stability Regions

The definition of stability that we developed above was based on the behavior of the method in the limit, as $h \to 0$. While perfectly valid, it is not immediately clear what it means for the practical implementation of methods, since we have to actually compute with $h > 0$. All we know, based on the material in Section 6.7.1, is that the Adams-Bashforth methods (for example) will be stable "for h sufficiently small."

To address this need for specific information about when a multistep method will work well, we have the notion of "stability regions."

Consider the model problem

$$y' = \lambda y, \quad y(0) = 1, \tag{6.63}$$

where we assume that λ is *complex*. If we apply, say, the second-order Adams-Bashforth method to this problem, then we have the recursion

$$y_{n+1} = \left(1 + \frac{3}{2}\lambda h\right) y_n - \frac{1}{2}\lambda h y_{n-1}.$$

If we look for the exact solution of this recursion as we have before, we are led to the polynomial

$$p(r) = r^2 - \left(1 + \frac{3}{2}\xi\right) r + \frac{1}{2}\xi$$

for $\xi = \lambda h$. The stability region is then defined as follows:

> *The stability region for a multistep method is that part of the complex plane where the method, when applied to the model problem (6.63), is absolutely stable; i.e., all components of the approximate solution decay to zero as $n \to \infty$.*

If we denote the roots of the polynomial as

$$r_0(\xi), r_1(\xi), \ldots, r_p(\xi),$$

then the stability region is that part of the complex plane[10] where $|r_k(\xi)| < 1$, for all k.

Figures 6.7–6.9 show the stability regions for three families of methods: the Adams-Bashforth family, the Adams-Moulton family, and the Adams-Bashforth-Moulton predictor-corrector family (in which we use the AB method as a predictor, and the AM method of the same accuracy as the corrector). Note that, in general, as the

---

[10] The reason we have to consider the complex plane is that when we consider *systems* of differential equations, the quantities corresponding to the λ in the differential equation are *eigenvalues* of a real matrix, and these can of course be complex.

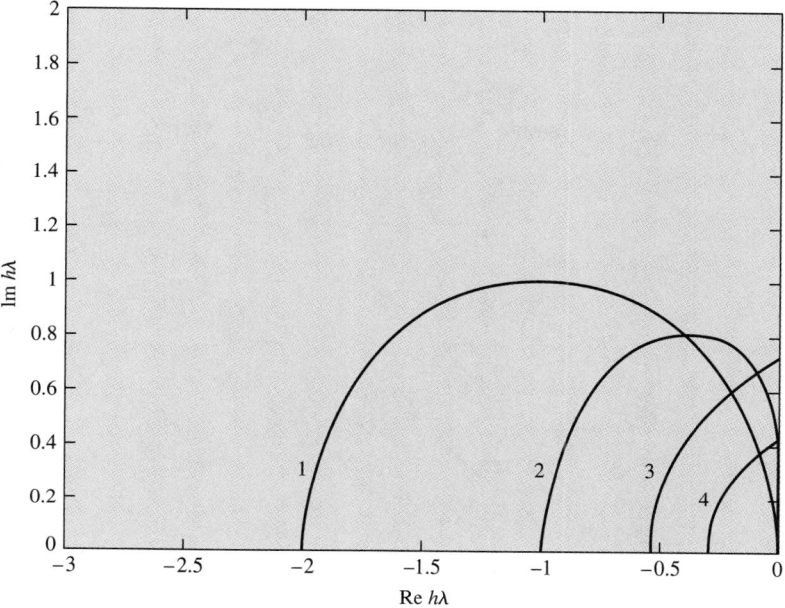

FIGURE 6.7 Regions of absolute stability for the Adams-Bashforth methods of orders 1–4.

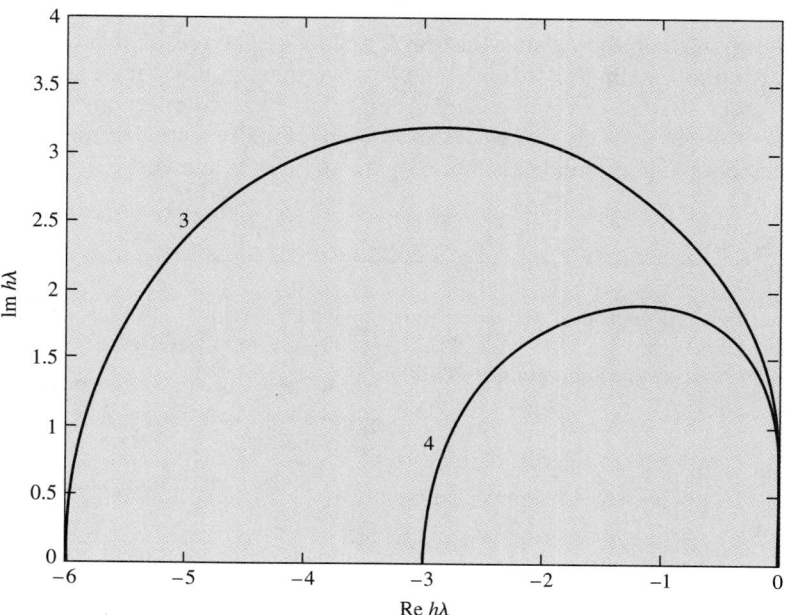

FIGURE 6.8 Regions of absolute stability for the Adams-Moulton methods of orders 3 and 4.

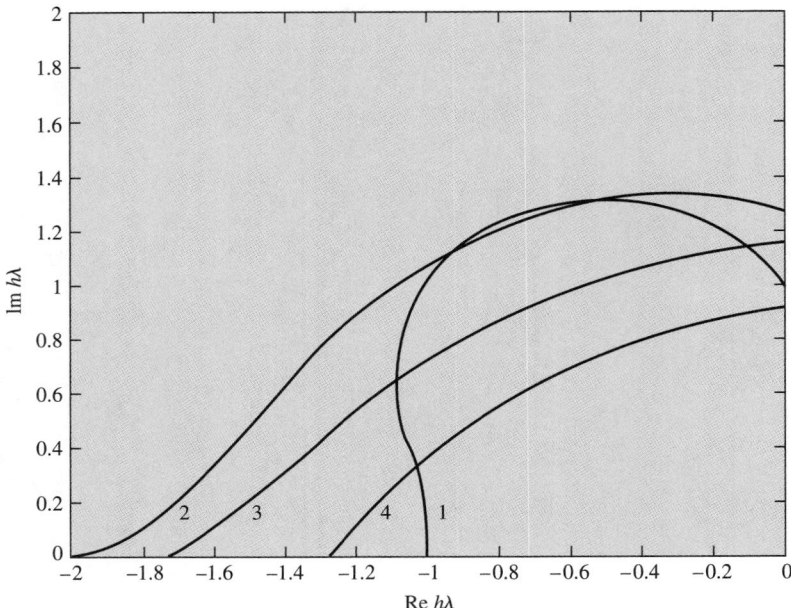

FIGURE 6.9 Regions of absolute stability for the Adams-Bashforth-Moulton predictor-corrector methods of orders 1–4.

accuracy of the method increases, the size of the stability region decreases. Note also that the implicit Adams-Moulton methods have larger stability regions than the Adams-Bashforth-Moulton predictor-corrector methods, which have larger stability regions than the Adams-Bashforth methods. The reason that only two of the Adams-Moulton methods (orders 3 and 4) have their stability regions plotted is that for the order 1 (backward Euler) and order 2 (trapezoid rule) schemes, the stability region is the entire left half-plane (see Exercise 4). Finally, note that the regions are symmetric about the horizontal axis, but only the top half is shown.

EXERCISES

1. Determine the residual and the truncation error for the method defined by $y_{n+1} = 4y_n - 3y_{n-1} - 2hf(t_{n-1}, y_{n-1})$. Try to use it to approximate solutions of the IVP $y' = -y \ln y$, $y(0) = 2$. Comment on your results.

2. Show that the method defined by $y_{n+1} = 4y_n - 3y_{n-1} - 2hf(t_{n-1}, y_{n-1})$ is unstable. Comment in light of the results above.

3. Consider the method defined by

$$y_{n+1} = y_{n-1} + \frac{1}{3}h\left[f(t_{n-1}, y_{n-1}) + 4f(t_n, y_n) + f(t_{n+1}, y_{n+1})\right],$$

which is known as Milne's method. Is this method stable or strongly stable?

4. Show for the backward Euler and trapezoid rule methods, that the stability region is the entire left half-plane.

5. Show that all four members of the BDF family are stable.

6. Show by experiment that the stability region for the fourth-order BDF method contains the entire negative real axis.

6.8   APPLICATION TO SYSTEMS OF EQUATIONS

So far in this chapter all of our examples of implementing methods have been for ordinary scalar equations. In this section we will look a bit at applying some of our methods to *systems* of equations.

6.8.1   Implementation Issues and Examples

Consider the system

$$y_1' = y_1 - y_1 y_2, \quad y_1(0) = y_{10} \tag{6.64}$$
$$y_2' = y_1 y_2 - y_2, \quad y_2(0) = y_{20}. \tag{6.65}$$

This is generally known as the Lotka-Volterra population model, since it was first used in the 1920s to model the population dynamics of two species, one of which feeds off the other. To apply any of the methods we have studied so far to this problem, it helps to formally write it out in terms of components and functions f_i, as follows:

$$y_1' = f_1(t, y_1, y_2), \quad y_1(0) = y_{10} \tag{6.66}$$
$$y_2' = f_2(t, y_1, y_2), \quad y_2(0) = y_{20}, \tag{6.67}$$

where

$$f_1(t, y_1, y_2) = y_1 - y_1 y_2 \tag{6.68}$$
$$f_2(t, y_1, y_2) = y_1 y_2 - y_2. \tag{6.69}$$

Then, for example, an Euler's method approximation would require that we compute using the recursions (note the plural)[11]

$$y_{1,n+1} = y_{1,n} + h f_1(t_n, y_{1,n}, y_{2,n})$$
$$y_{2,n+1} = y_{2,n} + h f_2(t_n, y_{1,n}, y_{2,n}),$$

whereas the second-order Runge-Kutta method (6.48) would require that we compute using the recursions

$$\bar{y}_{1,n+1} = y_{1,n} + \frac{1}{2} h f_1(t_n, y_{1,n}, y_{2,n})$$

$$\bar{y}_{2,n+1} = y_{2,n} + \frac{1}{2} h f_2(t_n, y_{1,n}, y_{2,n})$$

$$y_{1,n+1} = y_{1,n} + h f_1\left(t_n + \frac{1}{2} h, \bar{y}_{1,n+1}, \bar{y}_{2,n+1}\right)$$

$$y_{2,n+1} = y_{2,n} + h f_2\left(t_n + \frac{1}{2} h, \bar{y}_{1,n+1}, \bar{y}_{2,n+1}\right),$$

---

[11] Note also the use of double subscripts. The first one counts the element in the solution vector y_n, and the second one indicates the time step number.

and the second-order Adams-Bashforth method would use

$$y_{1,n+1} = y_{1,n} + \frac{1}{2}h\left(3f_1(t_n,y_{1,n},y_{2,n}) - f_1(t_n,y_{1,n},y_{2,n})\right)$$

$$y_{2,n+1} = y_{2,n} + \frac{1}{2}h\left(3f_2(t_n,y_{1,n},y_{2,n}) - f_2(t_n,y_{1,n},y_{2,n})\right).$$

EXAMPLE 6.20

To illustrate, let's do several example computations with the system

$$y_1' = y_1 - y_1y_2 + \sin \pi t, \quad y_1(0) = 2$$
$$y_2' = y_1y_2 - y_2, \quad y_2(0) = 1.$$

This is an instance of the Lotka-Volterra model in which the population of one of the species is driven by an external sinusoid. We have, in this case,

$$f_1(t,y_1,y_2) = y_1 - y_1y_2 + \sin \pi t$$
$$f_2(t,y_1,y_2) = y_1y_2 - y_2.$$

A single step of Euler's method, using $h = \frac{1}{4}$, would have us do the following computations:

$$y_{11} = y_{10} + hf_1(0,y_{10},y_{20}) = 2 + \frac{1}{4} \times (2 - (2)(1)) + \sin 0) = 2$$

$$y_{21} = y_{20} + hf_2(0,y_{10},y_{20}) = 1 + \frac{1}{4} \times ((2)(1) - 1) = \frac{5}{4}.$$

Now we can continue the computation using second-order Adams-Bashforth, using these Euler's method values as the starting values. In this case we get

$$y_{12} = y_{11} + \frac{1}{2}h\left(3f_1(t_1,y_{11},y_{2,1}) - f_1(t_0,y_{10},y_{20})\right)$$

$$= 2 + \frac{1}{2} \times \frac{1}{4}(3(2 - (5/4)2 + \sin \frac{1}{4}\pi) - (2 - (2)(1) + \sin 0))$$

$$= 2.078$$

$$y_{22} = y_{21} + \frac{1}{2}h\left(3f_2(t_1,y_{11},y_{21}) - f_2(t_0,y_{10},y_{20})\right)$$

$$= \frac{5}{4} + \frac{1}{2} \times \frac{1}{4}(3((2)(5/4) - (5/4)) - ((2)(1) - 1))$$

$$= 1.625.$$

∎

Generalizations to more accurate and more involved schemes simply involve a little more care and work.

An important special case, as with scalar equations, occurs when the system is a linear system. In this case we can write it in matrix-vector form as

$$y' = Ay + g, \quad y(0) = y_0,$$

where now A is an $N \times N$ matrix and y,g,y_0 are vectors of N components. It now becomes possible to (once again) employ implicit methods though the device of solving the linear equations explicitly. The difference between this case and what we had in Section 6.4.2 is that now we have a linear *system* to solve. For example, a direct solution of our general example via the trapezoid rule method involves the solution of the linear system

$$\left(I - \frac{1}{2}hA\right) y_{n+1} = \left(I + \frac{1}{2}hA\right) y_n + \frac{1}{2}h\left(g(t_n) + g(t_{n+1})\right)$$

at each step of the computation.

EXAMPLE 6.21

Consider the system

$$y_1' = -4y_1 + y_2 + \sin \pi t, \quad y_1(0) = 1$$
$$y_2' = y_1 - 4y_2, \quad y_2(0) = 2,$$

which we can write in matrix-vector form as

$$y' = Ay + g, \quad y(0) = y_0.$$

Here

$$A = \begin{bmatrix} -4 & 1 \\ 1 & -4 \end{bmatrix}, \quad y = \begin{bmatrix} y_1 \\ y_2 \end{bmatrix}, \quad g = \begin{bmatrix} \sin \pi t \\ 0 \end{bmatrix}, \quad y_0 = \begin{bmatrix} y_{10} \\ y_{20} \end{bmatrix}.$$

If we now do two steps of the trapezoid rule method for this system, using $h = \frac{1}{4}$, we get the following:

$$y_1 = \left(I - \frac{h}{2}A\right)^{-1} \left(I + \frac{h}{2}A\right) y_0 + \frac{1}{2}h\left(I - \frac{h}{2}A\right)^{-1} (g(t_0) + g(t_1))$$
$$= \begin{bmatrix} 0.6258 \\ 0.8021 \end{bmatrix}$$

$$y_2 = \left(I - \frac{h}{2}A\right)^{-1} \left(I + \frac{h}{2}A\right) y_1 + \frac{1}{2}h\left(I - \frac{h}{2}A\right)^{-1} (g(t_1) + g(t_2))$$
$$= \begin{bmatrix} 0.7690 \\ 0.8141 \end{bmatrix}.$$

$\blacksquare$

We will end this section by giving a pseudocode for approximating solutions of the system

$$y' = f(t,y), \quad y(0) = y_0$$

using the trapezoid rule predictor-corrector method (see Algorithm 6.3). The code assumes that the individual vector functions $f_j(t,y)$ can be called using an index passed

Algorithm 6.3: *Trapezoid rule predictor-corrector for systems*

```
input h,n,m,y0,t0
external f
!
t = t0
for j = 1 to m do
     y(j) = y0(j)
endfor
for k = 1 to n do
!
!      Euler predictor
!
       for j=1 to m do
           yb(j) = y(j) + h*f(j,t,y)
       endfor
!
!      Trapezoid rule corrector
!
       for j=1 to m do
           yc(j) = y(j) + (h/2)*(f(j,t,y) + f(j,t+h,yb))
       endfor
!
!      Update
!
       for j=1 to m do
           y(j) = yc(j)
       endfor
       t = k*h
endfor
```

into a single function name: `f(j,t,y)`. The number of unknowns in the code is taken to be *m*.

6.8.2 Stiff Equations

In some applications, we must deal with a large system of coupled differential equations, modeling processes that are occurring at substantially different rates. This is the essence of the notion of *stiffness* of a differential equation. Stiff systems can be very difficult to solve numerically.

To illustrate the phenomenon, consider the linear system

$$y_1' = 198y_1 + 199y_2, \quad y_1(0) = 1 \tag{6.70}$$

$$y_2' = -398y_1 + -399y_2, \quad y_2(0) = -1, \tag{6.71}$$

which we write in matrix-vector form as

$$y' = Ay, \quad y(0) = y_0, \tag{6.72}$$

TABLE 6.10 **Example of Stiff Approximation (Adams-Bashforth)**

$n = h^{-1}$	Error in y_1	Error in y_2
4	0.313335E–02	0.313335E–02
8	0.653059E–03	0.653059E–03
16	0.142067E–03	0.142069E–03
32	0.634997E+07	0.126999E+08
64	0.976963E+19	0.195393E+20
128	0.778921E+15	0.155784E+16
256	0.473265E–06	0.473265E–06
512	0.117631E–06	0.117631E–06
1024	0.293221E–07	0.293221E–07
2048	0.731982E–08	0.731982E–08

where

$$A = \begin{bmatrix} 198 & 199 \\ -398 & -399 \end{bmatrix}, \qquad y = \begin{pmatrix} y_1 \\ y_2 \end{pmatrix}, \qquad y_0 = \begin{pmatrix} y_1(0) \\ y_2(0) \end{pmatrix}.$$

The exact solution is

$$y_1(t) = e^{-t}, \quad y_2(t) = -e^{-t},$$

and we will compute approximations using second-order Adams-Bashforth, with Euler's method used to compute the starting value. The results are given in Table 6.10; note that we get decent accuracy at the outset, but then horrible accuracy until we have $h < 1/128$. Why is this?

The answer lies in the *eigenvalues* of the matrix A, which are $\lambda_1 = -1$, $\lambda_2 = -200$. This means that the general solution of the system will involve components of the form e^{-200t}, in addition to e^{-t}. The initial condition was cleverly chosen to make those components vanish. However, even though the e^{-200t} term is not present in the solution, we still have to use a value of h that is small enough to accurately resolve it. (To be precise, we have to take h small enough so that $-200h$ is in the development of how this comes about is deferred to Appendix A. For our part, we will call a linear system of differential equations *stiff* if the eigenvalues of the coefficient matrix A all have negative real parts, and if the ratio of the largest eigenvalue to the smallest, in absolute value, is "large."

One of the best ways to deal with a stiff system is to use an implicit method to compute the approximation. If the system is linear, the implicitness can be dealt with by solving a linear system; if the system is nonlinear, then some kind of root-finding scheme must be used to solve the nonlinear equations, or a predictor-corrector method must be used.

If we apply the trapezoid rule method to the solution of this example, we get the results shown in Table 6.11. Note that the error is much smaller and does not show

TABLE 6.11   **Example of Stiff Approximation (Trapezoid Method)**

$n = h^{-1}$	Error in y_1	Error in y_2
4	0.192913E–02	0.192913E–02
8	0.479822E–03	0.479822E–03
16	0.119803E–03	0.119803E–03
32	0.299413E–04	0.299413E–04
64	0.748472E–05	0.748472E–05
128	0.187114E–05	0.187114E–05
256	0.467784E–06	0.467784E–06
512	0.116946E–06	0.116946E–06
1024	0.292364E–07	0.292364E–07
2048	0.730911E–08	0.730911E–08

the erratic and undesirable behavior that we saw with the Adams-Bashforth method. The material in Appendix A gives some explanation for this.

6.8.3   A-Stability

Because of the phenomenon of stiffness, we are interested in numerical methods for ODEs that will always produce decaying approximate solutions when applied to a differential equation that has decaying solutions, regardless of the mesh size used. This is the concept of A-stability. (The "A" stands for "asymptotic.") The formal definition is as follows

Definition 6.7 (A-Stability)   Let y_k be the approximate solution to

$$y' = \lambda y, \quad y(0) = 1$$

for λ complex, as computed by some numerical method. If

$$\lim_{n \to \infty} y_n = 0$$

for all h and all λ such that $\mathrm{Re}(\lambda) < 0$, then we say that the numerical method is A-stable.

The trapezoid method and the backward Euler method are examples of A-stable methods. (See Exercise 3.) The advantage of an A-stable method is that the parasitic solutions will always decay, regardless of the mesh size; hence they can be used to solve stiff systems. Unfortunately, there are no A-stable multistep methods of greater than h^2 accuracy.

EXERCISES

1. Using a hand calculator and $h = \frac{1}{4}$, compute approximate solutions to the initial value problem

$$\begin{aligned} y_1' &= -4y_1 + y_2, \quad y_1(0) = 1 \\ y_2' &= y_1 - 4y_2, \quad y_2(0) = -1. \end{aligned}$$

Compute out to $t = 1$, using the following methods:

(a) Euler's method

(b) Trapezoid rule predictor-corrector

(c) AB2 with Euler's method for the starting values

(d) RK4

Organize your computations and results neatly to minimize mistakes.

2. Consider the second-order equation

$$y'' + \sin y = 0, \quad y(0) = \frac{1}{8}\pi, y'(0) = 0;$$

write this as a first-order system and compute approximate solutions using a sequence of grids and the following methods:

(a) Second-order Adams-Bashforth

(b) The method of Heun

(c) Fourth-order Runge-Kutta

The solution will be periodic; try to determine, experimentally, the period of the motion.

3. Show that the backward Euler and trapezoid rule methods are both A-stable.

4. Show that the second-order BDF method is A-stable.

5. Apply the third-order BDF method, using the trapezoid method to obtain the starting values, to the stiff example in the text. Compare your results with the exact solution and those obtained in the text.

6. Consider the following system of differential equations:

$$\begin{aligned} y_1'(t) &= ay_1(t) - by_2(t)y_1(t), \quad y_1(0) = y_{10} \\ y_2'(t) &= -cy_2(t) + dy_1(t)y_2(t), \quad y_2(0) = y_{20} \end{aligned}$$

Solve this system for a sequence of mesh sizes, using the data $a = 4, b = 1, c = 2, d = 1$, with initial values $y_{10} = 3/2$ and $y_{20} = 4$. Use (a) second-order Adams-

Bashforth, with Euler's method providing the starting value; and (b) fourth-order Runge-Kutta. Plot the solutions in two ways: as a single plot showing y_1 and y_2 versus t, and as a *phase plot* showing y_1 versus y_2. Note: The solutions should be periodic. Try to determine, experimentally, the period. (Note: This problem is an example of a *predator-prey* model, in which y_1 represents the population of a prey species and y_2 represents the population of a predator that uses the prey as its only food source. See Braun [3] for an excellent discussion of the dynamics of such systems, as well as the history behind the problem.)

7. One simple but important application of systems of differential equations is to the spread of epidemics. Perhaps the simplest model is the so-called *SIR* model, which models an infectious disease (such as measles) that imparts immunity to those who have had it and recovered. We divide the population into three categories:

- $S(t)$: These are the people who are susceptible; i.e., they are well at time t but might get sick at some future time,

- $I(t)$: These are the people who are infected, i.e., sick,

- $R(t)$: These are the people who have recovered and are therefore protected from getting the disease by being immune.

Under a number of simplifying assumptions,[12] the *SIR* model involves the following system, sometimes known as the continuous Kermack-McKendrick model.

$$\begin{aligned} \frac{dS}{dt} &= -rSI \\ \frac{dI}{dt} &= rSI - aI \\ \frac{dR}{dt} &= aI. \end{aligned}$$

The parameters r and a measure the rate at which susceptible people become sick upon contact with the infected, and the rate at which infected people are cured.

(a) For the case $I_0 = 1, S_0 = 762, R_0 = 0, a = 0.44$, and $r = 2.18 \times 10^{-3}$, solve the system using the numerical

---

[12] See J.D. Murray, *Mathematical Biology*, Springer-Verlag, Berlin, 1989, pp. 611ff.

method and step size of your choice, and plot the solutions. (This data is taken from a study of an influenza epidemic at an English boys' school; the time scale here is in days.)

(b) Now consider the case where $S_0 = 10^6$, $R_0 = 0$, $a = 1.74$, $r = 0.3 \times 10^{-5}$. Solve the system (again, using the method of your choice) for a range of values of $I_0 > 0$. Is there a critical value of I_0 beyond which the disease spreads to almost all of the population? (This data is taken, roughly, from a study of a plague epidemic in Bombay; the time scale is weeks.)

(c) In the previous data set, take $I_0 = 10^4$ and vary the value of r. Try to find the critical value of r such that, for $r < r_{\text{crit}}$ the number of infected drops monotonically, and for $r > r_{\text{crit}}$ the number of infected rises to a peak before falling off.

6.9 ADAPTIVE SOLVERS

Just as we did with numerical quadrature, we can devise methods for initial value problems that implement some kind of variable step size as well as some automatic error control to minimize the work while obtaining a user-defined accuracy. The basic principles are the same as we used in Section 5.8.3 for numerical integration: We do two computations of the same value, and from these derive a computable estimate of the error. If this computable estimate of the error is small enough, we accept the computed value and go on to the next point. If the error estimate is not small enough, then we go to a smaller mesh size and try again.

The "professional" codes use various schemes. One of the first such codes, DIF-SUB [8], used a variable order Adams-Bashforth-Moulton scheme. Others (such as the MATLAB routine `ode45`) use what is called a Runge-Kutta-Fehlberg scheme. What we will first describe here is a simpler but less accurate scheme that nonetheless contains the essential elements of the others, followed by a simple presentation of a scheme of Runge-Kutta-Fehlberg type, based on the more recent work of Dormand and Prince [6]. Consider the initial value problem

$$y' = f(t,y), \quad y(0) = y_0,$$

and assume we have computed accurate approximate solutions out as far as some t_n. Thus, we have $y_n \approx y(t_n)$. We have a user-defined error criterion τ. We define $u_n(t)$ as the exact solution of the IVP

$$y' = f(t,y), \quad y(t_n) = y_n;$$

thus $u_n(t)$ defines the solution of the ODE that passes through our most recent approximate value. We will design our algorithm to control the *local error*, i.e., the error between y_{n+1} and $u_n(t_{n+1})$. The *global error*, $y(t_{n+1}) - y_{n+1}$, can then be estimated from the local error and the differential equation.

Our first algorithm, which we will develop and justify in more detail, is based on using two approximate values generated from the trapezoid rule method, which of course is one instance of second-order Runge-Kutta. We require one preliminary lemma before proceeding with the design of the algorithm.

Lemma 6.1   *If the solution y is sufficiently smooth, then the truncation error for the second-order Runge-Kutta scheme*

$$y_{n+1} = y_n + \frac{1}{2}h\left(f(t_n,y_n) + f(t_{n+1},y_n + hf(t_n,y_n))\right)$$

satisfies

$$\tau(t,h) = -h^2 \left[\frac{1}{12} y'''(t) - \frac{1}{4} y''(t) f_y(t,y(t)) \right] + O(h^3). \qquad (6.73)$$

Proof   We have

$$\tau(t,h) = h^{-1} \left[y(t+h) - y(t) - \frac{1}{2} h \left(f(t,y(t)) + f(t,y(t) + hf(t,y(t))) \right) \right].$$

We proceed essentially as we did in deriving the Runge-Kutta methods, by expanding $y(t+h)$ and $f(t,y(t)+hf(t,y(t)))$ using Taylor's Theorem:

$$y(t+h) = y(t) + hy'(t) + \frac{1}{2} h^2 y''(t) + \frac{1}{6} h^3 y'''(t) + O(h^4)$$

and

$$\begin{aligned} f(t,y(t) + hf(t,y(t))) = {} & f(t,y(t)) + hf_t(t,y(t)) + hf(t,y(t))f_y(t,y(t)) \\ & + \frac{1}{2} \left(h^2 f_{tt}(t,y(t)) \right. \\ & + 2h^2 f(t,y(t)) f_{ty}(t,y(t)) \\ & + h^2 f^2(t,y(t)) f_{yy}(t,y(t)) \,) \\ & + O(h^3). \end{aligned}$$

Putting this all together gives us

$$\begin{aligned} \tau(t,h) = {} & \left(\frac{1}{6} y'''(t) - \frac{1}{4} \left(f_{tt}(t,y(t)) + 2f(t,y(t)) f_{ty}(t,y(t)) + f^2(t,y(t)) f_{yy}(t,y(t)) \right) \right) h^2 \\ & + O(h^3), \end{aligned} \qquad (6.74)$$

and the conclusion follows by computing $y'''(t)$ in terms of f and its derivatives. □

This lemma is the key to constructing our adaptive algorithm (see Algorithm 6.4). Note that (6.73) implies that—if the solution is smooth enough—we can use Taylor's Theorem to show that, for any real scalar a,

$$\tau(t_n + ah, h) = \tau(t_n, h) + O(h^3). \qquad (6.75)$$

The last two steps are a very simple way to control the step size;[13] more sophisticated methods actually use the information from one step to predict what the next value of h ought to be. These same ideas can be used to predict an initial value of h as well.

---

[13] We don't want the step size to become too small, because that would result in our doing too much work by taking many tiny steps where fewer, larger steps would be acceptably accurate.

Algorithm 6.4: *Simple Adaptive Solver for Initial Value Exercises*

Given t_n, y_n, a desired error tolerance, ϵ, the current value of the mesh, h, and a value $y_n \approx y(t_n)$, do the following:

1. Compute

$$Y_1 = y_n + \frac{1}{2} h \left(f(t_n, y_n) + f(t_{n+1}, y_n + hf(t_n, y_n)) \right);$$

   note that this is a single-step $O(h^2)$ approximation to $y(t_{n+1})$.

2. Compute

$$Y_2 = W + \frac{1}{4} h \left(f\left(t_n + \frac{1}{2}h, W\right) + f\left(t_{n+1}, W + \frac{1}{2}hf\left(t_n + \frac{1}{2}h, W\right)\right) \right),$$

   where

$$W = y_n + \frac{1}{4} h \left(f(t_n, y_n) + f\left(t_n + \frac{1}{2}h, y_n + \frac{1}{2}hf(t_n, y_n)\right) \right);$$

   note that Y_2 is a two-step $O(h^2)$ approximation to $y(t_{n+1})$.

3. Compute $Y_3 = (4Y_2 - Y_1)/3$; note that this is the Richardson extrapolated value for Y_1 and Y_2.

4. Compute $E = (Y_2 - Y_1)/3$; note that this is the Richardson estimated error.

5. If $\frac{1}{4}h\epsilon \le E \le h\epsilon$, then we set $y_{n+1} = Y_3$ and start all over with the next point.

6. If $E > h\epsilon$, then the step size is too large, so we cut h in half ($h \leftarrow h/2$) and repeat the computation.

7. If $E < \frac{1}{4}h\epsilon$, then the step size is too small, so we set $y_{n+1} = Y_3$ but we double the step size ($h \leftarrow 2h$) for the next step.

Because we do not accept an approximate value y_{n+1} unless the estimated truncation error is less than $h\epsilon$, we are confident that the local error satisfies

$$|u_n(t_{n+1}) - y_{n+1}| \le h\epsilon. \tag{6.76}$$

This is made precise in the following theorem.

Theorem 6.5   For Y_1, Y_2, Y_3, and E as defined above, we have (assuming that f is continuously differentiable in y)

$$u_n(t_{n+1}) - Y_3 = O(h^4)$$

and

$$u_n(t_{n+1}) - Y_2 = E + O(h^3).$$

Proof We prove this by simply going through the details of the error analysis, using the results of the lemma proved above. We have (since $u_n(t_n) = y_n$)

$$u_n(t_{n+1}) = y_n + \frac{1}{2}h\left(f(t_n,y_n) + f(t_{n+1},y_n + hf(t_n,y_n))\right) + h\tau(t_n,h)$$

and

$$Y_1 = y_n + \frac{1}{2}h\left(f(t_n,y_n) + f(t_{n+1},y_n + hf(t_n,y_n))\right),$$

so that

$$u_n(t_{n+1}) - Y_1 = h\tau(t_n,h).$$

Similarly, we can show that the error in the half-step is given by ($t_{n+1/2} = t_n + \frac{1}{2}h$)

$$u_n(t_{n+1/2}) - W = \frac{1}{2}h\tau\left(t_n,\frac{1}{2}h\right).$$

We then can get $u_n(t_{n+1}) - Y_2$ by looking at the next half-step:

$$u_n(t_{n+1}) = u_n(t_{n+1/2}) + \frac{1}{4}h\left(f(t_{n+1/2},u_n(t_{n+1/2})) + f(t_{n+1},u_n(t_{n+1/2})\right.$$
$$\left. + \frac{1}{2}hf(t_{n+1/2},u_n(t_{n+1/2})))\right) + \frac{1}{2}h\tau\left(t_{n+1/2},\frac{1}{2}h\right)$$

$$Y_2 = W + \frac{1}{4}h\left(f(t_{n+1/2},W) + f\left(t_{n+1},W + \frac{1}{2}hf(t_n,W)\right)\right).$$

Use subtraction and the Mean Value Theorem to write

$$f(t_{n+1/2},u_n(t_{n+1/2})) - f(t_{n+1/2},W) = f_y(t_{n+1/2},\upsilon)(u_n(t_{n+1/2}) - W),$$

where υ is a value between W and $u_n(t_{n+1/2})$. This yields

$$u_n(t_{n+1}) - Y_2 = \frac{1}{2}h\tau\left(t_n,\frac{1}{2}h\right)\left[1 + \frac{1}{2}hf_y + \frac{1}{8}h^2f_y^2\right] + \frac{1}{2}h\tau\left(t_{n+1/2},\frac{1}{2}h\right),$$

where $f_y = f_y(t_{n+1/2},\upsilon)$. Using (6.75) we can now simplify this to

$$u_n(t_{n+1}) - Y_2 = h\tau\left(t_n,\frac{1}{2}h\right) + O(h^4) \cdot \frac{1}{3}(u_n(t_{n+1}) - Y_1)$$

$$= \frac{4}{3}\left(h\tau\left(t_n,\frac{1}{2}h\right) + O(h^4)\right) - \frac{1}{3}h\tau(t_n,h)$$

$$= O(h^4)$$

and

$$
\begin{aligned}
E &= \frac{1}{3}(Y_2 - Y_1) \\
&= \frac{1}{3}\left(Y_2 - u_n(t_{n+1}) + u_n(t_{n+1}) - Y_1\right) \\
&= \frac{1}{3}\left(-h\tau\left(t_n, \frac{1}{2}h\right) + O(h^4) + h\tau(t_n, h)\right) \\
&= -\frac{1}{4}h\tau(t_n, \frac{1}{2}h) + O(h^3) \\
&= u_n(t_{n+1}) - Y_2 + O(h^3)
\end{aligned}
$$

and we are done. $\square$

Thus we can use E as a computable estimate to the local error in Y_2. If this estimate is small enough, we use the (more accurate) value Y_3 as the approximation; this is known as *local extrapolation*.

The use of $\frac{1}{4}h\epsilon$ as the lower bound for the error in a single step is somewhat arbitrary. Also, it is typical to have an absolute minimum step size; if the computation attempts to take a step size smaller than h_{min}, the computation terminates. Similarly, it is a good idea to have an absolute maximum step size, so that the code does not take too large a step.

This particular algorithm has a number of drawbacks compared with others, but it is simple enough to illustrate the ideas behind adaptive methods. The most glaring flaw—beyond the fact that it is a relatively low-order method—in this algorithm is that it requires a total of five different function evaluations in each step.

EXAMPLE 6.22

We can show how the algorithm works by using it to approximate, with a crude specified error, the solution to the initial value problem

$$
y' = 16y(1 - y), \quad y(0) = 1/1024,
$$

whose exact solution is $y(t) = 1/(1 + 1023e^{-16t})$. We take $\epsilon = 0.001$ and use $h = 0.125$ as the initial step size. We get

$$
\begin{aligned}
Y_1 &= 0.0048714 \\
W &= 0.0024385 \\
Y_2 &= 0.0060786,
\end{aligned}
$$

so the estimated error is

$$
E = 0.00040239,
$$

which is larger than the desired tolerance of $0.001h = 1.25 \times 10^{-4}$. Thus we cut h in half to $h = 0.0625$ and repeat the computation. This time we get

$$
\begin{aligned}
Y_1 &= 0.0024385 \\
W &= 0.0015860 \\
Y_2 &= 0.0025749,
\end{aligned}
$$

and the estimated error is now $E = 0.000045460 < \epsilon h = 6.25 \times 10^{-5}$. Therefore, we accept the *extrapolated* value

$$Y_3 = (4Y_2 - Y_1)/3 = 0.0026204$$

as the value of $y_1 \approx y(h)$; since the error is *not* less than $\frac{1}{4}\epsilon h$, we do not change h before continuing the computation.

We will show one more step. To compute y_2, we go through essentially the same computations as for y_1; we get

$$Y_1 = 0.0065304$$
$$W = 0.0042517$$
$$Y_2 = 0.0068921$$
$$E = 1.2056 \times 10^{-4}.$$

This estimated error is again too large, so we again cut h in half to $h = 0.03125$ and repeat the computation. This time we get

$$Y_1 = 0.0042517$$
$$W = 0.0033550$$
$$Y_2 = 0.0042946$$
$$E = 1.4296 \times 10^{-5}.$$

This time the estimated error is less than $\epsilon h = 3.125 \times 10^{-5}$, so we accept the extrapolated value $Y_3 = 0.0043089$ as the value for $y_2 \approx y(2h)$. Again, the estimated error is not *too* small, so we leave h alone and go on to the next step. ∎

A glance at Figure 6.10 should explain why we had to make two attempts each time to get an acceptable step size: The solution to this problem is very steep near the initial point. If we use the more reasonable tolerance of $\epsilon = 5 \times 10^{-6}$ and the same initial h of 0.25, we get the values in Table 6.12. Because of the large number of points generated, we have included pairs of consecutive results only when the step size changed. The leftmost column keeps track of the number of steps so far; the rightmost column shows K_f, the running total of calls to f. Finally, note that we are here reporting the *global* error, $\max_{k \leq n} |y(t_k) - y_k|$, even though it is the *local* error that is being controlled.

Our second, more accurate, method is also based on two Runge-Kutta schemes. This time we use two higher-order schemes chosen so as to minimize the number of function calls in using both of them together. This general idea is known as *Runge-Kutta-Fehlberg*.

Consider the two schemes

$$y_{n+1} = y_n + h\left(\frac{5179}{57,600}k_1 + \frac{7571}{16,695}k_3 + \frac{393}{640}k_4 - \frac{92097}{339,200}k_5 + \frac{187}{2100}k_6 + \frac{1}{40}k_7\right) \quad (6.77)$$

and

$$y_{n+1} = y_n + h\left(\frac{35}{384}k_1 + \frac{500}{1113}k_3 + \frac{125}{192}k_4 - \frac{2187}{6784}k_5 + \frac{11}{84}k_6\right); \quad (6.78)$$

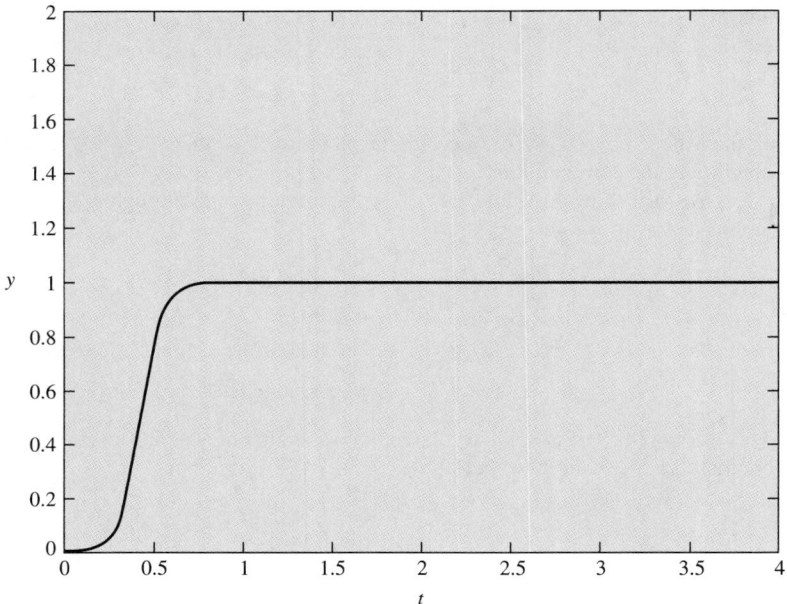

FIGURE 6.10   Plot of the exact solution for the adaptive algorithm example.

where

$$k_1 = f(t_n, y_n)$$

$$k_2 = f\left(t_n + \frac{1}{5}h, y_n + \frac{1}{5}hk_1\right)$$

$$k_3 = f\left(t_n + \frac{3}{10}h, y_n + \frac{3}{40}hk_1 + \frac{9}{40}hk_2\right)$$

$$k_4 = f\left(t_n + \frac{4}{5}h, y_n + \frac{44}{45}hk_1 - \frac{56}{15}hk_2 + \frac{32}{9}hk_3\right)$$

$$k_5 = f\left(t_n + \frac{8}{9}h, y_n + \frac{19{,}372}{6561}hk_1 - \frac{25{,}360}{2187}hk_2 + \frac{64{,}448}{6561}hk_3 - \frac{212}{729}hk_4\right)$$

$$k_6 = f\left(t_n + h, y_n + \frac{9017}{3168}hk_1 - \frac{355}{33}hk_2 + \frac{46{,}732}{5247}hk_3 + \frac{49}{176}hk_4 - \frac{5103}{18{,}656}hk_5\right)$$

$$k_7 = f\left(t_n + h, y_n + \frac{35}{384}hk_1 + \frac{500}{1113}hk_3 + \frac{125}{192}hk_4 - \frac{2187}{6784}hk_5 + \frac{11}{84}hk_6\right).$$

The first method (6.77) is fourth-order and the second one (6.78) is fifth-order, although we will not establish that here. It will be convenient, notationally, to define the two *Runge-Kutta-Fehlberg operators*

$$\text{RKF}_4(t_n, y_n) = y_n + h\left(\frac{5179}{57{,}600}k_1 + \frac{7571}{16{,}695}k_3 + \frac{393}{640}k_4 - \frac{92{,}097}{339{,}200}k_5 + \frac{187}{2100}k_6 + \frac{1}{40}k_7\right)$$

TABLE 6.12   Illustration of Second-Order Adaptive Algorithm

n	t_n	y_n	$\max_{k \leq n} \lvert y(t_k) - y_k \rvert$	h	K_f
0	0.00000000	0.00097656	0.000000E+00	0.250000E+00	0
1	0.00390625	0.00103948	0.316646E−09	0.390625E−02	29
11	0.04296875	0.00194025	0.648521E−08	0.390625E−02	79
12	0.04492188	0.00200171	0.672889E−08	0.195312E−02	88
57	0.13281250	0.00811817	0.340651E−07	0.195312E−02	313
58	0.13378906	0.00824496	0.346024E−07	0.976562E−03	322
154	0.22753906	0.03592021	0.150334E−06	0.976562E−03	802
155	0.22802734	0.03619173	0.151430E−06	0.488281E−03	811
999	0.64013672	0.96482762	0.108911E−05	0.488281E−03	5031
1000	0.64111328	0.96535403	0.108911E−05	0.976562E−03	5036
1098	0.73681641	0.99229815	0.108911E−05	0.976562E−03	5526
1099	0.73876953	0.99253335	0.108911E−05	0.195312E−02	5531
1144	0.82666016	0.99815985	0.108911E−05	0.195312E−02	5756
1145	0.83056641	0.99827114	0.108911E−05	0.390625E−02	5761
1167	0.91650391	0.99956231	0.108911E−05	0.390625E−02	5871
1168	0.92431641	0.99961373	0.108911E−05	0.781250E−02	5876
1179	1.01025391	0.99990231	0.108911E−05	0.781250E−02	5931
1180	1.02587891	0.99992393	0.108911E−05	0.156250E−01	5936
1185	1.10400391	0.99997821	0.108911E−05	0.156250E−01	5961
1186	1.13525391	0.99998681	0.108911E−05	0.312500E−01	5966
1188	1.19775391	0.99999516	0.108911E−05	0.312500E−01	5976
1189	1.26025391	0.99999829	0.108911E−05	0.625000E−01	5981
1190	1.32275391	0.99999939	0.108911E−05	0.625000E−01	5986
1191	1.44775391	1.00000000	0.108911E−05	0.125000E+00	5991
1192	1.69775391	1.00000000	0.108911E−05	0.250000E+00	5996

and

$$\mathrm{RKF}_5(t_n, y_n) = y_n + h \left(\frac{35}{384} k_1 + \frac{500}{1113} k_3 + \frac{125}{192} k_4 - \frac{2187}{6784} k_5 + \frac{11}{84} k_6 \right).$$

If we define

$$W_{n+1} = \mathrm{RKF}_4(t_n, y_n)$$

and

$$w_{n+1} = \mathrm{RKF}_5(t_n, y_n)$$

Algorithm 6.5 *High-Order Runge-Kutta-Fehlberg Adaptive Algorithm for Initial Value Problems*

Given t_n, y_n, a desired error tolerance, ϵ, the current value of the mesh, h, and a value $y_n \approx y(t_n)$, do the following:

1. Compute $Y_1 = \text{RKF}_4(t_n, y_n)$; note that this is a fourth-order approximation to $y(t_{n+1})$.

2. Compute $Y_2 = \text{RKF}_5(t_n, y_n)$; note that this is a fifth-order approximation to $y(t_{n+1})$.

3. Compute the error according to

$$E = |Y_1 - Y_2|.$$

4. If $\frac{1}{4}h\epsilon \leq E \leq h\epsilon$, then set $y_{n+1} = Y_2$ and continue on to the next point.

5. If $E > h\epsilon$, then the step size is too large, so replace h with $h/2$ and repeat the computation.

6. If $E < \frac{1}{4}h\epsilon$, then the step size has gotten too small, so set $y_{n+1} = Y_2$ but double the step size ($h \leftarrow 2h$) for the next step.

---

as two approximations to $y(t_{n+1})$, then it can be shown that the local error $u(t_{n+1}) - W_{n+1}$ can be estimated by a linear combination of the k_j values. In particular, we have

$$u(t_{n+1}) - W_{n+1} \approx w_{n+1} - W_{n+1}$$

$$= h\left(-\frac{71}{57,600}k_1 + \frac{71}{16,695}k_3 - \frac{71}{1920}k_4 + \frac{17,253}{339,200}k_5 - \frac{22}{525}k_6 + \frac{1}{40}k_7\right).$$

Again, this is a computable estimate of the error in a single step, and we can use it as a basis for controlling the step size, h. The algorithm is given, in outline form, as Algorithm 6.5.

Note that we use the more accurate fifth-order method as the actual approximate value, even though the error estimate is for the error in the fourth-order method. (Again, this is called *local extrapolation*.) Note also that the two Runge-Kutta methods use the same set of k_j values. This is deliberate, for it minimizes the need for excessive calls to the derivative function f. We have a fourth-order adaptive method that uses only seven calls to f; compare this with our first method (Algorithm 6.4), which uses five calls to f but is only second-order.

To illustrate the RKF method, we apply it to the same example as before, using the same parameters, with the results summarized in Table 6.13. Note that this method uses far fewer calls to f and far fewer steps than the crude algorithm outlined previously.

One drawback of adaptive methods is that they can "cycle" back and forth between h being too small and too large. This occurs in the RKF method in steps 22–26, for example. However, this is a minor issue considering the high accuracy that the adaptive methods deliver, for relatively low cost. Note also that the cost of the first step is high for both methods; this is because we deliberately set the initial step size rather high.

Recall that we tried to control only the *local* error, i.e., the error made in going from t_n to t_{n+1}. However, if the local error is controlled, then the global error is also

TABLE 6.13 Illustration of Runge-Kutta-Fehlberg Adaptive Algorithm

n	t_n	y_n	$\max_{k \leq n} \lvert y(t_k) - y_k \rvert$	h	K_f
0	0.00000000	0.00097656	0.000000E+00	0.250000E+00	0
1	0.03125000	0.00160906	0.164091E−05	0.312500E−01	25
5	0.15625000	0.01176855	0.827479E−05	0.312500E−01	65
6	0.17187500	0.01506077	0.828900E−05	0.156250E−01	78
21	0.40625000	0.39401030	0.829506E−05	0.156250E−01	225
22	0.43750000	0.51737154	0.829506E−05	0.312500E−01	232
23	0.45312500	0.57920584	0.829506E−05	0.156250E−01	245
24	0.48437500	0.69413320	0.829506E−05	0.312500E−01	252
25	0.50000000	0.74450451	0.829506E−05	0.156250E−01	265
42	0.76562500	0.99512871	0.829506E−05	0.156250E−01	432
43	0.79687500	0.99703971	0.829506E−05	0.312500E−01	439
48	0.95312500	0.99975633	0.829506E−05	0.312500E−01	486
49	1.01562500	0.99991024	0.829506E−05	0.625000E−01	493
52	1.20312500	0.99999551	0.829506E−05	0.625000E−01	520
53	1.32812500	0.99999922	0.829506E−05	0.125000E+00	527
54	1.45312500	0.99999987	0.829506E−05	0.125000E+00	534
55	1.70312500	0.99999956	0.829506E−05	0.250000E+00	541

controlled, although in general this depends somewhat on the nature of the differential equation. The precise statement is the following.

Theorem 6.6 (Global Error for Adaptive Methods, Version I)   *If f is Lipschitz continuous in y, if the local error at each step satisfies (6.76), and if $h_{\max}/h_{\min} < c$, then, for $h_{\max}$ sufficiently small, the global error is bounded by*

$$\lvert y(t_n) - y_n \rvert \leq C_0 \lvert y(t_0) - y_0 \rvert + C_1 \epsilon$$

where

$$C_0 = e^{cKt_n}, \quad C_1 = \frac{e^{cKt_n} - 1}{K}$$

and K is the Lipschitz constant for f.

Proof   We start with the triangle inequality, which gives us

$$\lvert y(t_{n+1}) - y_{n+1} \rvert \leq \lvert y(t_{n+1}) - u_n(t_{n+1}) \rvert + \lvert u_n(t_{n+1}) - y_{n+1} \rvert. \tag{6.79}$$

Note that the last term is bounded by the local error hypothesis (6.76). From the original IVP and the definition of u_n, we can get

$$y(t) = y(t_n) + \int_{t_n}^{t} f(s,y(s))\, ds$$

$$u_n(t) = y_n + \int_{t_n}^{t} f(s,u_n(s))\, ds.$$

Subtraction then yields

$$y(t) - u_n(t) = y(t_n) - y_n + \int_{t_n}^{t} (f(s,y(s)) - f(s,u_n(s)))\, ds. \tag{6.80}$$

Take absolute values and upper bounds, and use the Lipschitz condition to get

$$|y(t) - u_n(t)| \le |y(t_n) - y_n| + K(t - t_n) \max_{t_n \le s \le t} |y(s) - u_n(s)|. \tag{6.81}$$

Since this is true for all $t \in [t_n, t_{n+1}]$, it follows that

$$\max_{t_n \le s \le t_{n+1}} |y(s) - u_n(s)| \le |y(t_n) - y_n| + K(t_{n+1} - t_n) \max_{t_n \le s \le t_{n+1}} |y(s) - u_n(s)| \tag{6.82}$$

so that (assuming that $1 - K(t_{n+1} - t_n) > 0$)

$$\max_{t_n \le s \le t_{n+1}} |y(s) - u_n(s)| \le \frac{1}{1 - K(t_{n+1} - t_n)} |y(t_n) - y_n|. \tag{6.83}$$

Therefore,

$$|y(t_{n+1}) - u_n(t_{n+1})| \le \max_{t_n \le s \le t_{n+1}} |y(s) - u_n(s)|$$

$$\le \frac{1}{1 - K(t_{n+1} - t_n)} |y(t_n) - y_n|. \tag{6.84}$$

Hence, if $h_{n+1} = t_{n+1} - t_n$ satisfies $h_{n+1} < 1/K$, then

$$|y(t_{n+1}) - y_{n+1}| \le \frac{1}{1 - Kh_{n+1}} |y(t_n) - y_n| + |u_n(t_{n+1}) - y_{n+1}|. \tag{6.85}$$

The local error hypothesis allows us to write this as

$$|y(t_{n+1}) - y_{n+1}| \le \frac{1}{1 - Kh_{n+1}} |y(t_n) - y_n| + h_{n+1}\epsilon. \tag{6.86}$$

If we assume that h is small enough at each step (this amounts to making sure that $h_{\max}$ is small enough; i.e., $h_{\max} < 1/K$) and the local error hypothesis is valid at each step, then we can solve this recursion to get

$$|y(t_n) - y_n| \le \gamma^n |y(t_0) - y_0| + \epsilon \sum_{k=0}^{n-1} \gamma^{n-k} h_k,$$

where

$$\gamma = \frac{1}{1 - Kh_{max}} \leq 1 + Kh_{max} \leq e^{Kh_{max}}.$$

Thus

$$\begin{aligned} |y(t_n) - y_n| &\leq \gamma^n |y(t_0) - y_0| + \epsilon h_{max} \sum_{k=0}^{n-1} \gamma^{n-k} \\ &= \gamma^n |y(t_0) - y_0| + \epsilon h_{max} \frac{1 - \gamma^n}{1 - \gamma} \\ &= \gamma^n |y(t_0) - y_0| + \epsilon h_{max} \frac{1 - \gamma^n}{Kh_{max}}(1 - Kh_{max}) \\ &\leq \gamma^n |y(t_0) - y_0| + \epsilon \frac{\gamma^n - 1}{K} \\ &\leq e^{Knh_{max}} |y(t_0) - y_0| + \epsilon \frac{e^{Knh_{max}} - 1}{K}. \end{aligned}$$

Now, since we assume that $h_{max}/h_{min} < c$, then it follows that $nh_{max} < ct_n$; therefore,

$$|y(t_n) - y_n| \leq e^{cKt_n} |y(t_0) - y_0| + \epsilon \frac{e^{cKt_n} - 1}{K},$$

and we are done. □

If we assume that f is smooth and uniformly monotone decreasing in y, we can get a better result, and we actually get it more easily.

Theorem 6.7 (Global Error for Adaptive Methods, Version II) *If f is smooth and uniformly monotone decreasing in y, and if the local error at each step satisfies (6.76), then*

$$|y(t_n) - y_n| \leq |y(t_0) - y_0| + \epsilon t_n.$$

Proof We begin as we did for the previous theorem—writing the global error in terms of the local error, using the triangle inequality:

$$|y(t_{n+1}) - y_{n+1}| \leq |y(t_{n+1}) - u_n(t_{n+1})| + |u_n(t_{n+1}) - y_{n+1}|. \tag{6.87}$$

The local error hypothesis tells us that

$$|u_n(t_{n+1}) - y_{n+1}| \leq h_{n+1}\epsilon,$$

where $h_{n+1} = t_{n+1} - t_n$ is the step used in the most recent computation, so we need worry only about the first term on the right. Recall now the definition of u_n as the solution of the ODE with initial value $u_n(t_n) = y_n$. In other words, u_n satisfies

$$u_n' = f(t, u_n), \quad u_n(t_n) = y_n.$$

But we also have, of course, that y satisfies

$$y' = f(t, y), \quad y(t_n) = y(t_n).$$

We can therefore use the stability estimate for smooth and uniformly monotone equations (6.8) to get

$$|y(t) - u_n(t)| \leq \left(e^{-m(t-t_n)}\right) |y(t_n) - y_n|;$$

therefore,

$$|y(t_{n+1}) - u_n(t_{n+1})| \leq \left(e^{-m(t_{n+1}-t_n)}\right) |y(t_n) - y_n| = e^{-mh_{n+1}} |y(t_n) - y_n| = \gamma |y(t_n) - y_n|,$$

where $\gamma < 1$. We therefore have that

$$|y(t_{n+1}) - y_{n+1}| \leq \gamma |y(t_n) - y_n| + h_{n+1}\epsilon, \tag{6.88}$$

which is a recursive inequality in $e_k = |y(t_k) - y_k|$:

$$e_{n+1} \leq \gamma e_n + h_{n+1}\epsilon.$$

We can solve this recursion to get

$$e_n = |y(t_n) - y_n| \leq C_n |y(t_n) - y_n| + \epsilon \left(\sum_{k=1}^{n} \gamma^{k-1} h_n\right),$$

where

$$C_n = \gamma^n \leq 1$$

and

$$\epsilon \left(\sum_{k=1}^{n} \gamma^{k-1} h_n\right) \leq \epsilon t_n,$$

and we are done.   □

Note that this result is much better than the more general one that used only Lipschitz continuity of f. In either case, we see that the global error is thus controlled when the local error is controlled, but not in the same way or to the same degree.

EXERCISES

1. Use Algorithm 6.4 to solve the following IVP using the local error tolerance of $\epsilon = 0.001$:

$$y' = y^2, \quad y(0) = 1.$$

Compute four or five steps. Compare the computed solution to the exact solution $y = (1-t)^{-1}$.

2. Repeat the preceding using Algorithm 6.5. (Note:

Both of these exercises are designed to be done as hand calculations using a calculator rather than a computer program.)

3. Implement Algorithm 6.4 of the text and test it out on the problems listed in Exercise 4 of Section 6.2. Keep track of the total number of function evaluations.

4. Repeat the preceding using Algorithm 6.5.

6.10 BOUNDARY VALUE PROBLEMS

We have already briefly encountered boundary value problems, in Sections 2.7 and 4.9. In this section we will take a more in-depth look at the problem of approximating BVPs.

We will confine ourselves to the two broad classes of problems:

1. Linear BVPs in the form

$$
\begin{aligned}
-u'' + p(x)u' + q(x)u &= f(x), \quad 0 \le x \le 1, \\
u(0) &= g_0, \\
u(1) &= g_1,
\end{aligned}
$$

where p, q, and f are known functions.

2. Nonlinear BVPs that can be written in the form

$$
\begin{aligned}
-u'' &= F(x, u, u'), \quad 0 \le x \le 1 \qquad &(6.89)\\
u(0) &= g_0 &(6.90)\\
u(1) &= g_1. &(6.91)
\end{aligned}
$$

The first class of problems can be attacked by a modest extension of the ideas from Section 2.7 or 4.9. The second class we attack by adapting our IVP methods to the boundary value problem. In both cases we assume that a unique solution exists to the original BVP, and that this solution is smooth. Precisely how smooth will be apparent from the hypotheses to each result we obtain.

6.10.1 Simple Difference Methods

The basic ideas are the same as in Section 2.7. We first define a grid

$$
0 = x_0 < x_1 < x_2 < \cdots < x_{n-1} < x_n = 1
$$

with $x_k - x_{k-1} = h$ (therefore $x_k = kh$). We then use the derivative approximations derived from

$$
u'(x) - \frac{u(x+h) - u(x-h)}{2h} = \frac{1}{6} h^2 u'''(\xi_{x,h})
$$

$$
u''(x) - \frac{u(x-h) - 2u(x) + u(x+h)}{h^2} = \frac{1}{12} h^2 u^{(4)}(\eta_{x,h})
$$

to write the BVP as the difference equation

$$
-\left(1 + \frac{1}{2}p(x)h\right) u(x-h) + \left(2 + q(x)h^2\right) u(x) - \left(1 - \frac{1}{2}p(x)h\right) u(x+h)
$$
$$
= h^2 f(x) + R(x,h) \quad (6.92)
$$

where the remainder term is given by

$$
R(x,h) = \frac{1}{12} h^4 u^{(4)}(\eta_{x,h}) + \frac{1}{6} p(x) h^4 u'''(\xi_{x,h}). \quad (6.93)
$$

Algorithm 6.6 *Finite Difference Method for Linear Boundary Value Problems*

1. Given $h = (b-a)/n$, form the $(n-1) \times (n-1)$ tridiagonal matrix T, where

$$t_{ii} = \left(2+q_ih^2\right), \quad t_{i,i+1} = -\left(1-\frac{1}{2}p_ih\right), \quad t_{i,i-1} = -\left(1+\frac{1}{2}p_ih\right).$$

2. Form the vector $F \in \mathbf{R}^{n-1}$ given by

$$F_i = \begin{cases} \left(1+\frac{1}{2}p_1h\right)g_0 + h^2f_1 & i = 1 \\ h^2f_i & 2 \le i \le n-2 \\ \left(1-\frac{1}{2}p_{n-1}h\right)g_1 + h^2f_{n-1} & i = n-1. \end{cases}$$

3. Solve the system $TU = F$.

4. Then the approximate values are $U_i \approx u(x_i)$, $1 \le i \le n-1$.

Note that $R = O(h^4)$. Denote the approximate function as $U_h(x) \approx u(x)$, and use $U_i = U_h(x_i)$ for the approximation at the grid points. Then the vector of U_i values is defined by the system of (tridiagonal) linear equations

$$\left(2+q_1h^2\right)U_1 - \left(1-\frac{1}{2}p_1h\right)U_2 = \left(1+\frac{1}{2}p_1h\right)g_0 + h^2f_1 \qquad (6.94)$$

$$-\left(1+\frac{1}{2}p_ih\right)U_{i-1} + \left(2+q_ih^2\right)U_i - \left(1-\frac{1}{2}p_ih\right)U_{i+1} = h^2f_i, \quad 2 \le i \le n-2 \quad (6.95)$$

$$-\left(1+\frac{1}{2}p_{n-1}h\right)U_{n-2} + \left(2+q_{n-1}h^2\right)U_{n-1} = \left(1-\frac{1}{2}p_{n-1}h\right)g_1 + h^2f_{n-1}, \qquad (6.96)$$

where we have used $p_i = p(x_i)$ and similarly for q_i and f_i.

At this point it might be useful to summarize all this as an algorithm in outline form (see Algorithm 6.6).

EXAMPLE 6.23

Consider the BVP

$$-u'' + u' + u = \sin \pi x, \quad 0 \le x \le 1$$
$$u(0) = 1$$
$$u(1) = 0.$$

Therefore, $p(x) = 1$, $q(x) = 1$, and $f(x) = \sin \pi x$. If we take $h = \frac{1}{4}$, then $n = 4$ and the linear system is therefore 3×3. We have

$$T = \begin{bmatrix} 2.0625 & -0.875 & 0 \\ -1.125 & 2.0625 & -0.875 \\ 0 & -1.125 & 2.0625 \end{bmatrix}, \quad F = \begin{bmatrix} 1.1692 \\ 0.0625 \\ 0.04419 \end{bmatrix},$$

and the solution to the linear system is

$$U = \begin{bmatrix} 0.8423 \\ 0.6489 \\ 0.3754 \end{bmatrix},$$

thus giving us our approximate solution,

$$u(1/4) \approx 0.8423, \qquad u(1/2) \approx 0.6489, \qquad u(3/4) \approx 0.3754.$$

The following theorem guarantees that the tridiagonal linear system can be solved using Algorithm 2.6 from Section 2.6. ∎

Theorem 6.8   *Let p and q both be continuous on $[0,1]$, with $|p(x)| \le P_M$ and $0 < Q_m \le q(x)$ for all $x \in [0,1]$. Then, for h such that*

$$h \le 2/P_M \tag{6.97}$$

the discrete system (6.94)–(6.10.1) has a unique solution that can be computed using Algorithm 2.6.

Proof   The condition on h implies that

$$1 - \frac{1}{2} P_M h \ge 0;$$

therefore, $1 \pm \frac{1}{2} p_i h \ge 0$ regardless of the sign of p_i. Hence, we have

$$\left| 1 + \frac{1}{2} p_i h \right| + \left| 1 - \frac{1}{2} p_i h \right| \le 2$$
$$< 2 + h^2 Q_m$$
$$\le 2 + h^2 q_i$$

for all i (the arguments for the first and last equations are a little different); thus the matrix is diagonally dominant and we can invoke Algorithm 2.6. □

Now that we know that the approximate solution can be always computed, we can ask how accurate it is.

Theorem 6.9   *Under the same hypotheses as in Theorem 6.8, there exists a constant $C > 0$, independent of h, such that*

$$\max_{1 \le i \le n-1} |u(x_i) - U_i| \le C h^2 \left(\|u^{(4)}\|_\infty + \|u'''\|_\infty \right).$$

Proof   Define $e_i = u(x_i) - U_i$, with $e_0 = e_n = 0$. Subtracting (6.95) from (6.92) shows that the e_i satisfy the system of equations

$$\left(2 + q_1 h^2\right) e_1 - \left(1 - \frac{1}{2} p_1 h\right) e_2 = R_1$$

$$-\left(1 + \frac{1}{2} p_i h\right) e_{i-1} + \left(2 + q_i h^2\right) e_i - \left(1 - \frac{1}{2} p_i h\right) e_{i+1} = R_i, \quad 2 \le i \le n-2$$

$$-\left(1 + \frac{1}{2} p_{n-1} h\right) e_{n-2} + \left(2 + q_{n-1} h^2\right) e_{n-1} = R_{n-1},$$

where we have used $R_k = R(x_k, h)$ for the remainder as given in (6.93). Let J be the index for which e_i is largest in absolute value; $|e_J| \ge |e_i|$ for all i; assume for simplicity that $J \ne 1$ and $J \ne n-1$. We then have

$$(2 + q_J h^2)|e_J| \le \left|1 + \frac{1}{2} p_J h\right| |e_{J-1}| + \left|1 - \frac{1}{2} p_J h\right| |e_{J+1}| + |R_J|$$

$$\le \left(\left|1 + \frac{1}{2} p_J h\right| + \left|1 - \frac{1}{2} p_J h\right|\right) |e_J| + |R_J|$$

$$= 2|e_J| + |R_J|.$$

Now subtract $2|e_J|$ from both sides of the equation to get

$$q_J h^2 |e_J| \le |R_J| \quad \Rightarrow \quad |e_J| \le \frac{1}{h^2 Q_m} \max_{x \in [0,1]} |R(x,h)|.$$

But

$$\max_{x \in [0,1]} |R(x,h)| \le \frac{1}{6} h^4 \left(\|u^{(4)}\|_\infty + \|u'''\|_\infty\right),$$

and so we are done, with $C = (6Q_m)^{-1}$, since $|e_J| = \max_{1 \le i \le N-1} |u(x_i) - U_i|$ is the error we are trying to bound.   □

This result is an excellent example of what mathematicians would call a "non-sharp" result. It seems to suggest that the accuracy of the approximation would depend adversely on Q_m, when in fact we will retain second-order accuracy even if $q(x) \equiv 0$, i.e., $Q_m = 0$. This requires a different and more subtle mode of proof, though, which we choose to omit. It should be noted that the effect of P_M that is implied in the two theorems is very real; unless h is small compared with $p(x)$, we will not get the expected accuracy. Some of the exercises are designed to demonstrate this.

EXAMPLE 6.24

To illustrate, consider the example

$$-u'' + 25u' + 26u = 1 \tag{6.98}$$

$$u(0) = 0, \quad u(1) = 1, \tag{6.99}$$

TABLE 6.14   **Finite Difference Approximation to (6.98)–(6.99)**

$n = h^{-1}$	Error in approximating $u(x)$
4	0.410152632559E+00
8	0.226940305241E+00
16	0.795659554206E–01
32	0.185637947707E–01
64	0.437975179182E–02
128	0.110156098833E–02
256	0.274526544131E–03
512	0.685779319546E–04
1024	0.171423695873E–04

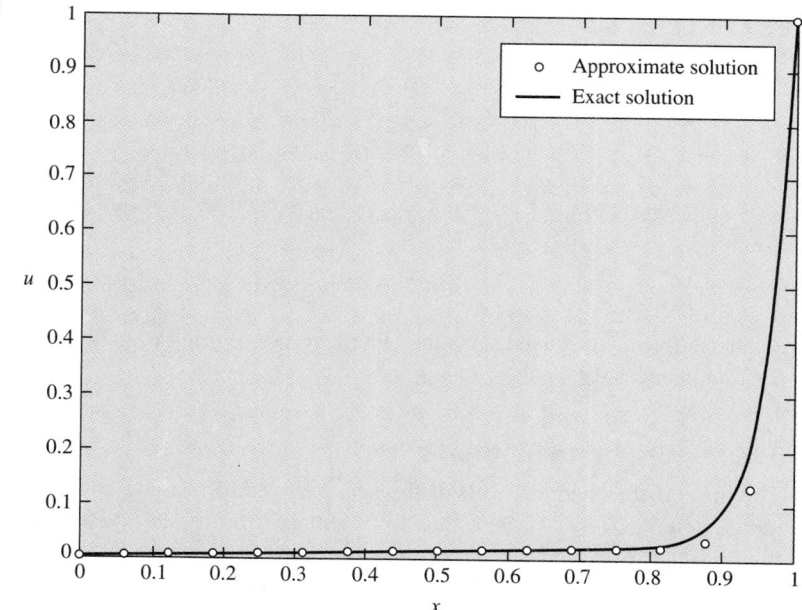

FIGURE 6.11   Solution for Example 6.24, together with the approximate solution (○) for $h = 1/16$.

for which the the exact solution is

$$u(x) = \frac{1}{26}\left(1 - \frac{e^{26} + 25}{e^{26} - e^{-1}}e^{-x} + \frac{25 + e^{-1}}{e^{26} - e^{-1}}e^{26x}\right).$$

We solve this with the preceding ideas, using a sequence of grids with mesh size $h = 2^{-k}$, $2 \leq k \leq 10$. Table 6.14 shows the maximum error over the entire interval for each value of h. Note that we do not begin to get the appropriate rate of convergence until $h \leq 1/16$, as we should expect. Figure 6.11 shows the exact solution, with the approximate solution for $h = 1/16$ marked by the circles. ∎

TABLE 6.15   Finite Difference Approximation to (6.100)–(6.101)

$n = h^{-1}$	Error in approximating $u(x)$
4	0.617591875069E–03
8	0.156383538506E–03
16	0.392871138047E–04
32	0.982751528289E–05
64	0.245793638298E–05
128	0.614467541404E–06
256	0.153625683064E–06
512	0.384063436609E–07
1024	0.960190649213E–08

EXAMPLE 6.25

A second example,

$$-u'' + u' = 2 \tag{6.100}$$
$$u(0) = 0, \quad u(1) = 1, \tag{6.101}$$

shows that the condition $q > 0$ is not required for second-order accuracy. Here the exact solution is

$$u(x) = 2x + \frac{1}{e-1}(1 - e^x),$$

and the problem was solved using the same sequence of grids as in the previous example. Table 6.15 summarizes the results. ∎

6.10.2   Shooting Methods

The finite difference methods of the previous section can be applied to nonlinear problems such as (6.89)–(6.91), but the result is a nonlinear system of equations rather than a linear one, and we are not yet prepared to approximate the solutions to such a problem. (But see Section 7.9.) What we do here is reformulate the problem so that we can apply our initial value problem methods.

Recall the problem of interest:

$$-u'' = F(x,u,u'), \quad 0 \le x \le 1$$
$$u(0) = g_0$$
$$u(1) = g_1.$$

Consider now the new problem

$$-y'' = F(x,y,y'), \quad 0 \le x \le 1 \tag{6.102}$$
$$y(0) = g_0 \tag{6.103}$$
$$y'(0) = p, \tag{6.104}$$

Algorithm 6.7 *Shooting Method for Nonlinear Boundary Value Problem*

1. Define $y_h(p)$ to be the (approximate) value of $y(1)$ that is found using a mesh spacing of h and an initial value of p for y', together with a reasonable initial value problem method applied to (6.102)–(6.104).

2. Use, for example, the secant method to update values of p:

$$p_{k+1} = p_k - y_h(p_k) \left(\frac{p_k - p_{k-1}}{y_h(p_k) - y_h(p_{k-1})} \right).$$

Note that each time we iterate on this, we have to solve the initial value problem for a new value of $p = p_k$; this is an example of a function whose root we are seeking, and that might take a substantial amount of time to compute.

3. Once the iteration converges, the approximate solution of the initial value problem is also an approximate solution of the boundary value problem.

---

and note that this is an *initial value problem*, although we are using x instead of t to denote the independent variable. We will view y as depending on the choice of initial value for y', and write $y(x; p)$ to denote this dependence.

The question then is, Can we find a value of p such that $y(1; p) = g_1 = u(1)$? If we can, then the fact that both the boundary and initial value problems have unique solutions implies that we have found the solution of the BVP (6.89)–(6.91). Essentially, if we can find the value of p that makes the function

$$f(p) = y(1; p) - g_1 \tag{6.105}$$

equal to zero, we will have found the correct value of p and therefore solved the BVP. But (6.105) is nothing more than a root-finding problem, just as we solved in Chapter 3!

The *shooting method* is based on making an initial guess for p, and then applying a root-finding method to an approximation to f. Note that (6.102)–(6.104) is a second-order differential equation, so we will have to recast it as a first-order system in order to proceed.

A naive approach to the problem might be as given in Algorithm 6.7.

Because of the high cost that is potentially involved in solving the IVP several times, it is usually more economical to initially iterate for p using a very coarse mesh, and then refine the value of p obtained in this way using the desired, finer mesh.

EXAMPLE 6.26

To illustrate, consider the nonlinear BVP defined by

$$-u'' = (u')^2 \tag{6.106}$$
$$u(0) = 0, \quad u(1) = 1, \tag{6.107}$$

which has exact solution $u = \ln((e - 1)x + 1)$. We used the trapezoid rule predictor-corrector method to solve the initial value problems, and we used the secant method with $p_0 = 1$ and $p_1 = 0.75$ to solve (iteratively) the root-finding problem. For the first several iterations we used

TABLE 6.16   Finite Difference Approximation to (6.106)–(6.107)

k	$n = h^{-1}$	p_k	$u_h(1) - 1$
0	4	0.1000000000E+01	−0.3126554719E+00
1	4	0.7500000000E+00	−0.4434700177E+00
2	4	0.1597516641E+01	−0.6162711822E−01
3	4	0.1734300664E+01	−0.1368453961E−01
4	4	0.1773343754E+01	−0.5578476751E−03
5	4	0.1775002976E+01	−0.5342620750E−05
6	4	0.1775019020E+01	−0.2114179276E−08
7	128	0.1717716619E+01	−0.2161118864E−03
8	128	0.1718310128E+01	0.2246518221E−05
9	128	0.1718304022E+01	0.2427305024E−09
10	128	0.1718304021E+01	−0.5551115123E−15

FIGURE 6.12   Approximate solution via shooting method for Example 6.26, using $h = 1/128$.

the very crude mesh size $h = \frac{1}{4}$, but on subsequent iterations we used the much smaller value $h = 1/128$. Table 6.16 shows the progress of the iteration for p, and Figure 6.12 shows a plot of the final approximate solution.

Let's look at the first iteration in more detail. The first-order system for this second-order equation is

$$w' = f(t, w), \quad w(0) = w_0,$$

where

$$w(t) = \begin{bmatrix} w_1(t) \\ w_2(t) \end{bmatrix} = \begin{bmatrix} u(t) \\ u'(t) \end{bmatrix}, \qquad f(t,w) = \begin{bmatrix} w_2(t) \\ -w_2^2(t) \end{bmatrix}, \qquad w_0 = \begin{bmatrix} 0 \\ p \end{bmatrix}$$

and we take the initial value of p to be 1. Then the first solution over the interval $[0,1]$, using $h = 1/4$ and the trapezoid rule predictor-corrector, goes like this:

$$\bar{w} = w_0 + hf(0,w_0) = \begin{bmatrix} w_{0,1} + hw_{0,2} \\ w_{0,2} - hw_{0,2}^2 \end{bmatrix} = \begin{bmatrix} 0.25 \\ 0.75 \end{bmatrix},$$

$$w_1 = w_0 + (h/2)(f(h,\bar{w}) + f(0,w_0)) = \begin{bmatrix} w_{0,1} + (h/2)(\bar{w}_2 + w_{0,2}) \\ w_{0,2} - (h/2)(\bar{w}_2^2 + w_{0,2}^2) \end{bmatrix} = \begin{bmatrix} 0.21875 \\ 0.8046875 \end{bmatrix};$$

$$\bar{w} = w_0 + hf(0,w_0) = \begin{bmatrix} w_{1,1} + hw_{1,2} \\ w_{1,2} - hw_{1,2}^2 \end{bmatrix} = \begin{bmatrix} 0.419921875 \\ 0.64280700683594 \end{bmatrix},$$

$$w_2 = w_1 + (h/2)(f(2h,\bar{w}) + f(h,w_1)) = \begin{bmatrix} w_{1,1} + (h/2)(\bar{w}_2 + w_{1,2}) \\ w_{1,2} - (h/2)(\bar{w}_2^2 + w_{1,2}^2) \end{bmatrix} = \begin{bmatrix} 0.39968681335449 \\ 0.67209714741330 \end{bmatrix};$$

$$\bar{w} = w_2 + hf(h,w_1) = \begin{bmatrix} w_{2,1} + hw_{2,2} \\ w_{2,2} - hw_{2,2}^2 \end{bmatrix} = \begin{bmatrix} 0.56771110020782 \\ 0.55916850352302 \end{bmatrix},$$

$$w_3 = w_2 + (h/2)(f(3h,\bar{w}) + f(2h,w_2)) = \begin{bmatrix} w_{2,1} + (h/2)(\bar{w}_2 + w_{2,2}) \\ w_{2,2} - (h/2)(\bar{w}_2^2 + w_{2,2}^2) \end{bmatrix} = \begin{bmatrix} 0.55359501972153 \\ 0.57654914855164 \end{bmatrix};$$

$$\bar{w} = w_3 + hf(3h,w_3) = \begin{bmatrix} w_{3,1} + hw_{3,2} \\ w_{3,2} - hw_{3,2}^2 \end{bmatrix} = \begin{bmatrix} 0.69773230685944 \\ 0.49344691837773 \end{bmatrix},$$

$$w_4 = w_3 + (h/2)(f(4h,\bar{w}) + f(3h,w_3)) = \begin{bmatrix} w_{3,1} + (h/2)(\bar{w}_2 + w_{3,2}) \\ w_{3,2} - (h/2)(\bar{w}_2^2 + w_{3,2}^2) \end{bmatrix} = \begin{bmatrix} 0.68734452808770 \\ 0.50456180080763 \end{bmatrix}.$$

Now, if we had used the correct value of $p = u'(0)$ to start the computation, we would have $w_{4,1} \approx u(1) = 1$. But clearly $p = 1$ was not a good enough approximation, so we want to do a secant iteration on the function $F(p) = w_{4,1} - 1$; this requires that we take a second initial guess and compute $w_{4,1}$ based on that guess. We take $p = 0.75$ as our second guess and repeat the computation outlined above, getting -0.4434700177 as our new value of $w_{4,1} - 1$. The secant step is then

$$p_2 = p_1 - F(p_1) \frac{p_1 - p_0}{F(p_1) - F(p_0)}$$

$$= 0.75 - (-0.4434700177) \left(\frac{0.75 - 1.00}{-0.4434700177 - (-0.3126554719)} \right) = 1.597516641,$$

and we would continue the computation using p_2 as the initial value for the first derivative; i.e., we would set $w_{0,2} = p_2$.

Table 6.16 summarizes the results we get by continuing this computation. Note that we use the crude mesh spacing $h = 1/4$ until we get a converged value for p, and then switch to the smaller, $h = 1/128$ mesh size. This is a more economical approach than doing the entire computation with the finer grid. ∎

EXERCISES

1. Use a mesh size of $h = \frac{1}{4}$ $(n = 4)$ to solve the boundary value problem

$$-u'' + u = \sin x, \quad 0 < x < 1$$
$$u(0) = u(1) = 0.$$

Don't bother to write a computer program; do this as a hand calculation.

2. Repeat the preceding for the BVP

$$-u'' + (2 - x)u = x, \quad 0 < x < 1$$
$$u(0) = u(1) = 0.$$

Don't bother to write a computer program; do this as a hand calculation.

3. Repeat the preceding for the BVP

$$-u'' + u = e^{-x}, \quad 0 < x < 1$$
$$u(0) = u(1) = 0.$$

4. Consider the linear boundary value problem

$$-u'' + (10 \cos 2x)u = 1$$
$$u(0) = u(\pi) = 0.$$

Solve this using finite difference methods and a decreasing sequence of grids, starting with $h = \pi/16, \pi/32, \ldots$. Do the approximate solutions appear to be converging to a solution?

5. Consider the nonlinear boundary value problem

$$-u'' + e^u = 1$$
$$u(0) = 0, u(1) = 1.$$

Use the shooting method combined with the trapezoid rule predictor-corrector to construct solutions to this equation, and then use a fourth-order Runge-Kutta scheme and compare the two approximate solutions. Are they nearly the same?

LITERATURE AND SOFTWARE DISCUSSION

The numerical solution of initial value problems for ordinary differential equations is one of the most important areas of scientific computation. The monographs [8], [10], [14], and [16] represent a good overview of the development of the subject in the last 40 years or so. More recent treatments are found in [7] and [12]. A selection of easily available automatic software—as well as a rich summary of recent results—is given in [15]. Other codes can be found by searching the Netlib repository at http://www.netlib.org.

Less is available about two-point boundary value problems. Much of what we have done here is based on Chapter 8 (Section 7) of [11]; a more comprehensive survey is given in [13]. One reason that less is written about BVPs for ordinary differential equations is that they can in almost all respects be considered as special cases of the Poisson partial differential equation, which we treat in Section 9.2. However, the ordinary differential equation case is *so* much simpler that it really deserves its own exposition, which can serve as an introduction to the PDE case.

BIBLIOGRAPHY

[1] ALLEN, MYRON B., & ISAACSON, ELI L. *Numerical Analysis for Applied Science*. John Wiley & Sons, New York, 1998.

[2] ATKINSON, KENDALL. *An Introduction to Numerical Analysis*. John Wiley & Sons, New York, 1987 (2nd ed.).

[3] BRAUN, MARTIN. *Differential Equations and Their Applications*. Springer-Verlag, New York, 1993 (4th ed.).

[4] BURDEN, RICHARD, & FAIRES, J. DOUGLAS. *Numerical Analysis*. Brooks/Cole, Pacific Grove, CA, 1997 (6th ed.).

[5] CODDINGTON, E.A., & LEVINSON, N. *Theory of Ordinary Differential Equations*. McGraw-Hill, New York, 1955.

[6] DORMAND, J.R., & PRINCE, P.J. "A family of embedded Runge-Kutta formulae," *Journal of Computational and Applied Mathematics*, vol. 6 (1980), no. 1, pp. 19–26.

[7] ERIKSSON, K., ESTEP, D., HANSBO, P., & Johnson, C.. *Computational Differential Equations*. Cambridge University Press, Cambridge, UK, 1996.

[8] GEAR, C. WILLIAM. *Numerical Initial Value Problems in Ordinary Differential Equations*. Prentice-Hall, Englewood Cliffs, NJ, 1971.

[9] GOLDSTINE, HERMAN H. *A History of Numerical Analysis from the 16th through the 19th Century*. Springer-Verlag, New York, 1977.

[10] HENRICI, PETER. *Discrete Variable Methods in Ordinary Differential Equations*. John Wiley & Sons, New York, 1962.

[11] ISAACSON, E., & KELLER, H.B. *Analysis of Numerical Methods*. Dover, New York, 1994 (originally published in 1966 by John Wiley & Sons.)

[12] ISERLES, ARIEH. *A First Course in the Numerical Analysis of Differential Equations*. Cambridge University Press, Cambridge, UK, 1996.

[13] KELLER, H.B. *Numerical Solution of Two Point Boundary Value Problems*. SIAM, Philadelphia, 1976.

[14] LAMBERT, J.D. *Computational Methods for Ordinary Differential Equations*. John Wiley & Sons, London, 1973.

[15] SHAMPINE, L.F. "The MATLAB ODE suite." *SIAM Journal of Scientific Computation*, vol. 18 (1997), pp. 1–22.

[16] SHAMPINE, L.F., & GORDON, M.K. *Computer Solution of Ordinary Differential Equations*. W.H. Freeman, San Francisco, 1975.

CHAPTER 7

NUMERICAL METHODS FOR THE SOLUTION OF SYSTEMS OF EQUATIONS

Pennsyltucky Pharmaceuticals makes a specialty line of three dietary supplements using different proportions of three raw materials: broccoli extract, gingko biloba, and ginseng oil. Product 1 is made by combining equal amounts of all three; product 2 is heavy on the gingko biloba, using four parts of it to one part each of broccoli extract and ginseng oil; and product 3 uses six parts ginseng oil to two parts gingko biloba and one part broccoli extract. These mixes can be modeled by a set of linear equations: If we let x_1 be the amount of broccoli extract, x_2 the amount of gingko biloba, and x_3 the amount of ginseng oil, and denote the amount of each product by p_1, p_2, and p_3, then we have the system of equations

$$x_1 + x_2 + x_3 = p_1$$
$$x_1 + 4x_2 + x_3 = p_2$$
$$x_1 + 2x_2 + 6x_3 = p_3.$$

Given desired production amounts of the p_i, $i = 1, 2, 3$, we can determine how much of each of the raw materials to buy by solving this system of equations.

This particular model is very simple, largely because of its small size, and does not really require any specialized mathematical techniques for its solution. However, the success of this line of dietary supplements has convinced Pennsyltucky Pharmaceuticals to expand this part of their business substantially, so that they are looking at as many as 50 products made from as many raw materials. The solution of a 50×50 linear system is not to be undertaken lightly in the absence of good computing software.

This chapter is about the techniques for solving linear and nonlinear systems of equations. One of the things we will learn early on is that the size of the system is an important consideration.

Here we will look at numerical methods for solving two important problems from linear algebra:

1. The linear systems problem: Given a matrix A and a vector b, both known, find the vector x such that

$$Ax = b.$$

2. The nonlinear systems problem: Given a vector-valued function F, find the vector x such that

$$F(x) = 0.$$

A third important problem—the algebraic eigenvalue problem—is deferred to the next chapter.

We begin with a review of the linear algebra concepts and notation that are necessary for this chapter.

7.1   LINEAR ALGEBRA REVIEW

A vector $x \in \mathbf{R}^n$ is defined to be an ordered n-tuple of real numbers:

$$x = (x_1, x_2, \ldots, x_n)^T$$

where the superscript T (denoting transpose) indicates that the vector is considered to be a column vector. A matrix $A \in \mathbf{R}^{m \times n}$ is a rectangular array of m rows and n columns:

$$A = \begin{bmatrix} a_{11} & a_{12} & a_{13} & \cdots & a_{1n} \\ a_{21} & a_{22} & a_{23} & \cdots & a_{2n} \\ a_{31} & a_{32} & a_{33} & \cdots & a_{3n} \\ \vdots & \vdots & \vdots & \ddots & \vdots \\ a_{m1} & a_{m2} & a_{m3} & \cdots & a_{mn} \end{bmatrix}.$$

We assume that the student is familiar with the basic operations of addition and multiplication of matrices.

We will use no diacritical marks to distinguish vectors from scalars. Matrices will, in general, be denoted by upper-case Roman (alphabet) letters, and vectors by lower-case Roman letters. The notational correspondence between the vector or matrix and its components will almost always be as in the above examples.

Our assumption that vectors are column vectors means we can regard them as matrices in $\mathbf{R}^{n \times 1}$, and this allows us to write the ordinary vector dot product as

$$(x,y) = x \cdot y = x^T y = \sum_{i=1}^{n} x_i y_i.$$

Note that reversing the order of multiplication results in a matrix, not a scalar:

$$xy^T = \begin{bmatrix} x_1 y_1 & x_1 y_2 & x_1 y_3 & \cdots & x_1 y_n \\ x_2 y_1 & x_2 y_2 & x_2 y_3 & \cdots & x_2 y_n \\ x_3 y_1 & x_3 y_2 & x_3 y_3 & \cdots & x_3 y_n \\ \vdots & \vdots & \vdots & \ddots & \vdots \\ x_n y_1 & x_n y_2 & x_n y_3 & \cdots & x_n y_n \end{bmatrix}.$$

Note also that this notation allows us to write

$$(x,Ay) = x^T Ay = (A^T x)^T y = (A^T x,y),$$

which is one of the important properties of the transpose of a matrix.

Given a square matrix $A \in \mathbf{R}^{n \times n}$, if there exists a second square matrix $B \in \mathbf{R}^{n \times n}$ such that $AB = BA = I$, then we say that B is the inverse of A and we write $B = A^{-1}$. *Not all square matrices have an inverse!* If a matrix $A \in \mathbf{R}^{n \times n}$ has an inverse, we say that A is *nonsingular*; otherwise we say that A is *singular*.

The following theorem summarizes the conditions under which a matrix is nonsingular, and also connects them to the solvability of the linear systems problem.

Theorem 7.1 *Given a matrix $A \in \mathbf{R}^{n \times n}$, the following statements are equivalent:*

1. *A is nonsingular.*
2. *The columns of A form an independent set of vectors.*
3. *The rows of A form an independent set of vectors.*
4. *The linear system $Ax = b$ has a unique solution for all vectors $b \in \mathbf{R}^n$.*
5. *The homogeneous system $Ax = 0$ has only the trivial solution $x = 0$.*
6. *The determinant is nonzero:* $\det A \neq 0$.

An important corollary to this theorem is the following.

Corollary 7.1 *If $A \in \mathbf{R}^{n \times n}$ is singular, then there exist infinitely many vectors $x \in \mathbf{R}^n$, $x \neq 0$, such that $Ax = 0$.*

There are a number of special classes of matrices. In Section 2.6 we looked at *tridiagonal* matrices, and later in this chapter we will look at *symmetric, positive definite* matrices. One class of special matrix that the student ought to be familiar with is the *triangular* matrix: A square matrix is lower (upper) triangular if all the elements above (below) the main diagonal are zero. Thus

$$U = \begin{bmatrix} 1 & 2 & 3 \\ 0 & 4 & 5 \\ 0 & 0 & 6 \end{bmatrix}$$

is upper triangular, and

$$L = \begin{bmatrix} 1 & 0 & 0 \\ 2 & 3 & 0 \\ 4 & 5 & 6 \end{bmatrix}$$

is lower triangular.

The student who is not familiar with the concepts of *independence/dependence*, *spanning, basis, vector space/subspace, dimension,* and *orthogonal/orthonormal* should review the appropriate sections of a linear algebra text. While we defer our treatment of the computation of eigenvalues and eigenvectors to the next chapter, the student should still be familiar with the basic properties and definition of these quantities, as they are mentioned in a couple of places in this chapter.

EXERCISES

1. Assume Theorem 7.1 and use it to prove Corollary 7.1.

2. Use Theorem 7.1 to prove that a triangular matrix is nonsingular if and only if the diagonal elements are all nonzero.

3. Suppose that we can write $A \in \mathbf{R}^{n \times n}$ as the product of two triangular matrices $L \in \mathbf{R}^{n \times n}$ and $U \in \mathbf{R}^{n \times n}$, where the diagonal elements of L and U are all nonzero. Prove that A is nonsingular.

7.2   LINEAR SYSTEMS AND GAUSSIAN ELIMINATION

In Section 2.6 we constructed an algorithm that solves tridiagonal linear systems by first reducing them to triangular form, and then solving the triangular system. In this section we will construct a general version of that algorithm. We begin by writing the linear system as a single augmented matrix:

$$A' = [A \mid b]$$

where the vertical bar is meant to separate the coefficient matrix from the right-side vector. The solution algorithm is then applied to A'.

The algorithm is the same one that is taught in the standard linear algebra course, and that we first saw in Section 2.6: Gaussian elimination. It works by systematically eliminating nonzero elements below the main diagonal of the coefficient matrix. This is accomplished by using only those operations that preserve the solution set of the system, known as *elementary row operations*:

1. Multiply a row by a nonzero scalar, c.

2. Interchange two rows.

3. Multiply a row by a nonzero scalar c, and add the result to another row.

If we can manipulate from one matrix to another using only elementary row operations, then the two matrices are said to be *row equivalent*.

The important theorem that connects the elementary row operations to the solution of linear systems is the following.

Theorem 7.2   *Let A' be the augmented matrix corresponding to the linear system $Ax = b$, and suppose A' is row equivalent to A'', which corresponds to the linear system $Tx = c$. Then the two linear systems have precisely the same solution sets.*

Our goal, then, is to use elementary row operations to reduce the augmented matrix A' to the new augmented matrix $A'' = [U \mid c]$, where U is *upper triangular*. This will mean that the new system $Ux = c$ is easy to solve.

EXAMPLE 7.1

To illustrate the process, we will work through a concrete example in detail. Consider the system of equations

$$4x_1 + 2x_2 - x_3 = 5$$
$$x_1 + 4x_2 + x_3 = 12$$
$$2x_1 - x_2 + 4x_3 = 12,$$

which can be written in matrix-vector form as

$$\begin{bmatrix} 4 & 2 & -1 \\ 1 & 4 & 1 \\ 2 & -1 & 4 \end{bmatrix} \begin{bmatrix} x_1 \\ x_2 \\ x_3 \end{bmatrix} = \begin{bmatrix} 5 \\ 12 \\ 12 \end{bmatrix}.$$

We write this as an augmented matrix:

$$A' = \begin{bmatrix} 4 & 2 & -1 & | & 5 \\ 1 & 4 & 1 & | & 12 \\ 2 & -1 & 4 & | & 12 \end{bmatrix}.$$

Then the elimination algorithm proceeds as follows:

$$A' = \begin{bmatrix} 4 & 2 & -1 & | & 5 \\ 1 & 4 & 1 & | & 12 \\ 2 & -1 & 4 & | & 12 \end{bmatrix} \sim \begin{bmatrix} 4 & 2 & -1 & | & 5 \\ 0 & \frac{7}{2} & \frac{5}{4} & | & \frac{43}{4} \\ 2 & -1 & 4 & | & 12 \end{bmatrix} \sim \begin{bmatrix} 4 & 2 & -1 & | & 5 \\ 0 & \frac{7}{2} & \frac{5}{4} & | & \frac{43}{4} \\ 0 & -2 & \frac{9}{2} & | & \frac{19}{2} \end{bmatrix}.$$

The first step was accomplished by multiplying the first row by $\frac{1}{4}$ and subtracting the result from the second row; the second step was accomplished by multiplying the first row by $\frac{1}{2}$ and subtracting the result from the third row. To finish the job, we have (by multiplying the second row by $-\frac{4}{7}$ and subtracting from the third row)

$$A' \sim \begin{bmatrix} 4 & 2 & -1 & | & 5 \\ 0 & \frac{7}{2} & \frac{5}{4} & | & \frac{43}{4} \\ 0 & -2 & \frac{9}{2} & | & \frac{19}{2} \end{bmatrix} \sim \begin{bmatrix} 4 & 2 & -1 & | & 5 \\ 0 & \frac{7}{2} & \frac{5}{4} & | & \frac{43}{4} \\ 0 & 0 & \frac{73}{14} & | & \frac{219}{14} \end{bmatrix} = A''.$$

This augmented matrix represents a triangular system—meaning the coefficient matrix is triangular—as follows:

$$A'' = [U \mid c] \implies Ux = c,$$

i.e.,

$$\begin{bmatrix} 4 & 2 & -1 \\ 0 & \frac{7}{2} & \frac{5}{4} \\ 0 & 0 & \frac{73}{14} \end{bmatrix} \begin{bmatrix} x_1 \\ x_2 \\ x_3 \end{bmatrix} = \begin{bmatrix} 5 \\ \frac{43}{4} \\ \frac{219}{14} \end{bmatrix};$$

and we can now solve by interpreting each row as follows:

Third row: $\frac{73}{14}x_3 = \frac{219}{14} \Rightarrow x_3 = 3$

Second row: $\frac{7}{2}x_2 + \frac{5}{4}x_3 = \frac{43}{4} \Rightarrow x_2 = 2$

First row: $4x_1 + 2x_2 - x_3 = 5 \Rightarrow x_1 = 1$

■

To render this example process into a general algorithm, we note the essential features:

> *Work down each column, eliminating (i.e., converting to a zero) each component below the main diagonal, and modifying the rest of the corresponding row appropriately.*

We start by presenting what we call "naive" Gaussian elimination, a version of the algorithm that is easy to understand yet is not completely general (see Algorithm 7.1). We then move quickly to a more robust and complete version.

Algorithm 7.1: Naive Gaussian elimination algorithm for $Ax = b$ (pseudocode)

```
for i=1 to n-1
      for j=i+1 to n
            m = a(j,i)/a(i,i)
            for k=i+1 to n
                  a(j,k) = a(j,k) - m*a(i,k)
            endfor
            b(j) = b(j) - m*b(i)
      endfor
endfor
```

Programming Notes: Let's talk about what this algorithm does and how it implements what was done in the example.

The outermost loop (the i loop) ranges over the columns of the matrix; the last column is skipped because we do not need to perform any eliminations there, since there are no elements below the diagonal. (If we were doing elimination on a nonsquare matrix with more rows than columns, we would have to include the last column in this loop.)

The middle loop (the j loop) ranges down the ith column, below the diagonal (hence j ranges only from $i+1$ to n). We first compute the *multiplier*, m, for each row. This is the constant by which we multiply the ith row in order to eliminate the a_{ji} element. Note that we *overwrite* the previous values with the new ones, and we do not actually carry out the computation that makes a_{ji} zero. Note also that this loop is where the right-side vector is modified to reflect the elimination step.

The innermost loop (the k loop) ranges across the jth row, starting after the ith column, modifying each element appropriately to reflect the elimination of a_{ji}.

Finally, we must be aware that the algorithm does not actually create the zeroes in the lower triangular half of A; this would be wasteful of computer time since we don't need to have these zeroes in place for the algorithm to work. If we were to apply our algorithm to the previous example, then the computer storage for A would look like this when the process was done:

$$B = \begin{bmatrix} 4 & 2 & -1 \\ 1 & \frac{7}{2} & \frac{5}{4} \\ 2 & -2 & \frac{73}{14} \end{bmatrix}$$

The process works because we work only with the upper triangular part of the matrix from this point forward, so the lower triangular elements need never be referenced. This will change somewhat in Section 7.4.

Algorithm 7.2: Backward solution algorithm for Ax = b (pseudocode)

```
x(n) = b(n)/a(n,n)
for i=n-1 to 1
     sum = 0
     for j=i+1 to n
          sum = sum + a(i,j)*x(j)
     endfor
     x(i) = (b(i) - sum)/a(i,i)
endfor
```

Programming Notes: This algorithm simply marches backward up the diagonal, computing each x_i in turn. Formally, we are computing

$$x_i = \frac{1}{a_{ii}} \left(b_i - \sum_{j=i+1}^{n} a_{ij}x_j \right),$$

which is what is necessary to solve a triangular system. The j loop is simply accumulating the summation term in this formula. To summarize, the algorithm is nothing more than a systematic application of the ordinary Gaussian elimination algorithm from basic linear algebra. There is one problem, however. In the computation of the multiplier, we divide by the current (note that the values of the matrix elements change throughout the elimination process) value of the diagonal element. Clearly, if this value is zero, then we cannot proceed any further and the algorithm breaks down. Note that this does not mean that the matrix is singular, since

$$A = \begin{bmatrix} 0 & 1 \\ 1 & 0 \end{bmatrix}$$

is an example of a nonsingular matrix for which the algorithm cannot even get started.
What do we do to fix this?
Recall that one of our allowed elementary row operations is the interchange of two rows. In the example above, if we switch the two rows, we get the new matrix

$$A' = \begin{bmatrix} 1 & 0 \\ 0 & 1 \end{bmatrix},$$

and from this point the elimination algorithm can obviously proceed to completion. (In fact, it is already finished!)

To finish the solution process we apply what is known as the backward solution or *back-solve* algorithm (see Algorithm 7.2) to the augmented matrix that results from the elimination step. It's called the *backward* solution step because we proceed backward up the diagonal from the bottom to the top.

The pivotal issue in this discussion is whether or not the diagonal elements become zero at any point in the process. Because they are so important in this regard, the diagonals are called *pivots* or *pivot elements*, and the process of swapping rows to avoid a zero element on the diagonal is called *pivoting*. There are actually several kinds of pivoting. *Partial* pivoting, in which only entries in the same column below the current diagonal are examined, is the most common and is the only type we will discuss here in any depth. *Complete* pivoting, which searches not only on the current column but also on all subsequent columns, is known to be more stable but is also

Algorithm 7.3: *Partial Pivoting (outline)*

1. Suppose we are about to work on the ith column of the matrix. Then we search that portion of the ith column below and including the diagonal, and find the element that has the largest absolute value. Let p denote the index of the row that contains this element.

2. Interchange rows i and p.

3. Proceed with the elimination.

---

much more expensive to implement. Finally, it is sometimes necessary to scale the rows before pivoting, and this is called *scaled* pivoting.

The partial pivoting algorithm is relatively easy to describe (see Algorithm 7.3).

Note that this algorithm does not merely search for the first nonzero element in the column to make the pivot, but it searches for the *largest* element in the column to use as the pivot. We will address this issue momentarily. It is not difficult to update our algorithm to include pivoting (see Algorithm 7.4).

Although the actual problem with naive Gaussian elimination is the potential division by a zero pivot, the entire algorithm will be less susceptible to rounding error if we choose to use the largest possible pivot element. To gain some insight into why this is so, consider the following example system:

$$\begin{bmatrix} \epsilon & 1 \\ 1 & 1 \end{bmatrix} \begin{bmatrix} x_1 \\ x_2 \end{bmatrix} = \begin{bmatrix} 1 \\ 2 \end{bmatrix}.$$

For any $\epsilon \neq 1$, the matrix is nonsingular and so a unique solution exists. The exact solution, expressed in terms of ϵ (see Exercise 7), is

$$x_1 = \frac{1}{1-\epsilon} = 1 + O(\epsilon), \quad x_2 = \frac{1-2\epsilon}{1-\epsilon} = 1 + O(\epsilon), \tag{7.1}$$

but what happens when we solve this system—under the assumption that ϵ is very small—using the naive versus the pivoting algorithm?

If we don't pivot, then the computation goes like this:

$$\left[\begin{array}{cc|c} \epsilon & 1 & 1 \\ 1 & 1 & 2 \end{array}\right] \sim \left[\begin{array}{cc|c} \epsilon & 1 & 1 \\ 0 & 1-1/\epsilon & 2-1/\epsilon \end{array}\right].$$

Suppose now that ϵ is so small that, within the machine arithmetic, $1 - 1/\epsilon = -\epsilon^{-1}$ and $2 - 1/\epsilon = -\epsilon^{-1}$. Then we have

$$\left[\begin{array}{cc|c} \epsilon & 1 & 1 \\ 0 & 1-1/\epsilon & 2-1/\epsilon \end{array}\right] \approx \left[\begin{array}{cc|c} \epsilon & 1 & 1 \\ 0 & -1/\epsilon & -1/\epsilon \end{array}\right] \Rightarrow x_2 \approx 1, \quad x_1 \approx 0.$$

While the value of x_2 is accurate, the value of x_1 is not. This error is caused by the rounding error associated with the large number $1/\epsilon$. On the other hand, if we first pivot by switching the rows, then the computation becomes

$$\left[\begin{array}{cc|c} \epsilon & 1 & 1 \\ 1 & 1 & 2 \end{array}\right] \sim \left[\begin{array}{cc|c} 1 & 1 & 2 \\ \epsilon & 1 & 1 \end{array}\right] \sim \left[\begin{array}{cc|c} 1 & 1 & 2 \\ 0 & 1-\epsilon & 1-2\epsilon \end{array}\right] \approx \left[\begin{array}{cc|c} 1 & 1 & 2 \\ 0 & 1 & 1 \end{array}\right] \Rightarrow x_2 \approx 1, \ x_1 \approx 1.$$

This example illustrates (but, of course, does not *prove*) the utility of partial pivoting even if we do not have a zero pivot. This is especially important when we consider that we cannot predict what the sizes of the pivot elements will be.

Algorithm 7.4: Gaussian elimination with partial pivoting (pseudocode)

```
for i=1 to n-1
!
!     This begins the pivoting code
!
      am = abs(a(i,i))
      p = i
      for j=i+1 to n
            if abs(a(j,i)) > am then
                  am = abs(a(j,i))
                  p = j
            endif
      endfor
      if p > i then
            for k = i to n
                  hold = a(i,k)
                  a(i,k) = a(p,k)
                  a(p,k) = hold
            endfor
            hold = b(i)
            b(i) = b(p)
            b(p) = hold
      endif
!
!     This ends the pivoting code
!
      for j=i+1 to n
            m = a(j,i)/a(i,i)
            for k=i+1 to n
                  a(j,k) = a(j,k) - m*a(i,k)
            endfor
            b(j) = b(j) - m*b(i)
      endfor
endfor
```

EXERCISES

For the sake of simplicity here, we will define at the outset several families of matrices, parameterized by their dimension. These will be referred to in several of the exercises throughout the chapter.

$$H_n = [h_{ij}], \quad h_{ij} = \frac{1}{i+j-1}.$$

$$K_n = [k_{ij}], \quad k_{i,j} = \begin{cases} 2 & i = j \\ -1 & |i-j| = 1 \\ 0 & \text{otherwise} \end{cases}$$

$$T_n = [t_{ij}], \quad t_{i,j} = \begin{cases} 4 & i = j \\ 1 & |i-j| = 1 \\ 0 & \text{otherwise} \end{cases}$$

$$A_n = [a_{ij}], \quad a_{i,j} = \begin{cases} 1 & i = j \\ 4 & i-j = 1 \\ -4 & i-j = -1 \\ 0 & \text{otherwise} \end{cases}$$

Even if we do not know the solution to a linear system, we can check the accuracy of a computed solution x_c by means of the residual $r = b - Ax_c$. If x_c is the exact solution, then each component of r will be zero; in floating-point arithmetic, there might be a small amount of rounding error, unless the matrix is "nearly singular," a concept that we will discuss in detail in Section 7.5.

1. Write a naive Gaussian elimination code and use it to solve the system of equations $Ax = b$, where

$$A = \begin{bmatrix} 14 & 14 & -9 & 3 & -5 \\ 14 & 52 & -15 & 2 & -32 \\ -9 & -15 & 36 & -5 & 16 \\ 3 & 2 & -5 & 47 & 49 \\ -5 & -32 & 16 & 49 & 79 \end{bmatrix}$$

and $b = [-15, -100, 106, 329, 463]^T$. The correct answer is $x = [0, 1, 2, 3, 4]^T$.

2. Write a naive Gaussian elimination code and use it to solve the system of equations

$$T_5 x = b,$$

where $b = [1, 6, 12, 18, 19]^T$. The correct answer is $x = [0, 1, 2, 3, 4]^T$.

3. Write a naive Gaussian elimination code and use it to solve the system of equations

$$H_5 x = b,$$

where $b = [5.0, 3.550, 2.81428571428571,$
$2.34642857142857, 2.01746031746032]^T$.
The correct answer is $x = [1, 2, 3, 4, 5]^T$.

4. Repeat the preceding problem, except now use $b_1 = 5.0001$. How much does the answer change?

5. Write your own naive Gaussian elimination code, based on the material in this chapter, and test it on the indicated families, over the range of $4 \leq n \leq 20$. Take b to be the vector of appropriate size, each of whose entries is 1.

(a) H_n
(b) K_n
(c) T_n

For each value of n, compute the value of $\max_{1 \leq i \leq n} |r_i|$, where $r = b - Ax$.

6. Modify the Gaussian elimination algorithm to handle more than a single right-hand side. Test it on a 5×5 example of your own design, using at least three right-side vectors.

7. Use the naive Gaussian elimination algorithm to solve (by hand) the following system. You should get the same results as in (7.1).

$$\begin{bmatrix} \epsilon & 1 \\ 1 & 1 \end{bmatrix} \begin{bmatrix} x_1 \\ x_2 \end{bmatrix} = \begin{bmatrix} 1 \\ 2 \end{bmatrix}$$

8. Use the naive Gaussian elimination algorithm to solve (by hand) the following system, using only three-digit decimal arithmetic. Repeat, using Gaussian elimination with partial pivoting. Comment on your results.

$$\begin{bmatrix} 0.0001 & 1 \\ 1 & 1 \end{bmatrix} \begin{bmatrix} x_1 \\ x_2 \end{bmatrix} = \begin{bmatrix} 1 \\ 2 \end{bmatrix}$$

9. Write a Gaussian elimination code that does partial pivoting and use it to solve the system of equations $Ax = b$, where

$$A = \begin{bmatrix} 9 & 3 & 2 & 0 & 7 \\ 7 & 6 & 9 & 6 & 4 \\ 2 & 7 & 7 & 8 & 2 \\ 0 & 9 & 7 & 2 & 2 \\ 7 & 3 & 6 & 4 & 3 \end{bmatrix}$$

and $b = [35,58,53,37,39]^T$. The correct answer is $x = [0,1,2,3,4]^T$.

10. Write a code to do Gaussian elimination with partial pivoting, and apply it to the system $A_5 x = b$, where $b = [-4, -7, -6, -5, 16]^T$ and the solution is $x = [0,1,2,3,4]^T$.

7.3 OPERATION COUNTS

Since many practical problems in numerical linear algebra involve very large matrices, an important issue is the number of operations that a specific algorithm requires in order to operate on a matrix of a given size. In this section we will go through the details of deriving the operation counts for Gaussian elimination and the back-solve step; in subsequent sections we will give the operation count without going into the details.

We first have to decide what constitutes an "operation." The usual convention is to count only the multiplications and divisions, since most multiplications (and divisions) are associated with a subsequent addition (or subtraction).[1] This is how we will count things, but the reader should be aware that other texts will occasionally count *all* operations, which will lead to different results (generally by a factor of 2) than what we get here.

There is little subtlety and no grace at all in the way we count the operations. For the naive Gaussian elimination algorithm, we simply write out the loop structure as a series of summations; inside each summation we put the number of operations that occur inside that loop. We thus have

$$C = \sum_{i=1}^{n} \sum_{j=i+1}^{n} \left(2 + \sum_{k=i+1}^{n} 1 \right).$$

We now proceed to add all this up. An exact answer will be more than a little ugly; fortunately, we are more interested in the gross size of the operation count than its exact value. What we really want to know is how the largest term in the count depends on the size of the matrix. We thus write

$$\sum_{j=i+1}^{n} \left(2 + \sum_{k=i+1}^{n} 1 \right) = \sum_{j=i+1}^{n} \left(2 + \sum_{m=1}^{n-i} 1 \right) =$$
$$\sum_{j=i+1}^{n} (n-i+2) = (n-i)(n-i+2);$$

---

[1] Traditionally, it was also true that multiplications and divisions were more costly than additions and subtractions, and thus it was more important to keep track of the more expensive operations. Divisions are still more costly than the other operations, but it is no longer the case that multiplications are substantially more expensive than additions and subtractions.

hence,[2]

$$C = \sum_{i=1}^{n-1}(n-i)(n-i+2) = \sum_{m=1}^{n-1}(m^2+2m) = \frac{1}{6}n(n-1)(2n-1)+n(n-1) = \frac{1}{3}n^3+O(n^2).$$

So we get an operation count for (naive) Gaussian elimination of

$$C = \frac{1}{3}n^3 + O(n^2).$$

If we do the same steps (Exercise 1) for the backward solution algorithm, we get

$$C = \frac{1}{2}n^2 + O(n),$$

so that the total cost of solving a linear system can be estimated by

$$\underbrace{C_T}_{\text{Total operations}} = \underbrace{\frac{1}{3}n^3 + O(n^2)}_{\text{Cost of elimination}} + \underbrace{\frac{1}{2}n^2 + O(n)}_{\text{Cost of forward solve}} + \underbrace{\frac{1}{2}n^2 + O(n)}_{\text{Cost of backward solve}}$$

$$= \tfrac{1}{3}n^3 + n^2 + O(n)$$
$$= \tfrac{1}{3}n^3 + O(n^2).$$

The significant thing here is that the solution step is a full power of n cheaper than the elimination step.

EXERCISES

1. Determine the operation count for the backward solution algorithm.

2. What is the operation count for computing the dot product $x \cdot y$ of two vectors?

3. What is the operation count for computing the matrix-vector product Ax?

4. Repeat the preceding, assuming that A is tridiagonal.

5. What is the operation count for a matrix-matrix product, AB?

6. Repeat the preceding, assuming that A is tridiagonal.

7. Repeat the preceding, assuming now that both A and B are tridiagonal.

8. What is the operation count for the *outer product* xy^T?

9. Determine the operation count for the tridiagonal solution algorithm of Section 2.6.

10. Assume that you are working on a computer that does one operation every 10^{-9} seconds. How long, roughly, would it take such a computer to solve a linear system for $n = 100,000$, using the cost estimates derived in this section for Gaussian elimination? What is the time estimate if the computer does only one operation every 10^{-6} seconds?

---

[2]The student should be aware that here we have used the formulas

$$\sum_{k=1}^{n}k = \frac{1}{2}n(n+1), \quad \sum_{k=1}^{n}k^2 = \frac{1}{6}n(n+1)(2n+1).$$

Sometimes these are part of the calculus discussion of Riemann sums.

7.4 THE *LU* FACTORIZATION

The Gaussian elimination algorithm developed in Section **??** is deficient in one respect: There is no way to "save" the work done in reducing A to upper triangular form. If we want to solve several linear systems having the same coefficient matrix, we can do it (see Exercise 6 of Section 7.2) only if all the right-side vectors are known ahead of time. However, we saw in Section 6.8.2 that it is possible to have a sequence of linear systems problems in which the right-side vector for one problem depends on the solution to the previous problem. Thus, unless we can somehow "save" the elimination step, we will have to repeat all that work, which is in fact the bulk of the work involved in the solution of the problem.

Our goal in this section is to develop a *matrix factorization* that allows us to save the work from the elimination step. Before proceeding with that effort, though, perhaps we should explain why an obvious way to proceed is not the best way to proceed. Why don't we just compute A^{-1}?

The answer is that it is not cost-effective to do so. The standard (and, with a little care, the best) way to compute the inverse is to solve the matrix-matrix system

$$AX = I, \qquad (7.2)$$

where I is the identity matrix. The solution X is the desired inverse: $X = A^{-1}$. Let's compute the cost of finding the inverse and of using it to solve linear systems.

The matrix equation (7.2) is a collection of n simple linear systems problems. We can easily modify the Gaussian elimination algorithm to find X, by adding a single extra loop to account for having more than one right-side vector. The elimination computation now takes $\frac{5}{6}n^3 + O(n^2)$ operations, and the triangular back-solves take $\frac{1}{2}n^3 + O(n^2)$; thus the total cost is $\frac{4}{3}n^3 + O(n^2)$. (Exercise 7 asks you to provide the details for all this.) A few operations can be saved by taking into account the fact that I is all 1s and 0s. When this is done, the total cost of producing the inverse is $n^3 + O(n^2)$. Compare this with the cost of Gaussian elimination alone, which is $\frac{1}{3}n^3 + O(n^2)$. If we can find some way to "save" the Gaussian elimination work, we will have a more efficient means of solving linear systems than computing and using the inverse.

What we will do is show that we can *factor* the matrix A into the product of a lower triangular and an upper triangular matrix:

$$A = LU.$$

This allows us to solve linear systems by instead solving two triangular systems:

$$Ax = b \Rightarrow Ux = y, \quad \text{where} \quad Ly = b.$$

Thus we first solve $Ly = b$ and then $Ux = y$ to get the solution:

$$Ux = y, \quad Ly = b \Rightarrow Ux = L^{-1}b \Leftrightarrow LUx = b \Leftrightarrow Ax = b.$$

This shows that we can solve linear systems by (1) computing the LU factorization of A and (2) then solving appropriate lower and upper triangular systems. An example might be useful at this point.

EXAMPLE 7.2

Consider the matrices

$$L = \begin{bmatrix} 1 & 0 & 0 \\ \frac{1}{2} & 1 & 0 \\ 0 & 0 & 1 \end{bmatrix}$$

and

$$U = \begin{bmatrix} 4 & 2 & 0 \\ 0 & 2 & 1 \\ 0 & 0 & 2 \end{bmatrix}.$$

These are lower and upper triangular, respectively, and their product is

$$LU = A = \begin{bmatrix} 4 & 2 & 0 \\ 2 & 3 & 1 \\ 0 & 1 & \frac{5}{2} \end{bmatrix}.$$

Consider now the linear systems problem

$$Ax = b,$$

where $b = (2,5,6)^T$. Since we have an LU factorization of A, we can go ahead and use it to compute the solution. We first solve

$$Ly = b$$

using a *forward* solution algorithm, which is perfectly analogous to the backward solution algorithm except it works by going forward *down* the diagonal. We have

$$\begin{bmatrix} 1 & 0 & 0 \\ \frac{1}{2} & 1 & 0 \\ 0 & 0 & 1 \end{bmatrix} \begin{bmatrix} y_1 \\ y_2 \\ y_3 \end{bmatrix} = \begin{bmatrix} 2 \\ 5 \\ 6 \end{bmatrix}$$

and we get $y = (2,4,4)^T$. Now solve

$$Ux = y$$

using our backward solution algorithm; we have

$$U = \begin{bmatrix} 4 & 2 & 0 \\ 0 & 2 & 1 \\ 0 & 0 & 2 \end{bmatrix} \begin{bmatrix} x_1 \\ x_2 \\ x_3 \end{bmatrix} = \begin{bmatrix} 2 \\ 4 \\ 4 \end{bmatrix}$$

and we get $x = (0,1,2)^T$, which is exactly the solution to the original linear system. ∎

Since it turns out (see below) that the LU factorization is nothing more than a properly organized Gaussian elimination, the total cost of this process is

$$C = \underbrace{\frac{1}{3}n^3 + O(n^2)}_{\text{Cost of elimination}} + \underbrace{\frac{1}{2}n^2 + O(n)}_{\text{Cost of back-solve}} + \underbrace{\frac{1}{2}n^2 + O(n)}_{\text{Cost of forward solve}} = \underbrace{\frac{1}{3}n^3 + O(n^2)}_{\text{Total cost}}.$$

If we already have done the factorization, then the cost of the two solution steps is simply

$$C = \frac{1}{2}n^2 + O(n) + \frac{1}{2}n^2 + O(n) = n^2 + O(n).$$

Constructing the LU factorization is surprisingly easy. In fact, we have already done it; we just don't know it!

We start by considering the matrix

$$E_1 = I - \begin{bmatrix} 0 & 0 & \dots & 0 \\ m_{21} & 0 & \dots & 0 \\ m_{31} & 0 & \dots & 0 \\ \vdots & 0 & \dots & 0 \\ m_{n1} & 0 & \dots & 0 \end{bmatrix},$$

where $m_{i1} = a_{i1}/a_{11}$. It is not hard to show that

$$E_1 A = \left[\begin{array}{c|ccc} a_{11} & a_{12} & \dots & a_{1n} \\ \hline 0 & & & \\ \vdots & & A' & \\ 0 & & & \end{array} \right]$$

where A' is an $n-1 \times n-1$ matrix. Thus we see that multiplication by E_1 accomplishes the same thing as the first pass through the i loop in the naive Gaussian elimination algorithm. Then define E_2 as

$$E_2 = I - \begin{bmatrix} 0 & 0 & 0 & \dots & 0 \\ 0 & 0 & 0 & \dots & 0 \\ 0 & m_{32} & 0 & \dots & 0 \\ 0 & m_{42} & 0 & \dots & 0 \\ \vdots & \vdots & 0 & \dots & 0 \\ 0 & m_{n2} & 0 & \dots & 0 \end{bmatrix}$$

where the m_{i2} values are defined from the elements of A' so that

$$E_2E_1A = \begin{bmatrix} a_{11} & a_{12} & a_{13} & \cdots & a_{1n} \\ 0 & a'_{12} & a'_{13} & \cdots & a'_{1n} \\ 0 & 0 & & & \\ \vdots & \vdots & & A'' & \\ 0 & 0 & & & \end{bmatrix}.$$

The general trend ought to be clear, but a little notation helps. Let's write $A^{(0)} = A$, $A^{(1)} = E_1A^{(0)}$, $A^{(2)} = E_2A^{(1)} = E_2E_1A$, etc., with a similar superscript placed on the individual matrix components: $a_{ij}^{(k)}$. Then we can define the E_k matrices rather easily as

$$E_k = I - R_k,$$

where R_k is almost entirely zeroes:

$$r_{ij}^{(k)} = \begin{cases} a_{ji}^{(k)}/a_{ii}^{(k)} & k = i \text{ and } j > i \\ 0 & \text{otherwise.} \end{cases}$$

Note that the nonzero elements here are *precisely* the multipliers for the Gaussian elimination step. It follows, then, that the matrix

$$U = E_{n-1}E_{n-2}\cdots E_2E_1A$$

will be upper triangular. In fact, it will be the same upper triangular matrix that is produced by naive Gaussian elimination. Moreover, each of the E_i matrices is lower triangular and nonsingular; therefore, their inverses all exist and are lower triangular. We thus can write

$$A = (E_{n-1}E_{n-2}\cdots E_2E_1)^{-1}U = LU,$$

where L is defined by

$$L = (E_{n-1}E_{n-2}\cdots E_2E_1)^{-1} = E_1^{-1}E_2^{-1}\cdots E_{n-1}^{-1}.$$

But the simplicity of the E_k matrices allows us to write their inverses explicitly (Exercise 8):

$$E_k^{-1} = I + R_k; \tag{7.3}$$

moreover, it is easy to show that the sequential products of these matrices are very simple (Exercise 8):

$$E_k^{-1}E_{k-1}^{-1} = I + R_k + R_{k-1}. \tag{7.4}$$

Thus, when all is said and done, we have (again, Exercise 8)

$$L = I + R_1 + R_2 + \cdots + R_{n-1}.$$

Algorithm 7.5: LU decomposition algorithm (no pivoting)

```
for i=1 to n-1
     for j=i+1 to n
          a(j,i) = a(j,i)/a(i,i)
          for k=i+1 to n
               a(j,k) = a(j,k) - a(j,i)*a(i,k)
          endfor
     endfor
endfor
```

But, since the nonzero elements of each R_k are nothing more than the multipliers, the lower triangular matrix is nothing more than the matrix of multipliers! Thus, the *LU* factorization is nothing more than a very slight reorganization of the same Gaussian elimination algorithm we studied earlier in this chapter.

This allows for a very simple and compact algorithm for computing the *LU* decomposition (see Algorithm 7.5).

The solution process is completed by using the forward-backward solution algorithm (Algorithm 7.6).

Note that the first half of the solution algorithm (the forward solve) makes explicit use of the fact that *L* is unit lower triangular; i.e., the diagonal elements are all 1s.

Algorithm 7.6: Forward-backward solution, using LU decomposition

```
!
!     First, solve Ly = b
!
x(1) = b(1)
for i=2 to n
     sum = 0.0
     for j=1 to i-1
          sum = sum + a(i,j)*x(j)
     endfor
     x(i) = b(i) - sum
endfor
!
!     Next, solve Ux = y
!
x(n) = x(n)/a(n,n)
for i=n-1 to 1
     sum = 0.0
     for j=i+1 to n
          sum = sum + a(i,j)*x(j)
     endfor
     x(i) = (x(i) - sum)/a(i,i)
endfor
```

It is important to understand how these algorithms manipulate the actual computer storage. Note that we use the lower triangular part of A to store L, and the upper triangular part to store U; separate arrays are not needed (and would be wasteful). Let's see how this would work in a particular case.

EXAMPLE 7.3

Take

$$A = \begin{bmatrix} 4 & 1 & 0 \\ 1 & 4 & 1 \\ 0 & 1 & 4 \end{bmatrix};$$

then the LU decomposition algorithm produces the following steps:

$$\begin{bmatrix} 4 & 1 & 0 \\ 1 & 4 & 1 \\ 0 & 1 & 4 \end{bmatrix} \sim \begin{bmatrix} 4 & 1 & 0 \\ (1/4) & 15/4 & 1 \\ (0) & 1 & 4 \end{bmatrix} \sim \begin{bmatrix} 4 & 1 & 0 \\ (1/4) & 15/4 & 1 \\ (0) & (4/15) & 56/15 \end{bmatrix}$$

where we have indicated the multipliers by putting parentheses around them. Thus, we have

$$L = \begin{bmatrix} 1 & 0 & 0 \\ 1/4 & 1 & 0 \\ 0 & 4/15 & 1 \end{bmatrix}, \quad U = \begin{bmatrix} 4 & 1 & 0 \\ 0 & 15/4 & 1 \\ 0 & 0 & 56/15 \end{bmatrix}.$$

Solving the system

$$\begin{bmatrix} 4 & 1 & 0 \\ 1 & 4 & 1 \\ 0 & 1 & 4 \end{bmatrix} \begin{bmatrix} x_1 \\ x_2 \\ x_3 \end{bmatrix} = \begin{bmatrix} 6 \\ 12 \\ 14 \end{bmatrix}$$

therefore requires that we solve the two triangular systems

$$\begin{bmatrix} 1 & 0 & 0 \\ 1/4 & 1 & 0 \\ 0 & 4/14 & 1 \end{bmatrix} \begin{bmatrix} y_1 \\ y_2 \\ y_3 \end{bmatrix} = \begin{bmatrix} 6 \\ 12 \\ 14 \end{bmatrix}$$

and

$$\begin{bmatrix} 4 & 1 & 0 \\ 0 & 15/4 & 1 \\ 0 & 0 & 56/15 \end{bmatrix} \begin{bmatrix} x_1 \\ x_2 \\ x_3 \end{bmatrix} = \begin{bmatrix} y_1 \\ y_2 \\ y_3 \end{bmatrix}.$$

The reader should check that we get the solution $x_1 = 1$, $x_2 = 2$, $x_3 = 3$. ∎

Pivoting and the LU Decomposition   The previous discussion and resulting algorithm both assumed that no pivoting was done. However, we have seen that pivoting is sometimes necessary and often desirable. Can we use pivoting in the LU decomposition without destroying the algorithm? And if so, how?

The answer is easier to explain than it is to prove. Because of the triangular structure of the LU factors, we can implement pivoting almost exactly as we did before. The difference is that we must keep track of how the rows are interchanged in order to properly apply the forward and backward solution steps.

The best way to see this is to look at an example.

EXAMPLE 7.4

Consider the 4×4 matrix

$$
A = \begin{bmatrix} 4 & 0 & 1 & 1 \\ 3 & 1 & 3 & 1 \\ 0 & 1 & 2 & 0 \\ 3 & 2 & 4 & 1 \end{bmatrix}.
$$

We will apply the LU decomposition to this, showing the values as stored in the computer. As in the previous example, to distinguish the multipliers that are stored in the lower triangular positions, we will enclose them in parentheses. Eliminating in the first column is very routine:

$$
\begin{bmatrix} 4 & 0 & 1 & 1 \\ 3 & 1 & 3 & 1 \\ 0 & 1 & 2 & 0 \\ 3 & 2 & 4 & 1 \end{bmatrix} \sim \begin{bmatrix} 4 & 0 & 1 & 1 \\ (3/4) & 1 & 9/4 & 1/4 \\ (0) & 1 & 2 & 0 \\ (3/4) & 2 & 13/4 & 1/4 \end{bmatrix}.
$$

To carry out the next step of elimination, we see that we need to pivot. Let's go ahead and swap the rows, just as we did before, and see what happens.

$$
\begin{bmatrix} 4 & 0 & 1 & 1 \\ (3/4) & 1 & 9/4 & 1/4 \\ (0) & 1 & 2 & 0 \\ (3/4) & 2 & 13/4 & 1/4 \end{bmatrix} \sim \begin{bmatrix} 4 & 0 & 1 & 1 \\ (3/4) & 2 & 13/4 & 1/4 \\ (0) & (1/2) & 3/8 & -1/8 \\ (3/4) & (1/2) & 5/8 & 1/8 \end{bmatrix}
$$

We have to pivot again (since $3/8 < 5/8$), so we end up with

$$
\begin{bmatrix} 4 & 0 & 1 & 1 \\ (3/4) & 2 & 13/4 & 1/4 \\ (3/4) & (1/2) & 5/8 & 1/8 \\ (0) & (1/2) & 3/8 & -1/8 \end{bmatrix} \sim \begin{bmatrix} 4 & 0 & 1 & 1 \\ (3/4) & 2 & 13/4 & 1/4 \\ (3/4) & (1/2) & 5/8 & 1/8 \\ (0) & (1/2) & (3/5) & -1/5 \end{bmatrix}.
$$

This suggests, then, that the L and U matrices are

$$
L = \begin{bmatrix} 1 & 0 & 0 & 0 \\ 3/4 & 1 & 0 & 0 \\ 3/4 & 1/2 & 1 & 0 \\ 0 & 1/2 & 3/5 & 1 \end{bmatrix}, \quad U = \begin{bmatrix} 4 & 0 & 1 & 1 \\ 0 & 2 & 13/4 & 1/4 \\ 0 & 0 & 5/8 & 1/8 \\ 0 & 0 & 0 & -1/5 \end{bmatrix}.
$$

If we multiply these out, we get

$$
\begin{bmatrix}
1 & 0 & 0 & 0 \\
3/4 & 1 & 0 & 0 \\
3/4 & 1/2 & 1 & 0 \\
0 & 1/2 & 3/5 & 1
\end{bmatrix}
\begin{bmatrix}
4 & 0 & 1 & 1 \\
0 & 2 & 13/4 & 1/4 \\
0 & 0 & 5/8 & 1/8 \\
0 & 0 & 0 & -1/5
\end{bmatrix}
=
\begin{bmatrix}
4 & 0 & 1 & 1 \\
3 & 2 & 4 & 1 \\
3 & 1 & 3 & 1 \\
0 & 1 & 2 & 0
\end{bmatrix}
\neq A.
$$

However, we do note that the product LU is the same as A, *once we have imposed the same row interchanges on A as we did in the pivoting process.* Thus, if we keep track of how the rows were swapped, we can impose this reordering on the right-side vector and carry on with the solution process. ■

So, how do we keep track of the row interchanges? One way is to use an *index array*, which is simply an integer-valued vector. It is initialized to the natural row ordering

$$
J =
\begin{bmatrix}
1 \\
2 \\
3 \\
\vdots \\
n
\end{bmatrix},
$$

and then every time two rows in the matrix are swapped, the corresponding two elements of the index array J are swapped. Thus, in our example, the final version of J is

$$
J =
\begin{bmatrix}
1 \\
4 \\
2 \\
3
\end{bmatrix}.
$$

The student should check that this is correct, of course. We interpret this as follows:

- The first row of A is the first row of the product matrix LU.
- The fourth row of A is the second row of the product matrix LU.
- The second row of A is the third row of the product matrix LU.
- The third row of A is the fourth row of the product matrix LU.
- In general, if we denote the product matrix LU by A', then

$$
a'_{i,j} = a_{J_i, j}.
$$

How do we use this information in the solution process? Since the product matrix $A' = LU$ does not equal A, we have to reorder the b vector accordingly. This is accomplished by the simple expedient of copying b into b', where

$$
b'_i = b_{J_i}.
$$

The factorization algorithm becomes the pseudocode given in Algorithm 7.7.

The forward-backward solve algorithm becomes the pseudocode given in Algorithm 7.8.

Algorithm 7.7: LU decomposition with partial pivoting (pseudocode)

```
for i=1 to n-1
!
!    This begins the pivoting code
!
     am = abs(a(i,i))
     p = i
     for j=i+1 to n
         if abs(a(j,i)) > am then
             am = abs(a(j,i))
             p = j
         endif
     endfor
     if p > i then
         for k = 1 to n
             hold = a(i,k)
             a(i,k) = a(p,k)
             a(p,k) = hold
         endfor
         ihold = indx(i)
         indx(i) = indx(p)
         indx(p) = ihold
     endif
!
!    This ends the pivoting code
!
     for j=i+1 to n
         a(j,i) = a(j,i)/a(i,i)
         for k=i+1 to n
             a(j,k) = a(j,k) - a(j,i)*a(i,k)
         endfor
     endfor
endfor
```

Algorithm 7.8: Forward-backward solution, using LU decomposition (pseudocode)

```
!
!      First, reorder the b vector
!
for k=1 to n
     x(k) = b(indx(k))
endfor
for k=1 to n
     b(k) = x(k)
endfor
```

```
!
!      Solve Ly = b
!
y(1) = b(1)
for i=2 to n
     sum = 0.0
     for j=1 to i-1
          sum = sum + a(i,j)*y(j)
     endfor
     y(i) = b(i) - sum
endfor
!
!      Next, solve Ux = y
!
x(n) = y(n)/a(n,n)
for i=n-1 to 1
     sum = 0.0
     for j=i+1 to n
          sum = sum + a(i,j)*x(j)
     endfor
     x(i) = (y(i) - sum)/a(i,i)
endfor
```

Programming Notes: Some comments on these algorithms are in order at this point. First, note that we used indx as the variable name for the index vector *J*. Second, note that the loop where the rows are interchanged now runs across the entire matrix (in other words, the loop starts at k = 1 instead of k = i). This is because the *LU* decomposition requires that we interchange the elements of *L* (the multipliers) as well as the elements of *U*. Finally, note that we used three separate vectors (y, b, and x) in the solution step. This really is not necessary but does make the algorithm easier to follow.

EXERCISES

1. Do, by hand, an *LU* factorization of the matrix

$$A = \begin{bmatrix} 2 & 1 \\ 1 & 2 \end{bmatrix}$$

and use it to solve the system $Ax = b$, where $b = (1,2)^T$. The exact solution is $x = (0,1)^T$. Verify that $LU = A$.

2. Repeat the preceding for

$$A = \begin{bmatrix} 4 & 1 \\ 1 & 5 \end{bmatrix}$$

and $b = (2,10)^T$; here the exact solution is $x = (0,2)^T$.

3. Write an *LU* factorization code and use it to solve the

system of equations $Ax = b$, where

$$A = \begin{bmatrix} 14 & 14 & -9 & 3 & -5 \\ 14 & 52 & -15 & 2 & -32 \\ -9 & -15 & 36 & -5 & 16 \\ 3 & 2 & -5 & 47 & 49 \\ -5 & -32 & 16 & 49 & 79 \end{bmatrix}$$

and $b = [-15, -100, 106, 329, 463]^T$. The correct answer is $x = [0,1,2,3,4]^T$.

4. Write an *LU* factorization code and use it to solve the system of equations

$$T_5 x = b,$$

where $b = [1,6,12,18,19]^T$. The correct answer is $x = [0,1,2,3,4]^T$.

5. Write an LU factorization code and use it to solve the system of equations

$$H_5 x = b$$

where $b = [5.0,3.550,2.81428571428571, 2.34642857142857,2.01746031746032]^T$. The correct answer is $x = [1,2,3,4,5]^T$.

6. Write your own LU factorization code, based on the material in this chapter, and test it on the following examples. In each case have your code multiply out the L and U factors to check that the routine is working.

(a) $K_5 x = b$, $b = [-1,0,0,0,5]^T$; the solution is $x = [0,1,2,3,4]^T$.
(b) $A_5 x = b$, $b = [-4,-7,-6,-5,16]^T$; the solution is $x = [0,1,2,3,4]^T$.

7. Determine the operation count for computing the inverse of a matrix, as outlined in this section.

8. Show that
(a)

$$E_k^{-1} = I + R_k \qquad (7.5)$$

for all k.

(b)

$$E_k^{-1} E_{k-1}^{-1} = I + R_k + R_{k-1} \qquad (7.6)$$

for all k.
(c) $L = I + R_1 + R_2 + \cdots + R_{n-1}$.

9. Modify the tridiagonal solution algorithm from Chapter 2 to produce an LU decomposition. Be sure to maintain the simple storage of the matrix that was used in Chapter 2, and assume that no pivoting is required.

10. Write an LU factorization code that does partial pivoting and use it to solve the system of equations $Ax = b$, where

$$A = \begin{bmatrix} 9 & 3 & 2 & 0 & 7 \\ 7 & 6 & 9 & 6 & 4 \\ 2 & 7 & 7 & 8 & 2 \\ 0 & 9 & 7 & 2 & 2 \\ 7 & 3 & 6 & 4 & 3 \end{bmatrix}$$

and $b = [35,58,53,37,39]^T$. The correct answer is $x = [0,1,2,3,4]^T$.

11. Write an LU factorization code with partial pivoting and apply it to the system $A_5 x = b$, where $b = [-4,-7,-6,-5,16]^T$ and the solution is $x = [0,1,2,3,4]^T$.

7.5 PERTURBATION, CONDITIONING, AND STABILITY

In this section we will study how the matrix solution process as outlined in the previous section is affected by changes to the problem. The main reason for doing this is to attain an understanding of why some linear system problems are difficult to solve. Usually this involves a concept known as "conditioning."

To motivate our discussion, let's consider a simple example.

EXAMPLE 7.5

Let A be given by

$$A = \begin{bmatrix} 1.002 & 1 \\ 1 & 0.998 \end{bmatrix}$$

and let b be given by

$$b = \begin{bmatrix} 2.002 \\ 1.998 \end{bmatrix}.$$

Then a direct computation shows that the solution to $Ax = b$ is given by $x = (1,1)^T$. Suppose now that we perturb the right-side vector a little bit, and define

$$b_c = \begin{bmatrix} 2.0021 \\ 1.998 \end{bmatrix}.$$

Note that this is a change of less than half a percent in one component of the vector b. Now a direct computation shows that the solution to $Ax = b_c$ is

$$x_c = \begin{bmatrix} -23.95 \\ 26.00 \end{bmatrix}.$$

Most of us would probably agree that this is a large change in the solution for such a small change in the problem. Why did this happen? ■

This is a serious issue in mathematics and computation; we like to know that small changes in the problem will produce correspondingly small changes in the solution, and we like to understand why the exceptions occur. And we especially want to know if the large change due to small perturbation is an artifact of the problem or of the computational scheme.

Before we can directly tackle the issues of this section, we need to introduce some notation and terminology that will allow us to measure the size of vectors and matrices. For the most part this is an extension of the function norm ideas introduced in Section 4.3, as well as a natural continuation of the idea of length of a vector that is part of an ordinary linear algebra course.

7.5.1 Vector and Matrix Norms

Because we will be concerned with measuring errors in vectors and matrices, it is necessary to introduce the concepts of *vector norm* and *matrix norm*. Informally, these norms are simply a generalization of the function norm concept that we introduced in Section 4.3. They are means of measuring the *size* of a vector or matrix. The formal definition of a vector norm is as follows.

Definition 7.1 (Vector Norm)   A vector norm *on* $\mathbf{R}^n$ *is any mapping* $\|\cdot\|$, *defined on* $\mathbf{R}^n$ *with values in* $[0,\infty)$, *that satisfies the three conditions*

  1. $\|x\| > 0$ *for any vector* $x \neq 0$.
  2. $\|ax\| = |a|\|x\|$ *for any scalar a.*
  3. $\|x+y\| \leq \|x\| + \|y\|$ *for any two vectors x and y.*

The most common examples of vector norms are the so-called infinity norm and the Euclidean 2-norm, respectively defined as

$$\|x\|_\infty = \max_{1 \leq i \leq n} |x_i|,$$

and

$$\|x\|_2 = \left(\sum_{i=1}^{n} x_i^2\right)^{1/2}.$$

We could use the same norm definition for matrix norms, but this would not give us a couple of important properties that we want to have. So we use a special kind of matrix norm, known as an *operator norm*.

Definition 7.2 (Matrix Norm)   *Let* $\|\cdot\|$ *be a given vector norm defined on* $\mathbf{R}^n$. *Define the corresponding* matrix norm, *for matrices* $A \in \mathbf{R}^{n \times n}$, *by*

$$\|A\| = \max_{x \neq 0} \frac{\|Ax\|}{\|x\|}. \tag{7.7}$$

One important consequence of this definition of matrix norm is that the norm of a product is always less than or equal to the product of the norms:

$$\|AB\| \leq \|A\|\|B\| \tag{7.8}$$

and

$$\|Ax\| \leq \|A\|\|x\|. \tag{7.9}$$

See Exercise 15.

As we have defined it here, matrix norms are always associated with a particular vector norm. Strictly speaking, this is not necessary; one can define a norm on a matrix that does not require any particular vector norm. However, such norms will not usually satisfy (7.8) and (7.9), and these conditions are necessary for what we want to do throughout this section.

The matrix infinity norm, $\|A\|_\infty$, can be shown to be equivalent to the following computation:

$$\|A\|_\infty = \max_{1 \leq i \leq n} \sum_{j=1}^{n} |a_{ij}|.$$

Thus $\|A\|_\infty$ is defined as the maximum row sum. The matrix 2-norm is more subtle. It can be shown that we have

$$\|A\|_2 = \sqrt{\Lambda(A^T A)}$$

where $\Lambda(B)$ is the largest (in absolute value) eigenvalue of the matrix B. Since this makes it much more difficult to compute using the matrix 2-norm, we will almost always use the infinity norm.

EXAMPLE 7.6

Let A be the simple matrix given by

$$A = \begin{bmatrix} 4 & -6 & 2 \\ 0 & 4 & 1 \\ 1 & 2 & 3 \end{bmatrix}.$$

Then $\|A\|_\infty = \max\{12, 5, 6\} = 12$. Furthermore,

$$A^T A = \begin{bmatrix} 17 & -22 & 11 \\ -22 & 56 & -2 \\ 11 & -2 & 14 \end{bmatrix},$$

which has eigenvalues $\lambda = 66.6816, 19.8065, 0.5118$, so that $\|A\|_2 = \sqrt{66.6816} = 8.1659$. ∎

7.5.2 The Condition Number and Perturbations

Suppose we want to look at the linear system

$$Ax = b$$

and suppose further that we are most interested in understanding how the solution x changes as the right-side vector b changes. Thus we might look at the two specific systems

$$Ax_1 = b_1$$

and

$$Ax_2 = b_2.$$

We have (assuming A is nonsingular)

$$x_1 - x_2 = A^{-1}(b_1 - b_2),$$

so that the relative error in x_2 as an approximation to x_1 is given by

$$\frac{\|x_1 - x_2\|}{\|x_1\|} \leq \|A^{-1}\| \frac{\|b_1 - b_2\|}{\|x_1\|}.$$

We would like to bound the change in solution by something that does not depend on the solution; thus we want to get rid of the x_1 in the denominator on the right. To do this, we note that $\|A\| \|x_1\| \geq \|b_1\|$, so that

$$\frac{1}{\|x_1\|} \leq \frac{\|A\|}{\|b_1\|};$$

hence we have

$$\frac{\|x_1 - x_2\|}{\|x_1\|} \leq \|A\| \|A^{-1}\| \frac{\|b_1 - b_2\|}{\|b_1\|}.$$

The multiplying coefficient $\|A\|\|A^{-1}\|$ is interesting. It depends entirely on the matrix in the problem, not on the right-side vector, yet it shows up as an amplifier to the relative change in the right-side vector. We will see this quantity often enough in the rest of this section that it is worth giving a name and special notation. We call it the *condition number*.

Definition 7.3 (Condition Number)   *For a given matrix $A \in \mathbf{R}^{n \times n}$ and a given matrix norm $\|\cdot\|$, the* condition number *with respect to the given norm is defined by*

$$\kappa(A) = \|A\|\|A^{-1}\|. \tag{7.10}$$

If A is singular, then we take $\kappa(A) = \infty$.

The justification for taking $\kappa(A) = \infty$ if A is given as singular is the following theorem.

Theorem 7.3   *Let $A \in \mathbf{R}^{n \times n}$ be given, and nonsingular. Then for any singular matrix $B \in \mathbf{R}^{n \times n}$, we have*

$$\frac{1}{\kappa(A)} \leq \frac{\|A - B\|}{\|A\|}.$$

Proof   We have

$$
\begin{aligned}
\frac{1}{\kappa(A)} &= \frac{1}{\|A\|\|A^{-1}\|} \\
&= \frac{1}{\|A\|}\left(\frac{1}{\displaystyle\max_{x \neq 0}\frac{\|A^{-1}x\|}{\|x\|}}\right) \\
&\leq \frac{1}{\|A\|}\left(\frac{1}{\dfrac{\|A^{-1}x\|}{\|x\|}}\right) \\
&\leq \frac{1}{\|A\|}\left(\frac{\|Ay\|}{\|y\|}\right),
\end{aligned}
$$

where $y \in \mathbf{R}^n$ is arbitrary. Now, let y be any nonzero vector such that $By = 0$; since B is singular, we know that such things exist. Then we can write

$$\frac{1}{\kappa(A)} \leq \frac{\|(A - B)y\|}{\|A\|\|y\|} \leq \frac{\|(A - B)\|\|y\|}{\|A\|\|y\|} = \frac{\|(A - B)\|}{\|A\|},$$

and we are done.   □

The importance of this result is that it tells us that if A is close to a singular matrix, then the reciprocal of the condition number will be near zero, i.e., $\kappa(A)$ itself will be "large." Thus, the condition number measures how close the matrix is to being singular; if $\kappa(A)$ is "large," then we know that A is close to being singular. One of the lessons of this section is that solving systems that are nearly singular can produce large errors.

Suppose we want to solve the system

$$Ax = b,$$

where $A \in \mathbf{R}^{n \times n}$ and $b \in \mathbf{R}^n$ are given. However, we know that the solution process will be affected by rounding error. While each instance of rounding error will no doubt be small, if n is large, then there will be an enormous number of individual rounding errors to add up. Is it possible that these will combine to dominate the computation and render the results meaningless? For example, if $n = 200$, which is not considered a large system, then ordinary Gaussian elimination will do about 2.7×10^6 operations to produce the solution. If each operation has error that is $O(10^{-7})$, and if they all accumulate, then the net error will be about 0.27—hardly small.

The main results to this section tell us how the solution of a linear system is affected by errors in the matrix or the right-side vector. Since some of the proofs can be rather involved, some of them are deferred to the appendix or are omitted entirely.

Theorem 7.4 (Effects of Perturbation in b)   *Let $A \in \mathbf{R}^{n \times n}$ (nonsingular) and $b \in \mathbf{R}^n$ be given, and define $x \in \mathbf{R}^n$ as the solution of the linear system $Ax = b$. Let $\delta b \in \mathbf{R}^n$ be a small perturbation of b, and define $x + \delta x \in \mathbf{R}^n$ as the solution of the system $A(x + \delta x) = b + \delta b$. Then*

$$\frac{\|\delta x\|}{\|x\|} \leq \kappa(A) \frac{\|\delta b\|}{\|b\|}.$$

Proof   This is a straightforward computation. We have $Ax = b$ and $A(x + \delta x) = b + \delta b$, so that $A\delta x = \delta b$. Thus

$$\delta x = A^{-1}\delta b \implies \|\delta x\| \leq \|A^{-1}\|\|\delta b\|$$

so that we have

$$\frac{\|\delta x\|}{\|x\|} \leq \frac{\|A\|\|A^{-1}\|\|\delta b\|}{\|A\|\|x\|}.$$

But $\|A\|\|x\| \geq \|b\|$ and the definition of condition number imply that

$$\frac{\|A\|\|A^{-1}\|\|\delta b\|}{\|A\|\|x\|} \leq \kappa(A) \frac{\|\delta b\|}{\|b\|},$$

and we are done. $\square$

The importance of this theorem is to show us that perturbations in the problem affect the final solution in a form that is amplified by the condition number of the matrix. Thus, if the matrix is ill conditioned—meaning that the condition number, $\kappa(A)$, is large—then a small change in the data could lead to a large change in the solution. A similar result involves the *residual* of a solution: Let x_c be a "computed" solution to $Ax = b$; then the residual is the vector $r = b - Ax_c$, i.e., r is the amount by which x_c fails to solve the system. If $r = 0$, then x_c is exact; one might think, then, that if r is small, then x_c is close to the exact solution. This isn't always so, as the following shows.

Theorem 7.5 (Effects of Residual on Accuracy)   *Let $A \in \mathbf{R}^{n \times n}$ (nonsingular) and $b \in \mathbf{R}^n$ be given, and define $x_c \in \mathbf{R}^n$ as a computed solution of the linear system $Ax = b$. Let $r \in \mathbf{R}^n$ be the residual $r = b - Ax_c$. Then*

$$\frac{\|x - x_c\|}{\|x\|} \le \kappa(A) \frac{\|r\|}{\|b\|}.$$

Proof   This is a small change from the previous proof, and so is left to the student. See Exercise 18.   □

The gist of these two results is that, in some sense, we cannot trust the solution to a problem involving an ill-conditioned matrix unless we have taken especial care in the solution. A similar but more involved result applies if we consider perturbations in the matrix itself.

Theorem 7.6 (Effects of Perturbation in A)   *Let $A \in \mathbf{R}^{n \times n}$ be given and nonsingular, and let $E \in \mathbf{R}^{n \times n}$ be a perturbation of A. Let $x \in \mathbf{R}^n$ be the unique solution of $Ax = b$. If $\kappa(A)\|E\| < \|A\|$, then the perturbed system $(A + E)x_c = b$ has a unique solution and*

$$\frac{\|x - x_c\|}{\|x\|} \le \frac{\theta}{1 - \theta},$$

where

$$\theta = \kappa(A) \frac{\|E\|}{\|A\|}.$$

Proof   See Appendix A.   □

The importance of these perturbation results is that we can establish that any computed (i.e., approximate) solution to the linear system problem is the *exact* solution to some "nearby" problem. The next result gives a sense of what is meant by this.

Theorem 7.7   *Let $A \in \mathbf{R}^{n \times n}$ (nonsingular) and $b \in \mathbf{R}^n$ be given, and define $x \in \mathbf{R}^n$ as the* exact *solution of the linear system $Ax = b$, and $x_c \in \mathbf{R}^n$ as a* computed *(that is, approximate) solution. Then there exists $\delta b \in \mathbf{R}^n$ such that x_c is the* exact *solution of the perturbed system*

$$Ax_c = b + \delta b,$$

where $\delta b = Ax_c - b = -r$ is the (negative) residual for x_c.

Proof   This is a straightforward computation:

$$Ax_c = A(x + (x_c - x)) = b + A(x_c - x) = b + \delta b.$$

On the other hand, we have

$$\delta b = A(x_c - x) = Ax_c - b = -r,$$

and we are done.   □

Thus, we can consider any computed solution to a linear system as the exact solution to a problem using slightly perturbed data. The idea of using this approach to estimating the error between two quantities—by establishing that they are both exact solutions to "nearby" problems—is called *backward error analysis* and is due to the British mathematician James H. Wilkinson,[3] widely regarded as the founder of modern numerical analysis.

A similar result shows that we can regard the computed solution as the exact solution to a problem involving the same right-side vector but a perturbed matrix. Considering how much more complicated the proof of Theorem 7.6 is compared with that of Theorem 7.4, it is surprising how easy this argument is. Note, however, that this result is restricted to the 2-norm, whereas our previous results have been independent of the choice of norm.

Theorem 7.8   *Let $A \in \mathbf{R}^{n \times n}$ and $b \in \mathbf{R}^n$ be given, with A nonsingular; define $x \in \mathbf{R}^n$ as the solution of $Ax = b$. Let x_c be an approximate solution to this system with residual $r = b - Ax_c$ and define*

$$E = rx_c^T / \|x_c\|_2^2. \tag{7.11}$$

Then x_c is the exact solution to $(A + E)x_c = b$ and

$$\frac{\|E\|_2}{\|A\|_2} \le \frac{\|r\|_2}{\|A\|_2 \|x_c\|_2}. \tag{7.12}$$

Proof   We have

$$(A + E)x_c = Ax_c + \frac{1}{\|x_c\|_2^2}(b - Ax_c)x_c^T x_c = Ax_c + (b - Ax_c)\frac{\|x_c\|_2^2}{\|x_c\|_2^2} = b.$$

Thus, x_c is the exact solution of the perturbed system. Next, for any $y \in \mathbf{R}^n$,

$$Ey = \frac{1}{\|x_c\|_2^2}rx_c^T y = \alpha_y r,$$

---

[3] James Hardy Wilkinson (1919–1986) was born in Kent, in England, and at the age of 16 won a scholarship to Trinity College, Cambridge, where he was widely regarded as one of the top students of his time. During World War II he did ballistics work, and soon after the end of the war he began work at Great Britain's National Physical Laboratory, on early computer designs. Out of this work came his studies of numerical algorithms, most notably on the effects of rounding errors on Gaussian elimination and methods for the algebraic eigenvalue problem, works that remain classics to this day. In 1969 he was elected to the Royal Society of London.

where

$$\alpha_y = \frac{x_c^T y}{\|x_c\|_2^2}.$$

Note that we have

$$|\alpha_y| \leq \frac{\|x_c\|_2 \|y\|_2}{\|x_c\|_2^2} = \frac{\|y\|_2}{\|x_c\|_2}.$$

Therefore,

$$\frac{\|Ey\|_2}{\|y\|_2} = \frac{|\alpha_y| \|r\|_2}{\|y\|_2} \leq \frac{\|r\|}{\|x_c\|_2},$$

from which

$$\|E\|_2 \leq \frac{\|r\|}{\|x_c\|_2}$$

follows, and we are done.   □

This last result allows us to estimate, via an upper bound, the size of the perturbation in A that our computed solution corresponds to, using the easily computable residual.

To use these perturbation results, we need to have one final set of results that estimates the effects of machine arithmetic—rounding error—on the Gaussian elimination process. In this connection we need one more definition.

Definition 7.4 (Gaussian Elimination Growth Factor)   Let $A^{(k)}$ be the matrix after k columns have been eliminated—thus $A^{(0)} = A$ and $A^{(n-1)} = U$—and let $a_{i,j}^{(k)}$ denote the individual elements of $A^{(k)}$. Then the growth factor *for Gaussian elimination applied to A is defined by*

$$\rho_n = \max_{i,j,k} \frac{|a_{i,j}^{(k)}|}{\|A\|_\infty}.$$

To illustrate this, let's go back to one of our computational examples from earlier in this chapter.

EXAMPLE 7.7

Consider the matrix from Example 7.4.

$$A = \begin{bmatrix} 4 & 0 & 1 & 1 \\ 3 & 1 & 3 & 1 \\ 0 & 1 & 2 & 0 \\ 3 & 2 & 4 & 1 \end{bmatrix}.$$

After eliminating in the first column, we get $A^{(1)}$:

$$A^{(1)} = \begin{bmatrix} 4 & 0 & 1 & 1 \\ 0 & 1 & 9/4 & 1/4 \\ 0 & 1 & 2 & 0 \\ 0 & 2 & 13/4 & 1/4 \end{bmatrix}.$$

Proceeding further, we get

$$A^{(2)} = \begin{bmatrix} 4 & 0 & 1 & 1 \\ 0 & 2 & 13/4 & 1/4 \\ 0 & 0 & 3/8 & -1/8 \\ 0 & 0 & 5/8 & 1/8 \end{bmatrix}$$

and

$$A^{(3)} = \begin{bmatrix} 4 & 0 & 1 & 1 \\ 0 & 2 & 13/4 & 1/4 \\ 0 & 0 & 3/8 & -1/8 \\ 0 & 0 & 0 & -1/5 \end{bmatrix}.$$

Now, it is easy to compute $\|A\|_\infty = 10$. Then, to compute the growth factor, we simply look at all the values in these four matrices to find the one that is largest in absolute value. In this case, we pick out $a_{1,1}^{(0)} = 4$, so, for this matrix the growth factor is

$$\rho_4 = 4/10.$$

The final piece of the puzzle, then, is the following theorem, taken from [3]. ■

Theorem 7.9 (Rounding Error Effects)   *Let L_c and U_c be the lower and upper triangular factors of A as computed on a machine with rounding unit u, using Gaussian elimination with either partial or complete pivoting.*

1. *Then there is a matrix E_1 such that $L_c U_c = A + E_1$ and*

$$\|E_1\|_\infty \le n^2 \rho_n u \|A\|_\infty.$$

2. *If x_c is the solution to the system $Ax = b$, computed using L_c and U_c, then there is a matrix E_2 such that $(A + E_2)x_c = b$ and*

$$\|E_2\|_\infty \le 1.01 n^2 (n+3) \rho_n u \|A\|_\infty.$$

Thus we see that Gaussian elimination produces the exact solution to a "nearby" problem, and we can combine Theorem 7.9 with Theorem 7.6 to get a bound on the error between the computed solution x_c and the exact solution, x:

$$\frac{\|x - x_c\|_\infty}{\|x\|_\infty} \leq \frac{1.01n^2(n+3)\rho_n \mathbf{u}\kappa_\infty(A)}{1 - 1.01n^2(n+3)\rho_n \mathbf{u}\kappa_\infty(A)} \tag{7.13}$$

Or can we? Just how big is ρ_n going to be? Note that we don't get the bound (7.13) unless $1.01n^2(n+3)\rho_n \mathbf{u}\kappa_\infty(A)\mathbf{u}$ is "sufficiently small," and we don't know (yet) how large ρ_n is going to be.

The bounds on ρ_n depend on what kind of pivoting we do. If we do complete pivoting—which is rather costly in terms of computer time—then it can be shown that

$$\rho_n \leq 1.8(n^{0.25 \ln n}),$$

which grows rather slowly: For $n = 100$, $\rho_n = O(400)$. However, if we do only partial pivoting, then it can be shown that there are matrices for which

$$\rho_n = 2^{n-1};$$

in other words, ρ_n grows exponentially. For comparison, we now have $\rho_{100} = O(10^{30})$.

So it is fair to ask, Is Gaussian elimination with partial pivoting a stable process?

Strictly speaking, (7.13) says that it is: For a sufficiently accurate computer (i.e., $\mathbf{u}$ small enough) and a sufficiently small problem (n small enough), then Gaussian elimination with partial pivoting will produce solutions that are stable and accurate. The problem is, we don't want to have to limit ourselves to computer architectures that use unrealistic word lengths to achieve extraordinary accuracy, nor do we want to restrict ourselves to very small problems in order to keep ρ_n small. So, as a practical matter, it appears that Gaussian elimination with partial pivoting is perhaps not stable.

However, the kinds of matrices that lead to large values of ρ_n do not appear to occur in practice. It can be shown that ρ_n is directly related to the growth of the elements in the upper triangular factor, U_c (this should be evident from the example we did above), and this growth is not observed in problems of interest. To quote from p. 166 of [10], "In fifty years of computing, no matrix problems that excite an explosive instability are known to have arisen under natural circumstances." This text also gives an interesting probabilistic argument to explain why this might be the case. Moreover, we should point out that ρ_n is relatively easy to compute as part of the elimination process, so if in any particular case we are worried about the size of ρ_n, we can compute it.

So, in the end, we conclude that, as a practical matter—meaning, for problems of interest that actually occur—Gaussian elimination with partial pivoting *is* a stable process.

It is possible to combine some of the other results to get a more meaningful *a posteriori* (after the fact) result. Note, again, that this result is specific to the 2-norm, because it uses Theorem 7.8.

Theorem 7.10   *Let x_c be a computed solution to the linear system $Ax = b$, and let $r = b - Ax_c$ be the corresponding residual. If $\kappa_2(A)\|r\|_2$ is sufficiently small, then we have*

$$\frac{\|x - x_c\|_2}{\|x\|_2} \leq \frac{\kappa_2(A)\dfrac{\|r\|_2}{\|b - r\|_2}}{1 - \kappa_2(A)\dfrac{\|r\|_2}{\|b - r\|_2}}.$$

Proof   This follows by cleverly combining several of the previous results in this section, and is left as an exercise for the student. (See Exercise 16.)   □

For practical purposes, the following "rule of thumb" is very useful.

Rule of Thumb   Let E and δb be the perturbations such that the computed solution exactly solves $(A + E)x_c = b + \delta b$. If we can assume

1. $\mathbf{u} \leq C_1 \times 10^{-s}$

2. $\|E\| \leq C_2\mathbf{u}\|A\|$

3. $\|\delta b\| \leq C_3\mathbf{u}\|A\|$

4. $\kappa(A) \leq C_4 \times 10^t$

then, for $t - s$ sufficiently negative, there is a constant $C > 0$ such that

$$\frac{\|x - x_x\|}{\|x\|} \leq C \times 10^{t-s}.$$

What this says is that, in practice, we observe that the error in a linear system computation goes like the condition number times the machine rounding unit.

7.5.3 Estimating the Condition Number

Singular matrices, as the terminology implies, are perhaps something of a rarity. This does not mean we don't have to worry about them, however. It can be shown, in a very mathematically precise sense, that all singular matrices are arbitrarily close to a nonsingular matrix. Once we begin doing numerical computations, then, the odds are very high that the inherent rounding error will perturb the matrix away from being singular. The resulting nonsingular matrix, however, will be very ill conditioned. This is why we have to be careful when doing matrix computations. The algorithms will fail (in the sense of having a zero pivot) if the matrix is singular, but this is only reliable in infinite-precision arithmetic. In floating-point arithmetic, we have to recognize that small pivots might mean that the matrix is *nearly singular*, or ill conditioned.

Generally speaking,

> *If the solution to a linear system changes a great deal when the problem changes only very slightly, then we suspect that the matrix is ill conditioned.*

Since the condition number is such an important indicator of how well we can compute with the matrix, it would be useful if it could be easily computed. Alas, this is asking too much, but we can *estimate* $\kappa_\infty(A)$ without too much more effort than is needed to solve the system $Ax = b$. A condition number estimator is a fairly standard feature of modern linear systems software.

Recall the definition

$$\kappa_\infty(A) = \|A\|_\infty \|A^{-1}\|_\infty.$$

Computing $\|A\|_\infty$ is not hard; computing $\|A^{-1}\|_\infty$ is.[4] However, the norm definition tells us

$$\|A^{-1}\|_\infty = \max_{x \neq 0} \frac{\|A^{-1}x\|_\infty}{\|x\|_\infty},$$

so that, for any fixed $w \in \mathbf{R}^n$, $w \neq 0$, we have

$$\|A^{-1}\|_\infty \geq \frac{\|A^{-1}w\|_\infty}{\|w\|_\infty}.$$

Now set $y = A^{-1}w$ and substitute into the condition number definition to get

$$\kappa_\infty(A) \geq \|A\|_\infty \frac{\|y\|_\infty}{\|w\|_\infty}, \tag{7.14}$$

and this holds for any $y \in \mathbf{R}^n$, $y \neq 0$, $w = Ay$. The trick, now, is to find the choice of y or w that maximizes the right side of (7.14). Since the inequality holds for any $y \in \mathbf{R}^n$, $w = Ay$, the larger the right side is, the more accurate is the estimate of $\kappa_\infty(A)$.

Recall that a singular matrix has at least one eigenvalue equal to zero. It therefore follows that a matrix that is "close to singular" will have at least one eigenvalue "close to zero." If we choose y to be the eigenvector of A corresponding to the smallest (in absolute value) eigenvalue of A, then

$$\frac{\|y\|_\infty}{\|w\|_\infty} = \frac{\|y\|_\infty}{\|Ay\|_\infty} = \frac{\|y\|_\infty}{\|\lambda y\|_\infty} = |\lambda|^{-1},$$

and this will tend to maximize the estimate of the condition number. We will see in Section 8.3 that the recursion

$$y^{(i+1)} = A^{-1}y^{(i)}/\|y^{(i)}\|_\infty, \quad i = 1, 2, \ldots$$

will produce a sequence of vectors that tend to be in the direction of the eigenvector corresponding to the smallest eigenvalue. Note that we would not compute this sequence by actually inverting A; rather, we would solve the sequence

---

[4] Anyway, if A is ill conditioned, any computation of A^{-1} will be unreliable.

of linear system problems

$$Ay^{(i+1)} = y^{(i)}/\|y^{(i)}\|_\infty$$

using an *existing LU* decomposition. Note also that if we are estimating the condition number as part of an *LU* decomposition routine, the factorization will already have been done, and all that has to be done are the relatively inexpensive triangular solves. The number of steps to take in the recursion is an issue; taking more steps is more costly but will lead to a more accurate estimate. The public-domain package LAPACK uses five iterations in the corresponding section of its condition number estimator, and that seems to be a reasonable choice.

In any event, with the recursion computed, we have

$$\kappa_\infty(A) \geq \|A\|_\infty \frac{\|y^{(5)}\|_\infty}{\|Ay^{(5)}\|_\infty} = \frac{\alpha v}{\omega}, \tag{7.15}$$

where

$$\alpha = \|A\|_\infty, \quad v = \|y^{(5)}\|_\infty, \quad \omega = \|Ay^{(5)}\|_\infty.$$

We thus could use the estimate

$$\kappa^* = \frac{\alpha v}{\omega}$$

for the condition number. However, the way we have set things up, we have

$$\omega = \|Ay^{(5)}\|_\infty = \left\| \frac{y^{(4)}}{\|y^{(4)}\|_\infty} \right\|_\infty = 1.$$

So we actually have

$$\kappa(A) \approx \kappa^* = \alpha v.$$

A complete algorithm is given in Algorithm 7.9.

Algorithm 7.9: *Condition number estimation*

Given an *LU* factorization of A, compute as follows:

1. Compute $\alpha = \|A\|_\infty$.

2. Take a random initial guess $y^{(0)}$.

3. Compute $y^{(5)}$ in the sequence defined by

$$y^{(i+1)} = A^{-1}y^{(i)}/\|y^{(i)}\|_\infty, \quad i = 0, 1, \ldots, 4, \tag{7.16}$$

by solving the systems using the extant factorization of A, and set $v = \|y^{(5)}\|_\infty$.

4. Set $\kappa^* = \alpha v$.

A more sophisticated process for choosing y is slightly better, but justifying it requires more linear algebra background than we have here, and this will work reasonably well.

EXAMPLE 7.8

To illustrate, consider the matrix

$$A = \begin{bmatrix} 1 & \frac{1}{2} & \frac{1}{3} \\ \frac{1}{2} & \frac{1}{3} & \frac{1}{4} \\ \frac{1}{3} & \frac{1}{4} & \frac{1}{5} \end{bmatrix}.$$

Since this is only 3×3, we can directly compute the exact condition number without too much effort, and we get $\kappa_\infty(A) = 748$. Applying the estimation algorithm here goes as follows. First, we compute an LU decomposition for A, getting

$$L = \begin{bmatrix} 1.0000 & 0 & 0 \\ 0.5000 & 1.0000 & 1.0000 \\ 0.3333 & 1.0000 & 0 \end{bmatrix}, \quad U = \begin{bmatrix} 1.0000 & 0.5000 & 0.3333 \\ 0 & 0.0833 & 0.0889 \\ 0 & 0 & -0.0056 \end{bmatrix}.$$

Choose a random vector for $y^{(0)}$—for example,

$$y^{(0)} = \begin{bmatrix} 0.2190 \\ 0.0470 \\ 0.6789 \end{bmatrix}.$$

Computing in the recursion (7.16), we get

$$y^{(5)} = \begin{bmatrix} 66.5558 \\ -372.1152 \\ 359.0420 \end{bmatrix}.$$

Then $\alpha = \|A\|_\infty = 1.833$ and $\upsilon = \|y^{(5)}\|_\infty = 372.1$, which gives us $\kappa^* = 682.2$, which is not a bad estimate. ∎

7.5.4 Iterative Refinement

Since Gaussian elimination can be adversely affected by rounding error, especially if the matrix is ill conditioned, it is sometimes worthwhile to improve the accuracy of a computed solution. This leads to the algorithm known as *iterative refinement* or *iterative improvement*.

We first observe that if x_c is a computed solution to $Ax = b$, then the error satisfies a very similar problem.

Algorithm 7.10: Iterative Refinement

For k from 0 until sufficient accuracy is achieved, do:

1. Compute $r^{(k)} = b - Ax^{(k)}$.

2. Solve $Ae = r^{(k)}$ using the existing LU factorization of A.

3. Define $x^{(k+1)} = x^{(k)} + e$.

Programming Hints: (1) Be sure to compute the residual using the original matrix A and not the product LU. (2) Because of the potential for subtractive cancellation, refinement works best when the residuals are computed in a higher precision than the rest of the calculation. (3) If $\kappa(A)\mathbf{u}$ is too large, then refinement may not work.

Theorem 7.11   *Let $A \in \mathbf{R}^{n \times n}$ be given and nonsingular, and let $x_c \in \mathbf{R}^n$ be a computed solution to the linear system $Ax = b$. Then the error $e = x - x_c$ satisfies $Ae = r$, where $r = b - Ax_c$ is the residual for x_c.*

Proof   This is a simple, direct computation:

$$Ae = A(x - x_c) = Ax - Ax_c = b - Ax_c = r. \qquad \square$$

Thus, given a computed solution $x_c \approx x$ and the LU factorization from which it was computed, we can then quickly solve the new problem

$$Ae = r$$

and set $x = x_c + e$.

If all the computations were carried out in exact arithmetic, then the "improved" value x computed above would be the exact solution. However, since the computations are almost surely not done exactly, the improved value will quite likely not be the exact solution. So it is worthwhile to continue the process iteratively. Describing this requires some better notation, though.

Let $x^{(0)}$ be the initial computed solution, and let $r^{(0)} = b - Ax^{(0)}$ be the corresponding residual. Then we compute the sequence of improved values $x^{(k)}$ according to Algorithm 7.10.

Note that the only additional work in the refinement algorithm is the computation of the residual and the two solution steps needed to complete the solution of the systems $Ae = r^{(k)}$; the factorization has already been done.

EXAMPLE 7.9

Let's consider the simple linear system

$$\begin{bmatrix} 1 & \frac{1}{2} & \frac{1}{3} \\ \frac{1}{2} & \frac{1}{3} & \frac{1}{4} \\ \frac{1}{3} & \frac{1}{4} & \frac{1}{5} \end{bmatrix} \begin{bmatrix} x_1 \\ x_2 \\ x_3 \end{bmatrix} = \begin{bmatrix} 3 \\ 23/12 \\ 43/30 \end{bmatrix},$$

which has exact solution $x = (1,2,3)^T$. The correct LU factorization of the coefficient matrix is

$$L = \begin{bmatrix} 1 & 0 & 0 \\ \frac{1}{2} & 1 & 0 \\ \frac{1}{3} & 1 & 1 \end{bmatrix}, \quad U = \begin{bmatrix} 1 & \frac{1}{2} & \frac{1}{3} \\ 0 & \frac{1}{12} & \frac{1}{12} \\ 0 & 0 & \frac{1}{180} \end{bmatrix},$$

but let's introduce some error by using only six digits to represent these:

$$L_c = \begin{bmatrix} 1 & 0 & 0 \\ 0.5 & 1 & 0 \\ 0.333333 & 1 & 1 \end{bmatrix}, \quad U_c = \begin{bmatrix} 1 & 0.5 & 0.333333 \\ 0 & 0.0833333 & 0.0833333 \\ 0 & 0 & 0.555556 \times 10^{-2} \end{bmatrix}.$$

Thus the solution we compute from the forward-backward solution steps will be inexact, however slightly. In fact, we get

$$x^{(0)} = \begin{bmatrix} 1.00002960003512 \\ 1.99982440014289 \\ 3.00017759985791 \end{bmatrix}.$$

Now apply the refinement algorithm. Our initial residual is

$$r^{(0)} = 10^{-5} \times \begin{bmatrix} -0.10000591998960 \\ -0.06666963332513 \\ -0.14866856774542 \end{bmatrix}.$$

We solve the system

$$L_c U_c e = r^{(0)}$$

to get

$$e = 10^{-3} \times \begin{bmatrix} -0.02960008024619 \\ 0.17559977427749 \\ -0.17759977587713 \end{bmatrix},$$

so the first improved solution is

$$x^{(1)} = x^{(0)} + e = \begin{bmatrix} 0.99999999995487 \\ 1.99999999991716 \\ 3.00000000008204 \end{bmatrix},$$

which is a substantial improvement over the original computed solution. Another step of refinement would continue to improve the solution, of course. ∎

EXERCISES

1. Let

$$A = \begin{bmatrix} 5 & 6 & -9 \\ 1 & 2 & 3 \\ 0 & 7 & 2 \end{bmatrix}.$$

Compute $\|A\|_\infty$.

2. Let

$$A = \begin{bmatrix} 1 & 2 & -7 \\ 4 & -8 & 0 \\ 2 & 1 & 0 \end{bmatrix}.$$

Compute $\|A\|_\infty$.

3. Let

$$A = \begin{bmatrix} -8 & 0 & -1 \\ 3 & 12 & 0 \\ 1 & 2 & 3 \end{bmatrix}.$$

Compute $\|A\|_\infty$.

4. Show that

$$\|A\|_\star = \max_{i,j} |a_{ij}|$$

does not define a matrix norm, according to our definition. (Hint: Show that one of the conditions fails to hold by finding a specific case where it fails.)

5. Let

$$A = \begin{bmatrix} 1 & \frac{1}{2} & 0 \\ \frac{1}{2} & \frac{1}{3} & \frac{1}{4} \\ 0 & \frac{1}{4} & \frac{1}{5} \end{bmatrix}.$$

Compute, directly from the definition, $\kappa_\infty(A)$. You should get $\kappa_\infty(A) = 18$.

6. Repeat the preceding for

$$A = \begin{bmatrix} 4 & 1 & 0 \\ 1 & 4 & 1 \\ 0 & 1 & 4 \end{bmatrix},$$

for which $\kappa_\infty(A) = 2.5714$.

7. Consider the linear system

$$\begin{bmatrix} 1.002 & 1 \\ 1 & 0.998 \end{bmatrix} \begin{bmatrix} x_1 \\ x_2 \end{bmatrix} = \begin{bmatrix} 0.002 \\ 0.002 \end{bmatrix},$$

which has exact solution $x = (1, -1)^T$ (verify this). What is the residual $b - Ax_c$ for the "approximate" solution $x_c = (0.29360067817338, -0.29218646673249)^T$? Explain.

8. Consider the linear system problem

$$Ax = b,$$

where

$$A = \begin{bmatrix} 4 & 2 & 0 \\ 1 & 4 & 1 \\ 0 & 2 & 4 \end{bmatrix}, \quad b = \begin{bmatrix} 8 \\ 12 \\ 16 \end{bmatrix},$$

for which the exact solution is $x = (1,2,3)^T$. (Check this.) Let $x_c = x + (0.002,0.01,0.001)^T$ be a computed (i.e., approximate) solution to this system. Use Theorem 7.8 to find the perturbation matrix E such that x_c is the exact solution to $Ax_c = b$.

9. Compute the growth factor for Gaussian elimination for the matrix in the previous problem.

10. Let

$$A = \begin{bmatrix} 2 & 1 & 0 \\ 1 & 2 & 1 \\ 0 & 1 & 2 \end{bmatrix}.$$

This has exact condition number $\kappa_\infty(A) = 8$. Use the condition number estimator in this section to approximate the condition number.

11. Repeat the preceding for

$$A = \begin{bmatrix} 1 & \frac{1}{2} & 0 \\ \frac{1}{2} & \frac{1}{3} & \frac{1}{4} \\ 0 & \frac{1}{4} & \frac{1}{5} \end{bmatrix},$$

for which $\kappa_\infty(A) = 18$.

12. Use the condition number estimator to produce a plot of κ^* versus n for each of the following matrix families.

(a) $T_n, 4 \le n \le 20$
(b) $K_n, 4 \le n \le 20$
(c) $A_n, 4 \le n \le 20$
(d) $H_n, 4 \le n \le 20$

13. Repeat the preceding, except this time produce a plot of the growth factor for Gaussian elimination for each matrix family, as a function of n.

14. Given a matrix $A \in \mathbf{R}^{n \times n}$, show that

$$\mu(A) = \frac{\max\limits_{x \neq 0} \frac{\|Ax\|}{\|x\|}}{\min\limits_{x \neq 0} \frac{\|Ax\|}{\|x\|}}$$

is equivalent to the condition number as defined in (7.10).

15. Prove that (7.8) and (7.9) follow from the definition of matrix norm.

16. Prove Theorem 7.10.

17. Give an argument to support the validity of the "rule of thumb" following Theorem 7.10.

18. Prove Theorem 7.5.

19. Consider the linear system problem $Ax = b$ for

$$A = \begin{bmatrix} 1 & \frac{1}{2} \\ \frac{1}{2} & \frac{1}{3} \end{bmatrix}, \quad b = \begin{bmatrix} 2 \\ 7/6 \end{bmatrix}.$$

Note that the exact solution is $x = (1,2)^T$.

(a) Represent the exact factorization of A using only four decimal digits, and solve the system.

(b) Do two steps of iterative refinement to improve your solution.

20. Do two more steps of refinement for the problem in Example 7.9.

7.6 *SPD* MATRICES AND THE CHOLESKY DECOMPOSITION

There are a number of classes of special matrices whose properties make the solution of linear systems easier. One important such class consists of the *symmetric, positive definite* matrices.

Definition 7.5 (SPD Matrices)   Let $A \in R^{n \times n}$ be given. If A satisfies

$$A = A^T$$

and

$$x^T A x > 0, \quad all \; x \neq 0,$$

then we say that A is symmetric positive definite, *abbreviated by writing $A \in SPD$.*

Like the diagonally dominant matrices, *SPD* matrices appear often in applications involving the approximation of functions and the solution of differential equations. What makes them important is that a special factorization and solution algorithm exists for them, which is about half as costly as ordinary Gaussian elimination. The result is due to the French mapmaker Cholesky.[5]

Theorem 7.12 (Cholesky Theorem)   Let $A \in R^{n \times n}$ be given, with $A \in SPD$. Then

  1. A is nonsingular.

  2. There exists a lower triangular matrix $G \in R^{n \times n}$, with $g_{ii} > 0$, such that $GG^T = A$.

---

[5] Andre-Louis Cholesky (1875–1918) was a French military officer who was killed in the final year of World War I. He was educated at École Polytechnique and served as a mapmaker for the French military, earning some distinction for his efforts in mapping Crete and some of the French colonies in North Africa. His method for the solution of positive definite systems of equations, which he developed for the solution of the so-called normal equations that arise in least squares problems, was published posthumously by a fellow officer, in 1924, with all due credit being given to Cholesky.

Proof   The nonsingularity is a straightforward consequence of the condition $x^T A x > 0$: If A is singular, there is a nonzero vector x such that $Ax = 0$; hence $x^T A x = 0$, and therefore a singular matrix cannot be *SPD*.

For the construction of G, we will use an argument that is based on induction on the size of the matrix. Let $n = 2$; then

$$A = \begin{bmatrix} a_{11} & a_{21} \\ a_{21} & a_{22} \end{bmatrix}$$

and the condition $x^T A x > 0$ implies that both diagonals a_{ii} must be positive (Exercise 1). Then it can easily be verified (by direct computation) that

$$G = \begin{bmatrix} \sqrt{a_{11}} & 0 \\ a_{21}/\sqrt{a_{11}} & \sqrt{a_{22} - a_{21}^2/a_{11}} \end{bmatrix}$$

is such that $A = GG^T$. We know that both of the square roots are defined because $A \in SPD$ implies (Exercises 1 and 2) that the diagonals must be positive and that $a_{22} - a_{21}^2/a_{11} > 0$. Thus G exists, at least for the special case of $n = 2$.

Assume now that G exists for $n = k - 1$; we will use this to show that it must exist for $n = k$, and this will complete the induction.

We write A (which is $k \times k$) in the partitioned form

$$A = \begin{bmatrix} A_{11} & a \\ a^T & a_{kk} \end{bmatrix},$$

where A_{11} is $(k - 1) \times (k - 1)$ and a is a vector in $\mathbf{R}^{k-1}$. The fact that $A \in SPD$ forces (Exercise 3) $A_{11} \in SPD$. Therefore, by the inductive hypothesis, there exists G_{11}, lower triangular, such that $A_{11} = G_{11}G_{11}^T$. Construct

$$G = \begin{bmatrix} G_{11} & 0 \\ g^T & g_{kk} \end{bmatrix},$$

where $g \in \mathbf{R}^{k-1}$ and $g_{kk} \in \mathbf{R}$ are yet to be determined. Note that if we can determine these two quantities, we will have finished the proof because we will have found the lower triangular matrix for the $k \times k$ case and the induction will be complete.

To see what g and g_{kk} have to be, we multiply out GG^T and set this equal to A:

$$GG^T = \begin{bmatrix} G_{11}G_{11}^T & G_{11}g \\ g^T G_{11}^T & g_{kk}^2 + g^T g \end{bmatrix} = \begin{bmatrix} A_{11} & a \\ a^T & a_{kk} \end{bmatrix}.$$

Note that the upper left element is correct: $G_{11}G_{11}^T$ does indeed equal A_{11} by the inductive hypothesis. To finish the job we have to have

$$G_{11}g = a \tag{7.17}$$

and

$$g_{kk}^2 + g^T g = a_{kk}.$$

Since the diagonal elements of G_{11} (which is upper triangular) are all positive, it is necessarily nonsingular; thus a unique solution exists to the system (7.17), and thus g is defined. With g defined, we simply set

$$g_{kk} = \sqrt{a_{kk} - g^T g}.$$

The argument to the square root has to be positive; otherwise there will be an x such that $x^T A x \leq 0$, which violates the positive definite condition (Exercise 4). The induction is complete, and hence the theorem is proved. □

There are a number of different ways of actually constructing the Cholesky decomposition, one of which can be deduced from the proof given above. All of these constructions are equivalent, because it can be shown that the Cholesky factorization is unique; they differ only in the way that the operations are organized. One common scheme uses the following formulas:

$$g_{ij} = \frac{1}{g_{jj}} \left[a_{ij} - \sum_{k=1}^{j-1} g_{ik} g_{jk} \right], \quad 1 \leq j < 1$$

$$g_{ii} = \left[a_{ii} - \sum_{k=1}^{i-1} g_{ik}^2 \right]^{1/2}.$$

This is a very efficient algorithm; the operation count (Exercise 5) is $\frac{1}{6}n^3 + O(n^2)$. Round-off error for a nearly singular *SPD* matrix can sometimes cause the algorithm to try to take a square root of a negative number. A variant exists that computes a factorization of the form $A = LDL^T$, where L is unit lower triangular and D is diagonal; this can be computed without using any square roots.

We will defer the detailed discussion of the Cholesky method to Section 9.22, where it is used to solve the large systems that arise in the discretization of partial differential equations.

Banded Systems   In many important applications (see Chapter 9 for examples) the coefficient matrix is very sparse (meaning most of the elements are zero), with the nonzero elements concentrated in a relatively narrow band around the main diagonal of the matrix. This is the case, for instance, for the matrices that arise in the discretization of partial differential equations. Note, also, that the tridiagonal matrices that were studied in Chapter 2 are a particularly simple example of banded matrices.

Banded systems can be factored "within the band"; that is, we do not need to store nor do we need to compute with the large number of zeroes that are located away from the diagonal ("outside the band"). For large problems this can result in a considerable savings in storage and execution time. Since the precise algorithm to use is almost always very specific to the problem being solved, we will defer a detailed discussion to Chapter 9, when we discuss the simple discretization of PDEs.

EXERCISES

1. Show that a matrix with one or more nonpositive diagonal elements cannot be *SPD*. (Hint: Try to find a vector x such that $x^T A x = a_{ii}$.)

2. For a 2×2 *SPD* matrix, show that $a_{22} - a_{21}^2 / a_{11} > 0$. (Hint: Consider the positive definite condition with $x = (a_{21}, -a_{11})^T$.)

3. Let $A \in \mathbf{R}^{n \times n}$ be partitioned as

$$A = \begin{bmatrix} A_{11} & A_{21} \\ A_{21} & A_{22} \end{bmatrix},$$

where A_{11} and A_{22} are square matrices. Show that both A_{11} and A_{22} must be *SPD* if $A \in SPD$.

4. In the proof of the Cholesky theorem, show that $a_{kk} - g^T g > 0$ must hold, or else there is a vector x such that $x^T A x \leq 0$. (Hint: Use Exercise 2 as a starting point.)

5. Derive the operation count given in the text for the Cholesky decomposition.

6. Prove the following: If $A \in SPD$, then there exists a unit lower triangular matrix L, and a diagonal matrix D with positive elements, such that $A = LDL^T$.

7. Derive an algorithm for the LDL^T factorization that does not require any square roots. (Hint: Look at the 3×3 case, explicitly. Multiply out LDL^T and set the result equal to A, and from this deduce the relationships

that are necessary. In other words, if

$$\begin{bmatrix} 1 & 0 & 0 \\ l_{21} & 1 & 0 \\ l_{31} & l_{32} & 1 \end{bmatrix} \begin{bmatrix} d_{11} & 0 & 0 \\ 0 & d_{22} & 0 \\ 0 & 0 & d_{33} \end{bmatrix} \begin{bmatrix} 1 & l_{21} & l_{31} \\ 0 & 1 & l_{32} \\ 0 & 0 & 1 \end{bmatrix}$$

$$= \begin{bmatrix} a_{11} & a_{12} & a_{13} \\ a_{21} & a_{22} & a_{23} \\ a_{31} & a_{32} & a_{33} \end{bmatrix},$$

what is the relationship between the components of L, D, and A? Then generalize to a problem of arbitrary size. Don't forget that A is symmetric!)

8. The T_n and K_n families of matrices are all positive definite. Write up a code to do the Cholesky factorization and test it on these matrices over the range $3 \leq n \leq 20$. Use a right-side vector that is all 1s, and confirm that the code is working by computing $\|r\|_\infty$, where r is the residual vector.

9. If A is tridiagonal *and SPD*, then the Cholesky factorization can be modified to work only with the *two* distinct nonzero diagonals in the matrix. Construct this version of the algorithm and test it on the T_n and K_n families, as above.

10. The H_n family of matrices is also *SPD*, but ill conditioned. Try to apply your Cholesky code to these, over the range $4 \leq n \leq 20$. What happens?

7.7   ITERATIVE METHODS FOR LINEAR SYSTEMS: A BRIEF SURVEY

If the coefficient matrix is *very* large and sparse—meaning that the number of nonzero elements is a small fraction of the total number of elements in the matrix—then Gaussian elimination may not be the best way to solve the linear system problem. In these circumstances, elimination techniques will be drastically slowed by the memory management issues associated with handling the very large matrix,[6] and they will also destroy the sparsity of the original matrix: Even though $A = LU$ is sparse, the individual factors L and U may not be *as* sparse as A.

We can illustrate this last point with a simple example.

---

[6]This is because most modern architectures use *virtual memory* concepts to enable a large amount of disk storage to substitute for a large amount of RAM storage. However, there is a time cost associated with writing the current RAM contents to disk and bringing new material into the RAM, and this time cost can be more than the cost of doing the actual computation.

EXAMPLE 7.10

Let

$$
A =
\begin{bmatrix}
-4 & 1 & 0 & 0 & 1 & 0 & 0 & 0 & 0 & 0 & 0 & 0 & 0 & 0 & 0 & 0 \\
1 & -4 & 1 & 0 & 0 & 1 & 0 & 0 & 0 & 0 & 0 & 0 & 0 & 0 & 0 & 0 \\
0 & 1 & -4 & 1 & 0 & 0 & 1 & 0 & 0 & 0 & 0 & 0 & 0 & 0 & 0 & 0 \\
0 & 0 & 1 & -4 & 0 & 0 & 0 & 1 & 0 & 0 & 0 & 0 & 0 & 0 & 0 & 0 \\
1 & 0 & 0 & 0 & -4 & 1 & 0 & 0 & 1 & 0 & 0 & 0 & 0 & 0 & 0 & 0 \\
0 & 1 & 0 & 0 & 1 & -4 & 1 & 0 & 0 & 1 & 0 & 0 & 0 & 0 & 0 & 0 \\
0 & 0 & 1 & 0 & 0 & 1 & -4 & 1 & 0 & 0 & 1 & 0 & 0 & 0 & 0 & 0 \\
0 & 0 & 0 & 1 & 0 & 0 & 1 & -4 & 0 & 0 & 0 & 1 & 0 & 0 & 0 & 0 \\
0 & 0 & 0 & 0 & 1 & 0 & 0 & 0 & -4 & 1 & 0 & 0 & 1 & 0 & 0 & 0 \\
0 & 0 & 0 & 0 & 0 & 1 & 0 & 0 & 1 & -4 & 1 & 0 & 0 & 1 & 0 & 0 \\
0 & 0 & 0 & 0 & 0 & 0 & 1 & 0 & 0 & 1 & -4 & 1 & 0 & 0 & 1 & 0 \\
0 & 0 & 0 & 0 & 0 & 0 & 0 & 1 & 0 & 0 & 1 & -4 & 0 & 0 & 0 & 1 \\
0 & 0 & 0 & 0 & 0 & 0 & 0 & 0 & 1 & 0 & 0 & 0 & -4 & 1 & 0 & 0 \\
0 & 0 & 0 & 0 & 0 & 0 & 0 & 0 & 0 & 1 & 0 & 0 & 1 & -4 & 1 & 0 \\
0 & 0 & 0 & 0 & 0 & 0 & 0 & 0 & 0 & 0 & 1 & 0 & 0 & 1 & -4 & 1 \\
0 & 0 & 0 & 0 & 0 & 0 & 0 & 0 & 0 & 0 & 0 & 1 & 0 & 0 & 1 & -4
\end{bmatrix}.
$$

This is a "small" example of the kind of sparse matrix we are talking about here; it has a total of 256 elements but only 64 are nonzero. If we factor this, using the LU factorization, then there will be a lot of "fill-in": Many of the elements that are zero in A will be nonzero in L or U. We can see this—without drowning in a sea of numbers—by creating a *sparsity plot* of the matrix and its LU factorization. This is a plot that puts a single mark (usually an asterisk or "x") at the location of every nonzero element and leaves blanks at the zero elements. Figure 7.1 shows the sparsity plots for A, the individual L and U factors, and the matrix A', which denotes the sparsity pattern of $A = LU$ as it would actually appear in the computer storage after the elimination process had run to completion. Note that the total number of nonzero elements has nearly doubled due to fill-in during the elimination step. (The captions give the number of nonzero elements in the matrix.)

Now, it was possible to store the original matrix using only five vectors (to represent the diagonals), each of length ≤ 16, and fewer if we took advantage of the symmetry of the matrix. On the other hand, storing the L and U factors will require much more storage. Moreover, we can easily write a code that will carry out matrix-vector multiplications of the form $z = Au$ by storing only the five diagonal vectors plus the u and z vectors. ∎

It is because of these issues that Gaussian elimination and factorization methods are sometimes not the best choice of solution technique. Instead, we are driven to consider *iterative* methods for solving the linear system. This might seem counterintuitive at first, since we are giving up a method (Gaussian elimination) that is exact in the absence of rounding error for a method that is, by definition, mathematically inexact. Nonetheless, the iterative methods are cost-effective for large, sparse problems, especially when A is symmetric and exhibits a strong structure in the placement of its zero and nonzero elements.

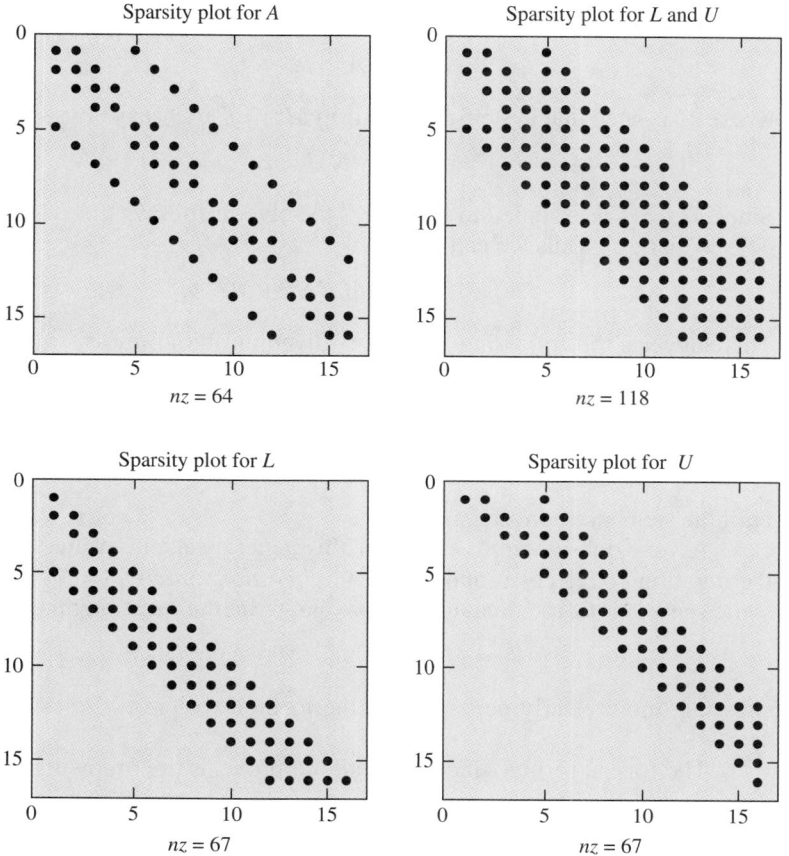

FIGURE 7.1 Sparsity plot for matrix A and its LU factors.

Unfortunately, a complete understanding of the mathematics underlying most iterative methods is beyond the intended level of this text. In addition, the most common realm of application for iterative methods is to solve the large, sparse, structured systems that come out of discretizing partial differential equations (PDEs). For this reason we will defer much of our discussion of iterative methods to Chapter 9; in this section we will outline some of the simpler ideas and show how they work on a simple 4×4 example. The most powerful technique for iteratively solving symmetric systems, the method of conjugate gradients, will be presented in Section 9.2.3. Still, we will be able to outline some significant ideas in this section.

An important class of matrix iterative methods come under the heading "splitting methods," because they are based on the notion of *splitting* the coefficient matrix into two parts.

Suppose we want to solve the linear system

$$Ax = b$$

where $A \in \mathbf{R}^{n \times n}$ is assumed to be nonsingular. We split A into a difference $M - N$:

$$A = M - N,$$

where M is such that systems of the form $Mz = f$ are "easy" to solve. Then we have

$$Ax = b \quad \Rightarrow \quad (M - N)x = b \quad \Rightarrow \quad Mx = Nx + b.$$

Since $Mz = f$ is assumed to be "easy" to solve, it follows that $M^{-1}f$ is easy to compute[7] for any f. Thus we can write

$$x = M^{-1}Nx + M^{-1}b.$$

This suggests the following iteration: Given an initial guess,[8] $x^{(0)}$, compute the sequence of vectors according to

$$x^{(k+1)} = M^{-1}Nx^{(k)} + M^{-1}b, \tag{7.18}$$

until the sequence converges.

A very rich literature exists for this general scheme of iterations, showing that the iteration converges if and only if the *iteration matrix* $T = M^{-1}N$ is less than 1 in some sense. Note, incidentally, that we can write the iteration matrix as

$$T = M^{-1}N = M^{-1}(M - A) = I - M^{-1}A,$$

showing, more clearly perhaps, that the method will work best when $M^{-1} \approx A^{-1}$ or $M \approx A$.

The formal result, which we shall not prove, is the following.

Theorem 7.13   *Let $A \in \mathbf{R}^{n \times n}$ be given. Define $T = M^{-1}N$, where $A = M - N$. Then the iteration (7.18) converges for all initial guesses $x^{(0)}$ if and only if there exists a norm $\|\cdot\|$ such that $\|T\| < 1$.*

A second, very important result relates the existence of such a norm to a fairly quantitative condition on the eigenvalues of the matrix T. First, a definition:

Definition 7.6 (Spectral Radius of a Matrix)   *Let $A \in \mathbf{R}^{n \times n}$ be given. Then the spectral radius of A, denoted $\rho(A)$, is the largest (in magnitude) of all the eigenvalues of A. Formally,*

$$\rho(A) = \max |\lambda|,$$

where the maximum is taken over all the eigenvalues of A.

---

[7]Note that this does not mean that we actually compute the inverse.

[8]We will, as a matter of convention, use parenthesized superscripts as iteration counters for vector quantities. This will also be an issue in Section 7.8, where we study nonlinear systems, and in Chapter 8, where we study eigenvalue-eigenvector approximations.

EXAMPLE 7.11

If A is the matrix

$$A = \begin{bmatrix} 4 & -3 \\ 2 & 1 \end{bmatrix},$$

then A has eigenvalues $2.5000 + 1.9365i$ and $2.5000 - 1.9365i$; thus the spectral radius of A is

$$\rho(A) = 3.1623 = |2.5000 \pm 1.9365i|.$$

■

Then the theorem connecting the spectral radius to the convergence of iterative methods is the following.

Therorem 7.14   *Let $A \in \mathbf{R}^{n \times n}$ be given. Then there exists a norm, $\| \cdot \|$, such that $\|A\| < 1$ if and only if the spectral radius satisfies $\rho(A) < 1$.*

From this we conclude that

The iteration (7.18) converges for all initial guesses $x^{(0)}$ if and only if $\rho(T) < 1$.

Having outlined the basic theory, let's now look at some specific methods. The simplest—and therefore, the slowest to converge—is known as the *Jacobi iteration*.[9] It might be more descriptive to call it "diagonal inversion," since it involves inversion of the diagonal part of A only, at each step of the iteration. We use the splitting

$$A = D - (D - A),$$

where D is the diagonal part of A. Then the iteration is

$$x^{(k+1)} = (I - D^{-1}A)x^{(k)} + D^{-1}b.$$

The most efficient implementation of the method—and this is the case for all of the splitting methods—depends greatly on the structure of the coefficient matrix, and so it is difficult to discuss in general. If we have a specialized routine for forming matrix-vector products involving A (using the structure of A to maximize efficiency), then we can write the iteration to take advantage of that:

$$x^{(k+1)} = x^{(k)} - D^{-1}(Ax^{(k)}) + D^{-1}b. \tag{7.19}$$

In Chapter 9 we will show how it would be done for the kinds of systems that arise from discretizing partial differential equations. Right now let's look at a simple example.

---

[9]Karl Gustav Jacob Jacobi (1804–1851) was born in Potsdam, Prussia (now part of Germany), and was educated at the University of Berlin. He earned his doctorate in 1825 and had a position at the University of Königsberg from 1826 until 1844, after which he returned to Berlin, where he remained until his death. Jacobi founded what we know as elliptic function theory and studied the determinant of functions known as the Jacobian. His iterative method for solving linear systems was first published in an astronomical journal in 1845.

EXAMPLE 7.12

Suppose we want to solve the system $Ax = b$, where

$$A = \begin{bmatrix} 4 & 1 & 0 & 0 \\ 1 & 5 & 1 & 0 \\ 0 & 1 & 6 & 1 \\ 1 & 0 & 1 & 4 \end{bmatrix}$$

and $b = (4,7,16,14)^T$, which means that the exact solution is $x = (0,1,2,3)^T$. Take $x^{(0)} = (0,0,0,0)^T$; then the first three iterations of the Jacobi iteration for this example go as follows:

$$x^{(1)} = x^{(0)} - D^{-1}Ax^{(0)} + D^{-1}b,$$

where $D = \text{diag}(4,5,6,4)$, meaning the diagonal matrix having elements (in order down the diagonal) 4,5,6,4. This yields

$$x^{(1)} = \begin{bmatrix} 1.0000 \\ 1.4000 \\ 2.6667 \\ 3.5000 \end{bmatrix};$$

$$x^{(2)} = x^{(1)} - D^{-1}Ax^{(1)} + D^{-1}b,$$

so that

$$x^{(2)} = \begin{bmatrix} -0.2250 \\ 0.6667 \\ 1.8500 \\ 2.5833 \end{bmatrix};$$

and

$$x^{(3)} = x^{(2)} - D^{-1}Ax^{(2)} + D^{-1}b,$$

so that

$$x^{(3)} = \begin{bmatrix} 0.1875 \\ 1.0750 \\ 2.1250 \\ 3.0938 \end{bmatrix}.$$

Even after only three iterations, it is becoming evident that the iteration is converging—and in fact we do converge, in the sense that $\|x^{(k)} - x^{(k-1)}\|_\infty < 10^{-6}$, in 19 iterations. ∎

The student should bear in mind that this example is meant to illustrate the barest workings of the method. There is very little reason to use an iterative method for solving a 4×4 linear system. For work with more realistic examples, see Section 9.2.

An obvious extension of the Jacobi iteration is to invert the entire lower triangular part of A, not just the diagonal. This is known as the Gauss-Seidel[10] iteration. We split A into[11]

$$A = L - (L - A),$$

so that the iteration becomes

$$x^{(k+1)} = (I - L^{-1}A)x^{(k)} + L^{-1}b,$$

or

$$x^{(k+1)} = x^{(k)} - L^{-1}(Ax^{(k)}) + L^{-1}b. \tag{7.20}$$

Again, we defer most of the implementation details until Chapter 9, when we can work with realistic examples. A simple example will illustrate and allow some comparison with the Jacobi method.

EXAMPLE 7.13

We use the same matrix A and the same initial vector as in the previous example. Then

$$x^{(1)} = x^{(0)} - L^{-1}(Ax^{(0)}) + L^{-1}b,$$

where

$$L = \begin{bmatrix} 4 & 0 & 0 & 0 \\ 1 & 5 & 0 & 0 \\ 0 & 1 & 6 & 0 \\ 1 & 0 & 1 & 4 \end{bmatrix}.$$

Of course, we do not actually compute the inverse; rather, we solve the system

$$Lz = Ax^{(k)}$$

for each index k. Since this is a lower triangular system, it is quite easy to solve. For this example we get

$$x^{(1)} = \begin{bmatrix} 1.0000 \\ 1.2000 \\ 2.4667 \\ 2.6333 \end{bmatrix}$$

---

[10]Ludwig Philipp von Seidel (1821–1896) was born in Zweibrücken, Germany, and studied at the University of Berlin beginning in 1840. Later he moved to Königsberg, where he studied under Jacobi. He received his doctorate from yet a third university, Munich, where he would eventually serve as a professor for many years. The bulk of his work was in optics and mathematical analysis. What we know as the Gauss-Seidel method was the result of a collaboration with Jacobi; Seidel published a memoir on the subject in 1874, but Gauss had described a similar method to a friend, who had published it in 1845.

[11]Note that L here is not the lower triangular matrix from the LU decomposition of A, but is simply the lower triangular part of A.

$$x^{(2)} = \begin{bmatrix} 0.0417 \\ 0.8983 \\ 2.0781 \\ 2.9701 \end{bmatrix}$$

$$x^{(3)} = \begin{bmatrix} 0.0329 \\ 0.9778 \\ 2.0087 \\ 2.9896 \end{bmatrix}.$$

Once again, the example displays evidence of convergence, and in fact the convergence appears to be faster than for the Jacobi iteration. If we carry on the computation, we converge (in the same sense as in the previous example) in 11 iterations.

∎

The two examples seem to indicate that Gauss-Seidel is converging faster, and this is consistent with our intuition, since Gauss-Seidel is based on inverting "more" of the matrix at each step. The formal theoretical results are the following.

Theorem 7.15 (Jacobi and Gauss-Seidel Convergence Theorem)
  1. *If A is diagonally dominant, then both Jacobi and Gauss-Seidel converge, and Gauss-Seidel converges faster in the sense that $\rho(T_{GS}) < \rho(T_J)$, where T_J is the iteration matrix for the Jacobi iteration, and T_{GS} is the iteration matrix for Gauss-Seidel.*
  2. *If $A \in SPD$, then both Jacobi and Gauss-Seidel will converge.*

While Gauss-Seidel is convergent for *SPD* matrices, it is still somewhat slow to converge. In the 1930s and 1940s, a number of schemes were proposed and developed to improve the convergence of Gauss-Seidel, usually by changing very slightly (or *relaxing*) the condition that defines the iteration. In the late 1940s and early 1950s this effort culminated with the development of the SOR (successive over-relaxation) iteration, which can be a substantial improvement over Gauss-Seidel. The problem is that this improved performance depends on the proper choice of a *relaxation parameter*.

The splitting for SOR is somewhat complicated. Let D be the diagonal of A, and let L and U be the *strictly* lower and upper triangular parts of A; i.e., L and U have zeroes on the diagonal. Then, for ω a real parameter, we split A as follows:

$$A = \left(\frac{1}{\omega} D + L \right) - \left(\left(\frac{1}{\omega} - 1 \right) D - U \right),$$

leading to the iteration

$$x^{(k+1)} = x^{(k)} - Q_\omega^{-1} A x^{(k)} + Q_\omega^{-1} b, \tag{7.21}$$

where

$$Q_\omega = \left(\frac{1}{\omega} D + L \right) = \omega^{-1}(D + \omega L),$$

which is lower triangular. The reason for this complicated splitting is, well, complicated. What is happening here is that we are *averaging* the result of a Gauss-Seidel step with the previous iterate, and ω is the weighting parameter in the average. What is interesting is that the best performance occurs when we take $\omega \in [1,2]$, which means that it is not a "true" average.

Let's look at an example, using the arbitrary value of $\omega = 1.05$.

EXAMPLE 7.14

The iteration, in general, is

$$x^{(k+1)} = x^{(k)} - Q_\omega^{-1} A x^{(k)} + Q_\omega^{-1} b,$$

where

$$Q_\omega = \begin{bmatrix} 3.8095 & 0 & 0 & 0 \\ 1.0000 & 4.7619 & 0 & 0 \\ 0 & 1.0000 & 5.7143 & 0 \\ 1.0000 & 0 & 1.0000 & 3.8095 \end{bmatrix}.$$

For the individual iterates we get

$$x^{(1)} = \begin{bmatrix} 1.0500 \\ 1.2495 \\ 2.5813 \\ 2.7218 \end{bmatrix}, \quad x^{(2)} = \begin{bmatrix} -0.0450 \\ 0.8749 \\ 2.0415 \\ 3.0148 \end{bmatrix}, \quad x^{(3)} = \begin{bmatrix} 0.0312 \\ 0.9910 \\ 1.9969 \\ 2.9919 \end{bmatrix}.$$

Again, it is apparent that the iteration is converging, and it is likewise apparent that it is converging faster than for Jacobi or Gauss-Seidel. If we run the iteration to completion, we get convergence (in the sense used in the two previous examples) in nine iterations, slightly faster than for Gauss-Seidel. ∎

As might be imagined, the convergence theory for SOR depends heavily on the choice of the parameter ω—for instance, taking $\omega = 1.75$ results in much *slower* convergence than for Gauss-Seidel. There exists a substantial theory for choosing the "best" value of ω, and this theory applies in a surprisingly wide range of important cases. Still, it and most of the theory for SOR is beyond the intended level of this text, so the reader is referred to the appropriate references at the end of this chapter. The most general theorem that can be succinctly stated is the following.

Theorem 7.16 (SOR Convergence Theorem) Let $A \in \mathbf{R}^{n \times n}$, SPD be given. Then, for any $b \in \mathbf{R}^n$, any initial guess $x^{(0)} \in \mathbf{R}^n$, and any value $\omega \in (0,2)$ the SOR iteration will converge to the exact solution of the linear system $Ax = b$. If $\omega < 0$ or $\omega > 2$, then the iteration will not converge.

It is worth noting that the methods outlined in this section have been, to a great extent, supplanted by more powerful iterative techniques, which we will outline in Section 9.2.3.

Since the important applications of these methods are to the solution of the kinds of linear systems that occur when partial differential equations are solved approximately, we will defer any detailed treatment of them to the appropriate sections of Chapter 9. The exercises ask the student to look at some very simple examples.

EXERCISES

1. Let

$$A = \begin{bmatrix} 4 & -1 & 0 & 0 \\ -1 & 4 & -1 & 0 \\ 0 & -1 & 4 & -1 \\ -1 & 0 & -1 & 4 \end{bmatrix}$$

and $b = (-4,2,4,10)^T$.

(a) Verify that the solution to $Ax = b$ is $x = (0,1,2,3)^T$.

(b) Do three iterations (by hand) of the Jacobi iteration for this matrix, using $x^{(0)} = (0,0,0,0)^T$.

(c) Do three iterations (by hand) of the Gauss-Seidel iteration for this problem, using the same initial guess.

2. Do three iterations of SOR for the previous exercise, using $\omega = 1.05$.

3. Solve the same system using SOR for a wide set of values of $\omega \in (0,2)$. For each ω, compute $r(\omega) = \|b - Ax^{(k)}\|_\infty$. Graph $r(\omega)$ for $k = 1,3,5,10$.

4. Write a computer code that does Jacobi for the previous exercise, for a specified number of iterations. How many iterations does it take to get convergence, in the sense that the consecutive iterates differ by less than 10^{-6}?

5. Repeat the preceding for Gauss-Seidel.

6. Repeat the preceding for *SOR*. Make sure your code can accept different values of ω as an input parameter.

7. Let A be the 16×16 matrix given at the beginning of this section. Take

$$b = (5,11,18,21,29,40,48,48,57,72,80,76,69,87,94,85)^T.$$

Write a computer code to do Jacobi, Gauss-Seidel, and SOR on this system of equations. Write the code to store only the nonzero diagonals of A, and make the code as efficient as possible.

8. Prove that the spectral radius of A, $\rho(A)$, is bounded above by $\|A\|$ for any norm such that $\|Ax\| \le \|A\|\|x\|$.

7.8   NONLINEAR SYSTEMS: NEWTON'S METHOD AND RELATED IDEAS

In Chapter 3 we spent considerable time and effort developing algorithms for the solution of a *single* nonlinear equation. In this section we will extend some of these ideas to tackle *systems* of nonlinear equations. Because we are talking about systems, we will use a lot of linear algebra to give structure and simplicity to the discussion.

Let f be a given function from a domain in $\mathbf{R}^k$ to a range also in $\mathbf{R}^k$, i.e., $f : \mathbf{R}^k \to \mathbf{R}^k$. We want to find the vector(s) $x \in \mathbf{R}^k$ such that $f(x) = 0$, where 0 here is understood to be the zero vector in $\mathbf{R}^k$.

For example, we might have

$$2x_1 - x_2 + \frac{1}{9}e^{-x_1} = 1 \tag{7.22}$$

$$-x_1 + 2x_2 + \frac{1}{9}e^{-x_2} = 0. \tag{7.23}$$

This implicitly defines $f = (f_1(x_1,x_2), f_2(x_1,x_2))^T$ via

$$f_1(x_1,x_2) = 2x_1 - x_2 + \frac{1}{9}e^{-x_1} - 1 \tag{7.24}$$

$$f_2(x_1,x_2) = -x_1 + 2x_2 + \frac{1}{9}e^{-x_2}. \tag{7.25}$$

How can we approximate the solution to this system?

This is a very active area of research, and many of the best ideas are relatively new. We will discuss how to derive and apply Newton's method for this kind of problem, and then show how certain variants might actually be superior to Newton's method.

One issue that is new here, in relation to the work we did in Chapter 3, is the *cost* of certain operations. This is entirely due to the "curse of dimensionality": The simple fact that we are working in $\mathbf{R}^k$ makes the size of the problem a *lot* bigger, and drives the cost up.

Because we are again doing iterations with a vector variable, we once again will use a parenthesized superscript for the iteration counter, in order to avoid confusion with the subscript denoting the element of the vector. Thus $x^{(k)}$ will refer to the vector at iteration k, and $x_j^{(k)}$ will refer to the jth component of that vector.

7.8.1 Newton's Method

To derive Newton's method for systems, take $x^{(0)} = (x_1^{(0)}, x_2^{(0)})$ as an initial approximation to the solution, and expand both component functions in a Taylor's series about that point:

$$f_1(x_1,x_2) = f_1(x_1^{(0)},x_2^{(0)}) + (x_1 - x_1^{(0)})\frac{\partial f_1}{\partial x_1}(x_1^{(0)},x_2^{(0)})$$

$$+ (x_2 - x_2^{(0)})\frac{\partial f_1}{\partial x_2}(x_1^{(0)},x_2^{(0)}) + R_1, \tag{7.26}$$

$$f_2(x_1,x_2) = f_2(x_1^{(0)},x_2^{(0)}) + (x_1 - x_1^{(0)})\frac{\partial f_2}{\partial x_1}(x_1^{(0)},x_2^{(0)})$$

$$+ (x_2 - x_2^{(0)})\frac{\partial f_2}{\partial x_2}(x_1^{(0)},x_2^{(0)}) + R_2. \tag{7.27}$$

Now, set $f_1(x_1,x_2) = f_2(x_1,x_2) = 0$ and drop the remainders; this yields

$$0 = f_1(x_1^{(0)},x_2^{(0)}) + (x_1 - x_1^{(0)})\frac{\partial f_1}{\partial x_1}(x_1^{(0)},x_2^{(0)}) + (x_2 - x_2^{(0)})\frac{\partial f_1}{\partial x_2}(x_1^{(0)},x_2^{(0)}), \tag{7.28}$$

$$0 = f_2(x_1^{(0)},x_2^{(0)}) + (x_1 - x_1^{(0)})\frac{\partial f_2}{\partial x_1}(x_1^{(0)},x_2^{(0)}) + (x_2 - x_2^{(0)})\frac{\partial f_2}{\partial x_2}(x_1^{(0)},x_2^{(0)}). \tag{7.29}$$

We want to solve this for the new values of x_1 and x_2, so we reorganize this as a matrix-vector equation:

$$0 = f(x^{(0)}) + A_f(x^{(0)})(x - x^{(0)}),$$

where

$$x^{(0)} = (x_1^{(0)}, x_2^{(0)})^T, \qquad x = (x_1, x_2)^T, \qquad A_f(x^{(0)}) = \begin{bmatrix} \dfrac{\partial f_1}{\partial x_1}(x_1^{(0)}, x_2^{(0)}) & \dfrac{\partial f_1}{\partial x_2}(x_1^{(0)}, x_2^{(0)}) \\ \dfrac{\partial f_2}{\partial x_1}(x_1^{(0)}, x_2^{(0)}) & \dfrac{\partial f_2}{\partial x_2}(x_1^{(0)}, x_2^{(0)}) \end{bmatrix}.$$

We can solve for x by multiplying by A_f^{-1} (assuming it exists):

$$x = x^{(0)} - A_f^{-1}(x^{(0)})f(x^{(0)}).$$

Generalizing the notation, then, we have

$$x^{(n+1)} = x^{(n)} - A_f^{-1}(x^{(n)})f(x^{(n)}),$$

where now $x^{(n)} = (x_1^{(n)}, x_2^{(n)})^T$ is the nth approximate (vector) value. This, of course, is almost exactly the classical Newton's method from Section 3.2, especially when we notice that the matrix A_f is the Jacobian matrix of first partial derivatives; essentially, this plays the role that the ordinary derivative did in the single-variable case. For this reason, we will introduce the notation of the derivative for this matrix.

Definition 7.7 (Gradient of a Function) *If f is a differentiable function from $\mathbf{R}^k$ to $\mathbf{R}^k$, then we will denote by f' the $k \times k$ matrix of first partial derivatives, also known as the gradient of f:*

$$f'(x) = \left\{ \frac{\partial f_i}{\partial x_j}(x) \right\}.$$

An alternative notation for the gradient is ∇f. Sometimes grad f is also used.

We will not go into detail on the theory of Newton's method for systems. Suffice it to say that the basic ideas of Chapter 3 carry over: If the initial guess $x^{(0)}$ is close enough to the actual solution, then convergence will occur and the convergence will be quadratic. However, because the implementation of Newton's method for systems is more problematic, we will spend some time on those issues.

For instance, while in Chapter 3 we had to worry about the possibility of dividing by a zero derivative in Newton's method, here we have to worry about a matrix being singular; moreover, we have to factor that matrix every step of the iteration. For the simple 2×2 example above, this is not going to be important, but for a larger system it is an issue. The scalar version of Newton's method cost two function evaluations each step; the k-dimensional version of Newton's method costs $k + k^2$ function evaluations, in addition to the roughly $\frac{1}{3}k^3$ cost of factoring and solving the linear system. Even for quadratic convergence, this can be a steep price to pay.

Table 7.1 shows the result of applying Newton's method to our example system, with the initial guess $x^{(0)} = (1,1)$.[12] Note the rapid convergence, which we expect with

---

[12] In this and all the other nonlinear system examples, the "estimated error" refers to the difference between consecutive iterates:

$$E_n = \max_{i=1,2} |x_i^{(n)} - x_i^{(n-1)}|.$$

TABLE 7.1   **Two-Dimensional Newton's Method Example**

n	$x_1^{(n)}$	$x_2^{(n)}$	Estimated error
0	1.000000000000	1.000000000000	NA
1	0.605042638147	0.267104846820	0.732895E+00
2	0.597220155135	0.255587267973	0.115176E−01
3	0.597216751762	0.255582530497	0.473748E−05
4	0.597216751761	0.255582530496	0.821260E−12

Newton's method when the initial guess is close to the actual solution. (The fact that this example is *nearly* linear helps.)

For comparison's sake, consider the chord method, which we introduced in Section 3.10.2. Here we compute according to

$$x^{(n+1)} = x^{(n)} - \left[f'(x^{(0)})\right]^{-1} f(x^{(n)}), \tag{7.30}$$

which means we do not need to reevaluate or refactor the gradient matrix at each step. Thus, each step of the chord iteration is substantially cheaper than Newton's method. As indicated in Section 3.10.2, we could periodically update the point on which the gradient is based, doing something like

$$x^{(n+1)} = x^{(n)} - [f'(x^*)]^{-1} f(x^{(n)}), \tag{7.31}$$

with x^* being updated every, say, p iterations. This reduces the cost compared with Newton's method but does speed up the convergence compared with the chord method. Tables 7.2 and 7.3 give the results of applying two chord iterations to our example problem. Table 7.2 uses the pure chord method (7.30); Table 7.3 uses (7.31), with $p = 3$. Compare the total cost to convergence with that for Newton's method, as given in Table 7.1; even though Newton's method took fewer iterations than either version of the chord method, the overall cost is cheaper for the chord methods. For Newton's method, the cost will be $k + k^2$ function evaluations plus $\frac{1}{3}k^3 + k^2$ operations in solving the linear system, per iteration. This works out to a total of about 24 function evaluations plus roughly 46 operations in solving the linear system. For (7.30), the pure chord method, the cost per iteration is only k function evaluations plus k^2 operations in solving the linear system; there is also an overhead cost of k^2 function evaluations and about $\frac{1}{3}k^3$ operations in factoring the linear system *once*. This works out to a total of about 22 function evaluations plus about 39 operations with the linear system. Finally, for the updated chord method (7.31), the cost per iteration is the same as for the pure chord method, with an additional k^2 function evaluations plus $\frac{1}{3}k^3$ operations for factoring the new gradient each time it is formed. This works out to about 18 function evaluations and 26 operations with the linear system. These differences are not gigantic, but they do show how the chord method can be cost-effective when compared with Newton's method, even though it is slower in terms of iterations to convergence.

TABLE 7.2   **Illustration of Two-Dimensional Chord Method (7.30)**

n	$x_1^{(n)}$	$x_2^{(n)}$	Estimated error
0	1.000000000000	1.000000000000	NA
1	0.605042638147	0.267104846820	0.732895E+00
2	0.597506380968	0.255993159449	0.111117E–01
3	0.597227336747	0.255597398559	0.395761E–03
4	0.597217136546	0.255583069746	0.143288E–04
5	0.597216765729	0.255582550060	0.519685E–06
6	0.597216752268	0.255582531206	0.188547E–07
7	0.597216751779	0.255582530522	0.684125E–09
8	0.597216751762	0.255582530497	0.248232E–10
9	0.597216751761	0.255582530496	0.900727E–12

TABLE 7.3   **Illustration of Modified Two-Dimensional Chord Method (7.31), Using $p = 3$**

n	$x_1^{(n)}$	$x_2^{(n)}$	Estimated error
0	1.000000000000	1.000000000000	NA
1	0.605042638147	0.267104846820	0.732895E+00
2	0.597506380968	0.255993159449	0.111117E–01
3	0.597216756247	0.255582536629	0.410623E–03
4	0.597216751761	0.255582530496	0.613292E–08
5	0.597216751761	0.255582530496	0.184247E–12

7.8.2   Fixed-Point Methods

The problem with many of the Newton-like methods is that they require that a linear system be solved at each iteration. Sometimes, this requirement is more of a computational burden than the need to evaluate the functions. Thus, a method that took more iterations (and therefore more function evaluations) but was significantly cheaper in cost for each iteration might well be more efficient. This brings us back to the notion of fixed-point iterations, which we first saw in Section 3.9.

Given the vector function f, mapping from $\mathbf{R}^k$ to $\mathbf{R}^k$, let us assume we have associated with f a second vector function g such that

$$\alpha = g(\alpha) \iff f(\alpha) = 0.$$

In other words, α is a root of f if and only if α is a fixed point of g. We can then think about employing the fixed-point iteration

$$x^{(n+1)} = g(x^{(n)}) \tag{7.32}$$

as a means of approximating α. We have a substantial theory for fixed-point methods in one dimension, developed in Chapter 3, and most of this carries over to the multidimensional case.

Theorem 7.17 (Multidimensional Fixed-Point Theorem) *Let g be a vector function defined on a domain $D \subset \mathbf{R}^k$, continuous, with the property that $g(x) \in D$ for all $x \in D$. Assume that there exists $\gamma \in [0,1)$ such that*

$$\|g(x) - g(y)\| \leq \gamma \|x - y\|$$

for all $x, y \in D$. Then there exists a unique fixed point $\alpha \in D$ and the iteration $x^{(n+1)} = g(x^{(n)})$ will converge to α for any initial value $x^{(0)} \in D$; furthermore, we have the error estimate

$$\|\alpha - x^{(n)}\| \leq \frac{\gamma^n}{1 - \gamma} \|x^{(1)} - x^{(0)}\|.$$

Proof   This is essentially Theorem 3.5 from Section 3.9, and the proof follows almost exactly the form there. $\square$

This tells us that if we can construct a function g from f such that the hypotheses of the theorem hold, then the iteration (7.32) can be used to compute approximate values of the solution to $f(x) = 0$. While there is no "general theory" for deriving g from f—it is something that is usually best done on a case-by-case basis—there are some general guidelines for broad classes of problems. We might note, for example, that one way to construct g is to use

$$g(x) = x - Af(x),$$

where A is a $k \times k$ matrix. To mimic the Newton scheme, we want $A \approx [f'(x)]^{-1}$. One such choice, very simplistic, would be to invert only the diagonal part of f', thus doing a kind of Jacobi iteration for a nonlinear problem; this requires no matrix factorizations and only k additional function evaluations. If we apply this idea to our example system, we find that it takes 43 iterations to reach convergence. This is slow, in terms of the number of iterations, compared with all our other methods. More sophisticated choices, however, work very well, as we shall see in the next section. The student might want to consider, for example, how to apply the Gauss-Seidel ideas to this setting.

EXERCISES

1. Consider the nonlinear system

$$2x_1 - x_2 + \frac{1}{9}e^{-x_1} = -1 \qquad (7.33)$$

$$-x_1 + 2x_2 + \frac{1}{9}e^{-x_2} = 1 \qquad (7.34)$$

Take $x^{(0)} = (1,1)^T$ and do two iterations of Newton's method; you should get

$$x^{(i)} = (-0.423172801349, 0.252702781305)^T.$$

2. Write a computer code to solve the system in the preceding problem, using

(a) Newton's method
(b) The chord method
(c) The chord method, updating every three iterations

3. Rewrite the system in Exercise 1 as

$$Kx + \phi(x) = b,$$

where K is the 2×2 matrix

$$K = \begin{bmatrix} 2 & -1 \\ -1 & 2 \end{bmatrix},$$

$\phi(x)$ is defined by

$$\phi(x) = \begin{pmatrix} \frac{1}{9}e^{-x_1} \\ \frac{1}{9}e^{-x_2} \end{pmatrix},$$

and $b = (-1,1)^T$.

(a) Do two iterations (by hand) of the fixed-point iteration

$$x^{(k+1)} = \frac{1}{2}(b - \phi(x^{(k)}) - Kx^{(k)} + 2x^{(k)})$$

for this system, using $x^{(0)} = (1,1)^T$.
(b) Do two iterations (by hand) of the fixed-point iteration

$$x^{(k+1)} = K^{-1}(b - \phi(x^{(k)}))$$

for this system, using $x^{(0)} = (1,1)^T$.
(c) Which one do you think is going to converge faster?

4. Write a computer code to implement the fixed-point iterations outlined in the previous exercise. Compare the total "cost to convergence" with your results for Newton's method and the chord iterations.

7.9 APPLICATION: NUMERICAL SOLUTION OF NONLINEAR BVPS

In Section 6.10, when we studied the approximate solution of boundary value problems, we were unable to apply the difference techniques of Section 6.10.1 to a nonlinear differential equation because we did not know how to solve the resulting system of (nonlinear) equations. Now we do, based on the material in Section 7.8.

Consider the problem

$$-u'' = (u')^2 \tag{7.35}$$
$$u(0) = 0, \quad u(1) = 1, \tag{7.36}$$

which we first looked at in Chapter 6. The exact solution is $u = \log((e-1)x+1)$. In Chapter 6, we had to use shooting methods to solve this problem, because the finite difference techniques would lead to a system of nonlinear equations:

$$2u_1 - u_2 - \frac{1}{4}u_2^2 = 0 \tag{7.37}$$

$$-u_{j-1} + 2u_j - u_{j+1} - \frac{1}{4}(u_{j+1} - u_{j-1})^2 = 0, \quad 2 \le j \le N-2 \tag{7.38}$$

$$-u_{N-2} + 2u_{N-1} - 1 - \frac{1}{4}(1 - u_{N-2})^2 = 0. \tag{7.39}$$

The student should verify (Exercise 1) that this is the correct system, assuming a grid using $h = 1/N$.

We can solve this nonlinear system using any one of the methods discussed in the previous section. Rather than go through all of those in detail, we elect to illustrate the choice of a fixed-point iteration, and leave the others to the exercises. (See Exercises 2–4.)

Recall that a common fixed-point iteration to solve the nonlinear equation $f(u) = 0$ is

$$u^{(n+1)} = u^{(n)} - Af(u^{(n)}),$$

where, ideally, the matrix A would be chosen to approximate the gradient of f. We can write our nonlinear system in the form

$$Ku + \phi(u) = b,$$

where K is the tridiagonal matrix with components given by

$$k_{ij} = \begin{cases} 2 & i = j \\ -1 & |i-j| = 1 \\ 0 & \text{otherwise,} \end{cases}$$

ϕ is the nonlinear function given by

$$\phi(u) = - \begin{pmatrix} \frac{1}{4}u_2^2 \\ \frac{1}{4}(u_3 - u_1)^2 \\ \vdots \\ \frac{1}{4}(u_N - u_{N-2})^2 \\ \frac{1}{4}(1 - u_{N-1})^2 \end{pmatrix},$$

and b is the constant vector

$$b = (0,0,\ldots,0,1)^T.$$

Since $f(u) = Ku + \phi(u) - b$, it is not hard to show (Exercise 5) that the gradient is given by $f'(u) = K + \phi'(u)$, where

$$\phi'(u) = - \begin{pmatrix} 0 & \frac{1}{2}u_2 & 0 & \cdots & 0 \\ \frac{1}{2}(u_1 - u_3) & 0 & \frac{1}{2}(u_3 - u_1) & \cdots & 0 \\ \vdots & \ddots & \ddots & \ddots & \vdots \\ 0 & \cdots & \frac{1}{2}(u_{N-2} - u_N) & 0 & (u_N - u_{N-2}) \\ 0 & \cdots & 0 & \frac{1}{2}(u_{N-1} - 1) & 0 \end{pmatrix} \tag{7.40}$$

is the gradient of ϕ. Thus a very efficient choice of A is $A = K$. Not only is K an important part of the gradient of f, but it is also independent of the iteration; thus we can factor it once and then we need only do the forward and backward solution steps each iteration. Table 7.4 shows the results that we get when we do this computation, using a sequence of meshes $h = \frac{1}{4}, \frac{1}{8}, \ldots$; for each mesh we printed out the number of iterations needed for convergence, and the error between the computed solution and the exact solution, as measured in the vector infinity norm. For the first case ($h = \frac{1}{4}$), we took as our initial guess the straight line connecting the boundary values. For the

TABLE 7.4 **Number of Iterations and Maximum Error for Approximate Solutions to the Nonlinear BVP (7.35)–(7.36)**

n	Iterations	Error
4	13	0.154547E–02
8	12	0.383951E–03
16	12	0.958279E–04
32	10	0.239854E–04
64	9	0.600085E–05
128	8	0.149813E–05
256	7	0.374157E–06
512	6	0.947443E–07
1024	5	0.242657E–07

subsequent cases, we took as our initial guess a vector based on the final solution from the previous case; thus, by starting off very close to the solution, we took fewer iterations to find it. For comparison's sake, if we had started out doing the $h = 1/1024$ case with the straight-line initial guess, it would have taken 15 iterations to converge.

EXERCISES

1. Set up the nonlinear system for the example (7.35)–(7.36), and verify that the result given by (7.37)–(7.39) is correct.

2. Apply Newton's method and the chord method to the approximate solution of the nonlinear BVP (7.35)–(7.36). Compare the number of iterations to convergence and the overall cost of convergence. Use the sequence of grids $h^{-1} = 4, 8, \ldots, 1024$.

3. Consider the nonlinear BVP

$$-u'' + e^{-u} = 1$$
$$u(0) = u(1) = 1.$$

Use finite difference techniques to reduce this (approximately) to a system of nonlinear algebraic equations,

and solve this system using several of the methods discussed in this chapter. Test the program on the sequence of grids $h^{-1} = 4, 8, \ldots, 1024$ (and further, if practical on your system). Compare the cost of convergence for each method, in terms of the number of iterations and the number of operations.

4. Now consider the nonlinear BVP

$$-u'' = \frac{u}{u+1}$$
$$u(0) = 0, \quad u(1) = 1.$$

Repeat the kind of study required in the previous exercise.

5. Verify that (7.40) gives the correct gradient for (7.37)–(7.39).

LITERATURE AND SOFTWARE DISCUSSION

The subject of numerical linear algebra has seen, in the last 25 years, an explosion of references and monographs. The standard references are the books by Stewart [9], Golub and van Loan [4], Watkins [11], Datta [1], and the very recent works by Trefethen and Bau [10] and Demmel [2]. Another classic, although dated, work is Householder [5]. For nonlinear systems the classic reference is still Ortega and Rheinboldt [7]; recent books by Kelley [6] and Saad [8] offer some more modern insight. The standard references on rounding error issues are Forsythe and Moler [3] and, of course, Wilkinson [12].

In terms of software, the most important developments have probably been the widespread use of the LAPACK library and the MATLAB package. MATLAB, in particular, has made it very easy for beginning students to use very sophisticated computational linear algebra techniques. Most of the example computations in this chapter were done using MATLAB.

BIBLIOGRAPHY

[1] DATTA, BISWA NATH. *Numerical Linear Algebra and Applications.* Brooks/Cole, Pacific Grove, CA, 1995.

[2] DEMMEL, JAMES. *Applied Numerical Linear Algebra.* SIAM, Philadelphia, 1997.

[3] FORSYTHE, GEORGE, & MOLER, CLEVE. *Computer Solution of Linear Algebraic Systems.* Prentice-Hall, Englewood Cliffs, NJ, 1967.

[4] GOLUB, G., & VAN LOAN, C. *Matrix Computations.* Johns Hopkins University Press, Baltimore, 1996 (3rd ed.).

[5] HOUSEHOLDER, ALSTON. *The Theory of Matrices in Numerical Analysis.* Dover, New York, 1975 (originally published in 1964 by Blaisdell).

[6] KELLEY, C.T. *Iterative Methods for Linear and Nonlinear Equations.* SIAM, Philadelphia, 1995.

[7] ORTEGA, J.M., & RHEINBOLDT, W.C. *Iterative Solution of Nonlinear Equations in Several Variables.* Academic Press, New York, 1970.

[8] SAAD, YOUSEF. *Iterative Methods for Sparse Linear Systems.* PWS, Boston, 1996.

[9] STEWART, G.W. *Introduction to Matrix Computations.* Academic Press, Orlando, 1973.

[10] TREFETHEN, L.N., & BAU, DAVID. *Numerical Linear Algebra.* SIAM, Philadelphia, 1997.

[11] WATKINS, DAVID. *Fundamentals of Matrix Computations.* John Wiley & Sons, New York, 1991.

[12] WILKINSON, J.H. *Rounding Errors in Algebraic Processes.* Prentice-Hall, Englewood Cliffs, NJ, 1963.

CHAPTER 8

APPROXIMATE SOLUTION OF THE ALGEBRAIC EIGENVALUE PROBLEM

Pennsyltucky Petroleum operates a number of chemical reaction plants, in which various complex reactions occur. Most of the reactions can be modeled by linear systems of differential equations:

$$\frac{dx}{dt} = Ax.$$

Here x is a *vector* of component concentrations, and the matrix elements $a_{i,j}$ model the rate at which the constituent j is turned into product i.

The simplest solution of such a system, given an initial vector $x(0)$, uses the *matrix exponential:*

$$x(t) = e^{At}x(0).$$

While this might look daunting, the matrix exponential is easily defined in terms of the *eigenvalues* and *eigenvectors* of the matrix A:

$$e^{At} = V^{-1}E(t)V,$$

where

$$E(t) = \begin{bmatrix} e^{\lambda_1 t} & & & \\ & e^{\lambda_2 t} & & \\ & & \ddots & \\ & & & e^{\lambda_n t} \end{bmatrix},$$

the λ_k are the eigenvalues of A, and

$$V = [v_1 \ v_2 \ \dots \ v_n]$$

is the matrix whose columns are the corresponding eigenvectors. (We are assuming here that the eigenvectors are independent.) For moderate-size problems this is a reasonable way to compute the solutions; thus Pennsyltucky Petroleum needs to understand how to approximate solutions to the algebraic eigenvalue problem.

In this chapter we will discuss, in some detail, some iterative methods for finding single eigenvalue-eigenvector pairs ("eigenpairs" is a common term) of a given real matrix A; we will also give an overview of more powerful and general methods that are commonly used to find *all* the eigenpairs of a given real A.

8.1 EIGENVALUE REVIEW

The algebraic eigenvalue problem is as follows: Given a matrix $A \in \mathbf{R}^{n \times n}$, find a nonzero vector $x \in \mathbf{R}^n$ and the scalar λ such that

$$Ax = \lambda x.$$

Note that this says that the vector Ax is parallel to x, with λ being an amplification factor, or *gain*. Note also that the expression above implies that

$$(A - \lambda I)x = 0,$$

thus showing (by Theorem 7.1) that $A - \lambda I$ is a singular matrix. Hence $\det(A - \lambda I) = 0$; it is easy to show that this determinant is a polynomial (of degree n) in λ, known as the *characteristic polynomial* of A, $p(\lambda)$, so that the eigenvalues are the roots of a polynomial. While this is *not* a good way to compute the eigenvalues, it does give us some insight into their properties. Thus we know that an $n \times n$ matrix has n eigenvalues, that the eigenvalues can be repeated, and that a real matrix can have complex eigenvalues, but these must occur in conjugate pairs. We summarize these and a number of other basic eigenvalue properties in the following theorem, presented without proof.

Theorem 8.1 (Basic Eigenvalue Properties)   *Let $A \in \mathbf{R}^{n \times n}$ be given. Then we have the following:*

1. *There are exactly n eigenvalues, counting multiplicities; complex eigenvalues will occur in conjugate pairs.*
2. *Eigenvectors corresponding to distinct eigenvalues are independent.*
3. *If an $n \times n$ matrix A has n independent eigenvectors, then there exists a nonsingular matrix P such that $P^{-1}AP = D$ is diagonal, and A is called diagonalizable. Moreover, the columns of P are the eigenvectors of A, and the elements $d_{ii} = \lambda_i$ are the eigenvalues of A.*
4. *If A is symmetric ($A = A^T$), then the eigenvalues are real and we can choose the eigenvectors to be real and orthogonal.*
5. *If A is symmetric, then there is an orthogonal matrix[1] Q such that $Q^T A Q = D$ is diagonal, where the elements $d_{ii} = \lambda_i$ are the eigenvalues of A.*
6. *If A is triangular, then the eigenvalues are the diagonal elements, $\lambda_i = a_{ii}$.*

EXAMPLE 8.1

Let

$$A = \begin{bmatrix} 4 & 1 & 0 \\ 1 & 4 & 1 \\ 0 & 1 & 4 \end{bmatrix}.$$

The characteristic polynomial for this matrix is

$$p(\lambda) = \lambda^3 - 12\lambda^2 + 46\lambda - 56,$$

---

[1]Recall that an orthogonal matrix is one such that $Q^T = Q^{-1}$.

and the eigenvalues are

$$\lambda_1 = 5.4142\ldots, \quad \lambda_2 = 4.000, \quad \lambda_3 = 2.5858\ldots \; .$$

Since the matrix is symmetric, we can find a set of orthonormal eigenvectors:

$$x_1 = \begin{bmatrix} 0.5000 \\ 0.7071 \\ 0.5000 \end{bmatrix}, \quad x_2 = \begin{bmatrix} -0.7071 \\ 0.0000 \\ 0.7071 \end{bmatrix}, \quad x_3 = \begin{bmatrix} -0.5000 \\ 0.7071 \\ -0.5000 \end{bmatrix}.$$

Since these vectors are orthonormal, when we arrange them as the columns of a matrix we get an orthogonal matrix:

$$Q = [x_1 | x_2 | x_3] = \begin{bmatrix} 0.5000 & -0.7071 & -0.5000 \\ 0.7071 & 0.0000 & 0.7071 \\ 0.5000 & 0.7071 & -0.5000 \end{bmatrix}.$$

Moreover, we have $Q^{-1}AQ = Q^{T}AQ = D$, diagonal:

$$(Q^{T}A)Q = \begin{bmatrix} 2.7071 & 3.8284 & 2.7071 \\ -2.8284 & 0.0000 & 2.8284 \\ -1.2929 & 1.8284 & -1.2929 \end{bmatrix} \begin{bmatrix} 0.5000 & -0.7071 & -0.5000 \\ 0.7071 & 0.0000 & 0.7071 \\ 0.5000 & 0.7071 & -0.5000 \end{bmatrix}$$

$$= \begin{bmatrix} 5.4142 & 0.0000 & 0.0000 \\ 0.0000 & 4.0000 & 0.0000 \\ 0.0000 & 0.0000 & 2.5858 \end{bmatrix}.$$

■

EXAMPLE 8.2

This time let

$$A = \begin{bmatrix} 5 & 4 & 3 \\ -1 & 0 & -3 \\ 1 & -2 & 1 \end{bmatrix}.$$

The characteristic polynomial is now

$$p(\lambda) = \lambda^3 - 6\lambda^2 + 32,$$

and the eigenvalues are

$$\lambda_1 = 4, \quad \lambda_2 = 4, \quad \lambda_3 = -2.$$

Since $\lambda_1 = \lambda_2$, we are not guaranteed that the corresponding eigenvectors are independent, and in this case they are not. An $n \times n$ matrix with $k < n$ independent eigenvectors is called *defective*. Note that having repeated eigenvalues is a *necessary* condition for a matrix to be defective, but it is not *sufficient;* there exist many nondefective matrices with repeated eigenvalues, the most obvious example being the identity matrix. Exercise 4 asks the student to explore a little bit more about defective matrices. ■

A certain amount of new terminology and notation will be useful. The collection of all eigenvalues of a matrix is typically called the *spectrum* of the matrix, which we will denote by $\sigma(A)$:
$$\sigma(A) = \{\lambda \in \mathbf{C}, Ax = \lambda x, x \neq 0\}.$$

The vector space spanned by all the eigenvectors corresponding to a single eigenvalue λ is called the *eigenspace* and is denoted by $E_\lambda(A)$:

$$E_\lambda(A) = \{x \in \mathbf{C}^n, Ax = \lambda x\}.$$

Note that the eigenspaces, in order to be vector spaces, must include the zero vector, which is not an eigenvector. If there exists a nonsingular matrix P such that $A = PBP^{-1}$, then A and B are said to be *similar*, which we will denote by $A \sim B$. Similar matrices have the same eigenvalues and related eigenvectors (Exercise 2).

An important tool in eigenvalue approximation is the ability to *localize* the eigenvalues, and the most important tool in eigenvalue localization is Gerschgorin's Theorem.

Theorem 8.2 (Gerschgorin's Theorem) Let $A \in \mathbf{R}^{n \times n}$ be given, and define the quantities

$$r_i = \sum_{\substack{j=1 \\ j \neq i}}^{n} |a_{ij}|$$

$$D_i = \{z \in \mathbf{C} \,|\, |z - a_{ii}| \leq r_i\}.$$

Then every eigenvalue of A lies in the union of the disks D_i, that is,

$$\lambda_k \in \bigcup_{i=1}^{n} D_i$$

for all $k = 1, 2, \ldots, n$. Moreover, if any collection of p disks is disjoint from the other $n - p$ disks, then we know that exactly p eigenvalues are contained in the union of the set of p disks, and exactly $n - p$ eigenvalues are contained in the set of $n - p$ disks.

EXAMPLE 8.3

To illustrate the theorem, consider the following examples. If we have

$$A = \begin{bmatrix} 2 & 1 & 0 \\ 1 & 2 & 1 \\ 0 & 1 & 2 \end{bmatrix},$$

then the disks are defined by the centers $a_{ii} = 2$, $1 \leq i \leq 3$, i.e., the disks are concentric, and the radii are

$$r_1 = 1, \; r_2 = 2, \; r_3 = 1.$$

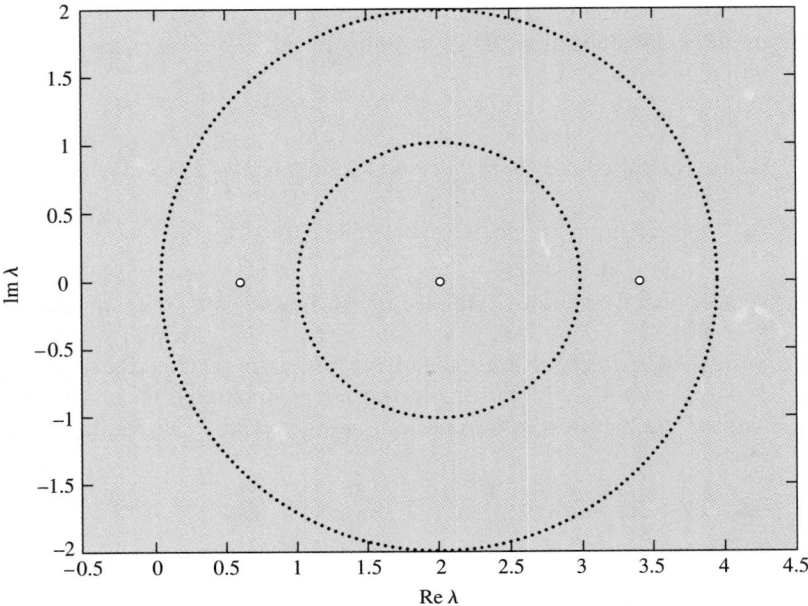

FIGURE 8.1 Illustration of Gerschgorin's Theorem.

Figure 8.1 shows the circles in the complex plane along with the exact eigenvalues (shown as small circles) $\lambda_1 = 3.1414$, $\lambda_2 = 2.000$, and $\lambda_3 = 0.5859$. Since the matrix is symmetric, we expect the eigenvalues all to be real; thus they lie on the x axis.

On the other hand, if

$$
A = \begin{bmatrix}
2 & -1 & -2 & 1 & 2 \\
2 & 2 & 1 & 0 & 0 \\
0 & -2 & -1 & 2 & -1 \\
2 & -2 & -2 & 0 & 1 \\
-2 & -1 & -2 & 0 & 2
\end{bmatrix},
$$

then we can no longer assume the eigenvalues are real. The disks are defined by the centers

$$a_{11} = 2, \ a_{22} = 2, \ a_{33} = -1, \ a_{44} = 0, \ a_{55} = 2,$$

and the radii

$$r_1 = 6, \ r_2 = 3, \ r_3 = 5, \ r_4 = 7, \ r_5 = 5.$$

The situation is depicted in Figure 8.2. Note that all of the eigenvalues are actually clustered within the smallest disk; they are all within all of the disks.

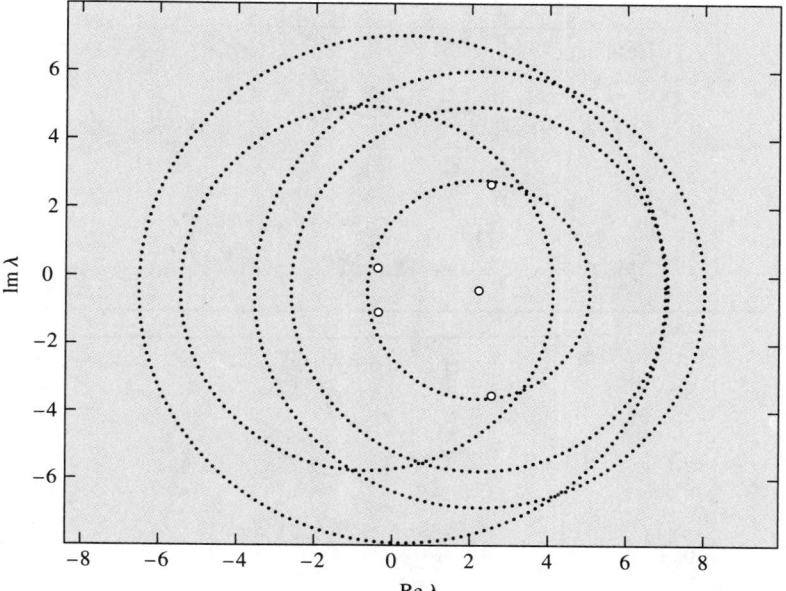

FIGURE 8.2   Illustration of Gerschgorin's Theorem.

Finally, we can illustrate the way Gerschgorin's Theorem can isolate eigenvalues by looking at the example

$$A = \begin{bmatrix} 0 & 2 & 0 \\ 2 & 7 & 1 \\ 0 & 1 & 4 \end{bmatrix};$$

then we have the centers

$$a_{11} = 0, \;\; a_{22} = 7, \;\; a_{33} = 4,$$

and the radii

$$r_1 = 2, \;\; r_2 = 3, \;\; r_3 = 1.$$

Note (see Figure 8.3) that we have a single eigenvalue within the isolated disk, and the other two eigenvalues are contained within the union of the other two disks.

If we change the matrix only slightly, to

$$A = \begin{bmatrix} 0 & 0 & 2 \\ 2 & 7 & 1 \\ 1 & 0 & 4 \end{bmatrix},$$

note that the disks remain the same but now the eigenvalues are positioned differently (see Figure 8.4). ∎

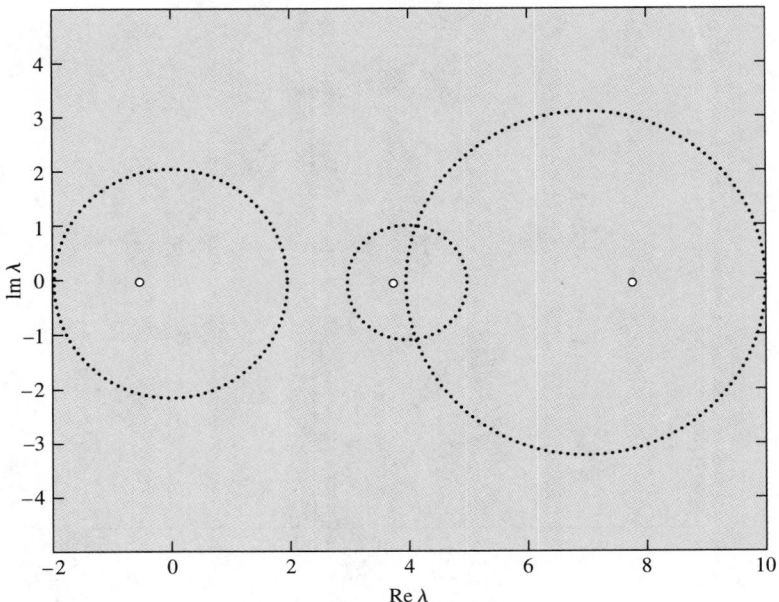

FIGURE 8.3 Illustration of Gerschgorin's Theorem.

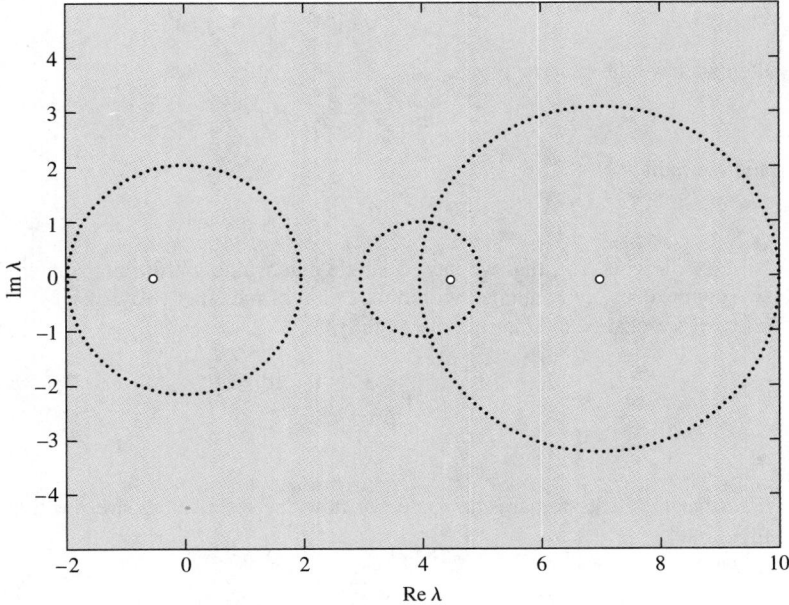

FIGURE 8.4 Illustration of Gerschgorin's Theorem.

EXERCISES

1. For each matrix below, find the characteristic polynomial and the eigenvalues by a hand calculation. For some of the exercises, the correct eigenvalues are given to four decimal places, so you can check your work.

(a)

$$A = \begin{bmatrix} 4 & 1 & 0 \\ 1 & 4 & 1 \\ 0 & 1 & 4 \end{bmatrix},$$

for which $\sigma(A) = \{5.4142, 4.0000, 2.5858\}$

(b)

$$A = \begin{bmatrix} 4 & 2 & 0 \\ 1 & 4 & 2 \\ 0 & 1 & 4 \end{bmatrix},$$

for which $\sigma(A) = \{2, 4, 6\}$

(c)

$$A = \begin{bmatrix} 6 & -2 & 0 \\ 2 & 6 & -2 \\ 0 & 2 & 4 \end{bmatrix},$$

for which $\sigma(A) = \{5.5698 \pm 2.6143i, 4.8603\}$

(d)

$$A = \begin{bmatrix} 6 & -1 & 0 \\ 2 & 6 & -1 \\ 0 & 2 & 4 \end{bmatrix},$$

(e)

$$A = \begin{bmatrix} 0 & 2 & 0 \\ 2 & 7 & 1 \\ 0 & 1 & 4 \end{bmatrix},$$

2. Prove that similar matrices have identical eigenvalues and related eigenvectors.

3. Apply Gerschgorin's Theorem to the matrices in Exercise 1 and determine the intervals or disks in which the eigenvalues must lie.

4. Show that the matrix

$$A = \begin{bmatrix} 3 & 2 & -1 \\ 0 & 2 & 2 \\ 0 & -2 & 7 \end{bmatrix}$$

has repeated eigenvalues and is defective, but that the matrix

$$A = \begin{bmatrix} -1 & -1 & 2 \\ -2 & 0 & 2 \\ -4 & -2 & 5 \end{bmatrix},$$

which also has repeated eigenvalues, is *not* defective.

5. Let A and B be similar matrices of size $n \times n$, with

$$B = \begin{bmatrix} \lambda & a^T \\ 0 & A_{22} \end{bmatrix},$$

where $\lambda \in \mathbf{R}$, $a \in \mathbf{R}^{n-1}$, and $A_{22} \in \mathbf{R}^{(n-1) \times (n-1)}$.
(a) Prove that $\lambda \in \sigma(A)$. (Hint: What is the product Be_1, where e_1 is the first standard basis vector?)
(b) Prove that each eigenvalue of A_{22} is also an eigenvalue of A.

6. Generalize the above: Let A and B be similar matrices of size $n \times n$, with

$$B = \begin{bmatrix} D & a^T \\ 0 & A_{22} \end{bmatrix},$$

where $D \in \mathbf{R}^{p \times p}$ is diagonal, $a \in \mathbf{R}^{n-p}$, and $A_{22} \in \mathbf{R}^{(n-p) \times (n-p)}$. Prove that each diagonal element of D is an eigenvalue of A: $d_{ii} \in \sigma(A), 1 \le i \le p$, and that each eigenvalue of A_{22} is also an eigenvalue of A.

8.2   REDUCTION TO HESSENBERG FORM

Since the computation of eigenvalues is equivalent to finding the roots of a polynomial, it follows that for $n \ge 5$ there will be no general algorithm that works in a finite number of steps.[2] Accordingly, we expect an eigenvalue solver to be an inherently iterative process, i.e., one that recursively computes better and better approximations. For this reason, it is usually more efficient to *preprocess* the matrix by making as much of a reduction as possible in a finite number of steps. This leads us to the notion of the *Hessenberg form* of a matrix.

Definition 8.1 (Hessenberg Form)   *A matrix $A \in \mathbf{R}^{n \times n}$ is in* Hessenberg form *if $a_{ij} = 0$ for all i, j such that $i - j > 1$.*

Thus, the matrix

$$A = \begin{bmatrix} 1 & 2 & 3 & 4 \\ 5 & 6 & 7 & 8 \\ 0 & 9 & 8 & 7 \\ 0 & 0 & 6 & 5 \end{bmatrix}$$

is in Hessenberg form. Note that one way to characterize Hessenberg form is that it is "almost" triangular. This is important, since the eigenvalues of a triangular matrix are the diagonal elements. Note, also, that a symmetric Hessenberg matrix is tridiagonal.

The important result, then, is the following.

Theorem 8.3   *Let $A \in \mathbf{R}^{n \times n}$ be given. Then there exists $A_H \in \mathbf{R}^{n \times n}$, Hessenberg, which can be computed from A in a finite number of steps and which has the same eigenvalues as A.*

Proof   The proof is constructive, but somewhat involved, and depends heavily on a preliminary result.

Claim   For any vector $x \in \mathbf{R}^n$, there exists an orthogonal matrix Q such that $Qx = \|x\|_2 e_1$, where e_1 is the first standard basis vector. •

We will prove this claim after the main proof of the theorem. To prove the theorem we will use the claim to construct a matrix P such that $B = PAP^{-1}$ is in Hessenberg form. This means A and B have the same eigenvalues (by virtue of similarity), and we will be done.

We start by writing A in partitioned form:

$$A = \left[\begin{array}{c|c} a_{11} & b^T \\ \hline a & A_{22} \end{array} \right],$$

---

[2]This is because there can be no algorithm for finding the roots of a polynomial of degree $n \geq 5$ in a finite number of steps. This follows from the work of Niels Henrik Abel (1802–1829) and Evariste Galois (1811–1832).

where $a \in \mathbf{R}^{n-1}$, $b \in \mathbf{R}^{n-1}$, and $A_{22} \in \mathbf{R}^{(n-1)\times(n-1)}$. Now let Q_1 be an orthogonal matrix such that $Q_1 a = \|a\|_2 e_1$, and use this to construct the matrix

$$P_1 = \left[\begin{array}{c|cc} 1 & \cdots & 0 & \cdots \\ \hline \vdots & & \\ 0 & & Q_1 \\ \vdots & & \end{array}\right].$$

(Note that Q_1 is in $\mathbf{R}^{(n-1)\times(n-1)}$.) Then P_1 is also orthogonal (thus $P_1^{-1} = P_1^T$) and

$$A_1 = P_1 A P_1^T = \left[\begin{array}{c|c} a_{11} & b^T Q_1^T \\ \hline \|a\|_2 e_1 & Q_1 A_{22} Q_1^T \end{array}\right].$$

Note that the first column of A_1 satisfies the criterion for being in Hessenberg form.

We can now continue the process in what is probably an obvious way. We write A_1 in the partitioned form

$$A_1 = \left[\begin{array}{cc|c} a_{11} & a_{12}^H & b_1^T \\ \|a\|_2 & a_{22}^H & b_2^T \\ \hline 0 & a_1 & A_{22}' \end{array}\right]$$

and define Q_2 to be the matrix such that $Q_2 a_1 = \|a_1\|_2 e_1$; define P_2 as

$$P_2 = \left[\begin{array}{cc|cc} 1 & 0 & \cdots & 0 & \cdots \\ 0 & 1 & \cdots & 0 & \cdots \\ \hline \vdots & \vdots & & \\ 0 & 0 & & Q_2 \\ \vdots & \vdots & & \end{array}\right],$$

so that $A_3 = P_2 A_1 P_2^T = P_2 P_1 A P_1^T P_2^T$ is in Hessenberg form. The process stops after $n-2$ such steps.

It remains only to prove the claim, which we state as a separate lemma. $\qquad\square$

Lemma 8.1 (Claim) *For any vector $x \in \mathbf{R}^n$, there exists an orthogonal matrix Q such that $Qx = \|x\|_2 e_1$, where e_1 is the first standard basis vector.*

Proof   The proof is constructive, using what are known as *Householder transformations*.[3] Define Q as

$$Q = I - \gamma w w^T,$$

where we regard w as a vector to be determined, and $\gamma = 2/\|w\|_2^2$. It is easily verified that Q is orthogonal since we have

$$
\begin{aligned}
Q^T Q &= (I - \gamma w w^T)(I - \gamma w w^T) \\
&= I - 2\gamma w w^T + \gamma^2 \|w\|_2^2 w w^T \\
&= I - 4\frac{w w^T}{\|w\|_2^2} + 4\frac{\|w\|_2^2 w w^T}{\|w\|_2^2} \\
&= I.
\end{aligned}
$$

Then[4]

$$Qx = x - \gamma(w^T x)w.$$

Since we want Qx to be a constant times e_1, we are going to want most of these components to vanish, so let's look for w in the form $w = x + c e_1$, where c is to be determined later. This gives us

$$
\begin{aligned}
Qx &= x - \gamma(x + c e_1)^T x (x + c e_1) \\
&= x - \gamma(\|x\|_2^2 + c x_1)(x + c e_1) \\
&= (1 - \gamma(\|x\|_2^2 + c x_1))x - \gamma c(\|x\|_2^2 + c x_1)e_1.
\end{aligned}
$$

We want the coefficient of x to vanish and the coefficient of e_1 to be $\|x\|_2$. Since $w = x + c e_1$, we know that

$$\gamma = \frac{2}{\|x\|_2^2 + 2c x_1 + c^2},$$

so we have (after a bit of manipulation) $c = -\|x\|_2$. This completes the proof. $\square$

Note: If we choose $c = \|x\|_2$, then we get $Qx = -\|x\|_2 e_1$. (See Exercise 1.) This is just as good a result for our purposes, and it can sometimes be more stable numerically. To minimize subtractive cancellation errors, it is best to use $c = \|x\|_2$ when $x_1 > 0$, and $c = -\|x\|_2$ when $x_1 < 0$.

---

[3] Alston Householder (1904–1993) was born in Rockford, Illinois, but raised in Alabama. He earned his Ph.D. in mathematics from the University of Chicago in 1937 and spent the first part of his career working in mathematical biology. In 1944 he joined the Division of Mathematics at Oak Ridge National Laboratory, where he stayed until his retirement in 1969. It was at Oak Ridge that Householder took up numerical analysis as his area of study, in which he made a number of important contributions. What we know as "Householder transformations" were first known as "elementary hermitian matrices" and are used in a variety of places in numerical linear algebra. Householder's early text [1] remains a classic in the field of numerical linear algebra.

[4] Note how easy it is to operate with these matrices. Matrix-vector multiplication is accomplished with a single dot product and a subtraction.

The special form of the Householder matrices allows us to accomplish the reduction to Hessenberg form very easily, with an algorithm that is very much akin to Gaussian elimination. If we are not interested in computing the eigenvectors of the original matrix, then very little additional storage is needed. On the other hand, if we do want to compute the eigenvectors, we will need to save the w vectors that are computed along the way so that we can reverse the transformation.

The operation count for the Hessenberg reduction is $\frac{5}{3}n^3 + O(n^2)$ for the general case, but only $\frac{2}{3}n^3 + O(n^2)$ if A is symmetric. In either case, this is inexpensive enough compared with the cost of operating with the full matrix that it is a worthwhile preliminary step for eigenvalue computations. We will consider only the symmetric case, for the sake of simplicity.

Assume that we have performed the necessary reductions on the first $i-1$ columns of A. Then the partially reduced matrix can be partitioned as follows:

$$P_{i-1}AP_{i-1} = \begin{bmatrix} A_{H,11} & B^T \\ B & A_{22} \end{bmatrix},$$

where $A_{H,11} \in \mathbf{R}^{i \times i}$ is in Hessenberg (therefore tridiagonal) form, all columns of B are zero except for the last one, and A_{22} is the remaining lower quadrant of the matrix. Let a be the nonzero column in B, and let w be the vector such that

$$Q = I - \gamma ww^T$$

is the Householder matrix for which $Qa = \pm\|a\|_2 e_1$. Then

$$P_i = \begin{bmatrix} I & 0 \\ 0 & Q \end{bmatrix},$$

so that the next stage of the reduction is

$$P_i P_{i-1} A P_{i-1} P_i = \begin{bmatrix} A_{H,11} & U^T \\ U & QA_{22}Q \end{bmatrix},$$

where all columns of U are zero except for the last one, which is $\pm\|a\|_2 e_1$. Thus we see that the work in each step of the reduction breaks down into three relatively simple steps:

1. Identify the vector a on which the Householder transformation is to operate.
2. Compute the vector w that defines the Householder transformation.
3. Update the lower quadrant of the matrix accordingly.

The first two steps are straightforward. The third step is equally so, using the equation

$$QA_{22}Q = (I - \gamma ww^T)A_{22}(I - \gamma ww^T) = A_{22} - \gamma(wp^T + pw^T) + \gamma^2 \alpha ww^T,$$

where

$$p = A_{22}w, \quad \alpha = w^T p = w^T A_{22}w.$$

The assumed symmetry of A and A_{22} is used in simplifying some of these expressions. If A is not symmetric, then there will be some changes.

The pseudocode given in Algorithm 8.1 on page 470 is for a symmetric reduction, but it is not organized in the most efficient manner; rather, it is organized to make following the code easy. One of the exercises asks the student to revise the code for increased efficiency.

EXAMPLE 8.4

Consider the matrix

$$A = \begin{bmatrix} 6 & 2 & 1 & 1 \\ 2 & 6 & 2 & 1 \\ 1 & 2 & 6 & 2 \\ 1 & 1 & 2 & 6 \end{bmatrix}.$$

Since this is only 4×4, we have to do only two steps to accomplish the reduction. In the first step, the vector a is defined by

$$a = \begin{bmatrix} 2 \\ 1 \\ 1 \end{bmatrix},$$

so that

$$w = \begin{bmatrix} 4.4495 \\ 1 \\ 1 \end{bmatrix}$$

and $\gamma = 0.0918$. Thus

$$Q_1 = I - \gamma ww^T = \begin{bmatrix} -0.8165 & -0.4082 & -0.4082 \\ -0.4082 & 0.9082 & -0.0918 \\ -0.4082 & -0.0918 & 0.9082 \end{bmatrix}$$

and

$$P_1 = \begin{bmatrix} 1 & 0 & 0 & 0 \\ 0 & -0.8165 & -0.4082 & -0.4082 \\ 0 & -0.4082 & 0.9082 & -0.0918 \\ 0 & -0.4082 & -0.0918 & 0.9082 \end{bmatrix}.$$

Therefore,

$$A_2 = P_1 A P_1 = \begin{bmatrix} 6.0000 & -2.4495 & -0.0000 & -0.0000 \\ -2.4495 & 8.6667 & -1.5749 & -0.7584 \\ -0.0000 & -1.5749 & 4.2584 & 0.6667 \\ -0.0000 & -0.7584 & 0.6667 & 5.0749 \end{bmatrix}.$$

Now, for the second (and final) step of the reduction, we have that the vector a is given by

$$a = \begin{bmatrix} -1.5749 \\ -0.7584 \end{bmatrix},$$

so that

$$w = \begin{bmatrix} -3.3229 \\ -0.7584 \end{bmatrix}$$

and $\gamma = 0.1722$. Thus

$$Q_2 = I - \gamma w w^T = \begin{bmatrix} -0.9010 & -0.4339 \\ -0.4339 & 0.9010 \end{bmatrix}$$

and

$$P_2 = \begin{bmatrix} 1 & 0 & 0 & 0 \\ 0 & 1 & 0 & 0 \\ 0 & 0 & -0.9010 & -0.4339 \\ 0 & 0 & -0.4339 & 0.9010 \end{bmatrix}.$$

Therefore,

$$A_H = P_2 A_2 P_2 = \begin{bmatrix} 6.0000 & -2.4495 & 0.0000 & -0.0000 \\ -2.4495 & 8.6667 & 1.7480 & 0 \\ 0.0000 & 1.7480 & 4.9333 & -0.7348 \\ -0.0000 & -0.0000 & -0.7348 & 4.4000 \end{bmatrix}.$$

∎

EXERCISES

1. Show that in the Householder construction, if we use $c = -\|x\|_2$, then we get $Qx = -\|x\|_2 e_1$, and that this will work just as well to construct the Hessenberg matrix.

2. Modify the algorithm in the text for the symmetric Hessenberg reduction to make it more efficient.

3. Modify the algorithm in the text for the symmetric Hessenberg reduction so it will work in the asymmetric case as well.

4. Compute (by hand) the Hessenberg form of the matrix

$$A = \begin{bmatrix} 6 & 1 & 1 & 1 \\ 1 & 6 & 1 & 1 \\ 1 & 1 & 6 & 1 \\ 1 & 1 & 1 & 6 \end{bmatrix}.$$

You should get

$$A_H = \begin{bmatrix} 6.0000 & 1.7321 & 0 & 0 \\ 1.7321 & 8.0000 & -0.0000 & 0 \\ 0 & -0.0000 & 5.0000 & 0 \\ 0 & 0 & 0 & 5.0000 \end{bmatrix}.$$

5. Write a computer code that takes an arbitrary $n \times n$ symmetric matrix A, $n \leq 20$, and produces the diagonal and subdiagonal of the corresponding Hessenberg matrix A_H. Test it on the example in the text and in the previous exercise.

Algorithm 8.1: Symmetric Hessenberg Reduction Algorithm

This code segment produces the Hessenberg form of the symmetric matrix a. Only the diagonal and one off-diagonal are saved, in the variable ah. None of the information needed to recover the original matrix from ah is saved; thus this could be used to compute only the eigenvalues of a, not the eigenvectors.

```
ah(2,1) = a(1,1)
ah(1,1) = 0
for i=2,n do
      for j=i,n do
            aa(j) = a(j,i-1)
            w(j) = aa(j)
      endfor
      sum = 0
      for j = i,n do
            sum = sum + aa(j)*aa(j)
      endfor
      anorm = sqrt(sum)
      sigma = 1
      if w(i) < 0 then sigma = -1
      w(i) = w(i) + sigma*anorm
      ah(1,i) = sigma*anorm
      sum = 0
      for j = i,n do
            sum = sum + w(j)*w(j)
      endfor
      gamma = 2/sum
      for j = i,n do
            sum = 0
            for k = i,n do
                  sum = sum + a(j,k)*w(k)
            endfor
            p(j) = sum
      endfor
      sum = 0
      for j = i,n do
            sum = sum + w(j)*p(j)
      endfor
      alpha = sum
      ga = gamma*gamma*alpha
      for j=i,n do
            for k = i,n do
                  u1 = ga*w(j)*w(k)
                  u2 = gamma*(w(j)*p(k) + w(k)*p(j))
                  a(j,k) = a(j,k) + u1 - u2
            endfor
      endfor
      ah(2,i) = a(i,i)
endfor
```

8.3 POWER METHODS

The simplest methods for approximating eigenvalues are based on the observation that the eigenvectors represent the directions in which the matrix operates, and the eigenvalues represent the gain along those directions. Thus, the quantity $A^N x$ should, eventually, begin to "line up" in the direction of the eigenvector associated with the largest (in absolute value) eigenvalue. While this is not an efficient approach for finding *all* the eigenvalues and eigenvectors of a matrix, it is useful for finding some of them, and much of the theory of more general methods is based on the essential ideas of the power methods.

We start with a formal theorem, from which we develop the first algorithm.

Theorem 8.4   *Let $A \in \mathbf{R}^{n \times n}$ be given, and assume that*

1. A has n linearly independent eigenvectors, x_k, $1 \leq k \leq n$.

2. The eigenvalues λ_k satisfy

$$|\lambda_1| > |\lambda_2| \geq |\lambda_3| \geq \cdots \geq |\lambda_n|.$$

3. The vector $z \in \mathbf{R}^n$ is such that

$$z = \sum_{k=1}^{n} \xi_k x_k$$

and $\xi_1 \neq 0$.

Then

$$\lim_{N \to \infty} \frac{A^N z}{\lambda_1^N} = c x_1 \tag{8.1}$$

for some $c \neq 0$, and

$$\lim_{N \to \infty} \frac{(z, A^N z)}{(z, A^{N-1} z)} = \lambda_1. \tag{8.2}$$

Proof   The key issue is that we can expand the vector z in terms of the eigenvectors, since the independence of the eigenvectors means they are a basis for $\mathbf{R}^n$. Thus we have

$$z = \xi_1 x_1 + \xi_2 x_2 + \cdots + \xi_n x_n = \sum_{k=1}^{n} \xi_k x_k,$$

where $\xi_1 \neq 0$. Then

$$A^N z = \sum_{k=1}^{n} \xi_k \lambda_k^N x_k,$$

Algorithm 8.2: Basic Power Method

For k from 1 until convergence, do

1. Compute $y^{(k)} = Az^{(k-1)}$.

2. $\mu_k = y_i^{(k)}$, where $\|y^{(k)}\|_\infty = |y_i^{(k)}|$.

3. Set $z^{(k)} = y^{(k)}/\mu_k$.

---

so that

$$\frac{A^N z}{\lambda_1^N} = \xi_1 x_1 + \sum_{k=2}^n \xi_k \theta_k^N x_k,$$

where

$$|\theta_k| = \left|\frac{\lambda_k}{\lambda_1}\right| < 1$$

for $k \geq 2$.

Thus, as $N \to \infty$, the θ_k^N vanish and we are left with (8.1).

The proof of (8.2) is very similar. We can write

$$\frac{(z, A^N z)}{(z, A^{N-1} z)} = \frac{\lambda_1^N \eta_1 + \lambda_2^N \eta_2 + \cdots + \lambda_n^N \eta_n}{\lambda_1^{N-1} \eta_1 + \lambda_2^{N-1} \eta_2 + \cdots + \lambda_n^{N-1} \eta_n},$$

where $\eta_k = (z, x_k)$. Now factor out the λ_1 terms in both numerator and denominator to get

$$\frac{(z, A^N z)}{(z, A^{N-1} z)} = \lambda_1 \frac{\eta_1 + \theta_2^N \eta_2 + \cdots + \theta_n^N \eta_n}{\eta_1 + \theta_2^{N-1} \eta_2 + \cdots + \theta_n^{N-1} \eta_n}.$$

As $N \to \infty$, the fraction goes to 1, since each of the θ_k terms goes to zero, and we are done. □

We can use the results of this theorem to construct an algorithm for finding the *dominant* eigenvalue (i.e., the one that is largest in absolute value) and the corresponding eigenvector. We can't use exactly the computations from the theorem, since in (8.1) we scaled the matrix powers by the exact eigenvalue, which we of course do not know. We also do not want to explicitly form the powers A^k. However, we can avoid forming A^k by using recursion, and it turns out (Exercise 3) that the precise form of the scaling is not all that important.

Consider Algorithm 8.2. Concerning this we can prove the following.

Theorem 8.5 (Basic Power Method)   *Let $A \in \mathbf{R}^{n \times n}$ be given, and assume the following three conditions:*

1. *A has n linearly independent eigenvectors, x_k, $1 \leq k \leq n$.*

2. *The eigenvalues λ_k satisfy*

$$|\lambda_1| > |\lambda_2| \geq |\lambda_3| \geq \cdots \geq |\lambda_n|.$$

3. *The vector $z^{(0)} \in \mathbf{R}^n$ is such that*

$$z^{(0)} = \sum_{k=1}^{n} \xi_k x_k$$

and $\xi_1 \neq 0$.

Then the basic power method converges in the sense that

$$\lim_{k \to \infty} \mu_k = \lambda_1, \quad |\lambda_1 - \mu_k| = O\left(\left(\frac{\lambda_2}{\lambda_1}\right)^k\right),$$

and there exists $c \neq 0$ such that

$$\lim_{k \to \infty} z^{(k)} = cx_1, \quad \|cx_1 - z^{(k)}\|_\infty = O\left(\left(\frac{\lambda_2}{\lambda_1}\right)^k\right).$$

Proof   An inductive argument (see Exercise 2) can be used to show that $z^{(k)} = c_k A^k z^{(0)}$, where c_k depends on all the scaling factors μ_k. We then expand in terms of the eigenvector basis, as we did before, to get

$$z^{(k)} = c_k \lambda_1^k \left(\xi_1 x_1 + \sum_{i=2}^{n} \xi_i \theta_i^k x_i \right),$$

where $\theta_i = \lambda_i / \lambda_1$ and $|\theta_i| < 1$ for all $i \geq 2$. Since each $z^{(k)}$ has infinity norm equal to 1, it follows that

$$\lim_{k \to \infty} c_k \lambda_1^k = \frac{1}{\xi_1 \|x_1\|_\infty} < \infty,$$

so that

$$\lim_{k \to \infty} z^{(k)} = \frac{1}{\|x_1\|_\infty} x_1.$$

Moreover, we clearly have that

$$\|z^{(k)} - \alpha_k x_1\|_\infty = \left\| c_k \lambda_1^k \sum_{i=2}^{n} \xi_i \theta_i^k x_i \right\|_\infty \leq C \theta_2^k,$$

which establishes the error estimate for the eigenvector.

As for the eigenvalue, we have

$$\mu_k z^{(k)} = y^{(k)} = A z^{(k-1)},$$

so that, taking dot products of both sides with $z^{(k)}$, we get

$$\mu_k = \frac{(z^{(k)}, A z^{(k-1)})}{(z^{(k)}, z^{(k)})} \quad \to \quad \frac{(x_1, A x_1)}{(x_1, x_1)} = \lambda_1.$$

The error estimate comes by manipulating with the ratio as was done in the proof of Theorem 8.4. See Exercise 4.   □

The power method requires several hypotheses for the theory to apply, and it is worth asking if all of them are necessary. What happens if some of them are violated?

Recall that if A does not have n linearly independent eigenvectors, then it is said to be *defective;* eigenvalue computations with defective matrices are unavoidably more difficult. However, just as all singular matrices are close to a nonsingular matrix, it is the case that all defective matrices are arbitrarily close to a nondefective matrix. Thus, applying the power method to a defective matrix usually results in, at worst, slow convergence (possibly *very* slow, if the dependent eigenvectors correspond to the two largest eigenvalues, λ_1 and λ_2).

If the two largest (in absolute value) eigenvalues are not separated, i.e., if $|\lambda_1| = |\lambda_2|$, then the iteration will cycle in to a family of vectors in the subspace spanned by x_1 and x_2.

The last hypothesis is actually the least important. Except in very simple situations (involving integer components), the rounding error associated with the computations will introduce some component in the direction of x_1 into the initial guess, and the iteration will then begin to converge. Making an initial guess that has absolutely no component in the direction of x_1 could result in convergence to the wrong eigenpair, however. For this reason, random, noninteger initial guesses are best.

Let's look at some examples.

EXAMPLE 8.5

Consider

$$A = \begin{bmatrix} 3 & 0 & 0 \\ -4 & 6 & 2 \\ 16 & -15 & -5 \end{bmatrix},$$

which has exact eigenvalues $\lambda_i = 3, 1, 0$, $i = 1, 2, 3$, and corresponding eigenvectors

$$x_1 = \begin{bmatrix} 1 \\ 0 \\ 2 \end{bmatrix}, \quad x_2 = \begin{bmatrix} 0 \\ 2 \\ -5 \end{bmatrix}, \quad x_3 = \begin{bmatrix} 0 \\ 1 \\ -3 \end{bmatrix}.$$

Note that the eigenvalues are all distinct; therefore, the eigenvectors are independent. We take the initial guess as (from a random number generator)

$$z^{(0)} = \begin{bmatrix} 0.21895918632809 \\ 0.04704461621449 \\ 0.67886471686832 \end{bmatrix}.$$

Then the first two iterations go as follows:

$$y^{(1)} = Az^{(0)} = \begin{bmatrix} 0.65687755898427 \\ 0.76416038571122 \\ -0.59664584630951 \end{bmatrix}, \ \mu_1 = 0.76416038571122,$$

$$z^{(1)} = y^{(1)}/\mu_1 = \begin{bmatrix} 0.85960692449779 \\ 1.00000000000000 \\ -0.78078615100441 \end{bmatrix};$$

$$y^{(2)} = Az^{(1)} = \begin{bmatrix} 2.57882077349338 \\ 1.00000000000000 \\ 2.65764154698677 \end{bmatrix}, \ \mu_2 = 2.65764154698677,$$

$$z^{(2)} = y^{(2)}/\mu_2 = \begin{bmatrix} 0.97034183425423 \\ 0.37627346740338 \\ 1.00000000000000 \end{bmatrix}.$$

If we continue the computation (letting a computer do the work), then we get convergence (using the difference in consecutive eigenvector approximations as our convergence criterion, with a tolerance of 10^{-6}) to $\lambda_1 = 3$ in 14 iterations. The approximate eigenvector is found to be

$$x_1 = \begin{bmatrix} 0.50000045604132 \\ 0.00000036483306 \\ 1.00000000000000 \end{bmatrix}.$$

It should be noted that we did not bother to put A in Hessenberg form for this example. ∎

EXAMPLE 8.6

We take as our second example the matrix

$$A = \begin{bmatrix} 4 & 1 & 0 \\ 1 & 4 & 1 \\ 0 & 1 & 4 \end{bmatrix},$$

whose eigenvalues are $\lambda_i = 5.414214, 4.0, 2.585786$, which are much closer together than was the case in the previous example; thus we expect slower convergence. With the initial guess

$$z^{(0)} = \begin{bmatrix} 0.67929640583661 \\ 0.93469289594083 \\ 0.38350207748986 \end{bmatrix}$$

(again, randomly generated), the first two iterations are

$$y^{(1)} = Az^{(0)} = \begin{bmatrix} 3.65187851928727 \\ 4.80157006708979 \\ 2.46870120590027 \end{bmatrix}, \quad \mu_1 = 4.80157006708979,$$

$$z^{(1)} = y^{(1)}/\mu_1 = \begin{bmatrix} 0.76055924796713 \\ 1.00000000000000 \\ 0.51414457592130 \end{bmatrix};$$

$$y^{(2)} = Az^{(1)} = \begin{bmatrix} 4.04223699186850 \\ 5.27470382388842 \\ 3.05657830368519 \end{bmatrix}, \quad \mu_2 = 5.27470382388842,$$

$$z^{(2)} = y^{(2)}/\mu_2 = \begin{bmatrix} 0.76634387954860 \\ 1.00000000000000 \\ 0.57947865998511 \end{bmatrix}.$$

If we continue the calculation we converge—using the same criterion as before—to $\lambda_1 = 5.414214$ in 37 iterations. ∎

EXAMPLE 8.7

We now want to consider two defective matrices:

$$A = \begin{bmatrix} -2 & 4 & -1 \\ -3 & 5 & -1 \\ -1 & 1 & 1 \end{bmatrix},$$

which has eigenvalues $\lambda_i = 2, 1, 1$, and

$$B = \begin{bmatrix} 5 & 4 & 3 \\ -1 & 0 & -3 \\ 1 & -2 & 1 \end{bmatrix},$$

which has eigenvalues $\lambda_i = 4, 4, -2$. When we apply the power method to A, we find that convergence occurs in about 23 iterations, using the initial guess

$$z^{(0)} = \begin{bmatrix} 0.51941637206795 \\ 0.83096534611237 \\ 0.03457211052746 \end{bmatrix}.$$

However, if we apply the power method to B, then we find that after 80 iterations the approximate eigenvalue is $\mu_k = 4.04888750871386$, which is not a good approximation to $\lambda_1 = 4$. The reason for the difference in performance is that for this matrix, it is the eigenvalue we are trying to find that is repeated, whereas for A the dominant eigenvalue $\lambda_1 = 2$ was not repeated. Note also that the iteration for the largest eigenvalue of B *was* converging, just very slowly. (The error in the eigenvector was going down by about 0.00001 each iteration.)   ■

EXAMPLE 8.8

Finally, consider trying to find the largest eigenvalue of

$$A = \begin{bmatrix} 3 & 0 & 0 \\ -4 & 6 & 2 \\ 16 & -15 & -5 \end{bmatrix},$$

which has eigenvalues $\lambda = 3,1,0$, using the initial guess

$$z^{(0)} = \begin{bmatrix} 0 \\ -0.375 \\ 1 \end{bmatrix}.$$

The exact eigenvectors are

$$x_1 = \begin{bmatrix} 1 \\ 0 \\ 2 \end{bmatrix}, \; x_2 = \begin{bmatrix} 0 \\ -2 \\ 5 \end{bmatrix}, \; x_1 = \begin{bmatrix} 0 \\ -1 \\ 3 \end{bmatrix},$$

and $z^{(0)}$ has been chosen to have no component in the direction of the dominant eigenvector, $x_1 = (1,0,2)^T$. If we use the power method, we converge to a multiple of the eigenvector x_2 in 6 iterations. By making a random choice of the initial vector, say

$$z^{(0)} = \begin{bmatrix} 0.6789 \\ 0.6793 \\ 0.9347 \end{bmatrix},$$

we minimize the chance of having absolutely no component in the x_1 direction. For this choice of initial vector, the power method converges to the correct eigenvector in 15 iterations.   ■

There are many variants to the basic power method. We first observe that if we use the power method on A^{-1}, then we get an algorithm that converges to the *smallest* eigenvalue of A (see Algorithm 8.3).

The theorem is now the following.

Theorem 8.6 (Inverse Power Method)   *Let $A \in \mathbf{R}^{n \times n}$ be given, nonsingular, and assume the following three conditions:*

1. *A has n linearly independent eigenvectors, x_k, $1 \le k \le n$.*

2. *The eigenvalues λ_k satisfy*

$$|\lambda_1| \ge |\lambda_2| \ge |\lambda_3| \ge \cdots \ge |\lambda_{n-1}| > |\lambda_n|.$$

Algorithm 8.3: Inverse Power Method

For k from 1 until convergence, do

   (a) Solve $Ay^{(k)} = z^{(k-1)}$.

   (b) $\mu_k = y_i^{(k)}$, where $\|y^{(k)}\|_\infty = |y_i^{(k)}|$.

   (c) Set $z^{(k)} = y^{(k)}/\mu_k$.

---

3. *The vector $z^{(0)} \in \mathbf{R}^n$ is such that*

$$z^{(0)} = \sum_{k=1}^{n} \xi_k x_k$$

and $\xi_n \neq 0$.

Then the inverse power method converges in the sense that

$$\lim_{k\to\infty} \mu_k = \lambda_n^{-1}, \quad |\lambda_n^{-1} - \mu_k| = O\left(\left(\frac{\lambda_n}{\lambda_{n-1}}\right)^k\right),$$

and there exists $c \neq 0$ such that

$$\lim_{k\to\infty} z^{(k)} = cx_n, \quad \|cx_n - z^{(k)}\|_\infty = O\left(\left(\frac{\lambda_n}{\lambda_{n-1}}\right)^k\right).$$

Proof   This is equivalent to applying the previous theorem to A^{-1}, so everything follows, with the largest (in absolute value) eigenvalue of A^{-1} being λ_n^{-1}.   □

Of course, we do not implement the inverse power method by constructing A^{-1}; rather, we compute the LU factorization of A and use it to solve the linear system defined in step 1 of the algorithm.

A single example at this point should illustrate.

EXAMPLE 8.9

Consider

$$A = \begin{bmatrix} 4 & 1 & 0 \\ 1 & 4 & 1 \\ 0 & 1 & 4 \end{bmatrix},$$

one of our previous example matrices. We factor this once to obtain

$$A = LU = \begin{bmatrix} 1 & 0 & 0 \\ -0.25 & 1 & 0 \\ 0 & -0.26666... & 1 \end{bmatrix} \begin{bmatrix} 4 & 1 & 0 \\ 0 & 3.75 & 1 \\ 0 & 0 & 3.7333... \end{bmatrix}$$

and then use this factorization to solve the systems in step 1 of the algorithm. Using the initial guess

$$z^{(0)} = \begin{bmatrix} 0.05346163504453 \\ 0.52970019333516 \\ 0.67114938407724 \end{bmatrix},$$

the first two iterations now look like this:

$$y^{(1)} = A^{-1} z^{(0)} = \begin{bmatrix} -0.01153083684992 \\ 0.09958498244420 \\ 0.14289110040826 \end{bmatrix}, \quad \mu_1 = 0.14289110040826,$$

$$z^{(1)} = y^{(1)} / \mu_1 = \begin{bmatrix} -0.08069667611890 \\ 0.69692921504333 \\ 1.00000000000000 \end{bmatrix};$$

$$y^{(2)} = A^{-1} z^{(1)} = \begin{bmatrix} -0.05353869646351 \\ 0.13345810973516 \\ 0.21663547256621 \end{bmatrix}, \quad \mu_2 = 0.21663547256621,$$

$$z^{(2)} = y^{(2)} / \mu_2 = \begin{bmatrix} -0.24713725702125 \\ 0.61604920078067 \\ 1.00000000000000 \end{bmatrix}.$$

If we continue, we get convergence to within 10^{-6} of x_3 in 39 iterations. The approximate eigenvalue is $\lambda_3 = 2.58578643767867$, correct to 11 places. ∎

It is worth repeating that the inverse power method is *not* implemented by actually multiplying by the inverse matrix. Rather, the system $A y^{(k+1)} = z^{(k)}$ is solved at each step, using, typically, an *LU* decomposition.

It is the inverse power method that opens up the possibility of efficient computation of more eigenvalues, because by introducing *shifts* we can converge to almost any eigenvalue we want. Moreover, we can increase the rate at which the method converges (see Algorithm 8.4).

The theorem now becomes the following.

Theorem 8.7 (Shifted Inverse Power Method) *Let $A \in \mathbf{R}^{n \times n}$ and a scalar λ_* be given, with λ_* not an eigenvalue of A. Define $A_* = A - \lambda_* I$, and assume the three conditions*

1. A has n linearly independent eigenvectors, x_k, $1 \le k \le n$.

2. There exists an index J such that λ_J is the eigenvalue of A that is strictly closest

Algorithm 8.4: Shifted Inverse Power Method

Let λ_* be given, with λ_* not an eigenvalue of A; define $A_* = A - \lambda_* I$. For k from 1 until convergence, do

(a) Solve $A_* y^{(k)} = z^{(k-1)}$.

(b) $\mu_k = y_i^{(k)}$, where $\|y^{(k)}\|_\infty = |y_i^{(k)}|$.

(c) Set $z^{(k)} = y^{(k)}/\mu_k$.

---

to λ_*:

$$|\lambda_J - \lambda_*| < |\lambda_k - \lambda_*|$$

for $k \neq J$.

3. *The vector $z^{(0)} \in \mathbf{R}^n$ is such that*

$$z^{(0)} = \sum_{k=1}^n \xi_k x_k$$

and $\xi_J \neq 0$.
Then the shifted inverse power method converges in the sense that

$$\lim_{k \to \infty} \mu_k = (\lambda_J - \lambda_*)^{-1}, \quad |(\lambda_J - \lambda_*)^{-1} - \mu_k| = O\left(\left(\frac{\lambda_J - \lambda_*}{\lambda_K - \lambda_*}\right)^k\right),$$

where K is the index of the eigenvalue that is second *closest to λ_*; and there exists $c \neq 0$ such that*

$$\lim_{k \to \infty} z^{(k)} = cx_J, \quad \|cx_J - z^{(k)}\|_\infty = O\left(\left(\frac{\lambda_J - \lambda_*}{\lambda_K - \lambda_*}\right)^k\right).$$

Proof   See Exercise 5.   □

The point of using shifts is that it allows us to control which eigenvector we converge to, as well as how fast we converge. Recall that the inverse power iteration converges to the eigenvector corresponding to the *smallest* eigenvalue. Using a shift λ_* means that, for some index J, $\lambda_J - \lambda_*$ is the smallest eigenvalue; thus we converge to the eigenvector corresponding to the eigenvalue *closest to* the shift. Furthermore, the convergence rate is affected by how close to an actual eigenvalue the shift is; taking $\lambda_* \approx \lambda_J$ will generally result in much faster convergence.

EXAMPLE 8.10

To illustrate the utility of shifts, we again look at

$$A = \begin{bmatrix} 4 & 1 & 0 \\ 1 & 4 & 1 \\ 0 & 1 & 4 \end{bmatrix}.$$

We now apply the shifted inverse power method, using the same initial guess as before. This time we use a shift of $\lambda_* = 2.5$, chosen because we know (from Gerschgorin's Theorem) that the smallest eigenvalue is at least as large as 2. From Example 8.9 the approximate eigenvalue, after two iterations, was $1/\mu_2 = 4.61604920078069$, which is not a very good approximation to the correct value of 2.58578643762691. This time the first two iterations (using the same initial vector as for the unshifted iteration) are

$$
y^{(1)} = (A - 2.5I)^{-1}z^{(0)} = \begin{bmatrix} 1.45298541901634 \\ -1.64486177807920 \\ 1.54400744143763 \end{bmatrix}, \quad \mu_1 = -1.64486177807920,
$$

$$
z^{(1)} = y^{(1)}/\mu_1 = \begin{bmatrix} -0.88334803469813 \\ 1.00000000000000 \\ -0.93868522085829 \end{bmatrix};
$$

$$
y^{(2)} = (A - 2I)^{-1}z^{(1)} = \begin{bmatrix} -9.44765403794921 \\ 13.28813302222568 \\ -9.48454549538931 \end{bmatrix}, \quad \mu_2 = 13.28813302222568,
$$

$$
z^{(2)} = y^{(2)}/\mu_2 = \begin{bmatrix} -0.71098430623377 \\ 1.00000000000000 \\ -0.71376057716502 \end{bmatrix}.
$$

The approximate eigenvalue is now

$$
\lambda \approx \lambda_* + \frac{1}{\mu_2} = 2.57525511660121,
$$

which is substantially better than we achieved in two iterations with the unshifted method. ∎

The performance of the power methods can be improved somewhat when A is symmetric by the use of the *Rayleigh quotient* as the approximate eigenvalue.

Definition 8.2 (Rayleigh Quotient) Given $A \in \mathbf{R}^{n \times n}$ and $x \in \mathbf{R}^n$, $x \neq 0$, the Rayleigh quotient *of A and x is defined by*

$$
R(A,x) = \frac{(x,Ax)}{(x,x)}.
$$

Note that if x is an eigenvector of A, then the Rayleigh quotient is the corresponding eigenvalue. If we define the approximate eigenvalue in the power methods via the Rayleigh quotient, i.e.,

$$
\mu_k = R(A,y^{(k)}),
$$

then the iteration converges to the correct eigenvalue *faster* than before, if A is symmetric. The reason for this is contained in the following theorem.

Theorem 8.8   *Let λ be an eigenvalue for the symmetric matrix A, with x the corresponding eigenvector. If $z \approx x$, $\mu = R(A,z)$, and $\|x\|_2 = \|z\|_2 = 1$, then*

$$|\lambda - \mu| \le C_A \|x - z\|_2^2.$$

Proof   We start by defining the error $e = x - z$, and we note that

$$x - e = z \;\Rightarrow\; \|x - e\|_2^2 = 1 \;\Rightarrow\; 1 - 2(x,e) + \|e\|_2^2 = 1,$$

so that $2(x,e) = \|e\|_2^2$. Then we have

$$\mu = R(A,z) = (x - e, A(x - e)) = (x, Ax) - (x, Ae) - (e, Ax) + (e, Ae) = \lambda - 2(e, Ax) + (e, Ae),$$

where we used the symmetry of A to write $(x, Ae) = (A^T x, e) = (Ax, e)$. Thus we have

$$\mu = \lambda - 2\lambda(e,x) + (e, Ae) = \lambda - \lambda(e,e) + (e, Ae) = \lambda + (e, (A - \lambda I)e),$$

so that

$$|\mu - \lambda| = |(e, (A - \lambda I)e)| \le \|A - \lambda I\|_2 \|e\|_2^2 = C_A \|e\|_2^2.$$

$\square$

Thus, if we write a power iteration for a symmetric matrix that uses the Rayleigh quotient to compute the approximate eigenvalue (and that scales the iteration so that $\|z^{(k)}\|_2 = 1$), then the eigenvalue error will be the square of the eigenvector error.

EXAMPLE 8.11

In Example 8.6 we considered the matrix

$$A = \begin{bmatrix} 4 & 1 & 0 \\ 1 & 4 & 1 \\ 0 & 1 & 4 \end{bmatrix},$$

which has eigenvalues $\lambda_i = 5.414214, 4.0, 2.585786$. With the initial guess

$$z^{(0)} = \begin{bmatrix} 0.67929640583661 \\ 0.93469289594083 \\ 0.38350207748986 \end{bmatrix},$$

we got the second approximate eigenvalue as $\mu_2 = 5.27470382388842$. If we use the same initial guess, but this time compute the approximate eigenvalue using the Rayleigh quotient while scaling the eigenvectors so that $\|z^{(k)}\|_2 = 1$, we get

$$y^{(1)} = Az^{(0)} = \begin{bmatrix} 3.65187851928727 \\ 4.80157006708979 \\ 2.46870120590027 \end{bmatrix},$$

so that the approximate eigenvalue is

$$r_1 = \frac{(z^{(0)}, Az^{(0)})}{(z^{(0)}, z^{(0)})} = \frac{(z^{(0)}, y^{(0)})}{(z^{(0)}, z^{(0)})} = 5.34045535705311.$$

The scaled approximate eigenvector is

$$z^{(1)} = \frac{y^{(1)}}{\|y^{(1)}\|_2} = \begin{bmatrix} 0.56026634732951 \\ 0.73665049610143 \\ 0.37874485692028 \end{bmatrix},$$

so for the second step we get

$$y^{(2)} = Az^{(1)} = \begin{bmatrix} 2.97771588541948 \\ 3.88561318865551 \\ 2.25162992378256 \end{bmatrix}$$

$$r_2 = \frac{(z^{(1)}, Az^{(1)})}{(z^{(1)}, z^{(1)})} = (z^{(1)}, y^{(2)}) = 5.38344613891082$$

and

$$z^{(2)} = \frac{y^{(2)}}{\|y^{(2)}\|_2} = \begin{bmatrix} 0.55261820977325 \\ 0.72110996710610 \\ 0.41786783744055 \end{bmatrix}.$$

Note that the approximate eigenvalue, r_2, is substantially more accurate than the second approximate eigenvalue obtained by the ordinary power iteration. ∎

Algorithm 8.5: Rayleigh Quotient Iteration

Given $A \in \mathbf{R}^{n \times n}$ and an initial guess $z^{(0)} \in \mathbf{R}^n$, compute as follows: For $k = 0$ to convergence, do

1. Define $\mu_k = R(A, z^{(k)})$.

2. Solve $(A - \mu_k I) y^{(k)} = z^{(k)}$.

3. Set $z^{(k+1)} = y^{(k)} / \|y^{(k)}\|_2$.

The Rayleigh quotient can be combined with the shifted inverse power iteration to produce a useful, if somewhat erratic, iteration called the *Rayleigh quotient iteration*. The basic idea is to update the shift at each step using the most recently computed approximate eigenvalue (see Algorithm 8.5).

Because we are updating the shift at each iteration, we have to do a new factorization of A, and this can be expensive. However, if A is symmetric, the Hessenberg reduction means we can assume that A is tridiagonal, and thus this is not a serious additional cost. Moreover, the rapid convergence that occurs because the shift keeps getting closer and closer to the correct eigenvalue makes the iteration very fast. The problem with the Rayleigh quotient iteration is that its performance is much akin to Newton's method; sometimes we have to start the iteration very close to the correct eigenvalue in order to converge to it. Note also that step 2 involves solving a linear system for a matrix that is getting closer and closer to singular. (Why is this?) This requires that some care be taken in doing the factorization; typically, the shift is no longer updated once it is a sufficiently accurate approximation to the eigenvalue.

EXAMPLE 8.12

Let's continue with the simple 3×3 example matrix that we have been using,

$$A = \begin{bmatrix} 4 & 1 & 0 \\ 1 & 4 & 1 \\ 0 & 1 & 4 \end{bmatrix},$$

which we note is symmetric. Using the same initial guess as in Examples 8.9 and 8.10, the first two iterations of Rayleigh quotient iteration are

$$\mu_0 = R(A, z^{(0)}) = 5.04601728854858, \quad y^{(1)} = (A - \mu_0 I)^{-1} z^{(0)} = \begin{bmatrix} 1.29838130282511 \\ 1.41159092492783 \\ 0.70786740234284 \end{bmatrix},$$

$$z^{(1)} = y^{(1)} / \|y^{(1)}\|_2 = \begin{bmatrix} 0.63510013864028 \\ 0.69047635711817 \\ 0.34625166304280 \end{bmatrix};$$

$$\mu_1 = R(A, z^{(1)}) = 5.35520043415499, \quad y^{(2)} = (A - 2I)^{-1} z^{(1)} = \begin{bmatrix} 8.18704621259896 \\ 11.73018872040136 \\ 8.40018699112711 \end{bmatrix},$$

$$z^{(2)} = y^{(2)} / \|y^{(2)}\|_2 = \begin{bmatrix} 0.49352861955263 \\ 0.70711508106093 \\ 0.50637709645942 \end{bmatrix}.$$

The next approximate eigenvalue—the next Rayleigh quotient—is

$$\mu_2 = R(A, z^{(2)}) = 5.41409682286229,$$

which is very close to the exact largest eigenvalue, $\lambda_1 = 5.41421356237310$, in only two iterations. ∎

Deflation   Once we have found a single eigenvalue, how do we keep from "recomputing" it when we try to find other eigenvalues? The answer, in general, is a process called *deflation*, in which we create a new matrix from the old matrix by "deleting" (in some sense) the eigenvalue and eigenvector that we have already found.

Let λ and x be an eigenvalue and corresponding eigenvector for A, and let Q be an orthogonal matrix such that $Qx = e_1$. (Here we have assumed that $\|x\|_2 = 1$.) Then it is not hard to show (Exercise 10) that

$$B = QAQ^T = \begin{bmatrix} \lambda & a^T \\ 0 & A_2 \end{bmatrix},$$

where $A_2 \in \mathbf{R}^{(n-1) \times (n-1)}$. Similarity implies that B and A have the same eigenvalues; thus A_2 has all the eigenvalues of A *except* for λ.

We can use this to compute eigenvalues in the obvious way. Once we have converged to λ and x, we define the vector w such that

$$Q = I - \gamma w w^T$$

satisfies $Qx = e_1$. This we already know how to do, from Section 8.2. Then compute B and pull out the lower submatrix to continue the computation. Although problems can develop due to the propagation of errors (since, in practice, neither λ nor x will be exact), this can be made an accurate procedure for eigenvalue computation. Note, however, that the deflation will usually destroy the Hessenberg form of the matrix, so that process will have to be repeated.

We close this discussion of the power methods by showing the details of a worked example.

EXAMPLE 8.13

Consider the matrix

$$A = \begin{bmatrix} 10 & -2 & 3 & 2 & 0 \\ -2 & 10 & -3 & 4 & 5 \\ 3 & -3 & 6 & 3 & 3 \\ 2 & 4 & 3 & 6 & 6 \\ 0 & 5 & 3 & 6 & 13 \end{bmatrix},$$

which has the spectrum

$$\sigma(A) = \{0.4000, 2.8156, 7.0186, 14.1179, 20.6479\}.$$

We will use the power methods to find *all* the eigenvalues of this matrix. We will do it two ways: first, using the ordinary power method with deflation, and second, using Rayleigh quotient iteration with deflation. The point of the exercise is to show that the power methods *can* be used to find all the eigenvalues (and eigenvectors, which we choose not to do here) and also to provide a basis for understanding that the power methods are not the most efficient way to do so.

The first step is to reduce A to Hessenberg form, using the algorithm discussed earlier. We get

$$A_{H,0} = \begin{bmatrix} 10.0000 & -4.1231 & 0 & 0 & 0 \\ -4.1231 & 9.2941 & 3.0920 & 0 & 0 \\ 0 & 3.0920 & 6.4587 & -5.3189 & 0 \\ 0 & 0 & -5.3189 & 17.9989 & 3.1371 \\ 0 & 0 & 0 & 3.1371 & 1.2483 \end{bmatrix}.$$

We now use the power method to find a single eigenpair. This takes 76 iterations, using the random initial guess

$$z^{(0)} = \begin{bmatrix} 0.7703 \\ 1.0000 \\ 0.8373 \\ 0.2883 \\ 0.0521 \end{bmatrix}.$$

We get the pair

$$(\lambda_1, x_1) = (20.6479, (0.0494, -0.1276, -0.4027, 1.0000, 0.1617)^T)$$

and now deflate to get the smaller matrix

$$A_1 = \begin{bmatrix} 8.2460 & 1.2308 & 4.6223 & 0.7475 \\ 1.2308 & 5.1488 & -2.0659 & 0.5260 \\ 4.6223 & -2.0659 & 9.9202 & 1.8307 \\ 0.7475 & 0.5260 & 1.8307 & 1.0370 \end{bmatrix},$$

which we have to convert to Hessenberg form:

$$A_{H,1} = \begin{bmatrix} 8.2460 & 4.8414 & 0 & 0 \\ 4.8414 & 8.9781 & -3.0113 & 0 \\ 0 & -3.0113 & 5.9943 & 1.7071 \\ 0 & 0 & 1.7071 & 1.1336 \end{bmatrix}.$$

We do another power iteration, this time lasting 43 iterations, using the random initial vector

$$z^{(0)} = \begin{bmatrix} 0.9731 \\ 0.4339 \\ 0.8364 \\ 1.0000 \end{bmatrix}$$

and resulting in the eigenpair

$$(\lambda, x_2) = (14.1179, (0.8245, 1.0000, -0.3812, -0.0501)^T).$$

Note that this vector is in $\mathbf{R}^4$, not $\mathbf{R}^5$. This is because it is an eigenvector for the 4×4 matrix $A_{H,1}$, *not* our original A. Some additional work will have to be done to recover the correct eigenvector for the original matrix.

In any event, we deflate again to get

$$A_2 = \begin{bmatrix} 3.2896 & -1.6907 & 0.1736 \\ -1.6907 & 5.8141 & 1.6834 \\ 0.1736 & 1.6834 & 1.1305 \end{bmatrix},$$

which has the Hessenberg form

$$A_{H,2} = \begin{bmatrix} 3.2896 & -1.6996 & 0 \\ -1.6996 & 5.4231 & -2.1243 \\ 0 & -2.1243 & 1.5215 \end{bmatrix}.$$

Using the initial vector

$$z^{(0)} = \begin{bmatrix} 1.0000 \\ 0.3686 \\ 0.2493 \end{bmatrix},$$

we use the power method for 34 iterations, getting the eigenpair

$$(\lambda, x_3) = (7.0186, (-0.4558, 1.0000, -0.3864)^T).$$

Deflation yields the 2×2 matrix

$$A_3 = \begin{bmatrix} 1.9061 & -1.1704 \\ -1.1704 & 1.3095 \end{bmatrix},$$

whose eigenvalues are found directly from the quadratic formula.

Now let's look at the Rayleigh quotient iteration. To make a fair comparison, we will use the same initial vector,

$$z^{(0)} = \begin{bmatrix} 0.7703 \\ 1.0000 \\ 0.8373 \\ 0.2883 \\ 0.0521 \end{bmatrix}.$$

The Hessenberg form is, of course, also the same. We now converge in only *three* iterations, to the eigenpair

$$(\lambda_3, x_3) = (7.0186, (-0.6578, -0.4756, -0.5271, -0.2210, -0.1201)^T).$$

We can deflate to get the new matrix

$$A_1 = \begin{bmatrix} 3.5922 & 3.1235 & 0.0132 & 0.0072 \\ 3.1235 & 13.5312 & -2.3536 & 1.6121 \\ 0.0132 & -2.3536 & 19.2422 & 3.8130 \\ 0.0072 & 1.6121 & 3.8130 & 1.6158 \end{bmatrix},$$

which has to be reduced to Hessenberg form:

$$A_{H,1} = \begin{bmatrix} 3.5922 & 3.1235 & 0 & 0 \\ 3.1235 & 13.5188 & 2.8192 & 0 \\ 0 & 2.8192 & 10.0064 & -9.6050 \\ 0 & 0 & -9.6050 & 10.8640 \end{bmatrix}.$$

Then with the initial vector

$$z^{(0)} = \begin{bmatrix} 0.3835 \\ 0.5194 \\ 0.8310 \\ 0.0346 \end{bmatrix}$$

we again take only three iterations to converge to the eigenpair

$$(\lambda_2, x_2) = (14.1179, (-0.2687, -0.9056, 0.1053, -0.3109)^T).$$

We again deflate and reconstruct the Hessenberg form, getting

$$A_2 = \begin{bmatrix} 5.1137 & 4.2466 & -4.2133 \\ 4.2466 & 9.7881 & -8.9606 \\ -4.2133 & -8.9606 & 8.9617 \end{bmatrix}$$

and

$$A_{H,2} = \begin{bmatrix} 5.1137 & 5.9821 & 0 \\ 5.9821 & 18.3385 & -0.3427 \\ 0 & -0.3427 & 0.4113 \end{bmatrix}.$$

Choose a random initial vector

$$z^{(0)} = \begin{bmatrix} 0.0535 \\ 0.5297 \\ 0.6711 \end{bmatrix}$$

and iterate, in this case five times, to get the eigenpair

$$(\lambda_4, x_4) = (2.8156, (-0.9323, 0.3581, -0.0511)^T).$$

We then deflate one last time to get

$$A_3 = \begin{bmatrix} 19.3308 & -4.9934 \\ -4.9934 & 1.7171 \end{bmatrix},$$

whose eigenvalues we can find from the quadratic formula.

The cost here is substantially less than for the ordinary power method, largely because the Rayleigh quotient iterations converged so much more quickly than the ordinary power iterations did. ■

The power methods are very useful for finding selected eigenvalues and corresponding eigenvectors, but to find all the eigenvalues and eigenvectors of a general matrix it is more efficient to make use of methods that work on all of the eigenvalues at once. In the next section we give a survey of the most commonly used such method, the QR iteration.

EXERCISES

1. Explain, in your own words, why it is necessary to scale the power iterations. (Hint: What happens if we don't scale the iteration?)

2. Use an inductive argument to show that, in the basic power method (Theorem 8.5), we have

$$z^{(k)} = c_k A^k z^{(0)}.$$

What is c_k?

3. Consider the iteration defined by

$$y^{(k+1)} = Az^{(k)}$$

$$z^{(k+1)} = y^{(k+1)}/\sigma_{k+1},$$

where σ_{k+1} is some scaling parameter. Assume that the iteration converges in the sense that $z^{(k)} \to z$; prove that σ_k must also converge, and that therefore (σ, z), where σ is the limit of the σ_k, must be an eigenpair of A.

4. Fill in the details of the eigenvalue error estimate in Theorem 8.5. In particular, show that there is a constant C such that

$$|\lambda_1 - \mu_k| \le C \left(\frac{\lambda_2}{\lambda_1} \right)^k.$$

5. Prove Theorem 8.7.

6. Write a program that executes the basic power method. Use it to produce a plot of the largest eigenvalue of the T_n family of matrices, as a function of n, over the range $2 \le n \le 20$.

7. Write a program that executes the inverse power method. Use it to produce a plot of the smallest eigenvalue of the H_n family of matrices, as a function of n, over the range $2 \le n \le 20$.

8. Write a program that does inverse power iteration on a symmetric matrix. Assume the matrix is tridiagonal, and store only the necessary nonzero elements of the matrix. Test it on the T_n and K_n families, and on your results from finding the Hessenberg forms for the H_n matrices. Produce a plot of the smallest eigenvalues of each of these matrices, as a function of n.

9. Add shifts to your inverse power iteration program. Test it on the same examples, using a selection of shifts.

10. Let λ and x be an eigenvalue and corresponding eigenvector for $A \in \mathbf{R}^{n \times n}$, and let Q be an orthogonal matrix such that $Qx = e_1$. Assume that $\|x\|_2 = 1$. Show that

$$B = QAQ^T = \begin{bmatrix} \lambda & a^T \\ 0 & A_2 \end{bmatrix}.$$

(Hint: Consider each of the inner products (e_i, Be_j), using the fact that $Qx = e_1$ and that Q is symmetric.)

11. An alternate deflation for symmetric matrices is based on the relationship

$$A' = A - \lambda_1 x_1 x_1^T.$$

If $A \in \mathbf{R}^{n \times n}$ is symmetric, $\lambda_1 \in \sigma(A)$ with corresponding eigenvector x_1, and if $\|x_1\|_2 = 1$, show that A' has the same eigenvalues as A, except that λ_1 has been replaced by zero. In other words, if

$$\sigma(A) = \{\lambda_1, \lambda_2, \dots, \lambda_n\},$$

then

$$\sigma(A') = \{0, \lambda_2, \dots, \lambda_n\}.$$

12. Show how to implement the deflation from the previous problem (known as Hotelling's deflation) without forming the product $x_1 x_1^T$. (Hint: What is $A'z$ for any vector z? Can we write this in terms of the dot product $x_1^T z$?) (According to Wilkinson [6], this deflation is prone to excessive rounding error and so is not of practical value.)

13. What is the operation count for one iteration of the basic power method applied to a symmetric matrix? Compute two values, one for the matrix in Hessenberg form and one for the full matrix.

14. What is the cost of the basic power method for a symmetric matrix if it runs for N iterations? Again, compute two values depending on whether or not a Hessenberg reduction is done. Is it always the case that a reduction to Hessenberg form is cost-effective?

15. Repeat Exercises 13 and 14 for the inverse power method.

16. Consider the family of matrices T defined by

$$t_{ij} = \begin{cases} a & i - j = 1 \\ b & i = j \\ c & i - j = -1 \\ 0 & \text{otherwise.} \end{cases}$$

Thus T is tridiagonal, with constant diagonal elements a, b, and c. It is commonplace to abbreviate such a matrix definition as

$$T = \text{tridiag}(a,b,c).$$

Write a computer program that uses the power method and inverse power method to find the largest and smallest (in absolute value) eigenvalues of such a matrix, for any user-specified values of a, b, and c. Test the program on the following examples; note that the exact values are given (to four places).

(a) $a = c = 4$, $b = 1$, $n = 6$, $\lambda_1 = -6.2708$, $\lambda_6 = $
-0.7802

(b) $a = b = c = 6$, $n = 7$, $\lambda_1 = 17.0866$, $\lambda_6 = 1.4078$

(c) $a = 2$, $b = 3$, $c = 4$, $n = 5$, $\lambda_1 = 7.8990$, $\lambda_6 = 0.1716$

(d) $a = c = 1$, $b = 10$, $n = 9$, $\lambda_1 = 11.9021$, $\lambda_6 = 8.0979$

(e) $a = c = 1$, $b = 4$, $n = 8$, $\lambda_1 = 5.8794$, $\lambda_6 = 2.1206$

17. Modify the program from the previous exercise to use the inverse power method with shifts. Test it on the same examples by finding the eigenvalue closest to $\mu = 1$.

8.4   AN OVERVIEW OF THE *QR* ITERATION

The power methods are very useful and efficient for finding single eigenvalues and eigenvectors; for finding many—or all—of the eigenvalues and vectors of a matrix, they are simply too costly. Since it was first published in the early 1960s, the standard algorithm for computing all the eigenvalues and eigenvectors of a general matrix has been the *QR* iteration, so named because it is heavily based on the *QR* factorization of a matrix.

It is an unfortunate fact—and no less true for being unfortunate—that motivating and explaining the *QR* iteration for computing eigenvalues is *very* difficult, especially in comparison with the simplicity of the algorithm. To give a complete explanation of why the method works or of how and why it might have evolved from earlier methods would require more background and more time than we want to devote to the subject. We will, however, make an attempt to provide some explanation.

One of the primary drawbacks to the power methods is that they work on a single vector at a time; if we want all the eigenpairs of a matrix, then we spend a lot of computation time in deflation and rereducing the new matrices to Hessenberg form. Why not do the power method on several vectors at once? Let Z_0 be an initial guess matrix, and do the computation

$$Y_{k+1} = AZ_k, \quad Z_{k+1} = Y_{k+1}D_{k+1},$$ (8.3)

where D_{k+1} is some properly chosen diagonal scaling matrix. The problem is that this won't work; the repeated multiplications by A will simply cause *all* of the columns of Y_{k+1} to line up in the direction of the dominant eigenvalue, and so we will converge to n copies of the dominant eigenpair—hardly an improvement on the power methods.

This algorithm does have something to recommend it, though. If the Z_k matrices are all *orthogonal* matrices, then we get something that might work, because the orthogonality of the columns of Z_k would prevent the columns of the Y_{k+1} from all lining up in the same direction. But how can we guarantee this?

An important tool in computational linear algebra, used in solving a number of important applied problems, is the *QR* factorization, in which we factor an arbitrary square matrix[5] A into the product of an orthogonal matrix Q and an upper triangular

---

[5]In fact, A does not even have to be square, and in many important applications it isn't. We choose to look at only the square case for simplicity.

matrix R.

Theorem 8.9   *Given a matrix $A \in \mathbf{R}^{n \times n}$, there exists an orthogonal matrix Q and an upper triangular matrix R such that $A = QR$.*

Proof   Let a_1 be the first column of A, and let Q_1 be the orthogonal matrix such that $Qa_1 = \|a_1\|_2 e_1$. Then

$$A_2 = Q_1 A = \begin{bmatrix} \|a_1\|_2 & b^T \\ 0 & A_{22} \end{bmatrix}.$$

Now let a_2 be the first column of A_{22}, and define Q_{22} as the orthogonal matrix such that $Q_{22}a_2 = \|a_2\|_2 e_1$. Define

$$Q_2 = \begin{bmatrix} 1 & 0 \\ 0 & Q_{22} \end{bmatrix}.$$

Then

$$A_3 = Q_2 A_2 = Q_2 Q_1 A,$$

and we continue on in this vein. After $n - 1$ steps, we have

$$A_{n-1} = Q_{n-1} Q_{n-2} \cdots Q_2 Q_1 A$$

and A_{n-1} is upper triangular. Thus,

$$R = Q_{n-1} Q_{n-2} \cdots Q_2 Q_1 A \quad \Rightarrow \quad A = QR$$

for

$$Q = Q_1^T Q_2^T \cdots Q_{n-1}^T. \qquad \qquad \square$$

Note: If A is nonsingular and we insist on the diagonal elements of R being positive, then the decomposition is unique. The QR factorization has a number of applications in other areas of computational mathematics, most notably in the solution of least squares problems and ill-posed problems.

EXAMPLE 8.14

Let

$$A = \begin{bmatrix} 4 & 1 & 0 \\ 1 & 4 & 1 \\ 0 & 1 & 4 \end{bmatrix}.$$

Then $a_1 = (4,1,0)^T$, and we use what we know about Householder matrices (Section 8.2) to find that

$$Q_1 = \begin{bmatrix} -0.9701 & -0.2425 & 0 \\ -0.2425 & 0.9701 & 0 \\ 0 & 0 & 1.0000 \end{bmatrix},$$

so that

$$A_1 = Q_1 A = \begin{bmatrix} -4.1231 & -1.9403 & -0.2425 \\ 0 & 3.6380 & 0.9701 \\ 0 & 1.0000 & 4.0000 \end{bmatrix}.$$

To continue on, we note that $a_2 = (3.6380, 1.0000)^T$, from which we get

$$Q_2 = \begin{bmatrix} 1.0000 & 0 & 0 \\ 0 & -0.9642 & -0.2650 \\ 0 & -0.2650 & 0.9642 \end{bmatrix}.$$

Thus

$$R = A_2 = Q_2 A_1 = Q_2 Q_1 A = \begin{bmatrix} -4.1231 & -1.9403 & -0.2425 \\ 0 & -3.7730 & -1.9956 \\ 0 & 0 & 3.5998 \end{bmatrix}$$

and

$$Q = Q_1^T Q_2^T = \begin{bmatrix} -0.9701 & 0.2339 & 0.0643 \\ -0.2425 & -0.9354 & -0.2571 \\ 0 & -0.2650 & 0.9642 \end{bmatrix}.$$

∎

Algorithm 8.6: Orthogonal Iteration

Given an initial orthogonal matrix, Q_0, for k from 1 until convergence, do

1. Set $Y_{k+1} = AQ_k$.

2. Define Q_{k+1} and R_{k+1} as the QR decomposition of Y_{k+1}: $Q_{k+1} R_{k+1} = Y_{k+1}$.

We use the QR decomposition to modify (8.3) in a slight but important fashion, producing what is known as orthogonal iteration (Algorithm 8.6).

One consequence of this algorithm is that we can write

$$R_{k+1} = Q_{k+1}^T A Q_k, \tag{8.4}$$

and it can be shown, under the proper conditions, that the diagonal elements of the triangular matrix R_{k+1} converge to the eigenvalues of A. Let's assume that the iteration does converge, and call the limiting upper triangular matrix R_∞ and the limiting orthogonal matrix Q_∞.

Algorithm 8.7: Basic QR Iteration

Given $A \in \mathbf{R}^{n \times n}$, set $A_0 = A$ and compute as follows: For k from 1 until convergence, do

1. Define Q_k and R_k as the *QR* decomposition of A_{k-1}: $Q_k R_k = A_{k-1}$.

2. Define $A_k = R_k Q_k$.

Let's look now at the slightly different iteration based on the computation

$$A_k = Q_k^T A Q_k, \tag{8.5}$$

where Q_k is exactly the same as in orthogonal iteration. Note that the A_k matrices are similar to A; thus they have the same eigenvalues. Moreover, if the orthogonal iteration converges, then so does this modified iteration (see Exercise 3), and it is not hard to show that $A_k \to R_\infty$. In addition, this is not much different from orthogonal iteration since it comes from merely shifting the index by 1 in the iteration (8.4). So the question arises, would it be possible to use the new iteration (8.5) instead of the one from orthogonal iteration (8.4)? If so, how could we efficiently organize the computation?

Note the following:

$$A_k = Q_k^T A Q_k = \left(Q_k^T A Q_{k-1} \right) Q_{k-1}^T Q_k = R_k \left(Q_{k-1}^T Q_k \right)$$

$$A_{k-1} = Q_{k-1}^T A Q_{k-1} = Q_{k-1}^T \left(A Q_{k-1} \right) = \left(Q_{k-1}^T Q_k \right) R_k.$$

Thus, since products of orthogonal matrices are themselves orthogonal (Exercise 1), there exists an orthogonal matrix $Q_k' = Q_{k-1}^T Q_k$ and an upper triangular matrix R_k such that

$$A_{k-1} = Q_k' R_k, \quad A_k = R_k Q_k'.$$

Thus, given A_{k-1}, we can get the next member of the sequence simply by doing a *QR* decomposition and then reverse-multiplying the two matrices. Now, suppose we start the iteration with the special orthogonal matrix $Q_0 = I$; then we have

$$A_0 = A, \quad A_1 = R_1 Q',$$

where $Q' R_1 = A$ is the *QR* decomposition of A. This is the final piece of the puzzle. We can do essentially the same computations as orthogonal iteration—which is an attempt to do the power methods on more than one vector at a time—simply by doing a *QR* decomposition on A and then reverse-multiplying the two factors. To be specific, we have Algorithm 8.7.

The important theorem is the one that tells us this is a sequence of similar matrices, and thus they all have the same eigenvalues.

Theorem 8.10   *Given $A \in \mathbf{R}^{n \times n}$, set $A_0 = A$ and compute the following sequence of matrices:*

$$A_{k+1} = R_k Q_k,$$

where $A_k = Q_k R_k$ is the QR factorization of A_k. Then the A_k matrices are all similar.

Proof We have

$$A_{k+1} = R_k Q_k,$$

so that

$$Q_k A_{k+1} = Q_k R_k Q_k = A_k Q_k \quad \Rightarrow \quad A_{k+1} = Q_k^{-1} A_k Q_k.$$

We know that the inverse exists since Q_k is orthogonal and hence $Q_k^{-1} = Q_k^T$. □

At this point it might be worth looking at an example.

EXAMPLE 8.15

Let

$$A = \begin{bmatrix} 10 & -2 & 3 & 2 & 0 \\ -2 & 10 & -3 & 4 & 5 \\ 3 & -3 & 6 & 3 & 3 \\ 2 & 4 & 3 & 6 & 6 \\ 0 & 5 & 3 & 6 & 13 \end{bmatrix},$$

which is the same matrix we looked at, extensively, at the end of the section on power methods. The Hessenberg form is

$$A_{H,0} = A_0 = \begin{bmatrix} 10.0000 & -4.1231 & 0 & 0 & 0 \\ -4.1231 & 9.2941 & 3.0920 & 0 & 0 \\ 0 & 3.0920 & 6.4587 & -5.3189 & 0 \\ 0 & 0 & -5.3189 & 17.9989 & 3.1371 \\ 0 & 0 & 0 & 3.1371 & 1.2483 \end{bmatrix}.$$

Since our discussion of the QR iteration is not going to get into implementation details, we will simply illustrate how the computation proceeds. The QR decomposition of A_0 gives us

$$Q = \begin{bmatrix} -0.9245 & -0.3488 & 0.1024 & 0.1072 & 0.0402 \\ 0.3812 & -0.8461 & 0.2485 & 0.2601 & 0.0974 \\ 0 & -0.4031 & -0.6102 & -0.6387 & -0.2393 \\ 0 & 0 & 0.7453 & -0.6244 & -0.2339 \\ 0 & 0 & 0 & -0.3508 & 0.9364 \end{bmatrix}$$

$$R = \begin{bmatrix} -10.8167 & 7.3546 & 1.1786 & 0 & 0 \\ 0 & -7.6715 & -5.2193 & 2.1438 & 0 \\ 0 & 0 & -7.1370 & 16.6596 & 2.3380 \\ 0 & 0 & 0 & -8.9417 & -2.3967 \\ 0 & 0 & 0 & 0 & 0.4351 \end{bmatrix},$$

so that the next matrix in the sequence is

$$A_1 = RQ = \begin{bmatrix} 12.8034 & -2.9242 & -0.0000 & -0.0000 & -0.0000 \\ -2.9242 & 8.5943 & 2.8766 & -0.0000 & -0.0000 \\ 0 & 2.8766 & 16.7709 & -6.6639 & 0 \\ 0 & 0 & -6.6639 & 6.4239 & -0.1526 \\ 0 & 0 & 0 & -0.1526 & 0.4074 \end{bmatrix}.$$

Some things are worth noting at this point.

1. A_1 is tridiagonal; i.e., the QR step and reverse multiplication preserve the Hessenberg form.

2. The $(5,4)$ element of A_1 is substantially smaller than it was in A_0. This is significant, because when this element reaches zero, it means that the $(5,5)$ element is an eigenvalue. (See Exercise 2.)

We now continue the computation, without showing all the details. After six iterations, we have

$$A_6 = \begin{bmatrix} 16.0164 & -2.9678 & -0.0000 & -0.0000 & -0.0000 \\ -2.9678 & 18.7347 & 0.3964 & -0.0000 & 0.0000 \\ 0 & 0.3964 & 7.0332 & -0.0167 & -0.0000 \\ 0 & 0 & -0.0167 & 2.8157 & 0.0000 \\ 0 & 0 & 0 & 0.0000 & 0.4000 \end{bmatrix}.$$

We again make some observations:

1. Since the $(5,4)$ element is zero (to four digits), we can take $\lambda_5 \approx 0.4000$ as our first approximate eigenvalue.

2. Note that the iteration found the smallest eigenvalue first.

3. Note that the other off-diagonal elements are generally smaller than in A_0. This is significant; one reason that the QR iteration is superior to the power methods is that each step contributes something toward finding *all* the eigenvalues, not just a single one.

4. We can deflate by simply extracting the upper left 4×4 submatrix; there is no need to do any special computation, and the Hessenberg form is preserved. This is one of the key cost-saving features of the QR algorithm over the power methods and orthogonal iteration. ∎

While this example suggests that the QR iteration would outperform the power methods, the method can still be improved. Just as shifts accelerated the convergence of the inverse power methods, they can also be used to improve the performance of the QR iteration (see Algorithm 8.8).

It can be shown (Exercise 4) that the A_k matrices are all similar, and it is possible to arrange the computation so that the shifted matrices are never explicitly formed (this is known in the literature as *implicit QR*).

What do we use as shifts? The general philosophy is the same as for the power methods: Choose shifts that are close to the eigenvalue you are trying to compute. Two

Algorithm 8.8: Shifted QR iteration

Given $A \in \mathbf{R}^{n \times n}$, set $A_0 = A$ and compute as follows: For k from 1 until convergence, do

1. Define Q_k and R_k as the QR decomposition of $A_{k-1} - \mu_{k-1}I$: $Q_k R_k = A_{k-1} - \mu_{k-1}I$.

2. Define $A_k = R_k Q_k + \mu_{k-1}I$.

---

common shift strategies are the following:
- *Rayleigh shift:* Set $\mu_k = $ the lower right diagonal element of A_k.
- *Wilkinson shift:* Let

$$S = \begin{bmatrix} \alpha_{11} & \alpha_{12} \\ \alpha_{21} & \alpha_{22} \end{bmatrix}$$

denote the lower right 2×2 submatrix in A_k. Then the Wilkinson shift is to set $\mu_k = $ the eigenvalue of S that is closest to α_{22}.

EXAMPLE 8.16

To illustrate the value of shifts in the QR iteration, we return to the same example matrix as we have used all along. The Hessenberg form is

$$A_{H,0} = A_0 = \begin{bmatrix} 10.0000 & -4.1231 & 0 & 0 & 0 \\ -4.1231 & 9.2941 & 3.0920 & 0 & 0 \\ 0 & 3.0920 & 6.4587 & -5.3189 & 0 \\ 0 & 0 & -5.3189 & 17.9989 & 3.1371 \\ 0 & 0 & 0 & 3.1371 & 1.2483 \end{bmatrix}$$

If we use the Rayleigh shift, then $\mu_0 = 1.2483$ and the initial QR factorization gives us

$$Q = \begin{bmatrix} -0.9046 & -0.3719 & 0.1068 & 0.1478 & 0.1005 \\ 0.4262 & -0.7893 & 0.2268 & 0.3137 & 0.2134 \\ 0 & -0.4886 & -0.4477 & -0.6192 & -0.4213 \\ 0 & 0 & 0.8583 & -0.4242 & -0.2886 \\ 0 & 0 & 0 & -0.5625 & 0.8268 \end{bmatrix}$$

and

$$R = \begin{bmatrix} -9.6743 & 7.1590 & 1.3178 & 0 & 0 \\ 0 & -6.3281 & -4.9864 & 2.5989 & 0 \\ 0 & 0 & -6.1967 & 16.7589 & 2.6927 \\ 0 & 0 & 0 & -5.5769 & -1.3307 \\ 0 & 0 & 0 & 0 & -0.9054 \end{bmatrix},$$

so that

$$A_1 = RQ + \mu_0 I = \begin{bmatrix} 13.0511 & -2.6970 & 0.0000 & 0.0000 & 0 \\ -2.6970 & 8.6795 & 3.0278 & 0.0000 & 0.0000 \\ 0 & 3.0278 & 18.4072 & -4.7869 & -0.0000 \\ 0 & 0 & -4.7869 & 4.3626 & 0.5093 \\ 0 & 0 & 0 & 0.5093 & 0.4997 \end{bmatrix}.$$

The next shift is then $\mu_1 = 0.4997$. If we continue the iteration, we find that

$$A_3 = \begin{bmatrix} 14.3388 & -1.5148 & -0.0000 & -0.0000 & -0.0000 \\ -1.5148 & 17.5527 & 5.4211 & 0.0000 & 0.0000 \\ 0 & 5.4211 & 9.8879 & -0.1601 & -0.0000 \\ 0 & 0 & -0.1601 & 2.8207 & 0.0000 \\ 0 & 0 & 0 & 0.0000 & 0.4000 \end{bmatrix};$$

thus, we have found the first eigenvalue in only three iterations, whereas the unshifted method took six. We can now deflate, by defining

$$A_3' = \begin{bmatrix} 14.3388 & -1.5148 & -0.0000 & -0.0000 \\ -1.5148 & 17.5527 & 5.4211 & 0.0000 \\ 0 & 5.4211 & 9.8879 & -0.1601 \\ 0 & 0 & -0.1601 & 2.8207 \end{bmatrix}$$

and continue the iteration. Thus $\mu_3 = 2.8207$ and A_4 is defined by

$$A_4 = \begin{bmatrix} 14.7851 & -2.0073 & -0.0000 & 0.0000 \\ -2.0073 & 19.7542 & 1.6637 & -0.0000 \\ 0 & 1.6637 & 7.2452 & -0.0002 \\ 0 & 0 & -0.0002 & 2.8156 \end{bmatrix}$$

and

$$A_5 = \begin{bmatrix} 15.5758 & -2.7220 & -0.0000 & -0.0000 \\ -2.7220 & 19.1737 & 0.4287 & -0.0000 \\ 0 & 0.4287 & 7.0349 & -0.0000 \\ 0 & 0 & -0.0000 & 2.8156 \end{bmatrix},$$

showing that we have found another eigenvalue. Hence, in fewer steps than it took to find a single eigenvalue in the unshifted algorithm, we have found two eigenvalues by using the very simple Rayleigh shift.[6] Using the Wilkinson shift on this example yields very similar results. ∎

Which shift do we use? We might expect that the simpler Rayleigh shift is less reliable than the more involved Wilkinson shift, and we would be right. While the Rayleigh shift can produce rapid convergence of the off-diagonal elements to zero,

---

[6]To within four decimals, that is. For greater accuracy we would need to take more iterations before deflating, and perhaps some more after deflating.

there exist examples for which it will not converge at all. On the other hand, it can be shown that QR using the Wilkinson shift will always converge [3]. Exercise 5 illustrates some of this, drawing on an example from [3].

Why does QR work? A complete discussion is beyond the intended level of this text. Good explanations are found in the works of Watkins ([4] and [5]) and Trefethen and Bau [3]; a brief answer is that the QR iteration is equivalent to a properly scaled and shifted inverse power iteration being done on an orthonormal set of vectors. Thus, we are working on *all* the eigenvectors at the same time. Some of this is suggested by the motivation for the QR iteration that we did at the beginning of this section, using orthogonal iteration.

Since the underlying theory for the QR iteration is similar to that for the power methods, it should have the same weak points, and this in fact is the case. Defective matrices or matrices that are nearly defective (meaning that two or more eigenvectors are nearly parallel) will slow the convergence of the iteration; fundamentally, this occurs because the eigenvalue problem is a polynomial root-finding problem, and defective matrices are caused by multiple roots for the polynomial (recall Section 3.10.4). But software such as the codes in LAPACK or the `eig` command in MATLAB is robust and reliable, and can be counted upon to deliver accurate results for almost all problems.

EXERCISES

1. Show that the product of two orthogonal matrices, Q_1 and Q_2, is also orthogonal.

2. Let $A \in \mathbf{R}^{n \times n}$ have the partioned form

$$A \begin{bmatrix} A_{11} & a \\ 0 & \lambda \end{bmatrix},$$

where $A_{11} \in \mathbf{R}^{(n-1) \times (n-1)}$ and $\lambda \in \mathbf{R}$. Show that λ must be an eigenvalue of A. How are the eigenvalues of A and A_{11} related?

3. Assume that the iteration in Algorithm 8.6 converges, in the sense that $R_k \to R_\infty$, which is upper triangular, and $Q_k \to Q_\infty$, which is orthogonal.
(a) Prove that the matrices A_k, defined in (8.5), must also converge to R_∞.
(b) Prove then that the eigenvalues of A can be recovered from the diagonal elements of A_k, in the limit as $k \to \infty$.

4. Show that the matrices in the shifted QR iteration all have the same eigenvalues.

5. Consider the matrix

$$A = \begin{bmatrix} 0 & 1 \\ 1 & 0 \end{bmatrix}.$$

(a) Find the exact eigenvalues of A.

(b) Show that

$$Q = \begin{bmatrix} 0 & 1 \\ 1 & 0 \end{bmatrix}, \quad R = \begin{bmatrix} 1 & 0 \\ 0 & 1 \end{bmatrix}.$$

is a valid QR decomposition of A.

(c) Use this to show that the shifted QR iteration for A, using the Rayleigh shift, will not converge.

6. The QR factorization is usually carried out for the QR iteration by means of *Givens transformations*, defined in part (a), below. In this exercise we will introduce the basic ideas of this kind of matrix operation.
(a) Show that the matrix

$$G(\theta) = \begin{bmatrix} \cos\theta & -\sin\theta \\ \sin\theta & \cos\theta \end{bmatrix}$$

is an orthogonal matrix for any choice of θ.
(b) Show that we can always choose θ so that

$$G(\theta)A = \begin{bmatrix} \cos\theta & -\sin\theta \\ \sin\theta & \cos\theta \end{bmatrix} \begin{bmatrix} a_{11} & a_{12} \\ a_{21} & a_{22} \end{bmatrix}$$

$$= \begin{bmatrix} r_{11} & r_{12} \\ 0 & r_{22} \end{bmatrix} = R.$$

(c) Use a Givens transformation to perform one step of

the basic *QR* iteration for the matrix

$$\begin{bmatrix} 4 & 1 \\ 1 & 4 \end{bmatrix}.$$

7. (a) If $G(\theta)$ is a Givens transformation, show that

$$\hat{G}(\theta) = \begin{bmatrix} 1 & 0 \\ 0 & G(\theta) \end{bmatrix}$$

is orthogonal.

(b) Show how to use a sequence of *two* Givens transformations to do a single *QR* step for the matrix

$$\begin{bmatrix} 4 & 1 & 0 \\ 1 & 4 & 1 \\ 0 & 1 & 4 \end{bmatrix}.$$

LITERATURE AND SOFTWARE DISCUSSION

Most of the works cited at the end of the previous chapter contain good discussions of the numerical solution of the algebraic eigenvalue problem. The classic work is by James Wilkinson [6]. An outstanding explanation of the *QR* algorithm is in the SIAM review article by Watkins [5] (the texts by Watkins [4] and Trefethen and Bau [3] contain some of the same material). Parlett's book [2] is a good discussion of the symmetric eigenvalue problem, which is substantially easier than the general case, as we have seen.

There exist other eigenvalue approximation ideas beyond what was discussed here, and Watkins [4] has a brief discussion of many of these, some of which are of mostly historic interest. Of particular interest is the "divide and conquer" algorithm due to Cuppen. The rise of parallel computing has meant that some of these older methods, such as the Jacobi method, are of more interest because they can be programmed in a manner that takes greater advantage of the parallel computing architecture.

BIBLIOGRAPHY

[1] HOUSEHOLDER, ALSTON. *The Theory of Matrices in Numerical Analysis*. Dover, New York, 1975 (originally published in 1964 by Blaisdell).

[2] PARLETT, BERESFORD. *The Symmetric Eigenvalue Problem*. SIAM Classics in Applied Mathematics, Philadelphia, 1998 (originally published by Prentice-Hall in 1980).

[3] TREFETHEN, L.N., and BAU, DAVID. *Numerical Linear Algebra*, SIAM, Philadelphia, 1997.

[4] WATKINS, DAVID. *Fundamentals of Matrix Computations*. John Wiley & Sons, New York, 1991.

[5] WATKINS, DAVID "Understanding the *QR* algorithm", *SIAM Review*, vol. 24 (1982) pp. 427–440.

[6] WILKINSON, J.H. *The Algebraic Eigenvalue Problem*. Oxford University Press, Oxford, 1965.

CHAPTER 9

A SURVEY OF FINITE DIFFERENCE METHODS FOR PARTIAL DIFFERENTIAL EQUATIONS

Pennsyltucky Pharmaceutical is modeling the absorption and diffusion of oxygen in living tissue. Building on a classic formulation,[1] they arrive at the partial differential equation

$$u_t = Du_{xx} - \frac{Vu}{K+u}, \, 0 \le x \le 1$$

$$u(0,t) = g_0(t),$$
$$u(1,t) = g_1(t),$$
$$u(x,0) = u_0(x).$$

Here $u(x,t)$ is the concentration of oxygen in the tissue; $D > 0$ is the diffusion coefficient, which measures how rapidly the oxygen diffuses through the tissue; V and K are positive parameters governing the absorption of the oxygen in the tissue; and $g_0(t)$, $g_1(t)$, and $u_0(x)$ are the known boundary and initial conditions.

This is an example of a nonlinear partial differential equation, and its solution is possible using some of the methods we discuss in this final chapter.

Partial differential equations (PDEs) are differential equations involving functions of more than one independent variable, such as the temperature at each point (x,y,z) in an iron bar. They are necessarily more complicated in some respects than ordinary differential equations, and it is beyond the intended scope of this text to provide an in-depth treatment of all the methods used for their approximate solution. Indeed, this is such an active area of research that it is best, in an introductory text, to confine ourselves to a brief survey of major ideas.

---

[1] See J. Crank and R. S. Gupta, "A moving boundary problem arising from the diffusion of oxygen in absorbing tissue," *Journal of Institutional Mathematics Applications*, vol. 10 (1972), pp. 19–33, and A. E. Berger, M. Ciment and J. C. W. Rogers, "Numerical solution of a diffusion consumption problem with a free boundary," *SIAM Journal on Numerical Analysis*, vol. 12 (1975), pp. 646–672.

9.1 DIFFERENCE METHODS FOR THE DIFFUSION EQUATION

9.1.1 The Basic Problem

Perhaps the simplest PDE is the *diffusion equation*, so called because it can be used to model a number of processes that are driven by diffusion, such as heat transfer and some kinds of slow mass transfer.

We seek the unknown function $u(x,t)$ such that

$$u_t = au_{xx} + f(x,t), t > 0, 0 < x < 1 \tag{9.1}$$

$$u(0,t) = g_0(t) \tag{9.2}$$

$$u(1,t) = g_1(t) \tag{9.3}$$

$$u(x,0) = u_0(x). \tag{9.4}$$

Here g_0 and g_1 are the (known) *boundary data*, f is a known source term, and u_0 is the (known) *initial data*. More general and involved boundary conditions are possible, and we can generalize the equation somewhat, but this is the standard form. For simplicity we will assume *homogeneous boundary data*, meaning that $g_0 = g_1 = 0$ for all time t. In the most commonplace examples, the unknown u gives the temperature at each point x along a thin rod, for each time t. The boundary conditions (9.3)–(9.4) give the temperature at which each end of the rod is maintained, and the initial condition (9.4) gives the temperature distribution at the start of the process. See a text such as [8] for background material on the diffusion equation.

The parameter a, which must be positive, is the *thermal diffusivity* of the material being studied. If a is large, it means that diffusion is rapid; if a is small, then diffusion is slow.

We will study two basic methods for the diffusion equation in this section: the so-called explicit method and the Crank-Nicolson method.

Before proceeding with the derivations and analysis, we need to deal with some notational issues. Our approximations will be constructed, as was the case for ODE problems in Chapter 6, only at discrete points on a grid defined over space and time. There will be times when we will want to consider the approximation as a discretely defined function, and there will be times when we want to consider it as a vector in a standard Euclidean space. At times we will also want or need to consider the collection of exact solution values defined at the same grid points. To accomplish all this with a minimum of excessive notation, we will use the following conventions.

If the dependent variable in a problem is u, then $u_{h,\Delta t}$ will denote the approximation, defined on a grid using h as the spatial step and Δt as the time step. Specific values can be denoted using the standard functional notation, $u_{h,\Delta t}(x_i,t_n)$, which we will occasionally abbreviate as u_i^n. If we wish to consider the Euclidean vector of $u_{h,\Delta t}$ values at a specific point in time, t_n, then we will write $u_{h,\Delta t}^n$. The similar notation u^n (note the lack of subscript) will be used for the vector of exact solution values at time t_n, using the same grid points. Some of this notation is abusive of standard conventions, but it enables us to work with a minimum of confusing new symbols. The need for most of this will not occur until we begin to discuss implicit methods and Crank-Nicolson, in Section 9.1.3.

9.1.2 The Explicit Method and Stability

The explicit method is constructed by using the most straightforward difference approximations for the derivatives in the PDE. We start by defining a grid in the spatial variable x:

$$0 = x_0 < x_1 < x_2 < \cdots < x_{N-1} < x_N = 1,$$

and we assume, for simplicity, that the mesh spacing is uniform: $x_k - x_{k-1} = h$, for all k. Define a time step $\Delta t > 0$ and a time grid by $t_n = n\Delta t$ for $n \geq 0$. Then at each point in our space-time grid we can define difference approximations to the derivatives:

$$\frac{\partial u}{\partial t}(x,t) = \frac{u(x,t+\Delta t) - u(x,t)}{\Delta t} - \frac{1}{2}\Delta t \frac{\partial^2 u}{\partial t^2}(x,\theta),$$

$$\frac{\partial^2 u}{\partial x^2}(x,t) = \frac{u(x-h,t) - 2u(x,t) + u(x+h,t)}{h^2} + \frac{1}{12}h^2 \frac{\partial^4 u}{\partial x^4}(\eta,t),$$

where $t \leq \theta \leq t + \Delta t$ and $x - h \leq \eta \leq x + h$. We use these in the PDE to replace $u_t(x,t)$ and $u_{xx}(x,t)$, and we get

$$\frac{u(x,t+\Delta t) - u(x,t)}{\Delta t} = a\left(\frac{u(x-h,t) - 2u(x,t) + u(x+h,t)}{h^2}\right) + f(x,t) + T_E(x,t),$$

$$(9.5)$$

where the truncation error $T_E(x,t)$ is given by

$$T_E(x,t) = \frac{1}{12}h^2 \frac{\partial^4 u}{\partial x^4}(\eta,t) + \frac{1}{2}\Delta t \frac{\partial^2 u}{\partial t^2}(x,\theta). \tag{9.6}$$

We define now our approximate solution $u_{h,\Delta t}(x,t)$ at the grid points by dropping the truncation error from (9.5) and replacing $u(x_i,t_n)$ with $u_i^n = u_{h,\Delta t}(x_i,t_n)$. We get

$$\frac{u_i^{n+1} - u_i^n}{\Delta t} = a\left(\frac{u_{i-1}^n - 2u_i^n + u_{i+1}^n}{h^2}\right) + f(x_i,t_n), \tag{9.7}$$

which can be simplified to

$$u_i^{n+1} = u_i^n + \frac{a\Delta t}{h^2}\left(u_{i-1}^n - 2u_i^n + u_{i+1}^n\right) + \Delta t f(x_i,t_n).$$

This is called the *explicit method* because it yields the value of the approximation at time t_{n+1} explicitly in terms of the approximate solution at time t_n.

Coding this is very easy, so it is worth looking at some examples right off. Consider the problem (9.1)–(9.4) with $a = \pi^{-2}$, $u_0(x) = \sin \pi x + \sin 2\pi x$, and $g_0 = g_1 = 0$, for which the exact solution is $u(x,t) = e^{-t}\sin \pi x + e^{-4t}\sin 2\pi x$. Figure 9.1 shows the solution profile for the approximate solution computed using $h = 1/16$, $\Delta t = 1/64$, which implies that

$$\frac{a\Delta t}{h^2} = \frac{4}{\pi^2},$$

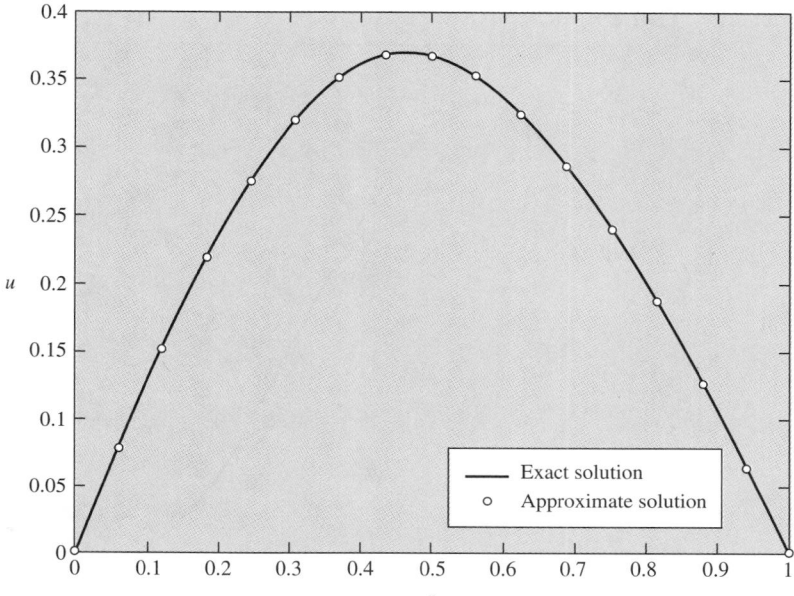

FIGURE 9.1 Solution profile at $t = 1$ for heat equation; explicit method, $h = 1/16, \Delta t = 1/64$.

and Figure 9.2 shows the error profile, both at time $t = 1$. The approximation appears to be doing a decent job. However, if we next look at the case defined by $h = 1/32$, $\Delta t = 1/128$, then we have

$$\frac{a \Delta t}{h^2} = \frac{8}{\pi^2},$$

and Table 9.1 on page XX shows how the values in the middle of the interval evolve for the first 45 time steps. Note that the approximate solution appears to be evolving normally for the first 40 or so steps, and then suddenly begins to "blow up." If we continue the computation, the solution values eventually get as large as 10^{29}. In fact, they will grow without bound. What is going on here?

The problem, in a nutshell, is that the time step is too large for the speed of diffusion and the spatial step. The explicit method requires that the time step be sufficiently small in order that the computation be stable and accurate. A formal result is the following.

Theorem 9.1   Let $u \in C^{4,2}(Q)$ for $Q = \{(x,t)|0 \leq x \leq 1, 0 < t \leq T\}$ for some $T > 0$. If h and Δt are such that

$$\Delta t \leq \frac{h^2}{2a},\tag{9.8}$$

then there exists a constant $C > 0$, independent of h and Δt, such that

$$\max_{(x_i,t_n) \in Q} \left| u(x_i, t_n) - u_{h,\Delta t}(x_i, t_n) \right| \leq \max_{0 \leq i \leq N} \left| u_0(x_i) - u_{h,\Delta t}(x_i, 0) \right| + C h^2 M,\tag{9.9}$$

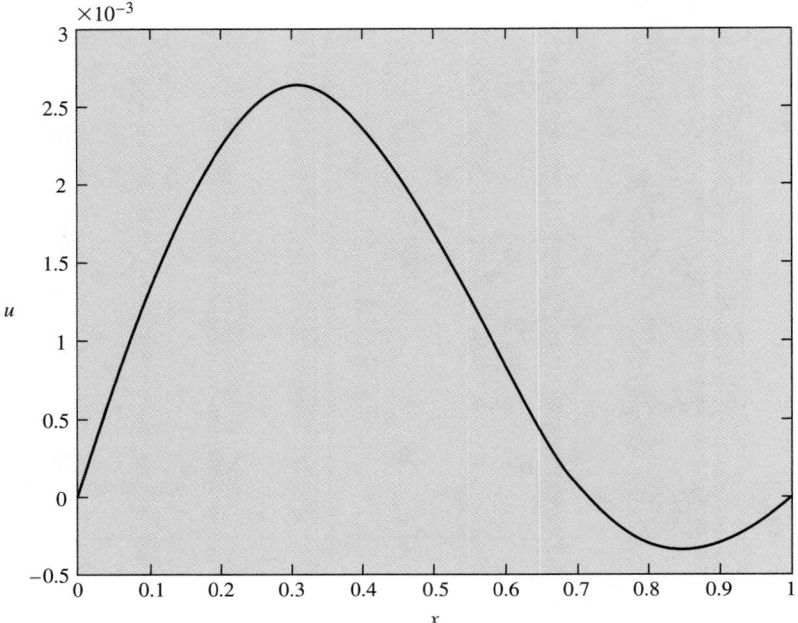

FIGURE 9.2 Error profile at $t = 1$ for heat equation; explicit method, $h = 1/16, \Delta t = 1/64$.

where

$$M = \max_{(x,t) \in Q} \{|u_{xxxx}(x,t)|, |u_{tt}(x,t)|\}.$$

Proof To construct an error estimate, we subtract (9.5) and (9.7); letting $e_i^n = u(x_i, t_n) - u_i^n$, we then have

$$e_i^{n+1} = e_i^n + \lambda \left(e_{i-1}^n - 2e_i^n + e_{i+1}^n \right) + \left(\frac{1}{12} h^2 \Delta t \frac{\partial^4 u}{\partial x^4}(\eta_i, t_n) + \frac{1}{2} \Delta t^2 \frac{\partial^2 u}{\partial t^2}(x_i, \theta_n) \right),$$

where $x_{i-1} \leq \eta_i \leq x_{i+1}$ and $t_n \leq \theta_n \leq t_{n+1}$, and we have used $\lambda = a\Delta t/h^2$. Taking absolute values and using the definition of M, plus some simplification, gives us

$$|e_i^{n+1}| \leq |1 - 2\lambda||e_i^n| + \lambda \left(|e_{i-1}^n| + |e_{i+1}^n| \right) + \left(\frac{1}{12} h^2 \Delta t + \frac{1}{2} \Delta t^2 \right) M, \quad 1 \leq i \leq N-1.$$

Now let

$$E_n = \max_{1 \leq i \leq N-1} |e_i^n|,$$

so that we have

$$|e_i^{n+1}| \leq (|1 - 2\lambda| + 2\lambda) E_n + \left(\frac{1}{12} h^2 \Delta t + \frac{1}{2} \Delta t^2 \right) M.$$

TABLE 9.1   Illustration of Instability for Explicit Method: $h = 1/32$, $\Delta t = 1/128$, $a = 1/\pi^2$

k	u_{14}^k	u_{15}^k	u_{16}^k	u_{17}^k	u_{18}^k
1	0.134389E+01	0.117643E+01	0.992194E+00	0.798403E+00	0.602366E+00
⋮	⋮	⋮	⋮	⋮	⋮
5	0.126978E+01	0.112348E+01	0.961573E+00	0.790403E+00	0.616417E+00
⋮	⋮	⋮	⋮	⋮	⋮
10	0.118573E+01	0.106234E+01	0.924624E+00	0.778003E+00	0.627984E+00
⋮	⋮	⋮	⋮	⋮	⋮
15	0.111007E+01	0.100618E+01	0.889094E+00	0.763449E+00	0.633948E+00
⋮	⋮	⋮	⋮	⋮	⋮
20	0.104172E+01	0.954414E+00	0.854929E+00	0.747210E+00	0.635278E+00
⋮	⋮	⋮	⋮	⋮	⋮
25	0.979764E+00	0.906558E+00	0.822080E+00	0.729670E+00	0.632816E+00
⋮	⋮	⋮	⋮	⋮	⋮
30	0.923295E+00	0.862416E+00	0.789955E+00	0.712294E+00	0.625040E+00
31	0.913011E+00	0.853028E+00	0.785740E+00	0.704519E+00	0.631205E+00
32	0.901453E+00	0.847107E+00	0.774446E+00	0.710929E+00	0.610682E+00
33	0.894185E+00	0.832261E+00	0.781858E+00	0.681157E+00	0.656032E+00
34	0.875812E+00	0.841599E+00	0.741088E+00	0.742416E+00	0.539034E+00
35	0.888149E+00	0.787860E+00	0.823636E+00	0.576484E+00	0.818067E+00
36	0.819241E+00	0.898150E+00	0.594303E+00	0.972637E+00	0.136538E+00
37	0.964831E+00	0.587900E+00	0.114726E+01	−0.117450E−01	0.176970E+01
38	0.552823E+00	0.134683E+01	−0.245593E+00	0.237169E+01	−0.214521E+01
39	0.157751E+01	−0.587537E+00	0.316667E+01	−0.341106E+01	0.717104E+01
40	−0.106303E+01	0.421043E+01	−0.520808E+01	0.104982E+02	−0.149249E+02
41	0.556995E+01	-0.769844E+01	0.151573E+02	-0.228400E+02	0.372474E+02
42	-0.110780E+02	0.215827E+02	-0.341683E+02	0.566645E+02	-0.855143E+02
43	0.303114E+02	−0.500811E+02	0.846480E+02	−0.132208E+03	0.202351E+03
44	−0.720490E+02	0.124290E+03	−0.200336E+03	0.314752E+03	−0.470614E+03
45	0.179537E+03	−0.297988E+03	0.480310E+03	−0.739356E+03	0.109814E+04

Note that if $1 - 2\lambda \geq 0$, then we can remove the absolute value symbol and write

$$|1 - 2\lambda| + 2\lambda = 1 - 2\lambda + 2\lambda = 1.$$

Therefore (if $1 - 2\lambda \geq 0$),

$$|e_i^{n+1}| \leq E_n + \left(\frac{1}{12}h^2\Delta t + \frac{1}{2}\Delta t^2\right)M,$$

and this holds for all i; hence

$$E_{n+1} \leq E_n + \left(\frac{1}{12} h^2 \Delta t + \frac{1}{2} \Delta t^2 \right) M. \tag{9.10}$$

This is a simple recursive inequality, very much akin to those we dealt with in Chapter 6, and applying the solution procedures from those sections gives us

$$E_n \leq E_0 + n \left(\frac{1}{12} h^2 \Delta t + \frac{1}{2} \Delta t^2 \right) M = E_0 + \left(\frac{1}{12} h^2 + \frac{1}{2} \Delta t \right) (M t_n)$$

$$\leq E_0 + \left(\frac{1}{12} h^2 + \frac{1}{2} \Delta t \right) (MT),$$

from which the conclusion (9.9) follows.

The condition $1 - 2\lambda \geq 0$ is equivalent to (9.8), and thus we are done.   □

If (9.8) does not hold, then (9.10) becomes

$$E_{n+1} \leq c E_n + \left(\frac{1}{12} h^2 \Delta t + \frac{1}{2} \Delta t^2 \right) M \tag{9.11}$$

for $c = 4\lambda - 1 > 1$. We then have

$$E_n \leq c^n E_0 + \left(\frac{1}{12} h^2 \Delta t + \frac{1}{2} \Delta t^2 \right) M \sum_{k=0}^{n-1} c^k.$$

This is important for a number of reasons. First of all, it shows that if the *stability condition* (9.8) does not hold, then any perturbation in the initial condition (which amounts to having $E_0 \neq 0$) might be amplified by a factor of c^n, making the error eventually very large. In addition, the presence of the c^k factors in the summation means that, even in the total absence of rounding error, the mathematical error is likewise amplified by exponentially growing factors.

The attentive reader might be concerned that (9.11), being an inequality, does not lead to the conclusion that E_n *must* grow as c^n does; it simply *allows* for such growth to occur. However, it is possible to establish a strict inequality showing that E_n will grow if the stability condition is not satisfied, and this growth is observed in practice, as our simple example showed.

The condition (9.8) is called a stability condition because it is both necessary and sufficient to ensure that the approximate solutions do not "blow up" during the course of the computation. It can be shown that if a numerical method for the approximate solution of the diffusion equation is stable and has a truncation error that goes to zero as the mesh parameters do, then the approximate solutions will converge to the exact solution. Note that (9.8) requires us to use a time step size that goes like the square of the spatial step; thus we will often be forced to use very tiny time steps. This is usually considered too much of a restriction for practical usage, although the simplicity of the explicit method makes it attractive for nonlinear problems and problems involving more than one space dimension.

9.1.3 Implicit Methods and the Crank-Nicolson Method

The time step restriction in the explicit method can be removed by using a very slightly different set of difference approximations in constructing the numerical method. If we use

$$\frac{\partial u}{\partial t}(x,t+\Delta t) = \frac{u(x,t+\Delta t) - u(x,t)}{\Delta t}$$

$$+ \frac{1}{2}\Delta t^2 \frac{\partial^2 u}{\partial t^2}(x,\theta)$$

$$\frac{\partial^2 u}{\partial x^2}(x,t+\Delta t) = \frac{u(x-h,t+\Delta t) - 2u(x,t+\Delta t) + u(x+h,t+\Delta t)}{h^2}$$

$$+ \frac{1}{12}h^2 \frac{\partial^4 u}{\partial x^4}(\eta, t+\Delta t),$$

where $t \le \theta \le t + \Delta t$ and $x_{i-1} \le \eta \le x_{i+1}$, then the numerical method becomes

$$u_i^{n+1} - \frac{a\Delta t}{h^2}\left(u_{i-1}^{n+1} - 2u_i^{n+1} + u_{i+1}^{n+1}\right) = u_i^n + f(x_i, t_{n+1}). \tag{9.12}$$

This is a tridiagonal system of $N-1$ equations in the $N-1$ unknowns $u_1^{n+1}, u_2^{n+1}, u_3^{n+1}$, $\ldots, u_{N-1}^{n+1}$, and we can solve it using Algorithm 2.6. Moreover, the method is unconditionally stable; i.e., it is stable for any value of Δt. The problem is that it is only first-order accurate in the time step.

Theorem 9.2   *For the implicit method (9.12) there exists a constant $C > 0$, independent of h or Δt, such that*

$$\max_{(x_i, t_n) \in Q}|u(x_i, t_n) - u_{h,\Delta t}(x_i, t_n)| \le \max_{0 \le i \le N}|u_0(x_i) - u_{h,\Delta t}(x_i, 0)| + C(h^2 + \Delta t)M, \tag{9.13}$$

where

$$M = \max_{(x,t) \in Q}\left\{|u_{xxxx}(x,t)|, |u_{tt}(x,t)|\right\}.$$

Proof   It is easy to show, using the same argument as for the explicit method, that the pointwise error $e_j^n = u(x_i, t_n) = u_i^n$ satisfies

$$e_i^{n+1} - \lambda\left(e_{i-1}^{n+1} - 2e_i^{n+1} + e_{i+1}^{n+1}\right) = e_i^n + \Delta t T_E(x_i, t_{n+1}), \tag{9.14}$$

where $\lambda = a\Delta t h^{-2}$ and $T_E(x_i, t_{n+1})$ is the truncation error:

$$T_E(x_i, t_{n+1}) = -\frac{1}{2}\Delta t^2 \frac{\partial^2 u}{\partial t^2}(x_i, \theta_{n+1}) + \frac{1}{12}h^2 \frac{\partial^4 u}{\partial x^4}(\eta_i, t_{n+1}).$$

Now, let J be the index such that

$$|e_J^{n+1}| = \max_{1 \le i \le n-1}|e_i^{n+1}|.$$

Then (9.14) implies that

$$(1+\lambda)|e_J^{n+1}| \le \lambda\left(|e_{J-1}^{n+1}| + |e_{J+1}^{n+1}|\right) + |e_J^n| + \Delta t \tau, \tag{9.15}$$

where τ is the maximum (in absolute value) truncation error:

$$\tau = \left| \frac{1}{2} \Delta t^2 + \frac{1}{12} h^2 \right| M.$$

Since $|e_J^{n+1}| \geq |e_j^{n+1}|$ for any j, we can write

$$(1 + 2\lambda)|e_J^{n+1}| \leq 2\lambda |e_J^{n+1}| + |e_J^n| + \Delta t \tau, \tag{9.16}$$

from which

$$E_{n+1} = \max_{1 \leq i \leq n-1} |e_i^{n+1}| = |e_J^{n+1}| \leq |e_J^n| + \Delta t \tau \leq \max_{1 \leq i \leq n-1} |e_i^n| + \Delta t \tau = E_n + \Delta t \tau \tag{9.17}$$

follows. Therefore, as in the previous proof, we solve the recursion $E_{n+1} \leq E_n + \Delta t \tau$ to get

$$E_n \leq E_0 + t_n \tau$$

from which the conclusion follows. □

This method is called *unconditionally stable* because there is no condition on the step size for the (approximate) solution to be bounded and accurate. The problem with this method, however, is that it is only $O(h^2 + \Delta t)$ accurate; the explicit method was $O(h^2)$ accurate, since $\Delta t = O(h^2)$ was required for stability. For comparison's sake, consider the same example we looked at for the explicit method. If we use the same grid sizes ($h = 1/16, \Delta t = 1/64$), then we get the error plot shown in Figure 9.3, which indicates that the implicit approximation was not as accurate as the explicit one was, by about a factor of 3. However, we can take much larger time steps and the solution remains stable. If we use $h = 1/32, \Delta t = 1/128$, then we get the error profile shown in Figure 9.4 on page XX; recall that the explicit approximation blew up for this case.

To regain $O(h^2)$ accuracy, we use an idea developed in the late 1940s by the British mathematicians Crank and Nicolson.[2]

There are a number of ways to derive the Crank-Nicolson scheme. Perhaps the best way is to first "solve" the PDE by integrating in time on both sides of the equation,

$$u(x,t + \Delta t) - u(x,t) = a \int_t^{t+\Delta t} (a u_{xx}(x,s) + f(x,s)) \, ds$$

---

[2]John Crank (1916–) was born in Lancashire, England, and received all of his education at Manchester University. From 1957 until his retirement in 1981, he was professor of mathematics at Brunel University in Acton.

Phyllis (Lockett) Nicolson (1917–1968) was born in Macclesfield, England, and received her early education (bachelor's and master's degrees) at Manchester University. In the immediate postwar period she was a research fellow at Girton College, Cambridge. In 1952 she was appointed lecturer in physics at the University of Leeds.

The Crank-Nicolson method resulted from a collaboration that took place in the immediate postwar period. It was first presented in a 1947 paper in the *Proceedings of the Cambridge Philosophical Society*, titled "A practical method for numerical evaluation of solutions to partial differential equations of the heat-conduction type."

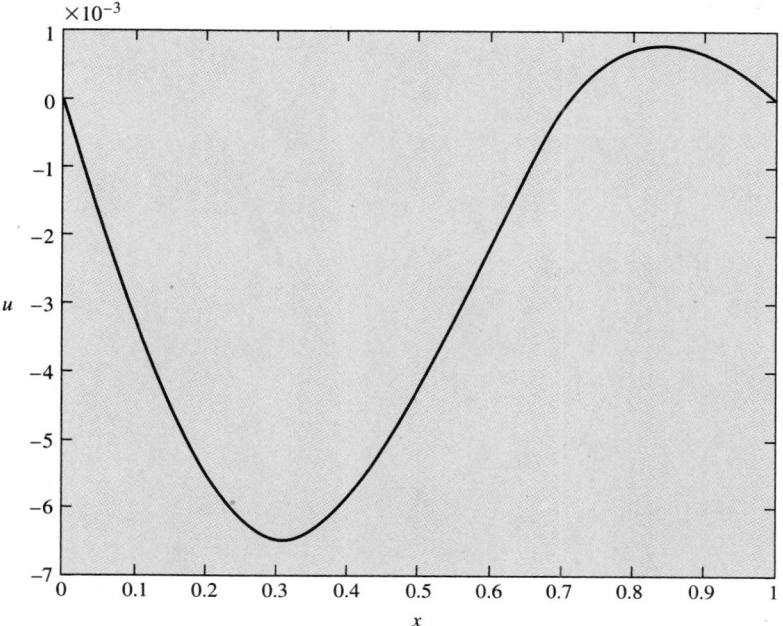

FIGURE 9.3 Error profile at $t = 1$ for heat equation; implicit method, $h = 1/16$, $\Delta t = 1/64$.

and then using the trapezoid rule to evaluate the integral on the right side:

$$u(x,t+\Delta t) - u(x,t) = \frac{1}{2}\Delta t a \left(u_{xx}(x,t) + u_{xx}(x,t+\Delta t) \right)$$
$$+ \frac{1}{2}\Delta t \left(f(x,t) + f(x,t+\Delta t) \right)$$
$$- \frac{1}{12}\Delta t^3 u_{xxtt}(x,\theta_t) - \frac{1}{12}\Delta t^3 f_{tt}(x,\eta_t).$$

Then, replacing the spatial derivatives with difference quotients, we get the lengthy expression

$$u(x,t+\Delta t) - u(x,t) = \frac{1}{2}\Delta t a \left[\frac{u(x+h,t+\Delta t) - 2u(x,t+\Delta t) + u(x-h,t+\Delta t)}{h^2} \right]$$
$$+ \frac{1}{2}\Delta t a \left[\frac{u(x+h,t+\Delta t) - 2u(x,t+\Delta t) + u(x-h,t+\Delta t)}{h^2} \right]$$
$$+ \frac{1}{2}\Delta t \left(f(x,t+\Delta t) + f(x,t) \right)$$
$$+ \frac{1}{24}h^2 \Delta t a \left(\frac{\partial^4 u}{\partial x^4}(\eta_1, t+\Delta t) + \frac{\partial^4 u}{\partial x^4}(\eta_0, t) \right) - \frac{1}{12}\Delta t^3 u_{xxtt}(x,\theta_t)$$
$$- \frac{1}{12}\Delta t^3 f_{tt}(x,\eta_t),$$

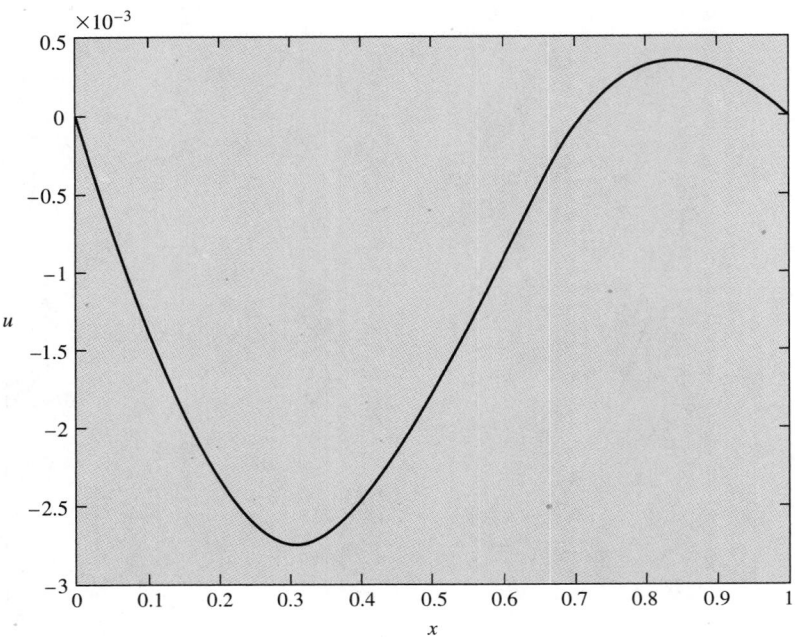

FIGURE 9.4   Error profile at $t = 1$ for heat equation; implicit method, $h = 1/32$, $\Delta t = 1/128$.

where $\eta_t \in [t, t+h]$, $\theta_t \in [t, t+h]$, $\eta_1 \in [x-h, x+h]$, and $\eta_0 \in [x-h, x+h]$. Finally, we drop the remainders to define the numerical method:

$$u_j^{n+1} - \mu(u_{j+1}^{n+1} - 2u_j^{n+1} + u_{j-1}^{n+1}) = u_j^n + \mu(u_{j+1}^n - 2u_j^n + u_{j-1}^n) \tag{9.18}$$

$$+ \frac{1}{2}\Delta t(f(x_j, t_{n+1}) + f(x_j, t_n)), \tag{9.19}$$

where $u_j^n = u_{h,\Delta t}(t_i, t_n) \approx u(x_j, t_n)$ is the approximate solution and $\mu = \frac{1}{2}a\Delta t h^{-2}$. This is again a system of $N-1$ equations in $N-1$ unknowns, but it is slightly more involved than the system for the implicit method (9.12). However, it is still unconditionally stable and it retains second-order accuracy in both space and time, as we shall soon see.

We can write the system (9.19) in matrix-vector form as

$$(I + \mu K)u_h^{n+1} = (I - \mu K)u_h^n + \frac{1}{2}\Delta t F^n,$$

where

$$F^n = (f(x_1, t_{n+1}) + f(x_1, t_n), \ldots, f(x_{N-1}, t_{n+1}) + f(x_{N-1}, t_n))^T$$

is the vector of source function values, and

$$K = \begin{bmatrix} 2 & -1 & 0 & \cdots & 0 & 0 \\ -1 & 2 & -1 & 0 & \cdots & 0 \\ \vdots & & \ddots & & & \vdots \\ 0 & \cdots & 0 & -1 & 2 & -1 \\ 0 & 0 & 0 & \cdots & -1 & 2 \end{bmatrix}.$$

Theorem 9.3 *Crank-Nicolson is unconditionally stable and second-order accurate in space and time, in the sense that*

$$\left(\sum_{i=1}^{N-1} (u(x_i,t_n) - u_{h,\Delta t}(x_i,t_n))^2 \right)^{1/2} \le C_0^n \left(\sum_{i=1}^{N-1} (u(x_i,0) - u_{h,\Delta t}(x_i,0))^2 \right)^{1/2} + C_1(h^2 + \Delta t^2)M,$$

where C_0 and C_1 are positive and independent of h and Δt, $C_0 < 1$, and M depends on the derivatives u_{xxxx}, u_{xxtt}, and f_{tt}.

Proof The proof involves a linear algebra argument that leans heavily on the fact that K is a symmetric, positive definite matrix. This means that the eigenvalues of K are real and positive, and that there is an orthogonal matrix Q that diagonalizes K (see Theorem 8.1).

We note that the vector of errors

$$e^n = (u(x_1,t_n) - u_1^n, \ldots, u(x_{N-1},t_n) - u_{N-1}^n)^T$$

satisfies the equation

$$(I + \mu K)e^{n+1} = (I - \mu K)e^n + \Delta t T^n, \tag{9.20}$$

where T^n is the truncation error vector at time t_n, given by

$$T_i^n = \frac{1}{24}h^2 a \left(\frac{\partial^4 u}{\partial x^4}(\eta_{1,i},t_{n+1}) + \frac{\partial^4 u}{\partial x^4}(\eta_{0,i},t_n) \right) - \frac{1}{12}\Delta t^2 u_{xxtt}(x_i,\theta_t) - \frac{1}{12}\Delta t^2 f_{tt}(x_i,\eta_t).$$

Since K is symmetric, there exists an orthogonal matrix Q such that

$$K = QDQ^T$$

where D is the diagonal matrix of the eigenvalues of K. If we substitute this into (9.20) and multiply by $Q^T = Q^{-1}$, we get

$$(I + \mu D)Q^T e^{n+1} = (I - \mu D)Q^T e^n + \Delta t Q^T T^n$$

so that

$$Q^T e^{n+1} = (I + \mu D)^{-1}(I - \mu D)(Q^T e^n) + \Delta t (I + \mu D)^{-1} Q^T T^n$$

or

$$Q^T e^{n+1} = D_1(Q^T e^n) + \Delta t D_2 Q^T T^n$$

for the diagonal matrices

$$D_1 = (I + \mu D)^{-1}(I - \mu D)$$

and

$$D_2 = (I + \mu D)^{-1}.$$

This recursion is easily solved to get

$$Q^T e^n = D_1^n(Q^T e^0) + \Delta t \sum_{k=0}^{n-1} D_1^k D_2 Q^T T^{n-1-k}$$

so that

$$e^n = Q D_1^n(Q^T e^0) + \Delta t \sum_{k=0}^{n-1} Q D_1^k D_2 Q^T T^{n-1-k}.$$

Now take the 2-norm of both sides to get

$$\|e^n\|_2 \leq \|Q\|_2 \|D_1^n\|_2 \|Q^T\|_2 \|e^0\|_2 + \Delta t \|Q\|_2 \sum_{k=0}^{n-1} \|D_1^k D_2 Q^T T^{n-1-k}\|_2.$$

But $\|Q\|_2 = 1$ for any orthogonal matrix Q (Exercise 5) and it is easily shown (Exercise 6) that

$$\|D\|_2 = \max_{1 \leq k \leq m} |d_{kk}|$$

for any $m \times m$ diagonal matrix. Thus, we have

$$\|e^n\|_2 \leq \|D_1^n\|_2 + t_n \|D_1 D_2\|_2 \max_{0 \leq k \leq n} \|T^k\|_2,$$

from which the conclusion follows. □

To illustrate the additional accuracy, consider the same simple example that we used to test the explicit and implicit methods. If we use Crank-Nicolson, we get better error for both the $h = 1/16, \Delta t = 1/64$ case and the $h = 1/32, \Delta t = 1/128$ case, as Figures 9.5 and 9.6 show. Of more importance is the fact that we can get the same accuracy that the explicit method had with a lot fewer time steps, as Figure 9.7 shows.

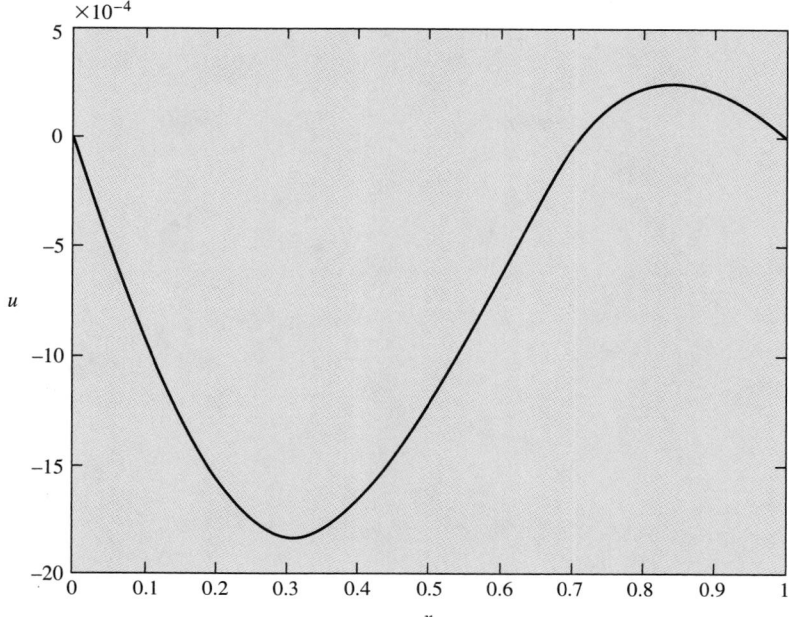

FIGURE 9.5 Error profile at $t = 1$ for heat equation; Crank-Nicolson method, $h = 1/16, \Delta t = 1/64$.

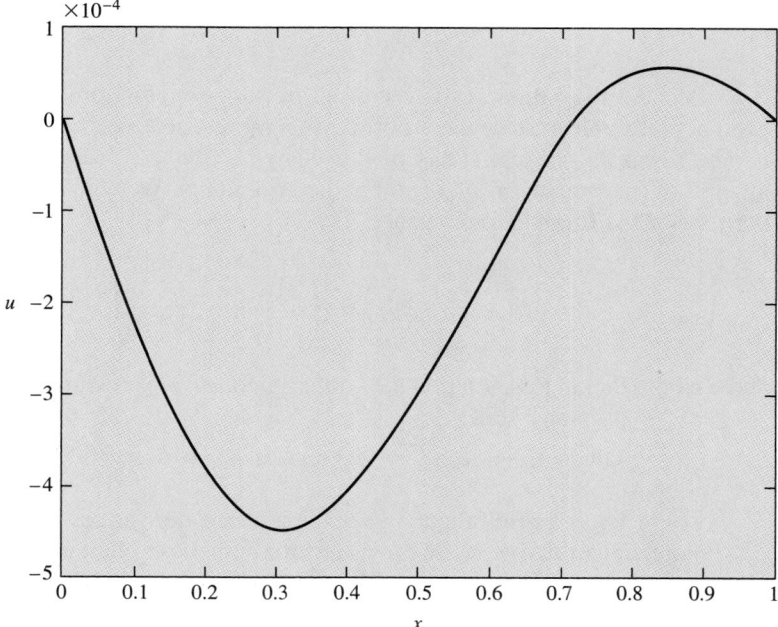

FIGURE 9.6 Error profile at $t = 1$ for heat equation; Crank-Nicolson method, $h = 1/32, \Delta t = 1/128$.

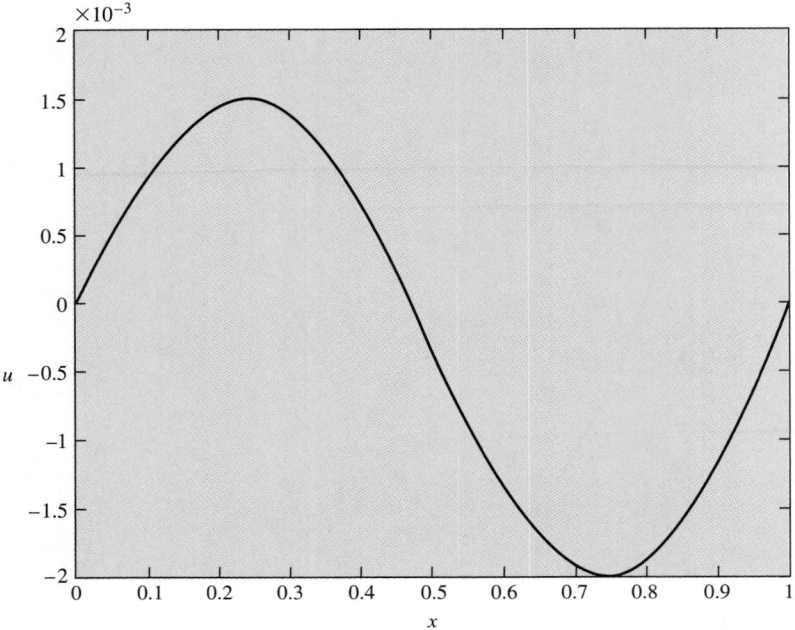

FIGURE 9.7   Error profile at $t = 1$ for heat equation; Crank-Nicolson method, $h = 1/16, \Delta t = 1/6$.

Because the values of the vector u_h^n are supposed to represent function values, it is often preferable to scale the 2-norm by the mesh size h. We will call this the *discrete integral 2-norm*, because it can be considered a trapezoid rule approximation to the integral of (the square of) $u_{h,\Delta t}(\cdot, t_n)$ at the grid points. We will write $\|u_{h,\Delta t}^n\|_{2,h}$ for this norm, which is formally defined as

$$\|u_{h,\Delta t}(\cdot, t_n)\|_{2,h} = \left(h \sum_{i=1}^{N-1} u_{h,\Delta t}^2(x_i, t_n) \right)^{1/2}, \tag{9.21}$$

and we note that in this notation the Crank-Nicolson error estimate can be written as

$$\|u^n - u_{h,\Delta t}^n\|_{2,h} \le C_0^n \|u_0 - u_{h,\Delta t}^0\|_{2,h} + C_1(h^2 + \Delta t^2) \|M_D\|_{2,h},$$

where M_D is a grid function dependent on the derivatives of the exact solution.

Note, however, that while the results for both the explicit method and the implicit method were in the infinity norm, the results for Crank-Nicolson are in the 2-norm. What does this mean, and what difference does it make?

It turns out to make quite a difference. The infinity norm gives us "pointwise" results, but the 2-norm does not; it gives average values. This allows the error in Crank-Nicolson to exhibit more "wiggles" than for the other methods; it also allows

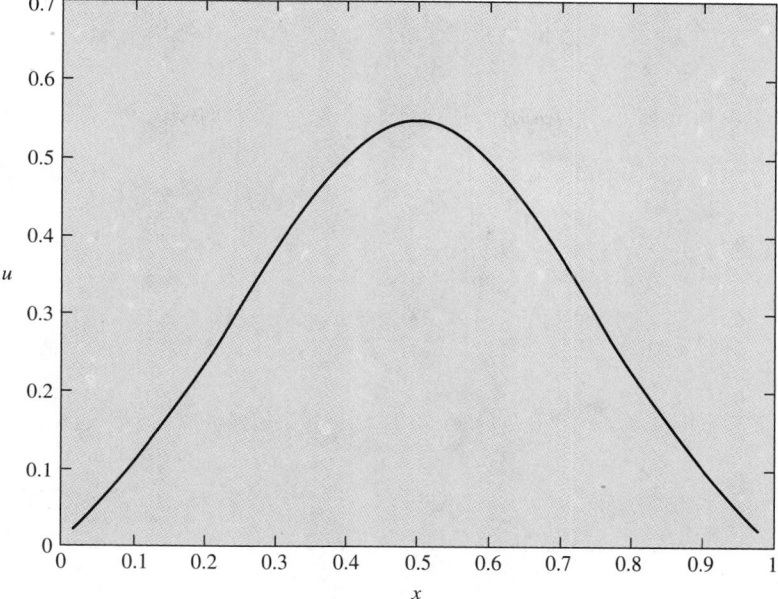

FIGURE 9.8   Solution profile at $t = \frac{1}{4}$ for discontinuous initial data; implicit method.

Crank-Nicolson to generate a negative approximation to a positive value, which will not happen with the other methods.

To illustrate what the difference between the two norms gives us, consider the new example defined by the initial data

$$u_0(x) = \begin{cases} 0 & 0 \le x \le \frac{1}{3} \\ 1 & \frac{1}{3} < x < \frac{2}{3} \\ 0 & \frac{2}{3} \le x \le 1. \end{cases}$$

This function is discontinuous at $x = \frac{1}{3}$ and $x = \frac{2}{3}$, but it can be shown that the exact solution will be very smooth for any $t > 0$. We approximated the solution to the heat equation using this data, with $a = \pi^{-2}, h = 1/64$, and $\Delta t = 1/32$, and computed out to $t = 1/4$ (thus using eight time steps) in both the implicit and Crank-Nicolson methods. As Figures 9.8 and 9.3 show, the implicit method handles the discontinuity in the initial data much better than Crank-Nicolson does. In order to remove the oscillations, we have to do one of two things: either take Δt very small in Crank-Nicolson—small enough to make the explicit method stable—or take the first step using the implicit method and then proceed with Crank-Nicolson (this smooths out the discontinuity enough to prevent the oscillations seen in Crank-Nicolson, without introducing too much error). Figures 9.8–9.10 illustrate this.

FIGURE 9.9 Solution profile at $t = \frac{1}{4}$ for discontinuous initial data; Crank-Nicolson.

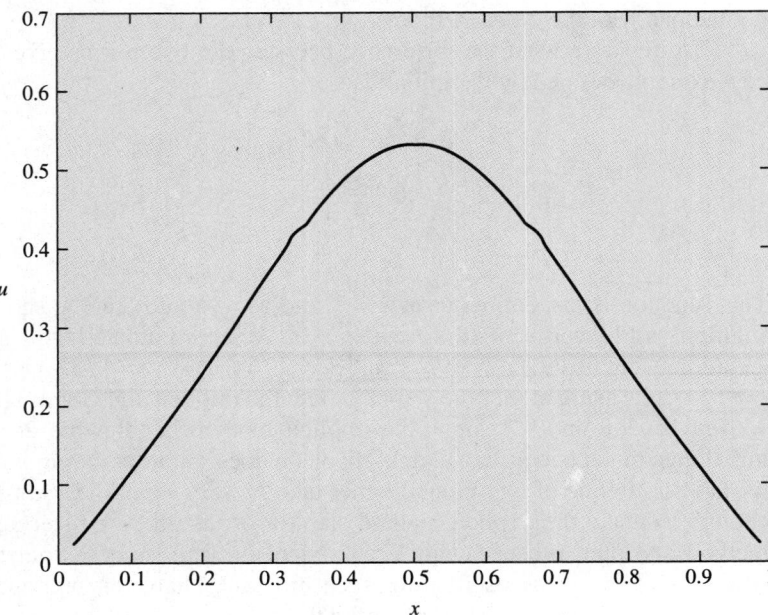

FIGURE 9.10 Solution profile at $t = \frac{1}{4}$ for discontinuous initial data; Crank-Nicolson with implicit first step.

EXERCISES

1. Write a program to use the explicit method to solve the diffusion equation

$$u_t = u_{xx}, \, t > 0, \, 0 < x < 1;$$
$$u(0, t) = 0;$$
$$u(1, t) = 0;$$
$$u(x, 0) = \sin \pi x.$$

which has exact solution

$$u(x, t) = e^{-\pi^2 t} \sin \pi x,$$

(The student should check that this is indeed the exact solution.) Use $h^{-1} = 4, 8, \ldots$, and take Δt as large as possible for stability. Confirm that the approximate solution is as accurate as the theory predicts. Compute out to $t = 1$.

2. Repeat the preceding exercise, but this time take Δt 10% too large to satisfy the stability condition, and attempt to compute solutions out to $t = 1$. Comment on what happens.

3. Apply Crank-Nicolson to the same PDE as in Exercise 1. For each value of h, adjust the choice of Δt to obtain comparable accuracy in the pointwise norm to what was achieved above. Comment on your results. Try to estimate the number of operations needed for each computation.

4. Modify the three algorithms for the diffusion equation to handle nonhomogeneous boundary conditions, i.e., to handle problems of the form

$$u_t = u_{xx} + f(x,t), \, t > 0, \, 0 < x < 1$$
$$u(0,t) = g_0(t)$$
$$u(1,t) = g_1(t)$$
$$u(x,0) = u_0(x),$$

where $g_0(t)$ and $g_1(t)$ are not identically zero. Test your work by applying it to the problem defined by

$$f(x,t) = 0, \, g_0(t) = \frac{1}{\sqrt{t+1}}, \, g_1(t) = \frac{e^{-1/4(t+1)}}{\sqrt{t+1}},$$

$$u_0(x) = e^{-x^2/4},$$

for which the exact solution is

$$u(x,t) = \frac{e^{-x^2/4(t+1)}}{\sqrt{t+1}}.$$

5. Let Q be an arbitrary $n \times n$ orthogonal matrix. Show that $\|Q\|_2 = 1$.

6. Recall the definition of the matrix 2-norm:

$$\|A\|_2 = \max_{u \neq 0} \frac{\|Au\|_2}{\|u\|_2}$$

where $\| \cdot \|_2$ is the usual vector 2-norm. If D is an $n \times n$ diagonal matrix, use this definition to show that

$$\|D\|_2 = \max_{1 \leq k \leq n} |d_{ii}|,$$

where d_{ii} are the diagonal elements of D.

7. Compute the number of operations need to compute out to $t = T$ using the explicit method, as a function of n, the number of points in the spatial grid. Assume that the time step is chosen to be as large as possible to satisfy the stability condition.

8. Compute the number of operations needed to compute out to $t = T$ using Crank-Nicolson, taking $\Delta t = ch$ for some constant $c > 0$, again as a function of n, the number of points in the spatial grid.

9. Use the explicit method to solve the nonlinear diffusion equation given at the beginning of the chapter. Take $g_0(t) = 1$, $g_1(t) = 0$, $u_0(x) = (1-x)^{15} + 9x^3(1-x)^2$ as the boundary and initial conditions. Use $D = 0.1$, $V = 1$, and $K = 0.1$ as the parameter values. Use $h = 1/16$ and $\Delta t = 1/320$ as the computational parameters; note that these values are appropriate for stability in the linear case. Compute your solutions out to $T = 0.1$ and plot the solution as a function of x for $t = T/4, T/2, 3T/4$, and T.

10. For the equation in the previous exercise, try to determine the "steady-state" solution by computing out for larger and larger t values until the solution stops changing. Plot the steady-state solution. How large a t value is needed to reach the steady state?

11. Repeat the previous two exercises with a smaller diffusion coefficient, $D = 0.05$. Keep everything else the same, and comment on what changes in your solutions.

9.2   DIFFERENCE METHODS FOR POISSON EQUATIONS

9.2.1   Discretization

In this section we will discuss methods for the approximate solution of PDEs of the form

$$-\Delta u = f, \; (x,y) \in D, \; u = g, \; (x,y) \in \Gamma = \partial D,$$

where Δ is the Laplace operator,

$$\Delta u = \frac{\partial^2 u}{\partial x^2} + \frac{\partial^2 u}{\partial y^2},$$

D is an open, connected subset of $\mathbf{R}^2$, and Γ is the boundary of D. The unknown function u therefore depends on two independent variables, x and y.

For simplicity's sake we will assume that D is the unit square $(0,1) \times (0,1)$, although much more general domains and, of course, much more general equations can be considered. It is also possible to consider problems posed in three dimensions.

In terms of the general theory, problems of this form have much in common with the boundary value problems studied in Section 6.10.1. The approximation is constructed by replacing the derivatives in the PDE with difference quotients, resulting in a system of linear equations. While we can apply Gaussian elimination and similar ideas to the solution of these systems, the size of the problem often mandates that we look at different methods.

Since the difference methods are fairly straightforward to apply, we will spend most of our time discussing efficient means for solving the large systems of linear equations that result from discretizing such problems. Thus, although the chapter is about approximating PDEs, this section has more to do with linear algebra than anything else.

We first define grids in both the x and y directions, in the usual way:

$$0 = x_0 < x_1 < x_2 < \cdots < x_{N-1} < x_N = 1$$

and

$$0 = y_0 < y_1 < y_2 < \cdots < y_{M-1} < y_M = 1.$$

For simplicity's sake, we will assume $N = M$, which implies an equivalent grid, with the same spacing h in each direction. We denote the approximate solution as the array of values $u_{i,j} \approx u(x_i, y_j)$, where x_i and y_j are points on a grid over D. Second, we use these in the usual difference approximation to the second derivative to write the approximate form of the PDE as

$$-\left[\frac{u_{i+1,j} - 2u_{i,j} + u_{i-1,j}}{h^2} \right] - \left[\frac{u_{i,j+1} - 2u_{i,j} + u_{i,j-1}}{h^2} \right] = f(x_i, y_j), \qquad (9.22)$$

which can be simplified somewhat to yield

$$-u_{i+1,j} - u_{i-1,j} + 4u_{i,j} - u_{i,j+1} - u_{i,j-1} = h^2 f_{i,j}, \; 1 \le i, j \le N-1,$$

where we have written $f_{i,j} = f(x_i, y_j)$. This is a system of $(N-1)^2$ equations in $(N-1)^2$ unknowns. To write it in a matrix-vector format, we define the vector u_h as a one-dimensional arrangement of the $u_{i,j}$ values:

$$u_h = \left(u_{1,1}, u_{1,2}, u_{1,3}, \ldots, u_{1,N-1}, u_{2,1}, \ldots, u_{N-1,N-1} \right)^T.$$

We could just as easily have ordered the $u_{i,j}$ in another way (by columns instead of rows, for example). The ordering of the unknowns can make a difference, especially in more complicated settings.

The resulting linear system can be quite large but has a very interesting structure. For $N = 4$ we have the following 9×9 system:

$$
\begin{bmatrix}
4 & -1 & 0 & -1 & 0 & 0 & 0 & 0 & 0 \\
-1 & 4 & -1 & 0 & -1 & 0 & 0 & 0 & 0 \\
0 & -1 & 4 & 0 & 0 & -1 & 0 & 0 & 0 \\
-1 & 0 & 0 & 4 & -1 & 0 & -1 & 0 & 0 \\
0 & -1 & 0 & -1 & 4 & -1 & 0 & -1 & 0 \\
0 & 0 & -1 & 0 & -1 & 4 & 0 & 0 & -1 \\
0 & 0 & 0 & -1 & 0 & 0 & 4 & -1 & 0 \\
0 & 0 & 0 & 0 & -1 & 0 & -1 & 4 & -1 \\
0 & 0 & 0 & 0 & 0 & -1 & 0 & -1 & 4
\end{bmatrix}
\begin{bmatrix}
u_{1,1} \\ u_{1,2} \\ u_{1,3} \\ u_{2,1} \\ u_{2,2} \\ u_{2,3} \\ u_{3,1} \\ u_{3,2} \\ u_{3,3}
\end{bmatrix}
= h^2
\begin{bmatrix}
f_{1,1} \\ f_{1,2} \\ f_{1,3} \\ f_{2,1} \\ f_{2,2} \\ f_{2,3} \\ f_{3,1} \\ f_{3,2} \\ f_{3,3}
\end{bmatrix}. \qquad (9.23)
$$

We can take advantage of the structure in this system by defining some notation. Let

$$
K = \begin{bmatrix}
4 & -1 & 0 \\
-1 & 4 & -1 \\
0 & -1 & 4
\end{bmatrix}
$$

and denote by u_i the column subvector of elements $u_{i,j}, 1 \leq j \leq N - 1$ (and similarly for f_i). Then we can write the large system more compactly as

$$
\begin{bmatrix}
K & -I & 0 \\
-I & K & -I \\
0 & -I & K
\end{bmatrix}
\begin{bmatrix}
u_1 \\ u_2 \\ u_3
\end{bmatrix}
= h^2
\begin{bmatrix}
f_1 \\ f_2 \\ f_3
\end{bmatrix}.
$$

There are a number of things that can be observed about this system of equations. The most important one, for our purposes, is that it contains a lot of zeroes. In fact, the coefficient matrix is "mostly" zeroes. Out of $O(N^4)$ elements in the matrix, only $O(5N^2)$ will be nonzero. Some of the other properties—the symmetry, the very nice and tidy block structure, the simplicity of the numbers—will be drastically affected by changes in the original PDE or the computational parameters or the shape of the domain D. But the *sparseness* of the matrix will remain.

We could solve this system by applying the standard techniques from Chapter 7, and for moderate-sized problems this is perhaps the best choice. The coefficient matrix is *SPD*, so the Cholesky algorithm is the natural first choice. However, for very large problems, so-called direct solution algorithms are not the best choice. They take too much computer time and use too much computer storage.

An alternative approach is to use an iterative technique like those studied in Section 7.7 to solve the system. This seems counterintuitive, perhaps, since we have available direct methods for exactly solving linear systems. Why would we use an iterative method that is only going to approximate the solution? The answer lies in the size of the problem. For large values of N—in other words, for small values of

Algorithm 9.1: Cholesky Algorithm

In this pseudocode segment, the (full) matrix is stored in the variable a and the dimension of the matrix is n2.

```
a(1,1) = sqrt(a(1,1))
for i=2,n2 do
    a(i,1) = a(i,1)/a(1,1)
end
for j=2,n2-1 do
    sum = 0.0d0
    for k=1,j-1 do
        sum = sum + a(j,k)*a(j,k)
    end
    a(j,j) = sqrt(a(j,j) - sum)
    for i=j+1,n2 do
        sum = 0.0d0
        for k=1,j-1 do
            sum = sum + a(i,k)*a(j,k)
        end
        a(i,j) = (a(i,j) - sum)/a(j,j)
    end
end
sum = 0.0d0
for k=1,n2-1 do
    sum = sum + a(n2,k)*a(n2,k)
end
a(n2,n2) = sqrt(a(n2,n2) - sum)
```

the mesh parameters—performing Gaussian elimination on a system like (9.23), but for larger N, will involve an enormous amount of computer storage, and the overhead in managing that storage can make the computation excessively expensive (even on modern computers). On the other hand, the iterative techniques that are used in practice converge very quickly, and the individual iterations are so cheap that the overall algorithms are cost-competitive with Gaussian elimination.

9.2.2 Banded Cholesky Solvers

Note that the coefficient matrix in our example above is entirely zero outside of three sub- or superdiagonals of the main diagonal. This is the essential feature of a *banded* matrix, that all the elements are zero if we are far enough away from the diagonal. Formally, we say a matrix A is banded, with *half-bandwidth p*, if

$$a_{i,j} = 0$$

for all i, j such that $|i - j| > p$. Thus our example is banded with half-bandwidth $p = 3$.

The important feature of banded systems is that the elimination algorithm can be carried out "within the band." That is, we don't need to reference, loop through, or even store those elements outside the band. This allows us to write elimination and solution routines that not only run faster but use much less storage.

To illustrate, consider a Cholesky factorization routine applied to systems like (9.23). A naive approach, ignoring the band structure, would yield a factorization algorithm much like Algorithm 9.1.

This requires storage of the full $n \times n$ matrix (in terms of the discretization parameters, then, we require about N^4 words of storage), and for larger values of n we will be doing a lot of meaningless computations with zeroes in the upper reaches of all the loops.

We can modify this algorithm to "stay within the band" simply by limiting some of the upper ranges on the loops, as Algorithm 9.2 shows.

Note, however, that this does not save us any storage; we have simply restricted where the program ranges in doing its work. To save storage, we have to decide on a scheme by which we put the band of the original matrix A into a smaller array, and then we have to modify the indexing of our banded algorithm accordingly.

Perhaps the simplest storage scheme is to store each column below (and including) the main diagonal as a column in a rectangular array. Thus we store the band of our example A as

$$
B_A = \begin{bmatrix}
4 & 4 & 4 & 4 & 4 & 4 & 4 & 4 & 4 \\
-1 & -1 & 0 & -1 & -1 & 0 & -1 & -1 & 0 \\
0 & 0 & 0 & 0 & 0 & 0 & 0 & 0 & 0 \\
-1 & -1 & -1 & -1 & -1 & 0 & 0 & 0 & 0
\end{bmatrix}
$$

Algorithm 9.2: Band-Limited Cholesky Algorithm

In this pseudocode segment, the (full) matrix is stored in the variable ab and the dimension of the matrix is n2. The loop indices are adjusted to range only over that portion of the matrix that lies within the band.

```
ab(1,1) = sqrt(ab(1,1))
for i=2,n do
     ab(i,1) = ab(i,1)/ab(1,1)
end
for j=2,n2-1 do
     sum = 0.0d0
     kmin = max(1,j-n+1)
     for k=kmin,j-1 do
          sum = sum + ab(j,k)*ab(j,k)
     end
     ab(j,j) = sqrt(ab(j,j) - sum)
     imax = min(n2,j+n-1)
     for i=j+1,imax do
          sum = 0.0d0
          kmin = max(1,j-n+1)
          for k=kmin,j-1 do
               sum = sum + ab(i,k)*ab(j,k)
          end
          ab(i,j) = (ab(i,j) - sum)/ab(j,j)
     end
end
sum = 0.0d0
for k=n2-n+1,n2-1 do
     sum = sum + ab(n2,k)*ab(n2,k)
end
ab(n2,n2) = sqrt(ab(n2,n2) - sum)
```

Instead of being a full 9×9, this is only 4×9. In terms of the discretization parameters, instead of storing a matrix of size $(N-1)^2 \times (N-1)^2$, we can get by with one that is $N \times (N-1)^2$. In large problems, this can be a substantial savings.

To finish the banded algorithm, we need an *index map* that takes us from the indices for the original matrix A and returns the appropriate indices for the band-only array B_A. One advantage of the band storage system we have chosen to use is that all of the second (i.e., column) indices remain the same. Thus all we need to do is determine the single-variable map that takes us from $a_{i,j}$ to $b^a_{k(i,j),j}$. A little experimentation with the indices ought to convince the reader that the map we want is

$$k(i,j) = i - j + 1,$$

so that we get our final band algorithm by replacing every occurrence of $a_{i,j}$ in Algorithm 9.2 with $b^a_{k(i,j),j}$. This gives us Algorithm 9.3.

The scheme we have used here to handle the banded matrix is somewhat crude but easy to understand. It is possible to construct much more sophisticated procedures which allow us to efficiently handle matrices that have "variable bandwidths" as we move down the diagonal. Some discussion of this is found in the text by Watkins [16]; a complete treatment is given in [5].

This is the kind of algorithm that is used in the comparison tests with the conjugate gradient and other iterative methods in the next section. While this is a very reasonable approach to take for problems where a fine grid is required to obtain the accuracy needed in the PDE approximation, the storage requirements will still be excessive on any but the largest computers; note also that even with the band storage system, as n gets larger we have to store a lot of elements that begin as zeroes but are overwritten with nonzeroes during the elimination process. (This is known as "fill-in.") This is why we are led to the use of iterative methods.

It is important to note that using the band-limited algorithm does not save us operations so much as *storage*. Algorithm 9.3 does exactly as many operations as Algorithm 9.2 but uses much less computer storage. Even with today's computers, this can be an issue.

9.2.3   Iteration and the Method of Conjugate Gradients

In Section 7.7 we looked at some very simple iterative methods for solving large sparse linear systems. In this section we will show how to implement them when applied to the kinds of systems generated by discretizing Poisson equations. We will also introduce a very powerful iterative technique, the method of conjugate gradients.

Recall the equation for the approximate value at (x_i, y_j):

$$-u_{i+1,j} - u_{i-1,j} + 4u_{i,j} - u_{i,j+1} - u_{i,j-1} = h^2 f_{i,j}. \tag{9.24}$$

We can solve this for $u_{i,j}$ as follows:

$$u_{i,j} = \frac{1}{4}\left(u_{i+1,j} + u_{i-1,j} + u_{i,j+1} + u_{i,j-1} + h^2 f_{i,j}\right). \tag{9.25}$$

Algorithm 9.3: Band-Limited Cholesky Algorithm (with Band Storage)

Here the matrix is stored in the variable ba, with only that part within the band being stored. The matrix thus has n rows and n2 columns. The function kk is the index map and is defined by kk(i,j) = i - j + 1.

```
ba(kk(1,1),1) = sqrt(ba(kk(1,1),1))
for i=2,n do
    ba(kk(i,1),1) = ba(kk(i,1),1)/ba(kk(1,1),1)
end
for j=2,n2-1 do
    sum = 0.0d0
    kmin = max(1,j-n+1)
    for k=kmin,j-1 do
        sum = sum + ba(kk(j,k),k)*ba(kk(j,k),k)
    end
    ba(kk(j,j),j) = sqrt(ba(kk(j,j),j) - sum)
    imax = min(n2,j+n-1)
    for i=j+1,imax do
        sum = 0.0d0
        kmin = max(1,j-n+1)
        for k=kmin,j-1 do
            sum = sum + ba(kk(i,k),k)*ba(kk(j,k),k)
        end
        ba(kk(i,j),j) = (ba(kk(i,j),j) - sum)/ba(kk(j,j),j)
    end
end
sum = 0.0d0
for k=n2-n+1,n2-1 do
    sum = sum + ba(kk(n2,k),k)*ba(kk(n2,k),k)
end
ba(kk(n2,n2),n2) = sqrt(ba(kk(n2,n2),n2) - sum)
```

Suppose now that we take $u_{i,j}^{(0)}$ as an initial guess, for $1 \leq i \leq N-1$ and similarly for j, and convert (9.25) to an iteration:

$$u_{i,j}^{(k+1)} = \frac{1}{4}\left(u_{i+1,j}^{(k)} + u_{i-1,j}^{(k)} + u_{i,j+1}^{(k)} + u_{i,j-1}^{(k)} + h^2 f_{i,j}\right). \qquad (9.26)$$

This is the same as the Jacobi iteration applied to the original system defined by (9.24), as can be seen by a modest manipulation (Exercise 2). Thus, the Jacobi iteration can be implemented by a very simple set of loops, as we can see in Algorithm 9.4.

The Gauss-Seidel iteration is not much more complicated than Jacobi. We have (Exercise 3)

$$u_{i,j}^{(k+1)} = \frac{1}{4}\left(u_{i+1,j}^{(k)} + u_{i-1,j}^{(k+1)} + u_{i,j+1}^{(k)} + u_{i,j-1}^{(k+1)} + h^2 f_{i,j}\right). \qquad (9.27)$$

Note that some of the u values on the right side are at iteration k, and some are at iteration $k+1$. This is how Gauss-Seidel differs from Jacobi: We are using the most current value of the approximate solution for those elements for which it has been

Algorithm 9.4: Jacobi Iteration for the Approximate Solution of Poisson Equations

This pseudocode assumes the initial guess is stored in u0, that the error tolerance is in eps, that the maximum number of iterations is itmax, and that the number of subintervals in each direction of the mesh is n.

```
input n, h, itmax, eps
for k = 1 to itmax do
      rmax = 0.0
      for i = 1 to n-1 do
            for j = 1 to n-1 do
                  uold = uo(i,j-1) + uo(i,j+1) + uo(i-1,j) + uo(i+1,j)
                  u(i,j) = 0.25*(uold + h*h*f(i,j))
            end
      end
      for i = 1 to n-1 do
            for j = 1 to n-1 do
                  u(i,j) = uo(i,j)
                  r = h*h*f(i,j) - 4*u(i,j) + (u(i,j-1) + u(i,j+1) + u(i-1,j)
                  + u(i+1,j))
                  if abs(r) > rmax then rmax = abs(r)
            end
      end
      if rmax < eps then stop
end
```

computed. This leads to an algorithm (Algorithm 9.5) that is actually more compact than for Jacobi.

The most complicated of the three splitting methods is SOR. The best way to organize the SOR iteration is to first take a Gauss-Seidel step,

$$\bar{u} = \frac{1}{4}\left(u_{i+1,j}^{(k)} + u_{i-1,j}^{(k+1)} + u_{i,j+1}^{(k)} + u_{i,j-1}^{(k+1)} + h^2 f_{i,j} \right),$$

and then average this value with the previous iterate value, using the relaxation parameter:

$$u^{(k+1)} = \omega \bar{u} + (1-\omega)u^{(k)}. \tag{9.28}$$

Exercise 4 asks the student to show that this is equivalent to the matrix formulation in Chapter 7. Note that if $\omega = 1$, then Gauss-Seidel and SOR are equivalent.

The implementation of SOR uses this two-stage approach (see Algorithm 9.6).

Of course, the convergence behavior of the SOR iteration will be heavily dependent on the choice of ω. A very complete theory for the choice of the "best" value of ω is available [16] for the kinds of *SPD* systems that come from discretizing Poisson equations. The precise details are more than we want to get into in this text, but a summary of the important results is the following.

Theorem 9.4   *If the system of equations $Au = f$ comes from discretizing a Poisson equation of the form*

$$-a_1 u_{xx} - a_2 u_{yy} + a_3 u = f,$$

Algorithm 9.5: Gauss-Seidel Iteration for the Approximate Solution of Poisson Equations

The parameters are the same as in the Jacobi algorithm, except that we do not use separate vectors for the new and old iterations; a single array (u) is used. The advantage of this is that we don't need to waste time with the final pair of nested loops.

```
input n, h, itmax, eps
for k = 1 to itmax do
     rmax = 0.0
     for i = 1 to n-1 do
          for j = 1 to n-1 do
               usave = u(i,j)
               uold = u(i,j+1) + u(i+1,j)
               unew = u(i,j-1) + u(i-1,j)
               u(i,j) = 0.25*(uold + unew + h*h*f(i,j))
          end
     end
     for i = 1 to n-1 do
          for j = 1 to n-1 do
               r = h*h*f(i,j) - 4*u(i,j) + (u(i,j-1) + u(i,j+1) + u(i-1,j)
                  + u(i+1,j))
               if abs(r) > rmax then rmax = abs(r)
          end
     end
     if rmax < eps then stop
end
```

Algorithm 9.6: SOR Iteration for the Approximate Solution of Poisson Equations

The parameters are the same as in the Gauss-Seidel algorithm, with the addition of the relaxation parameter, omega.

```
input n, h, itmax, eps, omega
om1 = 1.0 - omega
for k = 1 to itmax do
     rmax = 0.0
     for i = 1 to n-1 do
          for j = 1 to n-1 do
               uold = u(i,j+1) + u(i+1,j)
               unew = u(i,j-1) + u(i-1,j)
               ubar = 0.25*(uold + unew + h*h*f(i,j))
               u(i,j) = omega*ubar + om1*u(i,j)
          end
     end
     for i = 1 to n-1 do
          for j = 1 to n-1 do
               r = h*h*f(i,j) - 4*u(i,j) + (u(i,j-1) + u(i,j+1) + u(i-1,j)
                  + u(i+1,j))
               if abs(r) > rmax then rmax = abs(r)
          end
     end
     if rmax < eps then stop
end
```

where $a_1 > 0$, $a_2 > 0$, and $a_3 \geq 0$, and the unknowns are ordered within the vector u either row-wise or column-wise, then

1. *The spectral radius of the Jacobi iteration matrix, ρ_J, and the spectral radius of the Gauss-Seidel iteration matrix, ρ_{GS}, are related according to*

$$\rho_{GS} = \rho_J^2.$$

2. *The "best" value of ω for the SOR iteration ("best" in the sense that it leads to the smallest spectral radius for the SOR iteration matrix) is given by*

$$\omega_* = \frac{2}{1 + \sqrt{1 - \rho_J^2}}.$$

Note that to use the optimal value of ω, we apparently have to solve an eigenvalue problem first. In practice, an SOR iteration is usually started with $\omega = 1$, which is Gauss-Seidel, and from the way that the iterates change it is possible to estimate ρ_{GS} and therefore ρ_J, thus allowing us to use an estimate of the optimal ω (see Exercise 5). However, there is no denying that the need to come up with a value for ω makes effective use of SOR difficult, and is one reason it is no longer as widely used as it once was.

The most important modern technique for the solution of large, sparse symmetric linear systems is known as the *method of conjugate gradients*, or CG. The basic ideas go back to the mid-1950s, but the utility of the method was not completely recognized until the late 1980s. A full treatment of the theory underlying the method requires more background than we have assumed here, so we will simply present the algorithm and outline some of the major ideas. Complete treatments are in the references [4], [9], [12].

Theorem 9.5   Let $A \in \mathbf{R}^{n \times n}$ be given, SPD. For any $f \in \mathbf{R}^n$, the vector $u \in \mathbf{R}^n$ satisfies $u = A^{-1}f$ if and only if

$$(u, Au) - 2(u, f) < (v, Av) - 2(v, f)$$

for all $v \in \mathbf{R}^n$, $u \neq v$.

The point of the theorem is that we can find the solution of $Ax = b$ by finding the vector that minimizes

$$\phi(u) = (u, Au) - 2(u, f).$$

Many techniques exist for the solution of minimization problems, and many have the form

$$u^{(k+1)} = u^{(k)} + \alpha_k p^{(k)},$$

where α_k is a scalar (called the *step length*) and $p^{(k)}$ is a vector (called the *search direction*). Thus we go from the current approximate minimizer to the next approximate minimizer by determining a direction in which to move, and a distance to go along that

Algorithm 9.7: Conjugate Gradient Iteration

Assume $A \in SPD$ and compute as follows.

1. Given $u^{(0)}$, the initial guess, compute $r^{(0)} = f - Au^{(0)}$ and $p^{(0)} = r^{(0)}$. Let ϵ be the convergence tolerance.

2. For $k = 1$ to convergence do

(a) $w = Ap^{(k-1)}$

(b) $\alpha = (r^{(k-1)}, r^{(k-1)}) / (p^{(k-1)}, w)$

(c) $u^{(k)} = u^{(k-1)} + \alpha p^{(k-1)}$

(d) $r^{(k)} = r^{(k-1)} - \alpha w$

(e) If $\|r^{(k)}\|_2 \le \epsilon$, then stop.

(f) $\beta = (r^{(k)}, r^{(k)}) / (r^{(k-1)}, r^{(k-1)})$

(g) $p^{(k)} = r^{(k)} + \beta p^{(k-1)}$

direction. Often α_k is chosen to minimize ϕ along the search direction. Obviously, a lot will depend on how the search direction is chosen.

The CG method works by choosing the search direction so that the residual vectors $r^{(k)} = b - Ax^{(k)}$ are all mutually orthogonal:

$$(r^{(k)}, r^{(j)}) = 0, \ k \ne j.$$

It therefore follows that if A is $m \times m$, then $r^{(m)} = 0$, because otherwise we would have $m + 1$ mutually orthogonal vectors in an m-dimensional space. (See Exercise 6.) Thus, CG is not really an iterative method, since it finds the theoretically exact solution in a finite number of steps. However, as a practical matter, CG converges much faster than this would suggest, as some of the exercises will demonstrate.

The complete algorithm is presented as Algorithm 9.7. Note that we can execute this algorithm using only the storage for the matrix plus four additional vectors (u, w, r, and p), since we can overwrite the values from one iteration onto the next. In addition, we do not really need to store the entire matrix A; all we have to do is have a routine that can compute the product $w = Ap$. For a matrix as structured for the discretization of Poisson equations, this can be done very efficiently and with a minimum of storage.

EXAMPLE 9.1

To compare the performance of all four of these methods, we will try them each out on the linear system that we get from the PDE

$$-\Delta u = (2 + \pi^2 y(1-y)) \sin \pi x + (2 + \pi^2 x(1-x)) \sin \pi y, \ (x,y) \in R = (0,1) \times (0,1); \quad (9.29)$$
$$u = 0, \ (x,y) \in \partial R, \quad (9.30)$$

which has the exact solution

$$u(x,y) = y(1-y) \sin \pi x + x(1-x) \sin \pi y.$$

The discretization we are using is second-order accurate, so taking $h = 1/100$ gives us an expected error on the order of 10^{-4}; it also means that $N = 99^2 = 9801$. Doing a direct solution, even taking advantage of the symmetry and band structure of the matrix, would require working with a 99×9801 rectangular matrix; thus we would have to store 970,299 elements. Table 9.2 shows the results for each of the four iterations for a sequence of values of $h^{-1} = 4, 8, 16, \ldots, 128$; we show the number of iterations to convergence, an estimate of the number of "flops," or floating-point operations (additions, subtractions, multiplications, divisions, square roots) performed in doing the computation, and the error between the exact solution and the converged iteration, measured in the infinity norm. For SOR the optimal ω was precomputed and used, which is entirely unrealistic but allows us to see the best possible performance for SOR. The iterations were all stopped when the residual $f - Au$ was smaller than h^4; Jacobi, Gauss-Seidel, and SOR measured this in the infinity norm, but the 2-norm is much more natural for CG. This makes for something of an "apples and oranges" comparison, but it allows us to use the most natural norms for each method. Making the convergence tolerance dependent on h was done to ensure that the iteration error is much less than the $O(h^2)$ mathematical error in the approximation technique. For comparison's sake, the solution of the same system on the same computer, using a Cholesky factorization that took advantage of the band structure of the matrix, is summarized in Table 9.3. Note that CG was just as accurate as the Cholesky decomposition, but much cheaper in terms of the operation count. CG was also faster than the three splitting methods. CG, moreover, does not depend on a parameter like SOR does. Note that SOR and CG were the only methods for which we could run the $n = 256$ case. This last point deserves emphasizing, since one of the drawbacks to a direct solution technique such as the Cholesky decomposition is the memory requirement for storing even a band-limited form of the matrix. ∎

Conjugate gradients requires that the matrix $A \in SPD$; thus it will not work if the Poisson equation has a term such as $b(x,y)u_x$ in it (see Exercise 8). CG can be applied to asymmetric problems by multiplying by A^T; i.e., if we want to solve the linear system

$$Au = f$$

for $A \neq A^T$, we apply CG to the new system

$$A^T Au = A^T f.$$

If A is nonsingular, then $A^T A$ is SPD, and this can be implemented with only modest additional work in the algorithm (see Exercise 9). However, it can be shown that the condition number of $A^T A$ is the square of the condition number of A, and this can have drastic effects on the convergence of the CG algorithm. The development of an efficient version of CG that can be applied to asymmetric problems is a very active area of numerical analysis research.

The fundamental error and convergence theorem for CG can be stated once we define two notations, the first one of which we used in Chapter 8.

Definition 9.1 (Spectrum of a Matrix A)   *If A is a square matrix, the set $\sigma(A)$ is the set of all eigenvalues of A.*

Definition 9.2   *For any k the set Q_k will denote all polynomials p of degree k with the additional property that $p(0) = 1$.*

TABLE 9.2   Iterative solutions of (9.29)

Jacobi iteration

h^{-1}	Iterations	Error	Flops
4	15	1.1E–02	1908
8	82	2.5E–03	56,350
16	406	6.3E–04	1,279,350
32	1,917	1.6E–04	25,793,240
64	8,827	3.9E–05	490,489,020
128	39,921	9.8E–06	9,014,433,584

Gauss-Seidel iteration

h^{-1}	Iterations	Error	Flops
4	9	1.2E–02	1152
8	42	2.6E–03	28,910
16	204	6.3E–04	643,050
32	960	1.6E–04	12,917,762
64	4,415	3.9E–05	245,331,828
128	19,962	9.8E–06	4,507,571,630

SOR iteration with optimal ω

h^{-1}	Iterations	Error	Flops
4	6	1.3E–02	990
8	15	3.3E–03	13,328
16	33	8.0E–04	134,100
32	72	2.0E–04	1,247,378
64	158	4.9E–05	11,295,774
128	354	1.3E–05	102,806,246
256	831	3.2E–06	972,774,000

Conjugate gradient iteration

h^{-1}	Iterations	Error	Flops
4	3	6.8E–03	291
8	5	1.7E–03	3,045
16	9	4.2E–04	27,465
32	17	1.0E–04	232,593
64	39	2.6E–05	2,270,331
128	86	6.5E–06	20,596,860
256	178	1.6E–06	172,771,680

TABLE 9.3 Cholesky Solution of (9.29)

h^{-1}	Error	Flops
4	6.8298357260145E-03	322
8	1.6705855964484E-03	6,974
16	4.1526959889103E-04	122,230
32	1.0366786255403E-04	2,029,286
64	2.5907605235998E-05	33,021,382
128	6.4763166429495E-06	532,642,694

Then the basic theorem is as follows.

Theorem 9.6   Let the CG algorithm be applied to the solution of the linear system $Au = f$, *with* $A \in SPD$. *Let* $u^{(k)}$ *denote the* kth *iterate. Then the error satisfies*

$$\|u - u^{(k)}\|_2 \leq \sqrt{\kappa_2(A)} \left(\max_{\lambda \in \sigma(A)} \min_{p \in Q_k} |p(\lambda)| \right) \|u - u^{(0)}\|_2.$$

Stating the error in terms of a minimum over a set of polynomials allows us to use some advanced properties of Chebyschev polynomials to obtain the following very conservative result.

Theorem 9.7   The error in CG satisfies the conservative upper bound

$$\|u - u^{(k)}\|_2 \leq 2\sqrt{\kappa_2(A)} \left(\frac{\sqrt{\kappa_2(A)} - 1}{\sqrt{\kappa_2(A)} + 1} \right)^k \|u - u^{(0)}\|_2.$$

¿From this we see why the condition number of A is so important. For a poorly conditioned matrix, the method can converge very slowly, and this leads to the idea of *preconditioning*. If we want to solve the system

$$Au = f, \tag{9.31}$$

this is equivalent (assuming S is nonsingular) to solving

$$SAS^T S^{-T} u = Sf$$

or

$$SAS^T v = Sf \tag{9.32}$$

for $u = S^T v$. Note that the product SAS^T is always symmetric; moreover, if $(S^T S)^{-1} = M \approx A$, then it can be shown that

$$\kappa_2(SAS^T) \approx 1;$$

thus the CG iteration for solving (9.32) will converge more quickly than the CG iteration for solving the original system, and we recover u from v via $u = S^T v$. We can organize the CG iteration applied to the preconditioned system in such a way that we need to do only a few extra computations (see Algorithm 9.8).

Note that we do not actually compute the inverse of M; rather, we solve the system $Mz^{(k)} = r^{(k)}$. This means that the preconditioner must be a matrix for which the linear systems problem is *easier* to solve than it is for A.

The choice of preconditioners is an active area of research. Common choices amount to some kind of partial factorization of the original matrix A, or some direct approximation to A^{-1}.

One choice is the so-called incomplete Cholesky factorization; in this method, we compute only those elements of the Cholesky factorization of A that correspond to nonzero elements of A. (This avoids the massive fill-in that slows down the Cholesky factorization.) If L is the incomplete Cholesky factor, then the preconditioner is $M = LL^T$. The solution of the preconditioning equation is simple because M is already factored, and so the solution for $z^{(k)}$ is accomplished by a couple of triangular solution steps.

Another preconditioning scheme is *polynomial preconditioning*. Here we construct a matrix polynomial $P(A) \approx A^{-1}$ and precondition with $P(A)^{-1}$, so that the preconditioning steps are accomplished simply by evaluating $P(A)r^{(k)}$. For example, the polynomial

$$p(t) = \frac{176}{105} - \frac{86}{105}t + \frac{16}{105}t^2 - \frac{1}{105}t^3$$

is a decent approximation to t^{-1} on the interval $[1,8]$; it can be shown that this means that $p(A)$ will be a correspondingly decent approximation to A^{-1} for any matrix with real eigenvalues confined to the interval $[1,8]$. (The polynomial was computed by ordinary interpolation at the nodes 1, 3, 5, 7. These nodes were used because we know, from Gerschgorin's Theorem, that all the eigenvalues of the matrix A in (9.23)

Algorithm 9.8: Preconditioned Conjugate Gradient Algorithm

---

1. Given $u^{(0)}$, the initial guess, compute $r^{(0)} = f - Au^{(0)}$ and $p^{(0)} = z^{(0)} = M^{-1}r^{(0)}$. Let ϵ be the convergence tolerance.

2. For $k = 1$ to convergence, do
   (a) $w = Ap^{(k-1)}$
   (b) $\alpha = (z^{(k-1)}, r^{(k-1)})/(p^{(k-1)}, w)$
   (c) $u^{(k)} = u^{(k-1)} + \alpha p^{(k-1)}$
   (d) $r^{(k)} = r^{(k-1)} - \alpha w$
   (e) If $\|r^{(k)}\|_2 \le \epsilon$, then stop.
   (f) $z^{(k)} = M^{-1}r^{(k)}$
   (g) $\beta = (z^{(k)}, r^{(k)})/(z^{(k-1)}, r^{(k-1)})$
   (h) $p^{(k)} = z^{(k)} + \beta p^{(k-1)}$

TABLE 9.4 Incomplete Cholesky Preconditioned Conjugate Gradient Solution of (9.29)

h^{-1}	Iterations	Error	Flops
4	2	6.8E–03	799
8	3	1.7E–03	6,027
16	5	4.6E–04	42,667
32	9	1.6E–04	309,387
64	16	5.9E–05	2,195,175
128	30	1.7E–05	16,373,343
256	59	4.7E–06	128,241,523

TABLE 9.5 Polynomial Preconditioned Conjugate Gradient Solution of (9.29).

h^{-1}	Iterations	Error	Flops
4	1	6.9E–03	774
8	2	1.7E–03	6,324
16	4	4.2E–04	49,896
32	7	1.0E–04	347,742
64	14	2.6E–05	2,741,084
128	28	6.5E–06	21,750,168
256	55	1.6E–06	170,197,470

are in the interval $[0,8]$.) We thus define $M = p(A)^{-1}$ so that $M^{-1}r^{(k)} = P(A)r^{(k)}$, and use this in our preconditioned version of CG.

Tables 9.4 and 9.5 show the results; note that the number of iterations to convergence decreased markedly for both preconditioners, but the initial operation counts are actually greater, and it is only for the larger problems that the preconditioning pays off. This is because the preconditioning step involves a nontrivial number of additional operations. Note that polynomial preconditioning introduces many more matrix-vector multiplication into the computation (in computing $p(A)$), so we need to use the best possible polynomial, and it might not be a good idea to use too large a degree of a polynomial.

See Chapter 10 of Saad [12] for a good introduction to preconditioning ideas.

EXERCISES

1. What is the truncation error in the approximation defined by (9.21)?

2. Show that the Jacobi iteration (9.25) is equivalent to the matrix iteration (7.19) from Chapter 7.

3. Show that the Gauss-Seidel iteration (9.26) is equivalent to the matrix iteration (7.20) from Chapter 7.

4. Show that the SOR iteration (9.27) is equivalent to the matrix iteration (7.21) from Chapter 7.

5. For an iteration of the form

$$u^{(k+1)} = Tu^{(k)} + c,$$

show that

$$\frac{\|u^{(k+1)} - u^{(k)}\|_\infty}{\|u^{(k)} - u^{(k-1)}\|_\infty} \le \|T\|_\infty.$$

Can we use this to estimate ρ_J and therefore ω_*? (Hint: Recall Exercise 8 from Section 7.7.)

6. Let $x_1, x_2, \ldots, x_{n-1}, x_n$ be orthogonal vectors in a vector space V of dimension n. Show that if $z \in V$ is orthogonal to each one of the x_k, then $z = 0$.

7. Write a program to do each of the following:
(a) Banded Cholesky
(b) CG
(c) Preconditioned CG, using the polynomial

$$P(A) = \frac{1}{16}A^2 - \frac{25}{32}A + \frac{19}{8}I$$

as the preconditioner

Test each program on the example problem defined by

$$-\Delta u = \pi^2 \sin \pi x \sin \pi y, \ (x,y) \in (0,1) \times (0,1),$$

with $u = 0$ on the boundary, using $h = \frac{1}{8}$.

8. Discretize the Poisson equation

$$-u_{xx} - u_{yy} + bu_x = f, \ (x,y) \in (0,1) \times (0,1)$$
$$u(x,0) = u(x,1) = 0, \ x \in (0,1)$$
$$u(0,y) = u(1,y) = 0, \ y \in (0,1)$$

in the case $h = \frac{1}{3}$, $b \ne 0$. Is the resulting system symmetric?

9. Consider the linear system problem

$$Au = f,$$

where A is *not* symmetric. Modify the CG algorithm to solve the symmetrized system

$$A^T A u = A^T f$$

without explicitly forming the matrix $A^T A$.

10. Write a code to implement the algorithm you wrote in the previous exercise. Test it on the system obtained by discretizing the PDE

$$-u_{xx} - u_{yy} + u_x = \pi \cos \pi x \cos \pi y + 2\pi^2 \sin \pi x \sin \pi y,$$
$$(x,y) \in (0,1) \times (0,1)$$
$$u(x,0) = u(x,1) = 0, \ x \in (0,1)$$
$$u(0,y) = u(1,y) = 0, \ y \in (0,1)$$

for $h = \frac{1}{4}$ and $h = \frac{1}{8}$. The exact solution is $u(x,y) = \sin \pi x \sin \pi y$; use this to make sure your algorithm is working properly.

11. How does the discretization change when boundary data is nonhomogeneous (i.e., nonzero)? Demonstrate by writing down the discrete system for the PDE

$$-u_{xx} - u_{yy} = -2e^{x-y}, \ (x,y) \in (0,1) \times (0,1)$$
$$u(x,0) = e^x, \ u(x,1) = e^{x-1}, \ x \in (0,1)$$
$$u(0,y) = e^{-y}, \ u(1,y) = e^{1-y}, \ y \in (0,1)$$

for $h = \frac{1}{4}$. (Hint: It will help to write the values of the approximate solution at the grid points in two vectors— one for the interior grid points where the approximation is unknown, and one at the boundary grid points where the solution is known.)

12. Apply the following solution techniques to the system in the previous exercise, this time using $h = \frac{1}{8}$. Use the exact solution of $u(x,y) = e^{x-y}$ to verify that the code is working properly.
(a) Jacobi iteration
(b) Gauss-Seidel iteration
(c) SOR iteration, using $\omega = 1.4465$
(d) Conjugate gradient iteration

LITERATURE AND SOFTWARE DISCUSSION

The numerical solution of partial differential equations is one of the most active areas of numerical analysis research and is usually not treated extensively in undergraduate texts. Good sources for the basic difference methods outlined here are the books by Ames [1], Smith [13], Morton and Mayers [11], and Strikwerda [15]. A classic, though by now somewhat dated, source is the book by Forsythe and Wasow [6]. The question of numerical methods for the resulting linear systems gets extensive treatment. The best references for the CG method are Datta [4], Kelley [9], and Saad [12]. More classical iterative methods are found in the books by Young [17], Hageman and Young [7], and Birkhoff and Lynch [2].

One class of equations not treated here are first-order equations, often styled as "conservation laws." The general form of the equation in one space dimension is

$$u_t + (f(x,u))_x = 0.$$

These are important equations in areas such as fluid flow and gas dynamics; a good reference is the book by LeVeque [10].

Perhaps the most important technique for approximating solutions to PDEs is the finite-element method, which is sufficiently complicated that it deserves its own course. The classic reference is Strang and Fix [14]; a more modern treatment is the book by Brenner and Scott [3].

Easy-to-use software for approximating the solution of PDEs is difficult to write and therefore difficult to find. The generality of the problem and the need to specify solution techniques, data functions, problem geometries, boundary conditions, etc. all make it a challenging project.

BIBLIOGRAPHY

[1] AMES, W.F., *Numerical Methods for Partial Differential Equations*. Academic Press, New York, 1977.

[2] BIRKHOFF, GARRETT, & LYNCH, ROBERT. *Numerical Solution of Elliptic Problems*. SIAM, Philadelphia, 1984.

[3] BRENNER, SUZANNE, & SCOTT, RIDGWAY. *The Mathematical Theory of Finite Element Methods*. Springer-Verlag, New York, 1994.

[4] DATTA, BISWA NATH. *Numerical Linear Algebra and Applications*, Brooks/Cole, Pacific Grove, CA, 1995.

[5] DUFF, IAN, ERISMAN, A.M., & REID, J.K. *Direct Methods for Sparse Matrices*. University Press, Oxford, 1989.

[6] FORSYTHE, GEORGE, & WASOW, WOLFGANG. *Finite-Difference Methods for Partial Differential Equations*. John Wiley & Sons, New York, 1960.

[7] HAGEMAN, L.A., & YOUNG, DAVID. *Applied Iterative Methods*. Academic Press, New York, 1981.

[8] HUMI, MAYER, & MILLER, WILLIAM B. *Boundary Value Problems and Partial Differential Equations*. PWS-Kent, Boston, 1992.

[9] KELLEY, C.T. *Iterative Methods for Linear and Nonlinear Equations*. SIAM, Philadelphia, 1995.

[10] LEVEQUE, RANDALL. *Numerical Methods for Conservation Laws*. Birkhäuser, Basel, Switzerland, 1990.

[11] MORTON, K.W., & MAYERS, D.F. *Numerical Solution of Partial Differential Equations*, University Press, Cambridge, 1994.

[12] SAAD, YOUSEF. *Iterative Methods for Sparse Linear Systems*. PWS, Boston, 1996.

[13] SMITH, G.D. *Numerical Solution of Partial Differential Equations: Finite Difference Methods*. University Press, Oxford, 1985.

[14] STRANG, GILBERT, & FIX, GEORGE. *An Analysis of the Finite Element Method*. Prentice-Hall, Englewood Cliffs, NJ, 1973.

[15] STRIKWERDA, JOHN. *Finite Difference Schemes and Partial Differential Equations*. Wadsworth & Brooks/Cole, Pacific Grove, CA, 1989.

[16] WATKINS, DAVID. *Fundamentals of Matrix Computations*. John Wiley & Sons, New York, 1991.

[17] YOUNG, DAVID. *Iterative Solution of Large Linear Systems*. Academic Press, New York, 1971.

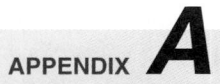

PROOFS OF SELECTED THEOREMS, AND OTHER ADDITIONAL MATERIAL

A.1 PROOFS OF THE INTERPOLATION ERROR THEOREMS

The interpolation error theorems all depend on applying a generalization of Rolle's Theorem to a very carefully constructed function. First, we will state and prove the general version of Rolle's Theorem.

Theorem A.1 (Generalized Rolle's Theorem)   Let $f \in C^n([a, b])$ be given, and assume there are n points $z_k, 1 \le k \le n$, in $[a, b]$ such that $f(z_k) = 0$. Then there exists at least one point $\xi \in [a, b]$ such that $f^{(n-1)}(\xi) = 0$.

Proof   By Rolle's Theorem, there exists at least one point η_k between each z_k and z_{k+1} such that $f'(\eta_k) = 0$. Thus there are $n - 1$ points where $g_1(x) = f'(x)$ is zero. By the same argument, then, there are $n - 2$ points where $g_2(x) = f''(x)$ is zero. Continuing onward, we end up with a single point where $g_{n-1}(x) = f^{(n-1)}(x)$ is zero.

This result allows us to prove, rather easily, Theorems 4.3 and 4.5.

Theorem A.2 (Lagrange Interpolation Error Theorem)   Let $f \in C^{n+1}([a, b])$ and let the nodes $x_k \in [a, b]$ for $0 \le k \le n$. Then, for each $x \in [a, b]$, there is a $\xi_x \in [a, b]$ such that

$$f(x) - p_n(x) = \frac{w_n(x)}{(n+1)!} f^{(n+1)}(\xi_x).$$

Proof   For any $t \in [a, b]$, $t \ne x_k$, define the function

$$G(x) = E(x) - \frac{w(x)}{w(t)} E(t),$$

where $E(x) = f(x) - p_n(x)$ and

$$w(x) = \prod_{k=0}^{n} (x - x_k).$$

535

Then $G(x_k) = 0$, for $0 \leq k \leq n$, and $G(t) = 0$; thus there are $n + 2$ points where G is zero. Therefore, by the generalized Rolle's Theorem, there is a point $\xi \in [a, b]$ such that $G^{n+1}(\xi) = 0$. But

$$G^{(n+1)}(x) = E^{(n+1)}(x) - \frac{w^{(n+1)}(x)}{w(t)} E(t)$$

$$= f^{(n+1)} - \frac{(n+1)!}{w(t)} E(t).$$

Therefore, $G^{n+1}(\xi) = 0$ implies that

$$f^{(n+1)} - \frac{(n+1)!}{w(t)} E(t) = 0 \;\Rightarrow\; f(x) - p_n(x) = \frac{w_n(x)}{(n+1)!} f^{(n+1)}(\xi_x),$$

and we are done. □

Theorem A.3 (Hermite Interpolation Error Theorem)   *Let $f \in C^{2n}([a, b])$ and let the nodes $x_k \in [a, b]$ for all k, $1 \leq k \leq n$. Then for each $x \in [a, b]$ there is a $\xi_x \in [a, b]$ such that*

$$f(x) - H_n(x) = \frac{\psi_n(x)}{(2n)!} f^{(2n)}(\xi_x), \tag{A.1}$$

where here

$$\psi_n(x) = \prod_{k=1}^{n} (x - x_k)^2.$$

Proof   The proof is essentially the same as for the Lagrange theorem. This time we define the auxiliary function as

$$G(x) = E(x) - \frac{\psi_n(x)}{\psi_n(t)} E(t),$$

where $E(x) = f(x) - H_n(x)$. Then the generalized Rolle's Theorem gives us a point $\xi \in [a, b]$ such that

$$0 = G^{(2n)}(\xi) = f^{(2n)} - \frac{(2n)!}{\psi_n(t)} E(t),$$

from which the result follows. □

A.2 PROOF OF THE STABILITY RESULT FOR SMOOTH AND UNIFORMLY MONOTONE DECREASING INITIAL VALUE PROBLEMS

The theorem as stated in the text (Theorem 6.2) is the following.

Theorem A.4   Let $f(t, y)$ be continuous for all (t, y) in a rectangle R, and smooth and uniformly monotone decreasing in y. Then for all $(t_0, y_0) \in R$ there exists a unique solution to the initial value problem

$$y' = f(t, y), \quad y(t_0) = y_0.$$

Moreover, if $z(t)$ is the solution to the same problem with initial data $z(t_0) = z_0$, then

$$|y(t) - z(t)| \le e^{-m(t-t_0)} |y_0 - z_0|. \tag{A.2}$$

Proof   The existence and uniqueness follows from Theorem 6.1 and was demonstrated in Section 6.1; all we need to do here is demonstrate the stability result (A.2). We have

$$y' = f(t, y), \quad y(t_0) = y_0,$$

and

$$z' = f(t, z), \quad z(t_0) = z_0.$$

Subtract and use the Mean Value Theorem on f to get

$$(y - z)' = f_y(t, \eta)(y - z), \quad y(t_0) - z(t_0) = y_0 - z_0.$$

Here η will be a function of y, z, and t, with value between $y(t)$ and $z(t)$. To simplify things let's introduce some notation:

$$p(t) = f_y(t, \eta), \quad \phi(t) = y(t) - z(t), \quad \phi_0 = y_0 - z_0.$$

We therefore have

$$\phi' - p(t)\phi(t) = 0, \quad \phi(t_0) = \phi_0,$$

which is a simple first-order linear differential equation. We can solve it by introducing the function

$$P(t) = -\int_{t_0}^{t} p(s) \, ds$$

and the integrating factor $e^{P(t)}$. We then have

$$\left(e^{P(t)}\phi(t)\right)' = 0,$$

so we integrate both sides (from t_0 to t) to get

$$e^{P(t)}\phi(t) = \phi_0.$$

Hence

$$\phi(t) = \phi_0 e^{-P(t)},$$

so we have

$$y(t) - z(t) = (y_0 - z_0)e^{-\int_{t_0}^t p(s)ds},$$

and therefore,

$$|y(t) - z(t)| \leq |y_0 - z_0| \max_{t_0 \leq s \leq t} e^{-P(t)}.$$

Maximizing the exponential on the right means we must maximize the exponent. But

$$-P(t) = \int_{t_0}^t p(s)\, ds \leq -m(t - t_0),$$

using the smooth and uniformly monotone decreasing condition to bound p (which is f_y, remember). $\qquad\Box$

A.3   STIFF SYSTEMS OF DIFFERENTIAL EQUATIONS AND EIGENVALUES

In this part of the appendix, we provide some more details on the relationship between the stiffness of a linear system of differential equations and the eigenvalues of the coefficient matrix. As in the text, we take as our example the IVP

$$y_1' = 198y_1 + 199y_2, \quad y_1(0) = 1 \tag{A.3}$$
$$y_1' = -398y_1 + -399y_2, \quad y_2(0) = -1, \tag{A.4}$$

which we write in matrix-vector form as

$$y' = Ay, \quad y(0) = y_0,$$

and we approximate its solution using the second-order Adams-Bashforth scheme. Thus we look at the recursion

$$y_{n+1} = y_n + \frac{1}{2}h\left(3Ay_n - \frac{1}{2}Ay_{n-1}\right) = \left(I + \frac{3}{2}hA\right)y_n - \frac{1}{2}hAy_{n-1}. \tag{A.5}$$

If A were a scalar instead of a matrix, it would be a simple matter to attack this recursion by looking for the solution in the form $y_n = r^n$ for some r, which we could determine by factoring a polynomial. But since A is a matrix, this approach appears fruitless.

If A were a *diagonal* matrix, then we could use the old approach. In that case, each row of (A.5) would represent a scalar recursion for that component of the vector y_{n+1}. In general, A will not be a diagonal matrix, but it almost always is *diagonalizable*.

Recall from linear algebra (see Theorem 8.1) that a matrix is diagonalizable if there exists an invertible matrix P such that $P^{-1}AP = D$, where D is diagonal. Recall further that if the eigenvectors of A are independent, then A is diagonalizable, with P being the matrix of eigenvectors and D the matrix of eigenvalues. We therefore have (if A is diagonalizable)

$$y_{n+1} = \left(I + \frac{3}{2}hP^{-1}DP\right) y_n - \frac{1}{2}hP^{-1}DPy_{n-1},$$

from which we get

$$Py_{n+1} = \left(I + \frac{3}{2}hD\right) Py_n - \frac{1}{2}hDPy_{n-1},$$

or, setting $z_n = Py_n$,

$$z_{n+1} = \left(I + \frac{3}{2}hD\right) z_n - \frac{1}{2}hDz_{n-1}.$$

Since D is diagonal, we can now break this apart into each separate scalar recursions:

$$z_{n+1,i} = \left(1 + \frac{3}{2}hd_i\right) z_{n,i} - \frac{1}{2}hd_iz_{n-1,i},$$

where d_i are the diagonal elements of D (and, therefore, the eigenvalues of A). For simplicity, let $hd_i = \eta_i$ and solve this using our usual procedure. We get

$$z_{n,i} = C_{1,i}r_{1,i}^n + C_{2,i}r_{2,i}^n,$$

where the $r_{j,i}$ values are the roots of the polynomial

$$q(r) = r^2 - \left(1 + \frac{3}{2}\eta_i\right) r + \frac{1}{2}\eta_i.$$

If we compute the roots of this polynomial as a function of the parameter η_i (for η_i real and negative), we find that the absolute value of both roots is not less than 1 until we have $|\eta_i| \leq 1$. Recall that $\eta_i = hd_i$, where d_i are the eigenvalues (-1 and -200) of the coefficient matrix A for the ODE. Thus, to be within the stability region for second-order Adams-Bashforth, we need to have $h \leq 1/200$. Note that this is the case even when the e^{-200t} component of the solution is not present, and even though this component would be negligibly small after the first time step.

In the text we saw that the stiffness of this system could be "defeated" by using an implicit method to compute the numerical solution. Let's see why that worked.

If we apply the trapezoid rule method to our example, we get the recursion

$$y_{n+1} = y_n + \frac{1}{2}h(Ay_n + Ay_{n+1}) = \left(I + \frac{1}{2}hA\right) y_n + \frac{1}{2}hAy_{n+1},$$

which can solve to get

$$\left(I - \frac{1}{2}hA\right) y_{n+1} = \left(I + \frac{1}{2}hA\right) y_n$$

or,

$$y_{n+1} = \left(I - \frac{1}{2}hA\right)^{-1} \left(I + \frac{1}{2}hA\right) y_n.$$

When we diagonalize the recursion, we get

$$z_{n+1} = \left(I - \frac{1}{2}hD\right)^{-1} \left(I + \frac{1}{2}hD\right) z_n,$$

so the scalar recursion of interest is

$$z_{n+1,i} = \frac{(1 + \frac{1}{2}\eta_i)}{(1 - \frac{1}{2}\eta_i)} z_{n,i}.$$

It can be shown that the fraction is always less than 1 in absolute value, so long as the real part of η_i is negative. Thus we do not require that h be sufficiently small to prevent the solution from growing without bound as n increases.

A.4   THE MATRIX PERTURBATION THEOREM

Our concern here is the proof of the following theorem from Chapter 7.

Theorem A.5   Let $A \in \mathbf{R}^{n \times n}$ be given, nonsingular, and let $E \in \mathbf{R}^{n \times n}$ be a perturbation of A. Let $x \in \mathbf{R}^n$ be the unique solution of $Ax = b$. If $\kappa(A)\|E\| < \|A\|$, then the perturbed system $(A + E)x_c = b$ has a unique solution and

$$\frac{\|x - x_c\|}{\|x\|} \leq \frac{\kappa(A)\frac{\|E\|}{\|A\|}}{1 - \kappa(A)\frac{\|E\|}{\|A\|}}.$$

Proof   We start by assuming that the perturbed matrix $A + E$ is nonsingular, and therefore the perturbed system has a unique solution; we will then prove this as a lemma.

We have, then, the two systems

$$Ax = b$$

and

$$(A + E)x_c = b.$$

We can subtract and rearrange to get

$$A(x - x_c) = -Ex_c$$

so that

$$\frac{\|x - x_c\|}{\|x\|} \leq \frac{\|A^{-1}\|\|E\|\|x_c\|}{\|x\|}$$

$$\leq \|A^{-1}\|\|E\| \left(\frac{\|x\| + \|x - x_c\|}{\|x\|} \right)$$

$$= \kappa(A)\frac{\|E\|}{\|A\|} \left(1 + \frac{\|x - x_c\|}{\|x\|} \right);$$

therefore,

$$\frac{\|x - x_c\|}{\|x\|} \leq \frac{\kappa(A)\frac{\|E\|}{\|A\|}}{1 - \kappa(A)\frac{\|E\|}{\|A\|}}.$$

The positivity of the denominator—which is required to preserve the direction of the inequality—is guaranteed by the assumptions of the theorem. □

Lemma A.1   *If $A \in \mathbf{R}^{n \times n}$ is nonsingular and $\kappa(A)\|E\| < \|A\|$, then $A + E$ is also nonsingular.*

Proof   Note that $A + E = A(I + A^{-1}E)$, so that $A + E$ is nonsingular if and only if $I + A^{-1}E$ is nonsingular. Note also that our hypothesis implies that $\|A^{-1}\|\|E\| < 1$. For simplicity, we write

$$F = -A^{-1}E$$

Assume, then, that $I + A^{-1}E = I - F$ is singular. Then there exists $z \neq 0$ such that $(I - F)z = 0$; thus $z = Fz$. Taking norms, then, we have that $\|z\| = \|Fz\| \leq \|F\|\|z\|$ so that $1 \leq \|F\|$, which is a contradiction. Thus $I - F$ is nonsingular; therefore, $A + E$ is nonsingular.

ANSWERS TO SELECTED EXERCISES

In this appendix we give short answers or brief hints to solving some of the exercises in the text, so that students can check their work, verify that they are doing things properly, or get some guidance on how to approach a problem. Generally, only the even-numbered exercises are treated here. When the text has several exercises of similar type, only a few are selected. For some of the computer exercises, we give solutions so that students can check the accuracy of their programs. For problems that ask for comments or plots or various comparisons, we have provided only the computational answer. the student still needs to provide the appropriate comment, plot, or comparison.

The later chapters of the text, because of the more involved nature of the material, do not have a lot of exercises that are well suited for a "short answer" section. This is why there are not many answers to exercises from Chapter 8 and none at all for Chapter 9.

1.1

2. $p_3(x) = 1 + (1/2)x - (1/8)x^2 + (1/16)x^3$

4. $|R(x)| \leq 0.00378$

6. $R(x)1/192$

8. $p_4(x) = 1 - x + x^2 - x^3 + x^4$

10(b). $p_3(x) = 1 - (1/2)(x - \pi/2)^2$

11(d). $p_3(x) = x - (1/2)x^2 - (1/3)x^3$. $|R_3(x)| \leq 1/4$

12(a). $p_6(x) = 1 + x + (1/2)x^2 + (1/6)x^3 + (1/24)x^4 + (1/120)x^5 + (1/720)x^6$

18. $p_4(x) = x^4 + 1$

20. $p_3(x) = x^3 + x$.

1.2

2. (hint) $e^x = 1 + x + (1/2)x^2 e^\xi$

4. (hint) $(1+x)^{-1} = 1 - x + x^2 - x^3(1+\xi)^{-4}$

8. (hint) Use the definition of the O notation.

1.3

2(a). Exact $= 4/15 = 0.26666666...$; chopped $= 0.266$; rounded $= 0.267$.
Abs. error (chopped) $= 6.67 \times 10^{-4}$; rel. error (chopped) $= 2.50 \times 10^{-3}$.

Abs. error (rounded) $= 3.33 \times 10^{-4}$; rel. error (rounded) $= 1.25 \times 10^{-3}$.

2(c). Exact $= 0.353968254...$, chopped $= 0.353$, rounded $= 0.354$.
Abs. error (chopped) $= 9.683 \times 10^{-4}$; rel. error (chopped) $= 0.002735$.
Abs. error (rounded) $= 3.175 \times 10^{-5}$; rel. error (rounded) $= 8.969 \times 10^5$.

4. Single precision requires $n = 8$. double precision requires $n = 13$.

6. You should get something like 2.25063 for the forward sum, but 2.27418 for the backward sum.

8. $a_2 = 50,000$ for all three cases.

12. $u_{\text{round}} = 0.596 \times 10^{-7}$.

1.5

2. $k = 3$.

4. $E_k(x) = O(1/(k+2)!)$

6. $p(x) = 1 - x^2 + (1/2)x^4 - (1/6)x^6$

10. $n = 7$ for 10^{-2}, $n = 14$ for 10^{-4}

1.6

2(a). $\ln 13 \approx 2.564948606$

2(c). $\ln(1/3) \approx -1.098608555$

4(a). $\ln 13 \approx 2.564949357461423$

4(c). $\ln(1/3) \approx -1.098612288664884$

2.1

1(b). $2 + x(3 + x^2)$

2(a). $1 + x^2((-1/2) + (1/24)x^2)$

3(a). $1 + x^2(-1 + x^2((1/2) - (1/6)x^2))$

7(a). $1 + x((1/3) - (1/60)(x - 3))$

2.2

2(a). $f'(1) \approx 0.35083360$ or $f'(1) \approx 0.35359656$

2(c). $f'(1) \approx 0.969994$ or $f'(1) \approx 1.00131$

4. (in part) $f'(1) \approx -0.48649$, $f'(1.5) \approx 0.072884$ $f'(1.8) \approx 0.30382$, $f'(1.9) \approx 0.38234$

6(a). $A = -3/2h$, $B = 2/h$, $C = -1/2h$

8. (hint) Expand $f(x \pm h)$ and $f(x \pm 2h)$ out to h^5, using Taylor's Theorem

10. (in part) $f'(1.2) \approx -0.2652536725$, $f'(1.5) \approx 0.3237018117E - 1$ $f'(1.6) \approx 0.1126454212$, $f'(1.8) \approx 0.2654434511$

12. $f(x) \approx (f(x+h) - 2f(x) + f(x-h))/h^2$

16. $\Delta_1(0.125) = 2.70271582$

20. $\Delta_2(0.125) = 2.718259666$

2.3

1. $y_1 = 1.210367746$, $y_2 = 1.458844986$, $y_3 = 1.690257280$, $y_4 = 1.851566895$

4. $y_1 = -0.320429543$, $y_2 = 0.121827147$, $y_3 = 0.486730952$, $y_4 = 0.812025140$

6. $y_8 = 0.7705812951$

2.4

1(c). $(0.34) \approx 0.3685329977$

2(a). $\Gamma(1.005) \approx 0.9975675385$

3(a). $\Gamma(1.290) - p_1(1.290) \approx$ $(1/8) \times (0.1)^2 \times \max\{1.2484, 1.0492\} = 1.5605 \times 10^{-3}$

4. $p_1(x) = 3 - 2x$, $|x^{-1} - p_1(x)| \leq (1/8) \times (1/2)^2 \times 16 = 1/2$

6. $p_1(x) = (4x + 3)/7$, $|x^{1/3} - p_1(x)| \leq 0.68056$

8. $h \leq 0.04472$ for 10^{-3} accuracy, $h \leq 1.414 \times 10^{-3}$ for 10^{-6}

2.5

2. $T_2(f) = 0.9480594490$, $h \leq 0.0874...$ for 10^{-3} accuracy, $h \leq 0.002764...$ for 10^{-6}

4. $T_8(f) = 0.2460937500$, $h \leq 0.0447$, $h \leq 0.00141$

6. $T_8(f) = 0.3856439099$, $h \leq 0.110$, $h \leq 0.00346$

8. $T_8(f) = 0.1361206304$, $h \leq 0.0483$, $h \leq 0.00153$

10. $I_4 = 0.8317002443$, $I_8 = 0.8346696206$, $I_R = 0.8356594123$

12(a). $h = 0.0018$, $h = 0.57E - 4$

12(c). $h = 0.11$, $h = 0.35E - 2$

14. For $h = 1/4$, we get $L = 2.791985108$; for $h = 1/16$, we get $L = 2.826154994$.

16(a). $h \leq 0.000245...$

18. $h \leq 0.2305940493E - 3$, so roughly 4337 points.

20. $T_5(f) = 0.48368...$, $T_{10}(f) = 0.48547...$

2.6

1. $x_1 = 0.008495$, $x_2 = 0.1575$, $x_3 = 0.2274$, $x_4 = -0.2009$

6. $x_1 = 1.0037$, $x_2 = 1.9925$, $x_3 = 3.3358$, $x_4 = 3.6082$

8. $x_1 = 1$, $x_2 = 2$, $x_3 = 3$, $x_4 = 4$

2.7

1. $U_1 = 0.0349$, $U_2 = 0.0563$, $U_3 = 0.0500$

2. $U_1 = 0.7149$, $U_2 = 1.2583$, $U_3 = 1.4521$, $U_4 = 1.1040$

5. (hint) Use $u'(x_k) \approx (u(x_{k+1}) - u(x_{k-1}))/(2h)$

6. The general form of the equation is $-(1 + 32h)U_{k-1} + (2 + h^2)U_k - (1 - 32h)U_{k+1} = h^2$. For $n = 4$ the solution is $U_1 = 0.0237$, $U_2 = 0.0019$, $U_3 = 0.0388$. note that the system in this case will no longer be diagonally dominant unless h is small.

8. For $n = 4$, $U_1 = 0.00001012$, $U_2 = 0.00001417$, $U_3 = 0.00001012$.

10. For $n = 4$, $U_1 = 0.0001620$, $U_2 = 0.0002268$, $U_3 = 0.0001620$.

3.1

1. $[a_0, b_0] = [0,2]$, $[a_1, b_1] = [1,2]$, $[a_2, b_2] = [1, 3/2]$, $[a_3, b_3] = [5/4, 3/2]$

2(c). We need to take $n = 4$ steps, and we get $[a_4, b_4] = [0.625, 0.6875]$.

4. For $C = 0.25$, $x = 3.807314332$; for $C = 0.50$, $x = 3.640022851$; for $C = 0.75$, $x = 3.504939653$

6. About 2.6%.

8(a). (hint) Set $\theta = \beta/\sqrt{k}$.

8(b). $\alpha = 0.0297, 0.2895, 1.4149, 2.4519$, for the sequence of θ values as given.

10. $[a_1, b_1] = [0.5, 2]$, $[a_2, b_2] = [0.8571, 2]$, $[a_3, b_3] = [1.0696, 2]$

3.2

2. $x_{n+1} = (2/3)(x_n + (1/x_n^2))$

4. $x_{n+1} = 2x_n - ax_n^2$

6. For $x_0 = 1$, we get $x_1 = 1.103638324$, $x_2 = 1.098624898$, $x_3 = 1.098612289$. For $x_0 = 2$, we get $x_1 = 1.406005850$, $x_2 = 1.141366984$, $x_3 = 1.099513383$. For $x_0 = 4$, we get $x_1 = 3.054946917$, $x_2 = 2.196322584$, $x_3 = 1.529956717$.

10. $\beta = 1.39174520027073$

3.3

4. For $t = T/4$, we get $x = 1.580795827$; for $t = T/8$, we get $x = 0.7925194063$.

6(a). About 6.8 feet

6(b). About 9.09 feet

8. About 2.6%

10. For $C = 0.25$, $x = 3.807314332$; for $C = 0.50$, $x = 3.640022851$; for $C = 0.75$, $x = 3.504939653$.

3.4

2. $x_1 = 1.481481481$, $x_2 = 1.499771376$, $x_3 = 1.499999965$, $x_4 = 1.500000000$, $x_5 = 1.500000000$.

4(b). $b = 5/8 = 0.625$, $x_0 = 1.75$, $x_1 = 0.158594E + 01$, $x_3 = 1.6$

4(c). $b = 7/8$, $x_0 = 1.25$, $x_4 = 1.142857143$, $1/7 \approx x_4/8 = 0.1428571429$.

6. $|\alpha - x_0| \le 0.0178$.

8. The initial error is now bounded by about 0.0025, and it takes three iterations, or about 15 total operations, to get the specified accuracy.

3.5

2. $|e_1| \le 1/9$, $|e_2| \le 1/81$, $|e_3| \le 1/6561$.

4. No, the method is not working.

6. Converges to the specified accuracy in six iterations, whereas bisection would take 29.

8. $|e_0| \le 0.1815712825E - 1$ is required for convergence.

3.6

2. Converges in four iterations.

4. We do not know if it will converge or not.

6. One step of bisection plus five steps of Newton's method suffice.

3.7

2. Five iterations

4. $|e_0| \le 0.00018803$

6. $|\sqrt{x} - p_2(x)| \le 0.006$ (experimentally), leading to convergence in only three iterations and thirteen total operations.

8. $|\sqrt{x} - q_1(x)| \le 0.03$ (experimentally), leading to convergence in four iterations and fourteen total operations.

10. $e_0 \le \max\{0.02441, 0.01418\} = 0.02441$; four iterations are required to get an absolute error of 10^{-16}, so this involves about 16 operations.

3.8

2. $x_2 = 2$, $x_3 = 0.5$, $x_4 = 0.8571$

6. Nine iterations

8. The barrels hit at a velocity of $v = 44.6041$ feet per second, so they will break.

10. For $k = -24$, $\alpha = 0.00094129631308$; for $k = -23$, $\alpha = 0.00125518056472$; for $k = 1$, $\alpha = 0.92775227544590$; for $k = 2$, $\alpha = 1.09242933874440$.

12. For $C = 0.25$, $x = 3.849238636$; for $C = 0.50$, $x = 3.714269213$; for $C = 0.75$, $x = 3.598701186$.

3.9

1((c). $x_1 = .4054651081$, $x_2 = 0.3403682857$, $x_3 = 0.2929444165$

2. $y_1 = 0.5433216988$, $y_2 = 0.5423768123$, $y_3 = 0.5423997893$

4. $\gamma = e^{-1} = 0.3679$, $|x_1 - x_0| \le 1$, so the iteration will converge in no more than 12 iterations.

10(a). Using $x_0 = 2$, the first three Newton iterates are $x_1 = 1.500000000$, $x_2 = 1.416666667$, $x_3 = 1.414215686$. the first three Steffenson iterates (using the same x_0) are $x_1 = 1.666666667$, $x_2 = 1.477477478$, $x_3 = 1.419177338$.

3.10

2. The basic iteration produces $x_1 = 0.2027$, $x_2 = 0.09230$, $x_3 = 0.04414$, $x_4 = 0.02160$; the first Aitken method produces $y_2 = 0.02702$, $y_3 = 0.006906$, $y_4 = 0.001754$; the second Aitken method produces $z_1 = 0.02702$, $z_2 = 0.1720E - 3$, $z_3 = 0.7392E - 8$, and $z_4 = 0.0000$. the exact value of the fixed point is $\alpha = 0$.

6(a). The first five iterates are $x_1 = 2.473684211$, $x_2 = 1.381181849$, $x_3 = 2.460662644$, $x_4 = 1.412737566$, $x_5 = 2.466610144$.

6(f). The first five iterates are $x_1 = 0.6487212714$, $x_2 = 0.7508805213$, $x_3 = 0.6986326250$, $x_4 = 0.7228530108$, $x_5 = 0.7109870517$.

6(i). The first five iterates are $x_1 = 0.3771477945$, $x_2 = 0.4879950020$, $x_3 = 0.3876940420$, $x_4 = 0.4717585386$, $x_5 = 0.3946163017$.

8(a). The first five iterates are $x_1 = 2.473684211$, $x_2 = 1.381181849$, $x_3 = 2.758444187$, $x_4 = -.054389925$, $x_5 = -2.510981319$.

8(f). The first five iterates are $x_1 = 0.6893502777$, $x_2 = 0.7092603344$, $x_3 = 0.7147863401$, $x_4 = 0.7148057746$, $x_5 = 0.7148059123$.

8(i). The first five iterates are $x_1 = 0.3771477945$, $x_2 = 0.4879950020$, $x_3 = 0.4145336696$, $x_4 = 0.4317713355$, $x_5 = 0.4262133737$.

10. $x_{n+1} = \frac{x_n(x_n^2 + 3a)}{3x_n^2 + a}$; converges to $\sqrt{5}$ in about five iterations.

20. Yes; $x_5 = 1.000000287$

22. $x_2 = 0.500000$, $x_3 = 0.250000$, $x_4 = 0.125069$, $x_5 = 0.0625346$, $x_6 = 0.0369239$, $x_7 = 0.0414641$.

4.1

2. $p_2(x) = 3x^2 - 2x$

4. $p_2(x) = -(32/105)x^2 + (22/21)x + (9/35)$

6. $p_3(x) = x^3$

10. (hint) Recall that the function $f(x) = 1$, for all x, is a polynomial of degree zero.

4.2

2(d). The divided difference coefficients are $a_0 = 0$, $a_1 = 1$, $a_2 = 1/6$.

4. $p_2(x) = -(12/35)x^2 + (23/21)x + (26/105)$; the empirical maximum error is about 0.01.

8. $p_5(x) = 0.1417E - 13x^5 - 0.3375E - 10x^4 + 0.2954E - 7x^3 - 0.00001066x^2 + 0.0002432x + 1.131$

10. $p_5 = -0.09270377605x^5 + 0.8632574013x^4 - 3.199846710x^3 + 6.287123228x^2 - 6.537486675x + 3.679656532$

12(a). $p_2(T) = 0.004165T^2 - 1.5434T + 155.15$

14(a).
$p_6(x) = 0.0001600x^6 + 0.00066298x^5 - 0.007638x^4 - 0.03204x^3 + 0.09548x^2 + 0.5895x + 0.7677$, $x = \log_{10} \theta$

4.3

2. $|x^{-1} - p_2(x)| \leq 1/(6\sqrt{3}) = 0.9622504490E - 1$

6. Let $q(\theta) = \sqrt{\theta^2 + \theta + 1}$; then $|f(x) - p_2(x)| \leq (1/162)p(\theta)h^3 \max_{x_0 \leq x \leq x_2} |f(\xi)|$, where $p(\theta) = (\theta + 2 - q(\theta))(\theta - 1 - q(\theta))(2\theta + 1 + q(\theta))$.

8. $n \geq 1166$ for the linear case., $n \geq 130$ for the quadratic case.

10. $n \geq 1111$ for the linear case, $n \geq 144$ for the quadratic case, $n \geq 45$ for the cubic case.

14(d). $\|f\|_\infty = 1$, $\|f\|_2 = 0.9101797207$

4.4

2. $x_3 = 0.7087486790$, $x_4 = 0.692849951$, $x_5 = 0.692849951$

4. $x_3 = 1.017357687$, $x_4 = 1.032876088$, $x_5 = 1.174008963$

6. $p_2(x) = -3.31370850x^2 + 3.656854248x$, $q_2(x) = y^2/2$

4.5

2(c). For $h = 1/4$, $f'(1) \approx 0.9742181938$, using (4.21). For $h = 1/4$, $f'(1) \approx 0.915162219$, using (4.23).

2(d). For $h = 1/8$, $f'(1) \approx 0.497464280$, using (4.21). For $h = 1/8$, $f'(1) \approx 0.497502407$, using (4.23).

4. $f'(x_1) = (-\eta/(h(h+\eta)))f(x_0) - ((h - \eta)/(h\eta))f(x_1) + (h/(\eta(h+\eta)))f(x_2) - (h\eta/6)f(\xi)$

6. $\Gamma'(1) \approx -0.5638283100$, $\Gamma'(1.5) \approx 0.3125765900E - 1$, $\Gamma'(2) \approx 0.4216022150$

4.6

2. $H_2(x) = -(1/\pi)x^2 + x$

6. $H_2(x) = 6 - 13x + 12x^2 - 4x^3$; the predicted error is $\leq 1/8$, the observed error is ≤ 0.022.

8. $H_2(x) = -2.227411278 + 4.364467670x - 3.046701500x^2 + 0.9096451100x^3$; the predicted error is $\leq 1/64$, and the observed error is ≤ 0.0037.

4.7

2.
$p_{2,1}(x) = -0.53875999x^2 + 1.24946568x + 0.218870860$,
$p_{2,2}(x) = -0.17822141x^2 + 0.84866259x + 0.329558822$.

4. $p_{3,1}(x) = -0.3427614666x^3 + 1.873584557x^2 - 3.283390574x + 2.752567484$,
$p_{3,2}(x) = 0.1596133333E - 2x^3 + 0.3766571466x^2 - 1.105518098x + 1.691638542$

4.8

2. no

4. $k = 3$

6. The spline coefficients are $c_{-1} = 0.4052$, $c_0 = 0.3333$, $c_1 = 0.2615$, $c_2 = 0.2207$, $c_3 = 0.1889$, $c_4 = 0.1667$, $c_5 = 0.1445$.

8. The spline coefficients are $c_{-1} = 0.0095$, $c_0 = 0.0140$, $c_1 = 0.0185$, $c_2 = 0.0212$, $c_3 = 0.0239$.

10. The spline coefficients are $c_{-1} = 0.1822$, $c_0 = 0.1667$, $c_1 = 0.1511$, $c_2 = 0.1470$, $c_3 = 0.1481$, $c_4 = 0.1542$, $c_5 = 0.1667$, $c_6 = 0.1792$.

18. The spline coefficients are $c_{-1} = -0.0029$, $c_0 = 0.0002$, $c_1 = 0.0032$, $c_2 = -0.0037$, $c_3 = 0.1056$, $c_4 = 0.3492$, $c_5 = 0.4793$, $c_6 = 0.5904$, $c_7 = 0.7016$.

24. The spline coefficients are $c_{-1} = 0.2002$, $c_0 = 0.1767$, $c_1 = 0.1532$, $c_2 = 0.1156$, $c_3 = 0.1053$, $c_4 = 0.1160$, $c_5 = 0.1267$.

4.9

2. The spline coefficients are $c_{-1} = 0.2021$, $c_0 = 0.1667$, $c_1 = 0.1312$, $c_2 = 0.0936$, $c_3 = 0.0513$, $c_4 = 0.0017$, $c_5 = -0.0582$.

3(a). For $h = 1/8$, the spline coefficients are $c_{-1} = -0.0647$, $c_0 = 0$, $c_1 = 0.0647$, $c_2 = 0.1195$, $c_3 = 0.1562$, $c_4 = 0.1690$, $c_5 = 0.1562$, $c_6 = 0.1195$, $c_7 = 0.0647$, $c_8 = 0$, $c_9 = -0.0647$.

3(e). For $h = 1/8$, the spline coefficients are $c_{-1} = -0.0650$, $c_0 = 0$, $c_1 = 0.0650$, $c_2 = 0.1201$, $c_3 = 0.1570$, $c_4 = 0.1699$, $c_5 = 0.1570$, $c_6 = 0.1201$, $c_7 = 0.0650$, $c_8 = 0$, $c_9 = -0.0650$.

4.10

2. $p_2(T) = 1.3873 - 0.00186222727273T + 0.00000088863636T^2$

5. $b = 149{,}670$, $m = 0.750825$

8(a). $(1/2)\sqrt{1 - 3e^{-2}}$

14. $p_2(x) = -2.279726631x^2 + 0.7599088770$; this is not especially accurate.

4.11

8. $p_1(x) = 1.7790534738 \times (x - 0.359834957) + 1.131111480 \times (0.890165043 - x)$

10. $\|f - p_n\|_\infty \le \frac{(b-a)^{n+1}}{2^{2n+1}(n+1)!}\|f^{(n+1)}\|_\infty$

5.1

3. (hint) Use the linearity that is established in Exercise 1 to reduce the problem of integrating an arbitrary polynomial exactly, to the problem of integrating each power exactly.

5.2

2. $T_4(f) = 0.9233310015$, $T_4^C(f) = 0.9270138493$

4. $T_4(f) = 0.8317002443$, $T_4^C(f) = 0.8356064943$

6(c). For $n = 16$, $T_n(f) = 1.29496946$, $T_n^C(f) = 1.29583752$

8. $T_8(f) = 1.570796328$, $T_8^C(f) = 1.570796328$

10. $T_{256}^C = 10.35673448$; for this problem, the correction is actually not an improvement until you take very small h.

5.3

2. $S_4(f) = 0.9271586870$

4. $S_4(f) = 0.8357855108$

6(c). $S_2 = 1.290400337$, $S_4 = 1.295321668$, $S_8 = 1.295798350$, $S_{16} = 1.295834312$

8. For $h = 1/4$, we get $\pi \approx 3.083595156$; for $h = 1/8$, we get $\pi \approx 3.121189170$

12. $S_2(f) = (h/6)(Af(a) + Cf(c) + Bf(b))$, $A = 2 + \theta - \theta^2$, $C = (3\theta^2 + 1 + \theta^3 + 3\theta)/\theta$, $B = (2\theta^2 + \theta - 1)/\theta$

14. (hint) Look at Exercise 1 of Section 5.1, and recall how a polynomial is written.

20. $S_{256}(f) = 11.39838665$

5.4

2. $M_4(f) = 0.9288993588$

4. $M_4(f) = 0.8376389970$

8(c). For accuracy to 10^{-3}, we need $h \le 0.1095445115$; for 10^{-6}, we need $h \le 0.3464101616\mathrm{E}-2$

5.5

2. For $\epsilon = 10^{-3}$, a conservative value is $n \approx 70$. For $\epsilon = 10^{-6}$, a conservative value is $n \approx 140$.

3. The limit is zero.

5.6

1(d). $G_4(f) = 1.493334622$

2(b). $G_4(f) = 0.9270388618$

2(d). $G_4(f) = 0.8356239692$

4(b). $G_4(f) = 0.6466426358$

6. $Q(x) = (1/3)(x + 1)$, $R(x) = (4/3)x - (2/3)$

12. (hint) Compute $\int_{-1}^1 x^k P_3(x)\,dx$ for $k = 0, 1, 2$.

18(b). $n = 10$

5.7

2. $S_2(f) = 1/4$, $S_4(f) = 1/4$, $R_4 = 1/4$, $E_4 = 0$; Simpson's rule is exact for this function.

4. $S_2(f) = 0.84$, $S_4(f) = 0.8358$, $R_4 = 0.83552$, $E_4 = 0.28E - 3$

10. The value of p is converging to 1, but for the midpoint rule it should converge to 2, so we conclude that the code has an error.

5.8

2(b). $\Theta_3(f) = 0.9270361380$

2(e). $\Theta_3(f) = 0.1352573547$

6(b). $I_4(f) = -0.7994$

10(b). $T_k(f)$ is exact for $k \geq 2$.

6.1

2(a). $f(t, y) = -4y$

3(b). $y_1' = y_2, y_2' = y_3, y_3' = y_4, y_4' = 1 - 4y_1$; $f(t, y) = (y_2, y_3, y_4, 1 - 4y_1)^T$

4(c). $K = 4$

5(a). Yes; $m = M = 3$

6.2

1(b). $y_4 = 0.7368351728 \approx y(1)$

2(c). $y'(0) \approx 3$, $y'(1/4) \approx 6.603978547$, $y'(1/2) \approx 17.88843470$, $y'(3/4) \approx 62.51297275$, $y'(1) \approx 95.66847529$

4(d). For $h = 1/4$, $y_1 = 2.173575402$, $y_2 = 2.379516214$, $y_3 = 2.552122473$, and $y_4 = 2.691102647$

6.3

2(a). $E(h) \leq 6.70h$

3(a). $M = m = 16$

6.4

2. The method is consistent, with truncation error $(h^2/3)y(\xi)$.

4(b). $y(1) \approx 0.7303809855$

8. By about $t = 5$ it is apparent that the limiting radius is about 0.416.

10. The limiting radius varies inversely with S_i, and directly with σ.

18. The method is $y_{n+1} = y_{n-1} + (h/3)(f(t_{n-1}, y_{n-1}) + 4f(t_n, y_n) + f(t_{n+1}, y_{n+1}))$; it is fourth order accurate, implicit, and multistep.

6.5

1(c). $y(1) \approx 37.62483136$

2(d). $y(1) \approx 5.151237020$

6(a). For $h = 1/4$, $y(1/4) \approx 0.53125$, $y(1/2) \approx 0.35546875$, $y(3/4) \approx 0.2895507812$, $y(1) \approx 0.2648315430$.

8. The limiting radius is now about 0.171.

6.6

1(a). $y(1) \approx 0.7226562500$

5(b). For $h = 1/8$, the maximum error is about 0.0265; for $h = 1/16$, it is about 0.0011.

6. $y_{n+1} = y_n + (h/2\theta)((1 + 2\theta)f(t_n, y_n) - f(t_{n-1}, y_{n-1}))$

8. The limiting radius is again about 0.1710.

6.7

2. The stability polynomial is $\sigma(r) = r^2 - 4r + 3 = (r - 3)(r - 1)$, so the roots are $r = 1, 3$; since none of the root conditions are satisfied, the method is unstable.

3. The method is stable, but not strongly stable.

5. For the two step method, we have $\sigma(r) = r^2 - (4/3)r + (1/3)$, which has roots $r = 1$ and $r = 1/3$. For the three step method, we have $\sigma(r) = r^3 - (18/11)r^2 + (9/11)r - (2/11)$, which has roots $r = 1$ and $r = 0.3181818182 \pm 0.2838635454i$.

6.8

1(c). $y_1(1) \approx 1.310058594$, $y_2(1) \approx -1.310058594$.

2(a). $dy_1/dt = y_2, dy_2/dt = -\sin y_1$. for $h = 1/4$, $y_1(1) \approx 0.3815058690$ and $y_2(1) \approx -0.9502278426E - 1$.

7(b). Using the trapezoid rule predictor-corrector with $h = 1/16$, the epidemic dies out in about 5 weeks, regardless of the size of I_0, and does not spread throughout the population.

6.9

2. If we start with an initial value of $h = 1/2$, then we go through two reductions to $h = 1/8$ and then get $y = (1.0000, 1.1468, 1.3448, 1.4704, 1.6222, 1.8093)$ for $t = (0, 0.1250, 0.2500, 0.3125, 0.3750, 0.4375)$. note that there is a further reduction to $h = 1/16$.

6.10

2. $u(1/4) \approx 0.0329$, $u(1/2) \approx 0.0538$, $u(3/4) \approx 0.0484$.

4. Using $h = \pi/8$, we get $(0.0601, 0.0316, -0.1511, -0.3233, -0.1511, 0.0316, 0.0601)^T$ as the solution vector.

7.2

4. $x = (1.0025, 1.9700, 3.1050, 3.8600, 5.0630)^T$

5(a). For $n = 4$, we get $x = (-4, 60, -180, 140)^T$.

8. If pivoting is not used, we get $x_2 = 1$, $x_1 = 0$. if partial pivoting is used, we get $x_2 = 1$ and $x_1 = 1$.

7.3

2. $C = n$

4. $C = 3n + O(1)$

6. $C = 9n + O(1)$

8. $C = n^2$

10. About 92.6 hours for the fast computer, and 92,592.6 hours (over 10 *years*) for the slow one.

7.4

2. $L = \begin{bmatrix} 1 & 0 \\ 0.25 & 1 \end{bmatrix}, U = \begin{bmatrix} 4 & 1 \\ 0 & 4.75 \end{bmatrix}$

7.5

2. $\|A\|_\infty = 12$

4. (hint) You ought to be able to find matrices A and B such that $\|AB\|_\star > \|A\|_\star \|B\|_\star$.

6. $\|A\|_\infty = 6$, $\|A^{-1}\|_\infty = 0.4286$

8. $E = \begin{bmatrix} -0.0020 & -0.0040 & -0.0060 \\ -0.0031 & -0.0062 & -0.0092 \\ -0.0017 & -0.0034 & -0.0051 \end{bmatrix}$

10. $\kappa^* = 6.7837$

12(b). $\kappa^*(K_3) = 6.8267$, $\kappa^*(K_6) = 20.1860$, $\kappa^*(K_9) = 40.8643$

7.6

2. (hint) For the suggested vector x, if $a_{22} - a_{21}^2/a_{11} \leq 0$, then $x^T A x \leq 0$, which violates the *SPD* condition.

6. (hint) We know there is a lower triangular G such that $A = GG^T$; let D_G be the diagonal part of G and consider the product $L = D_G^{-1} G$.

10. Depending on the machine precision, eventually the matrix fails to be "computationally positive definite" and the Cholesky algorithm breaks down.

7.7

1(b). $x^{(1)} = (-0.2500, 0.5000, 1.0000, 2.5000)^T$, $x^{(2)} = (-0.1250, 0.6875, 1.7500, 2.6875)^T$, $x^{(3)} = (-0.0781, 0.9062, 1.8438, 2.9062)^T$

2. $x^{(1)} = (-0.2625, 0.5939, 0.8941, 2.4592)^T$, $x^{(2)} = (-0.0935, 0.6167, 1.8007, 2.3855)^T$, $x^{(3)} = (-0.0959, 0.9430, 1.6624, 2.9908)^T$

7.8

2(a). $x^{(1)} = (-0.4232, 0.2527)^T$, $x^{(2)} = (-0.4831, 0.2136)^T$, $x^{(3)} = (-0.4834, 0.2134)^T$

3(a). $x^{(1)} = (-0.0204, 0.9796)^T$, $x^{(2)} = (-0.0669, 0.4689)^T$

7.9

2. For Newton's method with $h = 1/8$ and $u^{(0)} = 0$, we get convergence in 10 iterations to $u = (0.1267, 0.2529, 0.3787, 0.5039, 0.6287, 0.7529, 0.8767)^T$.

4. For Newton's method with $h = 1/8$ and $u^{(0)} = 0$, we get convergence in 7 iterations to $u = (0.1398, 0.2777, 0.4122, 0.5421, 0.6666, 0.7847, 0.8961)^T$.

8.1

1(a). $p(\lambda) = \lambda^3 - 12\lambda^2 + 46\lambda - 56$

3(b). $\lambda \in \{z \in \mathbf{C} \mid |z - 4| \leq 3\}$

4. For the first matrix, we get $\sigma(A) = \{3, 3, 6\}$, but the only eigenvector for $\lambda = 3$ is parallel to $x = (1, 0, 0)^T$. for the second matrix, we get $\sigma(A) = \{2, 1, 1\}$, and there are two independent eigenvectors for $\lambda = 1$.

8.2

4. $Q_1 = I - \gamma w w^T$ for $w = (-0.7312, 1, 1)^T$ and $\gamma = 0.7887$

8.3

6. For $4 \leq n \leq 10$, $\lambda_1 = $ 5.6180, 5.7321, 5.8019, 5.8478, 5.8794, 5.9021, 5.9190.

7. For $4 \leq n \leq 10$, $\lambda_n = 0.96702\text{E} - 4$, $0.32879\text{E} - 5$, $0.10828\text{E} - 6$, $0.34939\text{E} - 8$, $0.11116\text{E} - 9$, $0.34998\text{E} - 11$, $0.10931\text{E} - 12$.

10. (hint) Since Q is symmetric and orthogonal, $Q^{-1} = Q^T = Q$, so $Qx = e_1$ implies that $x = Qe_1$.

12. (hint) $A'z = Az - \lambda_1 x_1 (x_1^T z)$.

14. If a Hessenberg reduction is done, then the cost is roughly $C = 4nN + (2/3)n^3$. if the Hessenberg reduction is not done, then the cost is roughly $C = (n^2 + n)N$. If $(2/3)n^2 > (n-3)N$, then it might be cheaper not to do the reduction.

Index